Optimum Choice of Energy System Configuration and Storages for a Proper Match between Energy Conversion and Demands

Optimum Choice of Energy System Configuration and Storages for a Proper Match between Energy Conversion and Demands

Special Issue Editors

Andrea Lazzaretto
Andrea Toffolo

MDPI • Basel • Beijing • Wuhan • Barcelona • Belgrade

Special Issue Editors

Andrea Lazzaretto
University of Padova
Italy

Andrea Toffolo
Luleå University of Technology
Sweden

Editorial Office
MDPI
St. Alban-Anlage 66
4052 Basel, Switzerland

This is a reprint of articles from the Special Issue published online in the open access journal *Energies* (ISSN 1996-1073) 2019 (available at: https://www.mdpi.com/journal/energies/special_issues/Optimum_Choice_Energy_System_Configuration_Storages).

For citation purposes, cite each article independently as indicated on the article page online and as indicated below:

LastName, A.A.; LastName, B.B.; LastName, C.C. Article Title. *Journal Name* **Year**, *Article Number*, Page Range.

ISBN 978-3-03928-036-0 (Pbk)
ISBN 978-3-03928-037-7 (PDF)

Contents

About the Special Issue Editors

Andrea Lazzaretto (Professor), Ph.D., is professor of Energy Systems at the University of Padova, Italy. He studied at the same University, where he got the Master in Mechanical Engineering in 1986, and the Ph.D. in Energy Engineering in 1992 after a research period at the Tennessee Technological University. The research activities mainly focus on exergoeconomics, and simulation, optimization and experimental tests of energy systems and components. He is Fellow of the American Society of Mechanical Engineers, where he held several positions of responsibility, and is presently Awards and Honors chair of the Executive Committee of the Advanced Energy System Division. He is winner of the ASME Edward F. Obert Award in the years 1998, 2007 and 2018 for outstanding papers on Thermodynamics and Applied Thermodynamics. He has been Associate editor of the Journal of Energy Resources Technologies from 2006 to 2016, member of the editorial committees of other international journals and member of the scientific committee of the International Conference ECOS since 2005. He is author or co author of more than 220 papers, mostly published in international journals or conference proceedings.

Andrea Toffolo, Ph.D., has been Professor in Energy Engineering at the Division of Energy Science, Department of Engineering Sciences and Mathematics, Luleå University of Technology (Sweden), since 2011. He studied at the University of Padova (Italy), where he obtained his Master in Mechanical Engineering in 1996 and Ph.D. in Energy Engineering in 2003, subsequently serving as a Post-Doc and then Research Fellow at the Department of Mechanical Engineering of the same university (2003–2011). His main research activity is in the development and application of advanced artificial intelligence techniques (evolutionary algorithms, neural networks, fuzzy logic, expert systems) to solve optimization problems for the synthesis/design of energy conversion systems and their integration with industrial processes. Other research interests are related to the modeling, performance analysis, and diagnosis of malfunction in complex energy systems. He is the winner of the ASME Edward F. Obert Award in the years 2007 and 2018 for outstanding papers on Thermodynamics and Applied Thermodynamics. He is author or co-author of more than 90 papers published in international journals and conference proceedings.

Preface to "Optimum Choice of Energy System Configuration and Storages for a Proper Match between Energy Conversion and Demands"

Every sector of human activity, domestic, industrial, transportation, etc., requires energy in the different forms of electricity, heat, cooling, fuels, etc. However, each sector has the common need to match properly the generation of energy from the available sources with the users' demands. Traditional fossil sources can be easily stored to supply the energy in the desired form at any time it is required by the users. Renewable sources such as the Sun or wind are instead available depending on the meteorological conditions, and need to be stored properly after the conversion in the desired form to satisfy always the users' demand. In both cases, the search for a proper match between the available sources and users' demands requires a correct choice of type, number and size of the energy conversion and storage units. This is the topic of this Special Issue, which is certainly crucial for the development of a more sustainable and "smarter" society where the overall consumption of resources, costs and environmental impact are minimized.

Andrea Lazzaretto, Andrea Toffolo
Special Issue Editors

Editorial

Optimum Choice of Energy System Configuration and Storages for a Proper Match between Energy Conversion and Demands

Andrea Lazzaretto [1,*] and Andrea Toffolo [2]

[1] Department of Industrial Engineering, University of Padova, via Venezia 1, 35131 Padova, Italy

[2] Energy Engineering, Division of Energy Science, Department of Engineering Sciences and Mathematics, Luleå University of Technology, SE-971 87 Luleå, Sweden; andrea.toffolo@ltu.se

* Correspondence: andrea.lazzaretto@unipd.it; Tel.: +39-049-8276747

Received: 12 September 2019; Accepted: 12 October 2019; Published: 17 October 2019

Abstract: This Special Issue addresses the general problem of a proper match between the demands of energy users and the units for energy conversion and storage, by means of proper design and operation of the overall energy system configuration. The focus is either on systems including single plants or groups of plants, connected or not to one or more energy distribution networks. In both cases, the optimum design and operation involve decisions about thermodynamic processes, about the type, number, design parameters of components/plants, and storage capacities, and about mutual interconnections and the interconnections with the distribution grids. The problem is very wide, can be tackled with different methodologies and may have several, more or less valuable and complicated solutions. The twelve accepted papers certainly represent a good contribution to perceive its difficulty.

Keywords: smart power systems; multi-energy systems; optimization of energy systems design and operation

---

1. Introduction

Energy conversion systems convert a source of energy into the form desired by the users: Electricity, heat, fuel, etc. Users' demands vary with time and can often be predicted only with some uncertainty. On the other hand, the availability of the energy source may also vary depending, e.g., on uncertain weather predictions, as in the case of sun or wind, or on market availability, as in the case of fossil fuels. Thus, a good match between the availability of resources and users' demands is not easy to be found in terms of resource-saving, economic results, or sustainability. This match requires:

(1) A deep preliminary knowledge of the characteristics of the energy sources and the plants that will be included in the overall system, in addition to all the other constraints from the external environment that may limit system utilization (market costs and prices, transportation systems, and others). In this step, the definition of the optimum configuration of each single plant is crucial and involves a proper match among its components according to the purpose of the overall energy system.

(2) The definition of the final system configuration, which requires a proper match between sources, plants, and users. Storage capacities are often necessary to find this match, to store fuels or product energy streams.

The problem of defining a configuration and its design parameters is generally called the "synthesis/design problem" and applies both to single plants and to groups of plants. Plants or groups of plants can work in isolation or can be connected to the electric grid, or to other heat or fuel networks. Design and management of plants and networks are two parts of the same problem, i.e., the optimum coupling between users' demand and energy conversion systems, which is the topic of this Special Issue.

2. Overview of the Papers

To facilitate the organization of this Special Issue, the papers are subdivided into two main categories, about:

- Optimization of design and operation of groups of energy conversion and storage units;
- Optimization of the design of single energy conversion and storage units.

2.1. Optimization of Design and Operation of Groups of Energy Conversion and Storage Units

Sakalis et al. [1] supply very general and detailed information about the mathematical formulation, solution methods, and case studies of intertemporal static and dynamic optimization problems for the synthesis, design, and operation (SDO) of energy systems. The focus is on the energy systems aboard ships. The authors emphasize the imperative need to use optimization methods in this field because of the large variety of possible design configurations and time-changing operating conditions. The examples of application demonstrate that the optimal solution may differ significantly from the solutions suggested by the usual practice. Unlike other works in the literature, where the SDO optimization problems are solved by two- or three-level algorithms, single-level algorithms are applied here tackling all three aspects (S, D, O) concurrently. The methods can also be applied on land installations, e.g., power plants, cogeneration systems, etc., with proper modifications.

Wang et al. [2] review existing methodologies for the three general topics of: (1) System evaluation, (2) optimization of the system design, and (3) optimization of the system configuration (synthesis). The focus is on thermal power plants, which are still supposed to play a significant role in the near future. In particular, the authors present the potentialities of advanced exergy-based techniques in the evaluation and design improvement of thermal systems. In the optimization of system design, the focus is on the mathematical formulation, solution algorithms, and design and operating variables that are to be included in the decision variables set. In the optimization of the system configurations, after a general description of typical optimization problems and solving methods, the superstructure-based and -free concepts are reviewed and compared, showing the possible automatic generation of structural alternatives. In each topic, the authors identify the methods with higher possibilities of application and development.

Urbanucci et al. [3] present a methodology for the optimal design and operation of cogeneration systems with thermal energy storage. A two-level algorithm is proposed, which utilizes a genetic algorithm at the design level (upper level) to identify the components that are to be included in the energy system and their capacities, and combines it with a mixed-integer linear programming (MILP) formulation for the search of the optimal operation (lower level). The two problems are nested and solved simultaneously. For each individual solution (components sizes) generated by the upper level, the optimal annual operation cost is identified by the MILP solver, and the total equivalent annual cost is calculated. The procedure is repeated for each individual of each generation generated at the upper level until the stopping criteria are met. A rolling-horizon technique allows dividing the investigated period into smaller periods and optimizing each subproblem in sequence to reduce the computational time required without affecting the quality of the results. The optimized design of a cogeneration system for a secondary school in San Francisco is presented as an example of application. Results show that this system is able to meet around 70% of both the electric and thermal demands, while the thermal energy storage additionally covers 16% of the heat demand.

Rech [4] summarizes the basic theoretical and practical concepts that are required to simulate and optimize the design and operation of fleets of energy conversion and storage units, which have to match properly the availability of the resources with users' requirements. The paper is a sort of manual, which helps the reader define all steps of the optimization problem. The author supplies instructions to select variables and equations that are required to simulate the dynamic behavior of each conversion and storage unit included in the system, to define the operational constraints and to formulate the objective function (economic profit). A general combined heat and power (CHP) fleet of units is used as an example to present the construction of the dynamic model and the formulation of the optimization problem. The goal is to provide a "recipe" to choose the number and type of energy conversion and storage units that are able to exploit in an optimal way the available sources to fulfill the users' demands.

Vargas-Jaramillo et al. [5] emphasize the importance of properly taking into account the reliability of power networks, which depends on the uncertainties associated with generation, transmission and distribution, load demand, and the presence of unexpected catastrophic events. All these factors affect the sustainability of these networks, making their planning a difficult problem to solve. The authors propose a generation-transmission reliability approach to improve the sustainability assessment of power networks. The approach is based on a quasi-stationary multiobjective optimization problem which takes into account, at every node and instant of time, the propagation of uncertainties along the transmission lines, the uncertainties of the generation system (including the fluctuating effects associated with the wind and photovoltaic energy producers), and the uncertainties of the load demand. Six different objective functions are considered: The total daily costs for the economic aspects, SO2 and CO2 daily emissions for the environmental aspects, the disability-adjusted life year (DALY) for the social aspects, and the exergetic efficiency and expected energy not supplied (EENS) for the technical aspects. The sustainability–reliability approach is applied to the standard IEEE reliability test system [6] composed of several generation units and transmission lines, as well as nodes where only generation capacity or load demand are present. Results show that using a mixture of normals approximation (MONA) for the formulation of the EENS makes the reliability analysis simpler, and possible in large-scale optimization problems. Moreover, the authors emphasize the positive impact, in terms of sustainability, of including renewable energy producers in the optimal synthesis/design of power networks, counterbalanced by the negative impact on reliability.

Tran and Smith [7] propose a stochastic approach to minimize the operating cost of a district energy system (DES), including a CHP system, natural gas boilers, solar photovoltaics (PV), and wind turbines for generation of power, heating, and cooling. A district of buildings on a university campus is used as a case study. A Monte Carlo study is performed to analyze the stochastic power generation from the renewable energy resources in the DES. The optimization of the DES is carried out with a particle swarm optimization (PSO) algorithm in each hour of a day, to bid in the day-ahead market. The results suggest that the proposed DES can achieve approximately a 10% operating cost reduction with respect to the current system. The focus is on the importance of considering the uncertainty of energy loads and power generation from renewable energy sources to properly evaluate the operating costs of the DES, and in turn, plan correct management of the energy generation.

Mikolajková-Alifov et al. [8] tackle the problem of optimizing the supply chain of liquefied natural gas (LNG), compressed natural gas (CNG), or biogas for smaller regions. The task is to find the best supplier and the most efficient way to transport the gas to the customers to cover their demands, including the design of pipeline networks, truck transportation, and storage systems. To fulfil this task, the authors develop a mathematical model of a gas supply chain where gas may be supplied by pipeline, as compressed gas in containers or as LNG by tank trucks, with the goal to find the solution that corresponds to lowest overall costs. A mixed-integer linear programming approach is used to reduce the computational time. The model is applied to a gas distribution problem in western Finland and considers constraints and costs of the delivery and the investments required to realize the system. Therefore, it can be used to analyze the sensitivity of the design of the supply chain to changes in the parameters (e.g., constraints or costs) or in the gas supply and demand. Results show that the fuel price has a major effect on the optimal supply chain, including which fuel sources are to be used and how to deliver the gas to the customers. The costs of storage and pipes mainly influence the length of the pipeline and the number of storages to be constructed. Although the model is presented for a single-period problem (i.e., with fixed demands) it can be extended to multi-period problems. This will, however, increase the complexity of the numerical problem, restricting the size of the problems that can be solved without prohibitive computational burden.

Harahap et al. [9] look into the supply chain optimization of palm oil biomass residues in Sumatra island (Indonesia). A biomass supply chain includes biomass harvesting and collection, pretreatment, storage, transport and conversion into bio-based products, so its optimization heavily relies on the selection of plants, units, components, and technologies forming the configuration of the supply chain (also with reference to its spatial layout). In the paper, the optimization is performed using the BeWhere model. BeWhere is a geographically explicit techno-economic model, developed at the International Institute for Applied Systems Analysis (IIASA), which adopts a mixed-integer linear representation translated into GAMS equations and solved by CPLEX. The results from the maximization of supply chain profits show that palm oil biomass residues can be conveniently transformed into both energy and non-energy products, contributing to sustainable growth of the palm oil industry.

2.2. Optimization of the Design of Single Energy Conversion and Storage Units

Seki and Amano [10] propose a bottom-up procedure to build absorption systems configurations by a combination of elementary processes. The procedure brings ideas from the SYNTHSEP methodology [11–15] to form the so-called "basic configuration" of a system as a set of elementary thermodynamic cycles or part of them [16,17], using an original approach for the implementation. The authors present two examples of application under simplified/idealized operating conditions to show that existing absorption systems, otherwise designed on the basis of experience, could be obtained automatically. The final aim of the methodology is to allow engineers predicting all possible configurations of absorption systems and identifying simple yet feasible optimal ones.

Fiaschi et al. [18] propose a solar-integrated thermo-electric energy storage (TEES) system for one to two daily hours of operation to compensate for the mismatch between electricity generation and demand in small-to-medium-size photovoltaic systems (4 to 50 kWe). Given the drawbacks of alternative storage systems in this range of power, such as the limited life cycle of batteries or the low round-trip efficiency of chemical storage, the authors believe that TEES systems may represent an interesting solution to guarantee dispatchability to energy systems based on renewables. The proposed system consists of a power cycle, a solar-assisted heat pump, and a solar-assisted refrigeration cycle matched with properly sized hot and cold reservoirs of warm water at 120/160 °C and ethylene glycol at -10/-20 °C. In the storage mode, a supercritical heat pump restores sensible heat to the hot reservoir, while a cooling cycle cools the cold reservoir. Both the heat pump and cooling cycle operate on photovoltaic (PV) energy, and benefit from solar heat integration at low–medium temperatures (80–120 °C). The power cycle is a trans-critical CO_2 unit, including recuperation. The thermodynamic cycles are designed and optimized from an exergy and exergo-economic perspective to search for the highest possible performance for a variable heat input depending on the availability of the solar resource. Results show that the system can deliver electric energy with a marginal round-trip efficiency of around 50% without considering the solar heat input to solar-thermal collectors. The exergy round-trip efficiency is of the order of 35%. The levelized cost of electricity is around 0.7–0.75€/kWh, in line or slightly better than documented stand-alone renewable configurations. The authors observe that this cost is still pretty high because of the high costs of the solar collectors and of the refrigeration cycle, but it could be significantly improved working both on the reduction of equipment cost and on optimized control strategies.

Guewouo, et al. [19] optimize the design of a small-scale compressed-air energy storage (CAES) system operating without fossil fuel. To do that, they build a model of the system and use a modified real coded genetic algorithm (RCGA) to find the values of thirteen selected design parameters that maximize the global exergy efficiency. The model is partially validated (i.e., only for the filling and the discharge of the tank) with data from an experimental prototype existing in the authors' lab. The results of the optimization indicate that the electric energy consumed by the compressor is 103.83 kWh and the electric energy output is 25.82 kWh for the system charging and discharging times of about 8.7 and 2 h, respectively. This corresponds to an optimal round-trip efficiency of 79.07% and to a global exergy efficiency of 24.87%. The analysis of the variation of all design parameters during the evolution of the optimization process allows the authors to evaluate the effect of each design parameter on the global exergy efficiency. In particular, results show that a low mass flow of the pneumatic air motor coupled with a high mass flow rate of the compressor improves the efficiency of the storage system and the maximum value of air storage tank volume allowed by the constraints of space, cost, charge and discharge time should be preferred.

Margheritini and Kofoed [20] investigate the feasibility of a wave energy system made up of a number of Weptos wave energy converters (WECs) and sets of batteries, to provide the energy demands of a small island in Denmark. They simulate over one year the behavior of the combination of two different configurations of these machines, for a total installed power of 750 KW, supplemented by a 3 MWh battery bank and a backup generator. The goal consists in demonstrating that they are able to provide the total energy needs of the island. Due to the imbalance between demand and production, this goal is achieved only with the intervention of the backup generator, which covers approximately 5% to 7% of consumption, even if Weptos WECs supply much more energy than requested.

3. Conclusions

The twelve papers of this Special Issue show different approaches and applications to the synthesis of new configurations of energy systems made up of groups of plants and storage capacities, or single plants with or without storage capacities. All papers emphasize the criticalities involved in the search for the best match between production and demands, addressing both methodological and application aspects. Although far from being exhaustive, the presented overview is able to show the main problems and solutions in the search for new configurations of energy systems interconnected or not with the energy distribution grids.

Conflicts of Interest: The authors declare no conflict of interest.

References

1. Sakalis, G.N.; Tzortzis, G.J.; Frangopoulos, C.A. Intertemporal Static and Dynamic Optimization of Synthesis, Design, and Operation of Integrated Energy Systems of Ships. *Energies* **2019**, *12*, 893. [CrossRef]
2. Wang, L.; Yang, Z.; Sharma, S.; Mian, A.; Lin, T.E.; Tsatsaronis, G.; Maréchal, F.; Yang, Y. A Review of Evaluation, Optimization and Synthesis of Energy Systems: Methodology and Application to Thermal Power Plants. *Energies* **2019**, *12*, 73. [CrossRef]
3. Urbanucci, L.; D'Ettorre, F.; Testi, D. A Comprehensive Methodology for the Integrated Optimal Sizing and Operation of Cogeneration Systems with Thermal Energy Storage. *Energies* **2019**, *12*, 875. [CrossRef]
4. Rech, S. Smart Energy Systems: Guidelines for Modelling and Optimizing a Fleet of Units of Different Configurations. *Energies* **2019**, *12*, 1320. [CrossRef]
5. Vargas-Jaramillo, J.R.; Montanez-Barrera, J.A.; von Spakovsky, M.R.; Mili, L.; Cano-Andrade, S. Effects of Producer and Transmission Reliability on the Sustainability Assessment of Power System Networks. *Energies* **2019**, *12*, 546. [CrossRef]
6. Billinton, R.; Kumar, S.; Chowdhury, N.; Chu, K.; Debnath, K.; Goel, L.; Khan, E.; Kos, P.; Nourbakhsh, G.; Oteng-Adjei, J. A Reliability Test System for Educational Purposes. *IEEE Trans. Power Syst.* **1989**, *4*, 1238–1244. [CrossRef]
7. Tran, T.T.D.; Smith, A.D. Stochastic Optimization for Integration of Renewable Energy Technologies in District Energy Systems for Cost-Effective Use. *Energies* **2019**, *12*, 533. [CrossRef]
8. Mikolajková-Alifov, M.; Pettersson, F.; Björklund-Sänkiaho, M.; Saxén, H. A Model of Optimal Gas Supply to a Set of Distributed Consumers. *Energies* **2019**, *12*, 351. [CrossRef]
9. Harahap, F.; Leduc, S.; Mesfun, S.; Khatiwada, D.; Kraxner, F.; Silveira, S. Opportunities to Optimize the Palm Oil Supply Chain in Sumatra, Indonesia. *Energies* **2019**, *12*, 420. [CrossRef]

10. Seki, K.; Takeshita, K.; Amano, Y. Development of Complex Energy Systems with Absorption Technology by Combining Elementary Processes. *Energies* **2019**, *12*, 495. [CrossRef]

11. Toffolo, A. A synthesis/design optimization algorithm for Rankine cycle based energy systems. *Energy* **2014**, *66*, 115–127. [CrossRef]

12. Toffolo, A.; Rech, S.; Lazzaretto, A. Generation of complex energy systems by combination of elementary processes. *J. Energy Resour. Technol.* **2018**, *140*, 112005. [CrossRef]

13. Lazzaretto, A.; Manente, G.; Toffolo, A. SYNTHSEP: A general methodology for the synthesis of energy system configurations beyond superstructures. *Energy* **2018**, *147*, 924–949. [CrossRef]

14. Toffolo, A.; Lazzaretto, A. A Practical Tool to Generate Complex Energy System Configuration Based on the SYNTHSEP Methodology. *Int. J. Thermodyn.* **2019**, *22*, 45–53.

15. Toffolo, A.; Lazzaretto, A. Building the basic configuration of compression refrigeration systems with the SYNTHSEP method. In Proceedings of the ECOS 2019—The 32nd International Conference on Efficiency, Cost, Optimization, Simulation and Environmental Impact of Energy Systems, Wroclaw, Poland, 23–28 June 2019.

16. Lazzaretto, A.; Toffolo, A. A method to separate the problem of heat transfer interactions in the synthesis of thermal systems. *Energy* **2008**, *33*, 163–170. [CrossRef]

17. Morandin, M.; Toffolo, A.; Lazzaretto, A. Superimposition of Elementary Thermodynamic Cycles and Separation of the Heat Transfer Section in Energy Systems Analysis. *J. Energy Resour. Technol.* **2013**, *135*, 021602. [CrossRef]

18. Fiaschi, D.; Manfrida, G.; Petela, K.; Talluri, L. Thermo-Electric Energy Storage with Solar Heat Integration: Exergy and Exergo-Economic Analysis. *Energies* **2019**, *12*, 648. [CrossRef]

19. Guewouo, T.; Luo, L.; Tarlet, D.; Tazerout, M. Identification of Optimal Parameters for a Small-Scale Compressed-Air Energy Storage System Using Real Coded Genetic Algorithm. *Energies* **2019**, *12*, 377. [CrossRef]

20. Margheritini, L.; Kofoed, J.P. Weptos Wave Energy Converters to Cover the Energy Needs of a Small Island. *Energies* **2019**, *12*, 423. [CrossRef]

Review

A Review of Evaluation, Optimization and Synthesis of Energy Systems: Methodology and Application to Thermal Power Plants [†]

Ligang Wang [1],*, Zhiping Yang [2], Shivom Sharma [1], Alberto Mian [1], Tzu-En Lin [3], George Tsatsaronis [4], François Maréchal [1] and Yongping Yang [2]

[1] Industrial Process and Energy Systems Engineering, Swiss Federal Institute of Technology in Lausanne (EPFL), Rue de l'Industrie 17, 1951 Sion, Switzerland; shivom.sharma@epfl.ch (S.S.); alberto.mian@epfl.ch (A.M.); francois.marechal@epfl.ch (F.M.)

[2] National Research Center for Thermal Power Engineering and Technology, North China Electric Power University, Beinong Road 2, Beijing 102206, China; yzprr@163.com (Z.Y.); yyp@ncepu.edu.cn (Y.Y.)

[3] Laboratoire d'Electrochimie Physique et Analytique, Swiss Federal Institute of Technology in Lausanne (EPFL), Rue de l'Industrie 17, 1951 Sion, Switzerland; tzu-en.lin@epfl.ch

[4] Institute for Energy Engineering, Technical University of Berlin, Marchstraße 18, 10587 Berlin, Germany; tsatsaronis@iet.tu-berlin.de

* Correspondence: ligang.wang@epfl.ch or lgwangeao@163.com; Tel.: +41-21-69-34208

† This work is extended based on the doctoral thesis of Dr.-Ing. Ligang Wang entitled "Thermo-economic Evaluation, Optimization and Synthesis of Large-scale Coal-fired Power Plants" defended on July 2016 at the Technical University of Berlin.

Received: 16 September 2018; Accepted: 26 December 2018; Published: 27 December 2018

Abstract: To reach optimal/better conceptual designs of energy systems, key design variables should be optimized/adapted with system layouts, which may contribute significantly to system improvement. Layout improvement can be proposed by combining system analysis with engineers' judgments; however, optimal flowsheet synthesis is not trivial and can be best addressed by mathematical programming. In addition, multiple objectives are always involved for decision makers. Therefore, this paper reviews progressively the methodologies of system evaluation, optimization, and synthesis for the conceptual design of energy systems, and highlights the applications to thermal power plants, which are still supposed to play a significant role in the near future. For system evaluation, both conventional and advanced exergy-based analysis methods, including (advanced) exergoeconomics are deeply discussed and compared methodologically with recent developments. The advanced analysis is highlighted for further revealing the source, avoidability, and interactions among exergy destruction or cost of different components. For optimization and layout synthesis, after a general description of typical optimization problems and the solving methods, the superstructure-based and -free concepts are introduced and intensively compared by emphasizing the automatic generation and identification of structural alternatives. The theoretical basis of the most commonly-used multi-objective techniques and recent developments are given to offer high-quality Pareto front for decision makers, with an emphasis on evolutionary algorithms. Finally, the selected analysis and synthesis methods for layout improvement are compared and future perspectives are concluded with the emphasis on considering additional constraints for real-world designs and retrofits, possible methodology development for evaluation and synthesis, and the importance of good modeling practice.

Keywords: advanced exergy-based analysis; superstructure-based; superstructure-free; mathematical programming; flowsheet synthesis; multi-objective optimization; thermal power plants

1. Introduction

Thermal power plants are normally considered as the power stations, which produce electric power by various working-fluid based Rankine/combined cycles utilizing heat from different sources, e.g., fossil fuels, nuclear, solar and geothermal energy. Commonly-used working fluids for Rankine cycle are mainly water/steam for large-scale applications and high-temperature heat source, and various organic fluids for small-scale applications and intermediate-/low-grade heat. From the heat-source perspective, thermal power plants can be classified to coal-fired power, nuclear power, concentrated solar power, geothermal power, etc. However, as a usual term, thermal power plants mainly refer to those with fossil fuels (coal and natural gas). Particularly, coal-fired power will still contribute 40% to the total world electricity generation in 2020 [1], even with the current circumstance of fast growing of low-emission renewable power [2,3]. More importantly, to cope with the increasing injection of intermittent renewable power while maintaining stable and secure grid operation, thermal power plants are expected to operate flexibly by allowing faster load shifting [4], before large-scale technologies for electrical storage, e.g., power-to-gas [5], become widely available and affordable [6]. Therefore, in the foreseeable future, thermal power plants will continue to contribute the most in power generation sector. Regarding this context, state-of-the-art thermal power plants and trends of system development and integration are summarized by focusing on large-scale coal-fired power plants.

Coal-fired power plants have gone through nearly one hundred years of development. Key technology progress was mainly originated from the milestones of material improvement (Figure 1). Ferritic steel allows steam temperature below around 580 °C with the matched main steam pressure of around 250 bar. Austinite steel, about 20% of total steel applied to high-temperature components (final superheaters and reheaters, first stages of steam turbines) can push the temperatures of main and reheat steam up to 620 °C with the steam pressure of around 280 bar. Further using Ni-based steel (20%) together with austinite steel (25%) can enable plant operation with the steam temperature as high as 720 °C. The current trend of technology development is toward higher steam parameters (temperature and pressure) and larger generating capacity (over GW level). The next generation technology, advanced ultra-supercritical power plants, aiming at steam temperatures over 700 °C and pressures over 350 bar [7,8], has been under intensive R&D since the mid-1990s and promises to constitute a benchmark plant with a design efficiency of approximately 50%.

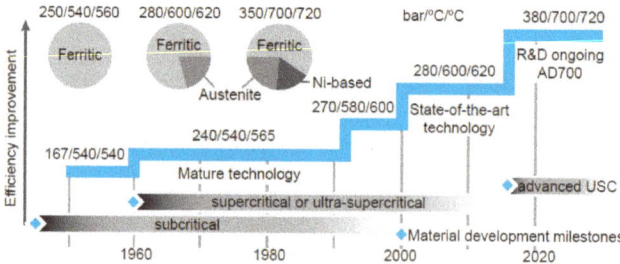

Figure 1. Technology development of pulverized coal power plants [9].

Pulverized-coal power plants are based on the classical Rankine cycle. The efficiency of an ideal Rankine cycle (η_{ideal}) is determined by average temperatures of heat absorption ($T_{\text{a,abs}}$) and heat release ($T_{\text{a,rel}}$) of the working fluid:

$$\eta_{\text{ideal}} = 1 - \frac{T_{\text{a,rel}}}{T_{\text{a,abs}}}, \tag{1}$$

The higher the average temperature of heat absorption and the lower the average temperature of heat release, the greater the cycle efficiency can be achieved. For condensing power plants, the average temperature of heat release depends on local ambient conditions. Thus, to achieve a higher cycle efficiency, the major means is to increase the average temperature of heat absorption, which can be achieved by increasing the temperatures of main and reheated streams, increasing the final feedwater preheating temperature, adding more feedwater preheaters and employing multiple reheating [10,11]. For real-world Rankine-cycle-based coal power plants, the increase of the pressure level of main steam and the reduction of thermodynamic inefficiencies occurring in real components (e.g., friction loss and steam leakage in steam turbines) can improve the plant efficiency as well. These design options for efficiency improvement have been considered during the development of future coal-fired power plants.

Although the temperature increase of main and reheated steams can improve the plant efficiency, it may lead to an overheating crisis of feedwater preheaters, especially those that extract superheated steam from the turbines after reheating. In addition, the superheat degree of steam extractions indicates incomplete steam expansion (i.e., the loss of work ability of the extracted steams). To address the potential overheat crisis of feedwater preheaters and ensure the complete expansion of extracted steams, a modified reheating scheme (Master Cycle [12]) has been proposed. The key idea of the Master Cycle is to employ a secondary turbine (ET) that receives non-reheated steam, drives the boiler feed pump, and supplies bled steam for feedwater preheaters, so that the superheat degrees of steam extractions can be significantly reduced. However, the impact of introducing a secondary turbine on the optimal design of the whole system has been limited studied [13,14].

New challenges lying ahead are associated with system-level integration. The integration opportunity flourishes, as multiple fluids are involved with wide temperature ranges (Figure 2), e.g., flue gas (130–1000 °C), steam (35–700 °C), feedwater (25–350 °C) and air (25–400 °C). On the one hand, there is a need to raise the heat utilization to the level of the overall system, which has not been achieved yet due to independent designs of the boiler and turbine subsystems. On the other hand, the integration of many available technologies or concepts, which deliver a significant improvement in overall plant efficiency, becomes possible. The options include topping or bottoming cycles (such as the CO_2-based closed Brayton cycle or the organic Rankine cycle [15]), low-grade waste heat recovery from flue gas [16], low-rank coal pre-drying [17], multiple heat sources (especially solar thermal energy [18–20]), etc. In addition, pollutant-removal technologies, particularly for CO_2 capture, should be considered as well.

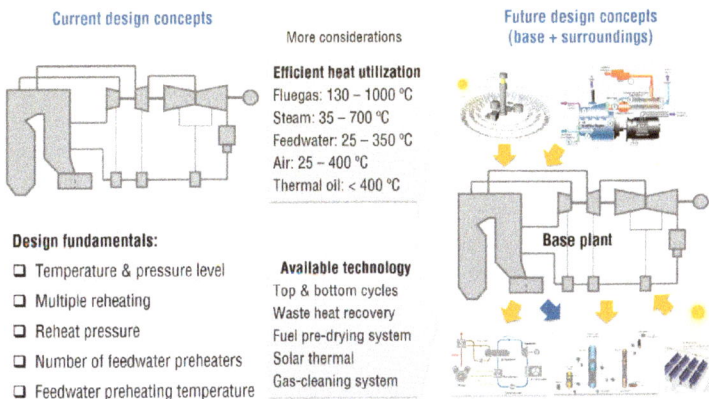

Figure 2. Fundamental considerations and new challenges for the design of thermal power plants [9].

Therefore, except for those fundamental considerations for the design of thermal power plants itself, such as employing more stages of reheating, increasing feedwater preheating temperature and implementing more feedwater preheaters, the future design concept of thermal power plants emphasizes system-level synthesis for integrating many available advantageous technologies (Figure 2). The question is then to find the best integration of multiple technologies considered by a systematic, effective synthesis and optimization method.

System synthesis and evaluation are at the heart of the overall system design of thermal power plants. The synthesis methods enable the engineers to create novel conceptual system designs, which are then evaluated with respect to various criteria for suggesting further improvements. In Sections 2–4, recent developments of thermodynamic evaluation methods (particularly exergy-based analysis method), optimization and synthesis approaches of both design/operating parameters and system layouts of energy systems are reviewed, respectively. The most influential methods, which are frequently used in literature and represent the state-of-the-art, are introduced with more details. To support comprehensive decision making with multiple objective functions, the techniques to handle multi-objective optimization are reviewed in Section 5. Therefore, this review provides a comprehensive and comparative view of these analysis and optimization methodologies with a summary and discussion of their applications to thermal power plants. A perspective for the future development, implementation, combination, and application of these methodologies is given in Section 6. Finally, some conclusions are given in Section 7.

2. Analysis of Energy Systems

The analysis of energy systems is a prerequisite for identifying the design imperfections and promoting improvement strategies, which is mainly based on energy analysis and exergy analysis. Energy analysis is obtained from the first law of thermodynamics and focuses on the quantity of energy, which has been carried out by many researchers over the past decades [21]. However, energy analysis only focuses on the quantity of energy and fails to identify any inefficiency in an adiabatic process [22]. While combing the concept of exergy, the exergy analysis considers also the quality of energy and then enhances the energy-based analysis. Detailed methods for physical and chemical exergies of different types of material flows, work and heat flows have been discussed in [23]. Here, the exergy-based analysis is mainly discussed for identifying the true performance of the considered components and systems.

This section is organized as follows: In Section 2.1, basic concept, indicators and short history of exergy analysis are given, which is further extended to exergoeconomic analysis in Section 2.2 by combining economic evaluation, and advanced exergy and exergoeconomic analyses in Section 2.3 by splitting exergy destruction (cost) based on their sources and avoidability. In Section 2.4, the application of exergy-based analysis to thermal power plants is summarized. Finally, the limitations of system evaluation are given in Section 2.5.

2.1. Exergy Analysis

All real processes are irreversible as their occurrence is driven by non-equilibrium forces, leading to thermodynamic inefficiencies inside the process boundaries (destruction (D) of exergy) and those across the process boundaries (loss (L) of exergy). An exergy analysis identifies the spatial distribution of thermodynamic inefficiencies within an energy system, pinpoints the components and processes with high irreversibilities, thus highlights the areas of improvement for the system [24].

The formulation of an exergy analysis usually includes exergy balance equations of the total system, a subsystem or a single component, which can be based on the incoming and outgoing exergy flows or the fuel (F) and product (P) definitions. In addition, by properly selecting the system boundaries, exergy losses occur only at the system level.

The key indicator of exergy analysis, exergetic efficiency, can be defined in many different ways [25], but the most accepted is introduced by Tsatsaronis in [26] as the following formulation:

$$\varepsilon = \frac{\dot{E}_P}{\dot{E}_F} = 1 - \dot{E}_D / \dot{E}_F, \tag{2}$$

where the subscripts F, P and D represent fuel exergy, product exergy and exergy destruction. The exergy destruction can identify the spatial and temporal distribution and magnitude of thermodynamic inefficiencies within an energy system.

The earliest contributions of exergy-based analysis can be dated back to the 1970s. Kotas et al. [27] pointed out that not all inefficiencies could be avoided due to the physical and economic constraints. Generally, the system analysis, particularly with exergy analysis, is the first step to understand the overall system performance. Singh and Kaushik [28] studied the optimization of Kalina cycle coupled with a coal-fired steam power plant by revealing the inherent mechanism on the impact of the ammonia mass fraction and turbine inlet pressure to the thermal efficiency. Some other applications can also be found in [29–32]. There are also several applications of exergy analysis for the next generation technology of advanced ultra-supercritical power plants, such as 700 °C-advanced plants, e.g., [33].

2.2. Exergoeconomic Analysis

Exergoeconomic analysis provides a deep understanding of costs related to equipment and thermodynamic inefficiencies as well as their interconnections and considers the interaction between the components and the whole system by unit costs of exergy flows and those of exergy destructions, thus tells us how we could iteratively improve the efficiency and cost-effectiveness of the system [26]. More importantly, in an exergoeconomic optimization, individual optimization of system components decomposed from the whole optimization problem is made possible. This decomposition relies on the statement that exergy is the only rational basis for the costs of energy flows and the inefficiencies within a system [26].

Major theoretical fundamentals of exergoeconomics have been established during the 1980s and 1990s. The term exergoeconomics was coined by Tsatsaronis [26], referred to as an exergy-aided cost-reduction method [34]. Key contributions of exergoeconomics came from a number of researchers, such as Tsatsaronis and Winhold [35,36], Tsatsaronis and Pisa [37], Tsatsaronis et al. [38], Lazzaretto and Tsatsaronis [39,40], Valero et al. [41–43], Valero and Torres [44], Valero et al. [45], Lozano and Valero [46], Frangopoulos [47–50], von Spakovsky [51], von Spakovsky and Evans [52], von Spakovsky [53], etc. These works can be classified as accounting and calculus methods [54].

2.2.1. Accounting Methods

The accounting approaches aim at understanding the formation of product costs, evaluating the performance of components and the system, and improving the system iteratively. To obtain unknown costs of all exergy flows, a set of algebraic equations are built. The equation set consists of cost balance equations associated with each unit (a component or a set of components of the system) and auxiliary cost equations that are needed for the units, of which the number of output streams is larger than the number of input streams. Evaluation of the equation set starts from the known costs of all input resources. With the costs of all exergy flows known, several exergoeconomic variables associated with each unit are calculated for performance evaluation and system improvement [37,38].

The allocation of costs to internal flows and products are mostly performed on the monetary basis (sometimes on exergetic cost basis [43]). The monetary cost of an exergy flow usually is accounted by the average cost associated with different exergy forms (thermal, mechanical and chemical) [40,55]. A systematic, generic and easy-to-use methodology, the specific exergy costing (SPECO) method, has been proposed by Lazzaretto and Tsatsaronis [56], which has been the milestone of the accounting methods. In the SPECO method, cost balance equations of each unit include the cost flow rates associated with capital amortization from an economic accounting, while fuel and product definitions and auxiliary cost equations are developed at the component level and in the most complex case considering the separate components of exergy. This approach has become the most widely accepted exergoeconomic analysis method even for complex energy systems (e.g., [57–60]) and has combined with mathematical algorithms for iterative optimization (e.g., [61–63]).

2.2.2. Calculus Methods

The calculus methods serve directly for mathematical cost minimization. The central idea is to closely approach thermoeconomic isolation, by means of thermoeconomic decomposition, for quickly and accurately assessing the effect of a certain parameter on the system performance without optimizing the whole problem (local optimization) [50]. Different decomposition approaches, i.e., the thermoeconomic functional analysis [47,48,50,64], Engineering Functional Analysis [51–53] and Three-Link Approach [65,66], have been developed for energy systems of different levels of design detail.

When the method of Lagrange multipliers is applied to the optimization algorithm, such as in the thermoeconomic functional analysis, the system is first decomposed by a functional analysis into units (the functional diagram [50], which is, in fact, the productive structure), each one of which has one specific function with a single exergy product. Then, the cost objective function is reformulated by adding a summation of Lagrange multipliers-weighted exergy products of all units. Thus, the multipliers do have their physical meaning: marginal costs of the exergy flow in the functional diagram. Introducing the marginal costs makes the problem readily solved by sequential algorithms.

However, the marginal costs are difficult to interpret regarding the process of cost formation [67], thus these methods are unable to reveal the physical and economic interrelationships among the components [47]. In addition, thermoeconomics decomposition becomes limited when complex systems are considered and less necessary due to the rapid developments of direct mathematical optimization tools and computation ability. Therefore, there have been no new developments or interesting applications of these calculus methods in recent years.

2.2.3. Recent Developments

In general, the maturity of exergoeconomics is marked by the SPECO method [56]; however, methodological and fundamental discussions have still been continued. One recent focus is the cost accounting associated with dissipative components, i.e., those whose productive purpose is neither intuitive nor easy to define. Torres [68] and Seyyedi et al. [69] discussed the mathematical basis and different criteria for cost assessment and formation process of the residues, and suggested that the costs entering a dissipative component should be charged to the productive component responsible for the residue. Piacentino and Cardona [70] introduced the Scope-Oriented Thermoeconomics, which identified cost allocation criteria for dissipative components, based on a possible non-arbitrary concept of Scope, and classified the system components by Product Maker/Product Taker but not by the classical dissipative/productive concepts. The subsequent optimization application, i.e., [71], presented that the method enabled to disassemble the optimization process and to recognize the formation structure of optimality, i.e., the specific influence of any thermodynamic and economic parameter in the path toward the optimal design. Banerjee et al. [72] proposed an extended thermoeconomics to allow for revenue-generating dissipative units and discussed the true cost of electricity for systems with such potential. Despite these, it seems that the choice of the best residue distribution among possible alternatives is still an open research line.

Efforts were also made to enhance the ability of exergoeconomics. Paulus and Tsatsaronis [73] formulated the auxiliary equations for specific exergy revenues based on SPECO, and presented "the highest price one would be willing to pay per unit of exergy is the value of the exergy". Cardona and Piacentino [74] extended exergoeconomics to analyze and design energy systems with continuously varying demands and environmental conditions. Moreover, an advanced exergoeconomic analysis, developed by the research group of Tsatsaronis [75–78], is capable of identifying the sources and availability of capital investments and exergy-destruction costs.

With these fundamental research, exergoeconomic analysis had a wide application on the thermal power plant recently. Rashidi and Yoo [79] analyzed a power-cooling cogeneration system from an exergoeconomic point of view to obtain the unit cost of power-cooling generation and the most exergy destruction location of the system. Sahin et al. [80] carry out exergoeconomic analysis for a combined cycle power plant. Different weighting factors were applied to energy efficiency, exergy efficiency, levelized cost and investment cost in three different scenarios; namely, the conventional case, the environmental conscious case, and the economical conscious case. Thus, the optimization of the size and configuration is depended on the user priorities. Ahmadzadeh et al. [81] applied the SPECO approach to evaluate the cost of a solar driven combined power and ejector refrigeration system. A genetic algorithm was used in their optimization process with the total cost rate as the objective function. Baghsheikhi [82] used a soft computing system to realize the real-time exergoeconomic optimization of a steam power plant, which was developed based on experts' knowledge and experiences regarding the exergoeconomic performance and features of the proposed power plant. It is proved to be an efficient method for real-time optimal response to the variation of operating condition. In [83], the exergoeconomic analysis was conducted to an existing ultra-supercritical coal-fired power plant for giving a promising solution for future design by using total revenue requirement (TRR) and the specific exergy costing (SPECO) methods for economic analysis and exergy costing.

2.3. Advanced Exergy-Based Analysis

When attempting to reduce thermodynamic inefficiencies within a system, additional factors must be taken into account: (a) Not all inefficiencies can be avoided [27], due to physical and economic constraints. The technical possibilities of exergy savings (i.e., the avoidable inefficiencies) of a component or system are always lower than the corresponding theoretical limit of thermodynamic exergy savings [46]. (b) The components in an energy system are not isolated whereas interactions among them always exist. Thus, part of the exergy destruction within a component is, in general, caused by the inefficiencies of the remaining components of the system [84]. (c) The same amount of exergy destruction within different components is not equivalent [27], because of different fundamental mechanisms of irreversibility and the component-system interactions. In other words, the same amount of decrease in exergy destruction within two different components has different impacts on the overall fuel consumption of the system [46]. These issues, however, cannot be addressed by the conventional exergy-based analysis.

Conventional exergy-based analysis can only identify the location and magnitude of inefficiencies, while an advanced exergy analysis can further reveal the source and avoidability of the inefficiency [85]. Thus, as one solution, an advanced exergy (exergoeconomic) analysis has been developed continuously since the last decade by Tsatsaronis and his coworkers [34,75–77,84–90], in which the exergy destruction (and cost) within a system component are further split: the avoidable (AV) and unavoidable (UN) parts, the endogenous (EN) and exogenous (EX) parts, and their combinations. Similarly, in the advanced exergoeconomic analysis, not only the exergy destruction but also the investment cost for each system component is split into avoidable/unavoidable and endogenous/exogenous parts [91].

2.3.1. Avoidable/Unavoidable Exergy Destruction and Cost

By employing technically feasible designs and/or operational enhancement, part of exergy destruction and costs associated with a system or component can be avoided, thus this part is considered as avoidable.

The estimation procedure has been initially discussed in [84,86]. Practically, the cost behavior exhibited by most components is that the investment cost ($\dot{Z}$) per unit of product exergy increases with decreasing exergy destruction ($\dot{E}_D$) per unit of product exergy or with increasing efficiency [86]. Thus, for the kth component, which is considered in isolation, if two limit states (Figure 3), one with extremely large investment cost and one with extremely high thermodynamic inefficiency, can be estimated with reasonably, then the unavoidable exergy destruction ratio $(\dot{E}_D/\dot{E}_P)^{\text{UN}}$ and the unavoidable investment cost ratio $(\dot{Z}/\dot{E}_P)_k^{\text{UN}}$ with respect to per unit of product exergy could be determined:

$$\dot{E}_{D,k}^{\text{UN}} = \dot{E}_{P,k} \cdot \left(\frac{\dot{E}_D}{\dot{E}_P}\right)_k^{\text{UN}}, \tag{3}$$

$$\dot{Z}_k^{\text{UN}} = \dot{E}_{\text{P},k} \cdot \left(\frac{\dot{Z}}{\dot{E}_{\text{P}}} \right)_k^{\text{UN}}.$$ (4)

Once the exergy destruction $\dot{E}_{\text{D},k}^{\text{UN}}$ and the cost $\dot{Z}_k^{\text{UN}}$ are known, the avoidable parts can be obtained:

$$\dot{E}_{\text{D},k}^{\text{AV}} = \dot{E}_{\text{D},k} - \dot{E}_{\text{D},k}^{\text{UN}},$$ (5)

$$\dot{Z}_k^{\text{AV}} = \dot{Z}_k - \dot{Z}_k^{\text{UN}}.$$ (6)

In general, both extreme states for the ratios $(\dot{E}_{\text{D}}/\dot{E}_{\text{P}})_k^{\text{UN}}$ and $(\dot{Z}/\dot{E}_{\text{P}})_k^{\text{UN}}$ are not industrially achievable; however, they can be simulated by adjusting a set of thermodynamic parameters associated with the considered component, including the parameters of incoming and outgoing streams, and the key design parameters of the component itself.

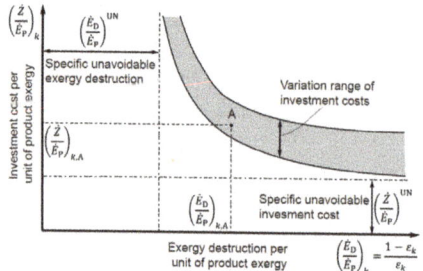

Figure 3. Definition of specific unavoidable exergy destruction $(\dot{E}_{\text{D}}/\dot{E}_{\text{P}})_k^{\text{UN}}$ and specific unavoidable investment cost $(\dot{Z}/\dot{E}_{\text{P}})_k^{\text{UN}}$ based on the expected relationship between investment cost and exergy destruction (or exergetic efficiency) for the k th component. (Reproduced from [86]).

2.3.2. Endogenous/Exogenous Exergy Destruction and Cost

The endogenous exergy destruction within the kth component $(\dot{E}_{\text{D},k}^{\text{EN}})$ is that part of the entire exergy destruction within the same component $(\dot{E}_{\text{D},k})$ that would still appear when all other components in the system operate in an ideal (or theoretical) way while the kth component operates with its real exergetic efficiency [75,76]. The exogenous exergy destruction within the kth component $(\dot{E}_{\text{D},k}^{\text{EX}})$ is the remaining part of the entire exergy destruction $(\dot{E}_{\text{D},k})$ and is caused by the simultaneous effects of the irreversibilities occurred in the remaining components. The exergy destruction $\dot{E}_{\text{D},k}^{\text{EX}}$ can also be expressed by a sum of the exogenous parts directly caused by the rth component $(\sum \dot{E}_{\text{D},k}^{\text{EX},r})$ plus a mexogenous (MX) exergy destruction term $(\dot{E}_{\text{D},k}^{\text{MX}})$ [89,92], caused by simultaneous interactions of other components. The endogenous and exogenous concepts are different from malfunction/dysfunction, which are used in thermoeconomic diagnosis based on the structural theory (for more details, see [75,76]).

To calculate the exergy destruction $\dot{E}_{\text{D}}^{\text{EN}}$, an ideal thermodynamic cycle needs to be defined first and then irreversibility of each component is introduced by turn [88,93,94]. This approach, however, is only appropriate for the system without chemical reactors and heat exchangers, of which the ideal operations are hard to define.

New calculation approach for $\dot{E}_{\text{D}}^{\text{EN}}$ has been proposed recently by Penkuhn et al. [95]. The basis of the new concept is that the nature of an ideal reversible process or system defines the relation between the exergy input and output. This feature pinpoints that the details on how the exergy is transferred or converted within a reversible process is not significant when constructing the simulation with only the considered component under real condition and all remaining components under their theoretical conditions: The considered component under its real condition is connected with a thermodynamically-reversible operated black-box, which makes the determination of each endogenous exergy destruction fairly easy. Note that the ideal operation of the black-box scales the mass flow rates of all streams and may change the thermodynamic properties of streams flowing into and out of the considered component.

The endogenous investment cost of the kth component ($\dot{Z}_k^{\text{EN}}$) is reasonably determined by exergy product at the theoretical condition and the investment cost per unit exergy product at the real condition:

$$\dot{Z}_k^{\text{EN}} = \dot{E}_{\text{P},k}^{\text{EN}} \cdot (\dot{Z}/\dot{E}_{\text{P}})_k \tag{7}$$

Subsequently, the endogenous part is obtained:

$$\dot{Z}_k^{\text{EX}} = \dot{Z}_k - \dot{Z}_k^{\text{EN}}. \tag{8}$$

2.3.3. Combination of the Two Exergy-Destruction Splits

All possible splits of exergy destructions within each component as well as the related costs are given in Figure 4. The primary splits are endogenous/exogenous (split 1) and avoidable/unavoidable (split 2). Considering the endogenous/exogenous split for unavoidable exergy destruction/cost yields the split 3b with unavoidable-endogenous and unavoidable-exogenous parts calculated as follows:

$$\dot{E}_{\text{D},k}^{\text{UN,EN}} = \dot{E}_{\text{P},k}^{\text{EN}} \cdot \left(\dot{E}_{\text{D}}/\dot{E}_{\text{P}}\right)_k^{\text{UN}}, \tag{9}$$

$$\dot{E}_{\text{D},k}^{\text{UN,EX}} = \dot{E}_{\text{D},k}^{\text{UN}} - \dot{E}_{\text{D},k}^{\text{UN,EN}}, \tag{10}$$

$$\dot{Z}_k^{\text{UN,EN}} = \dot{E}_{\text{P},k}^{\text{EN}} \cdot \left(\dot{Z}^{\text{UN}}/\dot{E}_{\text{P}}\right)_k, \tag{11}$$

$$\dot{Z}_k^{\text{UN,EX}} = \dot{Z}_k^{\text{UN}} - \dot{Z}_k^{\text{UN,EN}}. \tag{12}$$

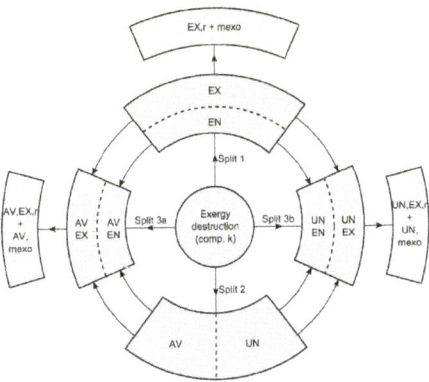

Figure 4. Complete splits of the exergy destruction in an advanced exergetic analysis [96].

Similarly, the avoidable exergy destruction/cost can be further split into avoidable-endogenous and avoidable-exogenous parts (split 3a):

$$\dot{E}_{\text{D},k}^{\text{AV,EN}} = \dot{E}_{\text{D},k}^{\text{EN}} - \dot{E}_{\text{D},k}^{\text{UN,EN}}, \tag{13}$$

$$\dot{E}_{\text{D},k}^{\text{AV,EX}} = \dot{E}_{\text{D},k}^{\text{EX}} - \dot{E}_{\text{D},k}^{\text{UN,EX}}, \tag{14}$$

$$\dot{Z}_k^{\text{AV,EN}} = \dot{Z}_k^{\text{EN}} - \dot{Z}_k^{\text{UN,EN}}, \tag{15}$$

$$\dot{Z}_k^{AV,EX} = \dot{Z}_k^{EX} - \dot{Z}_k^{UN,EX}.$$ (16)

Further insights can be obtained via the splits to consider the interaction between any two components ($\dot{E}_k^{UN,EX,r}$ and $\dot{E}_k^{AV,EX,r}$, $\dot{Z}_k^{UN,EX,r}$ and $\dot{Z}_k^{AV,EX,r}$) and the effects of the remaining components to the considered component ($\dot{E}_k^{UN,mexo}$ and $\dot{E}_k^{AV,mexo}$, $\dot{Z}_k^{UN,mexo}$ and $\dot{Z}_k^{UN,mexo}$).

An evaluation should consider all available data and be conducted in a comprehensive way. In general, improvement efforts should be made to those components with relatively high avoidable exergy destructions or costs. Besides, the sources of the avoidability are more reasonably identified and the improvement or optimization will not be misguided.

2.4. Applications

2.4.1. Conventional Exergy-Based Analysis

There has been a misuse of the term "exergy analysis" for its application in literature: Some references named with "exergy analysis" only calculated an overall exergy efficiency but did not perform a component-based analysis. Component-based exergy analysis has been intensively applied to various (coal-fired and gas-fired) thermal power plants with different capacities and operating parameters since 1980s. We summarize below the major findings related to major types of thermal power plants.

For coal-fired power plants ranging from 50–1440 MW, the overall exergy efficiency is reported from 25–37%, for which the exergy efficiency of the turbine subsystem over 80% and that of the boiler subsystem mostly below 50% [97]. All component-based analyses, e.g., [85,98], concluded similarly that the overall exergy dissipation is mostly contributed by the boiler subsystems, followed by the turbine subsystem and exergy losses. For modern coal-fired power plants, their exergy destruction ratios are over around 70%, 10% and 10%, respectively [85]. The boiler subsystem is mainly contributed by the combustion (around 70%) process and heat transfer (around 30%) process. The turbine system is dominated by the turbine (around 50%), followed by the condenser (around 20%) and other components. It is also obtained that along the improvement of the operating pressure and temperature, the overall efficiency is enhanced from 35 to over 40% for modern power plants, with the exergy destruction ratio of the boiler subsystem greatly reduced.

For gas-fired power plants, the overall exergy efficiency, over 50% depending on the operating parameters [99], is much higher than that of the coal-fired power plants. The major exergy destruction comes from the reformer and combustor with their overall exergy destruction ratio over 65%, followed by turbine, heat recovery system and air compressor, which contributed similarly by 4–8%. Varying the flue gas temperature at the gas turbine inlet can significantly enhance the overall exergy efficiency, almost 1 percentage point for each 50 °C increment.

For solar thermal power plants, the investigation of a 50 MWe parabolic trough plant [100] showed that the major exergy destruction is dominated by the collector-receiver (over 80%), whose exergy efficiency is as low as 39%. The remaining components, e.g., the boiler and turbine, contribute minor to the overall exergy dissipation. Increasing turbine inlet pressure from 90 bar to 105 bar enhances the overall exergy efficiency from 25.8% to 26.2%. The analysis of a solar tower power plant [101] showed that the overall exergy destruction is mainly contributed by the collector (heliostat field, 33%) and the central receiver (44%), whose exergy efficiency is around 75% and 55%, respectively. The overall efficiency of the considered solar tower power plants is around 24.5%, slightly lower than those reported for the parabolic trough plant evaluated in [100]. It should be noted that the performance of different types of solar collectors depends not only on the design itself but also the local solar irradiation, which might be one reason for the efficiency difference mentioned above.

The component-based exergoeconomic analyses have been applied to various steam cycles including subcritical or supercritical coal- and gas-fired power plants with the plant capacity ranging from 150 MWe to 1000 MWe, as summarized in [102]. These analyses clearly reveal the formation process of the cost of the final product, e.g., Figure 5 for coal-fired power plants [102]. For coal-fired power plants as detailed analyzed in [83,102], The air preheater and furnace have far less exergoeconomic factor indicating the related costs of these two components due to large exergy destruction rates, while the relative cost differences of the heat surfaces in the boiler subsystem are much larger than those of the turbine subsystem, mainly due to their high investment costs. The exergoeconomic performance of the turbine stages can be improved by enhancing the stage design and that of the feedwater preheater has a relatively small contribution from the investment costs.

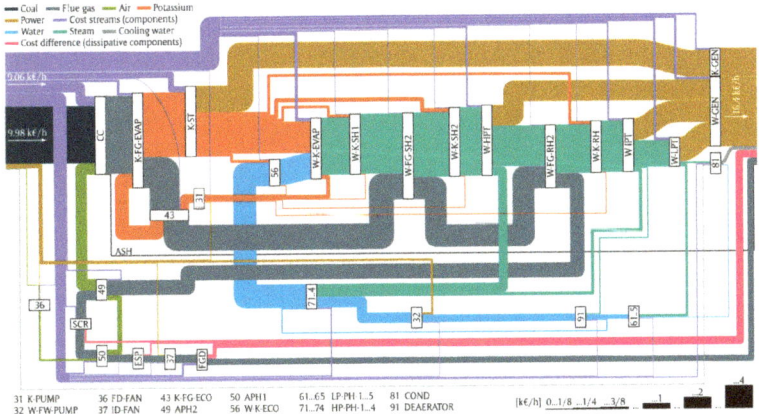

Figure 5. Cost formation process for coal-fired power plants [102]. The readers kindly refer [102] to interpret the involved abbreviations.

2.4.2. Advanced Exergy-Based Analysis

As summarized in Table 1, advanced exergy-based analysis has been initially (from 2006 to 2010) applied to simple systems (e.g., refrigeration system [88] and liquefied natural gas fed cogeneration system [89]) to assist the methodology development, particularly, proposing and comparing different calculation methods. The developed advanced analysis methods have been intensively applied to many different energy systems for various purposes, e.g., evaluating comparatively various power plants with CO_2 capture technologies [90,103–106], coal-fired power plants [85,107] with the anomalies diagnosis [108,109], gas-fired power plants [106,110], and concentrated solar thermal and geothermal power plants [98,111]. Most of them perform only advanced exergy analysis and only limited references have done advanced exergoeconomic and exergo-environmental analyses.

For coal-fired thermal power plants reported in [85,103–107], the major findings from advanced exergy analysis are (1) The contribution of the exogenous exergy destruction to the overall exergy destruction differs significantly from one component to another from 10% (e.g., turbine stages and boiler's component) up to 30% (feedwater preheater). However, in [98], it is mentioned that the exogenous exergy destruction obtained for the considered plant is directly proportional to the association degree, which might be due to an improper calculation procedure. (2) A large part (35–50%) of exergy destructions within heat exchangers and 30–50% within turbo-machines may be avoided; while this number for feedwater preheater is around 20%. (3) It is also found that most of the avoidable exergy destructions are endogenous; however, for some components, this number can be as high as 70%. The advanced exergoeconomics showed that around 10% of both total investment and exergy destruction costs of the system are avoidable. The boiler contributes the largest avoidable investment cost, while ST contributes the largest avoidable exergy destruction cost. For boiler's heating surfaces, steam turbine, most (over 60%) of the avoidable costs are endogenous, while for pumps and fans the most parts are exogenous.

Table 1. Summary of major applications of advanced exergy-based analysis for power plants.

Year	Authors	Applications	Component-Based	Advanced Exergy Analysis	Advanced Exergoeconomic Analysis	Advanced Exergoenvironmental Analysis
2006–2009	Morosuk and Tsatsaronis [88,93–95], Kelly et al. [76]	Absorption refrigeration machine, gas-turbine power plant	✓	✓		
2010	Tsatsaronis [89]	Liquefied natural gas fed cogeneration system	✓	✓		
2010–2012	Petrakopoulou et al. [90,103–106,112]	Power plants with CO_2 capture	✓	✓	✓	✓
2013	Yang et al. [85,107,113,114]	Ultra-supercritical coal-fired power plants	✓	✓		
2013	Manesh [115]	Cogeneration system	✓	✓	✓	✓
2014	Acikkalp et al. [110]	Natural gas fed power-generation facility	✓	✓		
2015	Tsatsaronis [116]	Gas-turbine-based cogeneration system	✓	✓	✓	✓
2015	Bolatturk [117]	Coal-fired power plants	✓		✓	
2016	Zhu et al. [98]	Solar tower aided coal-fired power plant	✓	✓		
2016	Gökgedik et al. [111]	Degradation analysis of geothermal power plant	✓	✓		
2017	Wang and Fu et al. [108,109]	Anomalies diagnosis of thermal power plants	✓	✓		

For gas-fired thermal power plants/facility, it is reported in [104,110,115] that the combustion chamber, the high-pressure steam turbine and the condenser have high improvement potentials and the interactions between components are weak reflected by a contribution of the endogenous exergy destruction of 70%, which seems quite different from that identified for coal-fired power plants. The total avoidable exergy destruction is calculated as around 38% of the total.

2.5. Limitations

Analysis methods can evaluate thermodynamic inefficiencies of a specific system and potentially guide parametric optimization of the analyzed system. These methods can assist the improvement of system flowsheet if combining with engineers' experience and judgments. However, they cannot, at least until now, optimize the design and operating variables and generate structural alternatives automatically and algorithmically, for which mathematical programming is usually needed for system optimization and synthesis to be discussed in the following sections.

3. Optimization of Energy Systems

System analyses introduced in Section 2 cannot realize systematic and automatic design and operational improvement of energy systems, which can be achieved via mathematical optimization. A general optimization problem consists of an objective function to be minimized or maximized, equality and/or inequality constraints, and the considered independent decision variables. For energy systems, there are usually three types of decision variables [118], i.e., binary structural variables (s) associated with the structure of the system, continuous or discrete design variables (d) related to nominal characteristics and sizes of the system and the components, and continuous or discrete operational variables (o) determining operation strategies at the system and/or component levels. Note that structural variables (s) refers to the degrees of freedom in the system structure and will be discussed in detail in Section 4 (synthesis of energy systems).

The optimization model discussed in this section can be formulated as follows:

$$\min_{d,o} f(d,o), \tag{17}$$

$$\text{s.t.} h(d,o) = 0, \tag{18}$$

$$g(d,o) \leq 0, \tag{19}$$

where f is the objective function, h and g represent the equality and inequality constraints.

Generally, the algorithms for different optimization problems can be divided into deterministic algorithms and metaheuristic algorithms [119], most of which have been well developed with various solving methods and solvers. Deterministic methods are usually solved by mathematical approaches with or without the aid of special speed-up techniques associated with thermodynamics or thermo-economics (e.g., [120]).

This section is organized as follows: Mathematical optimization is introduced in Section 3.1, focusing on deterministic (Section 3.1.1) and meta-heuristic (Section 3.1.2) methods. Then, the application to thermal power plants is summarized in Section 3.2 with insights on nonlinearity and integrity in Section 3.2.1, scope and key results in Section 3.2.2, and limitations in Section 3.2.3.

3.1. Mathematical Optimization

Depending on whether discrete (i.e., integer) decision variables are incorporated, the optimization problems are first classified as continuous and discrete. Then, considering the nature of functions involved, important subclasses are further identified: (continuous) linear programming (LP), (continuous) nonlinear programming (NLP), integer programming (IP), mixed integer linear programming (MILP), mixed integer nonlinear programming (MINLP), generalized disjunctive programming (GDP), etc.

The algorithms for different optimization problems, either deterministic or metaheuristic [119], have been well developed and exhaustively reviewed in many references, e.g., a comprehensive description of the most effective methods in continuous optimization [121], an extensive review on mathematical optimization for process engineering [122,123], recent advances in global optimization [124], derivative-free algorithms for bound-constrained optimization problems [125,126], and a broad coverage of the concepts, themes and instrumentalities of metaheuristics [119]. According to these, the basis of commonly used deterministic and metaheuristic optimization algorithms associated with the scope of this review are briefly introduced below.

3.1.1. Deterministic Algorithms

For a specific input, a deterministic algorithm always passes through the same sequence of the search pattern and converges potentially fast to the same result. The algorithms usually take advantage of the analytical properties of the optimization problems; thus, the problems need to be well formulated to avoid misguiding the search. However, for good formulations, particularly of complex problems, the user may have to manually address some trivial issues [127], e.g., scaling of (intermediate) variables and functions. In addition, the search may end up with bad local optimal solutions for complex problems. The optimization of LP, if no global solution algorithm is used, is a relatively mature field. For a well-conditioned linear problem with the abounded objective function, the feasible region is geometrically a convex polyhedron, which implies a local extremum is always globally optimal. The optimal solution, possibly not unique, is always attained at the boundary of the feasible region. The optimality can be reached with a finite steps, from any feasible solution either at the boundary (primal-dual simplex algorithms [128]) or at the interior (interior point algorithms [129]) of the feasible region. Several modern solvers, e.g., XPRESS, CPLEX, and IPOPT, are capable of handling LP with an unlimited number of variables and constraints, subject to available time and memory.

For NLP problems, the optimal solution can basically occur anywhere in the feasible region. Most NLP algorithms require derivative information of the objective function and constraints for efficiently determining effective searching directions. Commonly used solvers are usually based on successive quadratic programming (SQP), e.g., IPOPT, KNITRO, and SNOPT, which generate Newton-like steps and need the fewest function evaluations, or generalized reduced gradient (GRG), e.g., GRG2 and CONOPT, which work efficiently when function evaluations are relatively cheap.

MILP problems have a combinatorial feature and are usually NP-hard [130]. The solving algorithms are mostly based on a branch-and-bound idea, which incorporates a systematic rooted-tree enumeration of candidate solutions by "branch" and efficient eliminations of non-promising solutions by "bound". The algorithm can be further enhanced, as branch and cut, by introducing cutting planes (linear inequities) to tighten the lower bound of LP relaxations. The best-known MILP solvers include CPLEX, XPRESS.

Mixed integer nonlinear problems are also NP-hard. The solving idea is similar by generating and tightening the bounds of the optimal solution value. The algorithms, generally branch-and-bound or branch-and-cut like, rely on relaxations of the integrity to yield NLP subproblems and (linear) relaxations of the nonlinearity [131].

There is another problem of the above-mentioned MINLP methods: when fixing certain discrete variables as zero for branching or approximation, the redundant equations and intermediate variables may cause singularities and poor numerical performance [132]. To circumvent this, GDP methods have been developed as an alternative and receive increasing attention (see [133]). In GDP, the combination of algebraic and logical equations is allowed, thus the representation of discrete decisions is simplified. However, the algorithms for GDP are mostly under development (see [134]) and currently only the LOGMIP software [135] is available.

In addition, state-of-the-art solvers for deterministic optimization have been highly integrated with several well-developed high-level algebraic modeling environments, e.g., GAMS and AMPL, tailored for complex, large-scale applications.

3.1.2. Metaheuristic Algorithms

Metaheuristic algorithms are capable of escaping from local optima and robustly exploring a decision space. Although the metaheuristics are still not able to guarantee the global optimality for some classes of problems, e.g., MILP and MINLP, they can generally find sufficiently good solutions. Commonly used algorithms mainly include single-solution based, e.g., simulated annealing, tabu search, and population-based, e.g., evolutionary algorithms, ant colony optimization, and particle swarm optimization. Moreover, metaheuristic algorithms can be applied to highly nonlinear (even ill-conditioned) or black-box problems. The major disadvantages, however, include potential slow speed of convergence, unclear termination criterion, incapability of certifying the optimality of the solutions, and the potential need for designing problem-specific searching strategies.

In the following, the basis of population-based evolutionary algorithms is briefly introduced. Evolutionary algorithms (EAs), inspired by biological evolution, are generic, stochastic, derivative-free, population-based, direct search techniques. EAs can often outperform derivative-based deterministic algorithms for complex real-world problems, even with multi-modal, non-continuous objective function, incoherent solution space, and discrete decision variables; moreover, the global optimality, although not guaranteed, can be closely approached by a limited number of function evaluations.

The basic run (Figure 6) of an evolution algorithm (EA) starts from an initialization, in which a set of candidate solutions (population and individuals) are proposed and evaluated for assigning the fitnesses (the objective function value, if feasible; otherwise, a penalty value). Afterward, for evolving the current parent population to an offspring population, the algorithm starts an iteration loop: parent selection, recombination (crossover), mutation, evaluation and offspring selection. To produce each new individual, based on the fitness values, one or more parents are selected for crossover and mutation: A crossover operation randomly takes and reassembles parts of the selected parents, whereas a mutation operation performs a small random perturbation of one individual. The newly born offsprings are then evaluated; finally, a ranking of offspring (and parent) individuals is performed, so that those individuals with the larger possibility of leading to the optimality survive and are selected as the offspring population. The iteration continues until certain termination criterion, e.g., a limit of computation time, fitness-evaluation number, or generation number, is reached.

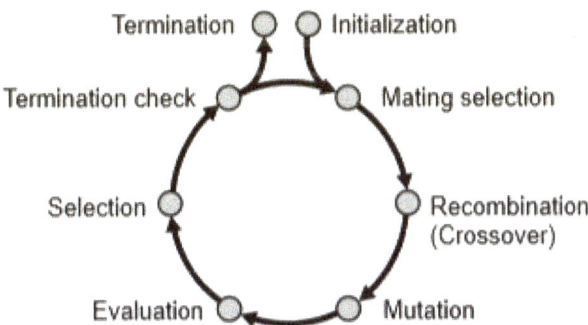

Figure 6. Flowchart of an evolutionary algorithm (adapted from [136]).

Selection, crossover and mutation are three genetic operators of evolutionary algorithms for maintaining local intensification and diversification of the search. Different strategies on these three aspects lead to a variety of evolutionary algorithms. Selection strategy mainly exerts influence on population diversity. One commonly used strategy of selection is the $(\mu + \lambda)$-selection proposed in evolution strategies [137], where μ and λ, satisfying $1 \leq \mu \leq \lambda$, denote the sizes of parent and offspring populations, respectively. Selection ranks the fitness of all $\mu + \lambda$ individuals and takes the μ best individuals. Depending on the search space and objective function, the crossover and/or the mutation may or may not occur in specific instantiations of the algorithm [119,137]. There are different mechanisms of crossover and mutation. For example, genetic algorithm [138] usually employs bit strings to represent variables. Besides, differential evolution (DE [139]), mentioned as the fastest evolution algorithm [139], does not rely on any coding but directly manipulates real-valued or discrete variables. Basically, for mutation DE adds the weighted difference between two parents' variable vectors to a third vector, thus the scheme remains completely self-organizing without using separate probability distribution and has no limitations for implementation compared to other evolutionary algorithms.

3.2. Applications to Thermal Power Plants

3.2.1. Nonlinearity and Integrity

The optimization problems of thermal energy systems are usually highly constrained and nonlinear, thus belong to NLP or MINLP. The nonlinearity and integrity may be led to by thermodynamic properties of working fluids, design and operational characteristics of components, the investment cost functions of components, energy balance equations, etc. These need to be well addressed, so that the problems, in the best case, can be transformed to LP or MILP for deterministic optimizations.

For the properties of working fluids, particularly water and steam (IAPWS-IF97 [140]), the highly nonlinear exact mathematical formulations can hardly be employed. One direct means incorporates polynomial approximations of low degrees of nonlinearity at the expense of accuracy [141–145]. However, inaccurate regressions may result frequently in non-applicable "optimal" solutions.

Another approach evaluates the property's value and associated derivatives of high accuracy based on reformulated exact formulations or reprocessed steam tables, e.g., TILMedia Suite [146] and freesteam [147] library. in these libraries, the discontinuities and even jumps of the thermodynamic properties are smoothed, and the integer variables indicating the state zones are encapsulated.

The nonlinear (or perhaps discrete) thermodynamic (operational) behavior of components can be properly reformulated. For example, for modeling turbine, alternatives include constant entropy efficiency model, Willan's Line [148], Turbine Hardware Model [149] and Stodola ellipse [150]. In those models, the set of variables which the isentropic efficiency depends on differs, thus the predictions of the off-design behavior are also different in accuracy. For heat exchangers, the logarithmic mean temperature difference can be replaced by a refinement of the arithmetic mean [151]. While for mixers, the discrete equality nonlinear relationship of the flow pressures between inlets and outlet can be either relaxed as an inequality nonlinear constraint [152] or linearized by introducing additional integer variables [153].

The investment cost functions are always needed if an economic objective is involved in the optimization. A cost function links the purchased equipment cost of one component with its key characteristic variables and associated flow parameters; thus, the function may be of high nonlinearity. To cope with this, cost functions are usually reformulated with separable terms of each variable, which are subsequently piecewise linearized with the aid of integer SOS2 variables [154].

Continuous nonconvex bilinear term ($v_1 \cdot v_2$) is another common source of nonlinearity, e.g., the term $\dot{m} \cdot h$ involved in energy balance equations. This nonconvex nonlinearity is usually handled by a convex/concave McCormick relaxation [155] or a quadratic reformulation. For the latter approach, two new variables $z_1 = (v_1 + v_2)/2$ and $z_1 = (v_1 - v_2)/2$ are introduced to replace the bilinear term with $z_1^2 - z_2^2$. The quadratic term can also be further linearized by SOS2 variables.

3.2.2. Scope and Key Results

Given a specific structure of an energy system, the application of optimization on the energy systems becomes an easy task, since integer variables are seldom involved for a given system layout. Dated back to half century ago, the first applications of mathematical optimization to thermal power plants or steam cycles, i.e., [156,157], were realized by analytical deduction to find the optimal heat-load distribution among feedwater preheaters, which derived the two well-known methods of equal increase in feed water enthalpy or temperature. Nowadays, the optimization methods are seldom used to optimize only the continuous variables in literature, but they are mostly combined with the optimization of non-continuous or integer variables to be discussed in Section 4, which can be optimized to bring larger benefits for performance improvement. Thus, the limited relevant references are summarized in Table 2.

Parametric optimization of steam cycles can be performed by mathematical optimization with thermodynamic, economic or environmental objectives, e.g., [158], or combining with thermoeconomic techniques for an economic optimization, e.g., [159,160]. The cost-optimal design of a dry-cooling system for power plants was investigated in [161] with SQP and relevant decomposition methods, which showed that with well-structured optimization problem and solving strategy, the direct optimization of complex problems is not necessary to be time-consuming and difficult. Similar optimization problem for modern coal-fired power plants is solved in [162] considering more comprehensively the off-design performance of the whole plant calibrated with historical operating data, thus potentially yielding practical operating strategies to cope with different operating scenarios of power plants. The SQP algorithms are also employed in [158] to optimize the steam cycles considering its interaction with boiler cold-end, which took the steam-extraction pressures as independent variables to optimize the overall plant efficiency. An efficiency gain of 0.7 percentage points was achieved. The implementation of the optimization utilized Aspen Plus to simulate the plant performance with given decision variables.

Combining thermoeconomic techniques for economic optimization, Uche et al. [159] performed global optimization of a dual-purpose power and desalination plant with cost savings of approximately 11% of the total cost at nominal operating conditions. Similarly, Xiong et al. [160] optimized the operation of a 300 MW coal-fired power plant using the structural theory of thermoeconomics and obtained a 2.5% reduction in total annual cost.

Using heuristic methods, particularly genetic algorithms and artificial neural network (ANN), to optimize thermal power plants is quite late since 2010. In [163], these two algorithms were employed to optimize the plant efficiency considering 9 design parameters, including the pressure of main and reheated steam, the pressure of steam extractions. The optimizer employed professional process simulator for evaluating the plant efficiency at the lower level, while the upper level with GA and ANN varied the decision variables and optimize the plant efficiency. In this case, the nonlinearity involved can be handled more efficiently via professional simulators. It is also concluded that the coupled GA-ANN algorithm can greatly improve the computational performance without loss of accuracy, thus is suitable for online applications. The optimal plant efficiency from the GA-ANN algorithm is slightly better than that obtained from mathematical programming approach, indicating that the heuristics methods may achieve the global optimum. More (ten) decision variables were considered in [164] to maximize plant efficiency and minimize the total cost rate. One design point identified showed a 3.76% increase in efficiency and a 3.84% decrease in total cost rate simultaneously, compared with the actual data of the running power plant. A correlation between two optimum objective functions and 15 decision variables were investigated with acceptable accuracy using ANN for decision making.

It should also be mentioned that the "optimization" term has been widely misused in literature. In many references, e.g., [165,166] for solar thermal power plants, the "optimization" was achieved by sensitivity analysis.

Table 2. Summary of the application of optimization to thermal power plants or steam cycles.

Year	Authors	Application	Objective Function	Method
1949, 1960	Haywood [156] and Weir [157]	Steam cycles	Optimal heat-load distribution of feedwater preheating system	Analytical deductions
1998, 2018	Conradie et al. [161], Li et al. [162]	Cooling systems for thermal power plant	Cost or net-power increment	SQP algorithms
2014	Espatolero et al. [158]	Layouts of feedwater preheating and flue-gas heat recovery system	Steam-extraction pressures	SQP algorithms
2001, 2012	Uche et al. [159] and Xiong et al. [160]	Steam cycles	Local cost optimization	Quadratic programming (QP) approximation
2011, 2012	Suresh et al. [163] and Hajabdollahi et al. [164]	Coal-fired power plant	Plant efficiency and/or cost	Genetic algorithm and artificial neural network

3.2.3. Limitations

As mentioned above, without the consideration of structural variables, the parametric optimizations only explore a limited number of design structures. More importantly, the structural options are generated not in a systematic way. Consequently, the best solutions searched may be far away from the optimal solution. In the following Section 4, we introduce the optimal synthesis of energy systems, which specifically copes with such an issue.

4. Synthesis of Energy Systems

The optimization discussed in Section 3 handles only parametric optimization to find the best design and operational variables; however, the optimization of a process structure (topology), process synthesis, may contribute more to the improvement of system performances. Process synthesis, namely complete flowsheet synthesis when performed at an overall system level, deals with the selection of process structure (topology), i.e., the set of technical components employed and their interconnections. The optimal synthesis phase usually contributes a major part to achieving the predefined goal or finding the globally optimal design option [167]. However, optimal synthesis tends to be a tough task compared to a simple design or operation optimization: It normally takes the design and/or operation optimization into account in a sequential or simultaneous fashion; moreover, the design space of structural alternative is basically not known a priori for a complex system, thus a complete, exact mathematical formulation of the synthesis problem seems not possible [168]. To systematically address the synthesis of energy and process systems, a vast number of research has been conducted in this field and methodologically reviewed by many researchers, e.g., [169–172]. Accordingly, the synthesis methodologies can be basically categorized into three groups, which are complementary to each other: (a) heuristic methods, (b) targeting or task-oriented methods, and c) mathematical optimization-based methods.

The heuristic and targeting methods are knowledge-based. The heuristic methods incorporate rules derived from long-term engineering knowledge and experience. The aims are to propose "reasonable" initial solutions and improve them sequentially. One influential method in this group is the hierarchical decision procedure for process synthesis [173], which introduces common concepts for almost any systematic synthesis method proposed afterward, such as [174,175]. The method explores the process nature by sequential decomposition and aggregation for further improvement [176] and has been extended for synthesizing complete flowsheet of the separation system [177]. Other heuristic rules based methods and practices can be found elsewhere, e.g., [171].

The targeting methods integrate physical principles to obtain, approach and even reach the targets for the optimal process synthesis. The most widely applied targeting method is the pinch methodology [178], which is fundamentally developed for the systematic synthesis of HEN. The method has been extended for complete flowsheet synthesis of total site utility systems [148,179].

To realize automatic and computer-aided synthesis using these guidelines, a number of knowledge-based expert systems have been developed for various processes and systems, such as chemical processes [180–182], thermal processes [183–185] and renewable energy supply systems [186]. Expert systems apply various logical inference procedures, e.g., means-end analysis [187] and case-based reasoning [188], to reproduce engineers' design maps, thus suggest the best-suited process for a particular application.

The heuristic and targeting methods are generally effective to quickly identify suboptimal structural alternatives [171]. However, they are unable to guarantee the optimality, mainly because of the sequential nature and mathematically non-rigorousness. Thus, much more comprehensive methods, the mathematical optimization-based methods, have been greatly developed.

The optimization-based methods consider simultaneously the structural options, design and operation conditions, and perform rigorously with any objective function. In these methods, a synthesis task is formulated as a mathematical optimization problem with an explicit (superstructure-based) or implicit (superstructure-free) representation of considered structural alternatives, among which the optimal structure is identified.

In the following, the optimization-based synthesis methods are reviewed in more details. In Section 4.1, superstructure-based synthesis is discussed with superstructure representation, superstructure generation, modeling and solving methods and strategies. Then, superstructure-free methods are reviewed in Section 4.2. Finally, the application to thermal power plants are summarized in Section 4.3.

4.1. Superstructure-Based Synthesis

The superstructure explicitly defines a priori structural space to mathematically formulate the synthesis problems. The superstructure concept was first proposed by Duran and Grossmann [189] to describe the outer approximation algorithm for solving MINLP, and was initially illustrated for addressing process synthesis issues in HEN [190]. Later, the synthesis concept was generalized as a systematical superstructure-based synthesis method [132,191,192], which has been widely applied to a multitude of process synthesis with different levels of detail, such as HEN [193,194], separation and distillation sequences [195], water networks [196], polygeneration process [197], steam utility systems [142,198], and thermal power plants [199–202].

The superstructure-based synthesis aims at locating the optimal solution from all possible alternatives embedded in the superstructure, which represents all considered components and the possible links. The fundamental basis of the superstructure-based synthesis involves three aspects: superstructure representation and generation, superstructure modeling and mathematical optimization of the problem.

4.1.1. Superstructure Representation

A (super)structure can be presented in forms of string, connectivity matrix or graph, such as digraph, signal-flow graph, P-graph (for these three types, see [203]) and S-graph [204]. The string-based representation is favorable for applying replacement rules (grammars), such as in a string rewriting system [205] for HEN [206]; however, the grammars tend to be too complicated for presenting detailed flowsheets. The connectivity matrix, digraph and signal-flow graph are only suitable for process analysis, e.g., the matrix representation in the structural theory of thermoeconomics [67], but may become ambiguous for variable structures. P-graph [196] represents the structure of a process (system) in a unique and mathematically rigorous form, while S-graph is more suitable for representing a detailed flowsheet. Current software status (see [207]) allows for the modular graphical representation of a flowsheet, e.g., [208,209].

4.1.2. Superstructure Generation

For most applications, the superstructure considers only a limited number of promising alternatives, which may be generated by knowledge-based methods, such as heuristic rules [175,177] and thermodynamic insights [174,210]. Great efforts, e.g., [211], have been made to enhance user-friendly generation. However, the generation procedure usually requires trivial manual interactions and specifications. More importantly, many good alternatives may be left out of the solution space spanned by the superstructure.

In principle, an excessively large superstructure can include as many good alternatives as possible. However, it may encompass also a large number of meaningless or even infeasible alternatives, which potentially lead to the forbiddingly large computational effort, as the computation complexity and difficulty of the optimal synthesis problems almost always increase exponentially with the number of components considered in the superstructure [185,203,212]. In addition, for realistic problems, the number of structural alternatives tends to be very large, e.g., over 10^9 structural enumerations for the feed-water preheating train of thermal power plants [213]. Considering the current computation capability of mathematical programming, it is basically not possible to take all possible alternatives into account.

To cope with some of these fundamental problems, many systematic or even algorithmic generation of superstructure have been developed. Toward systematic generation, there are stage-wise synheat superstructure for HEN [214], multi-level hierarchical aggregation [175,215,216], state-task and state-equipment network [217] for process systems, or decision tree [213,218] for power plants.

The algorithmic generation of superstructure automatically and systematically ignores structurally infeasible structures. The most prominent algorithms [219,220] are based on the P-graph representation. The P-graph framework explores the combinatorial nature of considered technical components and minimizes the number of components in the maximal structure [219]. Therefore, the complexity of the superstructure is reduced. The P-graph framework was originally proposed for synthesizing chemical processes and has been deployed to a wide range of synthesis problems (see [221]). The detailed implementation of the framework is introduced by Bertok et al. [209]. The disadvantage of the original framework, however, is that multiple redundant instances of one type of technical components are not considered. Recently, a combinatorial algorithm was proposed to add necessary redundancy of supply chains [222].

To enable the automated synthesis of distributed energy supply systems, Voll et al. [223] proposed a superstructure generation algorithm based on the P-graph framework. The algorithm first generates a maximal structure considering all feasible types of components. Then, the maximal structure is successively expanded by adding multiple redundant components, which is achieved by manipulating the connectivity matrix. However, limited by the matrix representation, the connections of the newly added redundant component are identical to the already existing component of the same type.

Although these methods make superstructure generation an easy task for certain processes, there are more challenges for complex energy systems: A complex flowsheet comprises only several types of components, which indicates that multiple redundant components are always involved with different connections. Additionally, one task may be fulfilled by several sequentially or parallelly connected components of the same or different types. Thus, it seems these generation methods are not adequate for such applications.

4.1.3. Superstructure-Based Modeling and Solving

The superstructure is usually modeled by introducing binary selection variables to allow the activation/deactivation of each considered component, as reviewed in [122,131,168]. Such superstructure-based problems are generalized as MILP, MINLP or GDP:

$$\min_{s,d,o} f(s,d,o), \tag{20}$$

$$\text{s.t.} h(s,d,o) = 0, \tag{21}$$

$$g(s,d,o) \le 0, \tag{22}$$

$$s \in \{0,1\}^n, \tag{23}$$

where the vector s contains n binary structural variables indicating the (non-)existence of components for design synthesis and the on/off-state of components (when involving operation synthesis). Note that a superstructure can be formulated at different levels of details [131]: (a) aggregate models concerning only major features like energy balance [190] (b) short-cut models considering simple nonlinear models for component performance and (c) rigorous models involving detailed modeling of component performance [143,224]. The solving algorithms have been introduced in Section 3.1.

Since the whole model is usually difficult and expensive to solve, many speedup techniques have been developed for different applications. For instance, several decomposition methods, e.g., [48,214,225,226], can partition the superstructure into several subproblems of smaller size. Another approach implicitly indicates the existence of considered components by using continuous variables, e.g., use zero mass flow rate to bypass the components for non-existence [227,228]. In this way, the discrete decision variables are eliminated, and the synthesis problems are reformulated to continuous optimization problems; while, the quality of local solutions highly depends on initial specifications.

For addressing the global optimization with many discrete decision variables, hybrid algorithms combining metaheuristic algorithms and mathematical programming (memetic algorithm) become popular. For example, Urselmann et al. [229] proposed a two-level memetic algorithm, where the upper level the integrity constraints and discontinuous cost functions are handled by genetic algorithm, while in the lower level continuous sub-problems are efficiently solved by robust solvers of mathematical programming for state variables [230].

4.2. Superstructure-Free Synthesis

The fundamental problems of superstructure-based optimization remain: On the one hand, good alternatives (particularly, the optimal solution) might be excluded from the superstructure, while on the other hand, many meaningless or even infeasible alternatives may be considered. To overcome these problems, superstructure-free approaches apply metaheuristic algorithms to explore a practically unconstrained solution space, which is not limited a priori by a superstructure model.

In fact, back to 1970s, Stephanopoulos and Westerberg [231] have outlined a crucial view of the evolutionary synthesis: Given an initial structure and rules to systematically adjust the structure with small changes, an effective strategy applying the rules produces neighbor structures and thus "enumerates" all feasible structures, in which the optimal structure lies. Based on this idea, Seader and Westerberg [232] synthesized a simple separation sequence. Modern superstructure-free approaches apply metaheuristic algorithms, which perform "intelligently" and stochastically, thus many unpromising structures are not considered. Two-level hybrid algorithms are always involved: the upper level manipulates the structural representation (e.g., S-graph, see Section 4.1.1) for generating structurally feasible structures, while the lower-level evaluates the generated structures.

$$\min_{\sigma} f(s(\sigma), d, o) \quad \sigma \in \Sigma, \tag{24}$$

$$\text{s.t. } \min_{d,o} f(s(\sigma), d, o), \tag{25}$$

where the term of σ is solution structure evolved by mutation. The term of Σ is the space of all structure alternatives that can be possibly reached by repeated structural mutation. To exploit the bi-level formulation, the superstructure-free optimization employs a hybrid algorithm combining an evolutionary algorithm for the upper level with deterministic optimization for the lower level. The upper-level evolutionary algorithm generates structural alternatives s, i.e., units-selection and interconnections among the employed units, while each alternative generated by the upper level is then optimized deterministically in the lower level, i.e., identification of optimal sizing d and operation o of the employed units. The structural decisions s are not explicitly modeled in a superstructure, but the structures are evolved with the new structural alternatives σ generated by an evolutionary algorithm.

For HEN synthesis, Fraga [206] proposed a set of grammars for string representations to add heat exchangers and split streams. With a string rewriting system, the genetic algorithm can generate complex networks. Toffolo [233] proposed a more flexible graph representation, with which genetic algorithms were used to perform the insertion and deletion of the heat exchanger, and swaps of hot and cold sides of two heat exchangers. However, these approaches are tailored to HEN synthesis.

Wright et al. [234] performed both mutation and crossover to heating, ventilating and air conditioning system for an evolutionary synthesis. The mutation swaps two randomly selected components or their interconnections, while the crossover allows the offsprings inheriting structural properties and technical specifications from two parents either separately or in an equal measure. However, this approach is basically incapable of being extended to other applications.

Toffolo [235] proposed a hybrid algorithm, which is further developed as *SYNTHSEP* [236] approach for complete flowsheet synthesis of thermal power plants. The approach decomposes a thermal system into the heat transfer section and the remaining parts (basic configuration) by a heat separation decomposition [237]. The algorithm sequentially synthesizes the basic configuration by genetic algorithm and SQP, and the heat transfer section by pinch method.

Emmerich et al. [204] proposed an S-graph based genetic algorithm making total flowsheet synthesis of energy and process systems more flexible. A set of symmetric replacement rules for generating the closest neighboring structures are defined as minimal moves, such as insert a heater parallel to an existing heater or swap a by-product stream with a recycle stream. The minimal move mutation operator recognizes existing patterns, such as one component or a set of components, and replaces them with similar patterns according to the replacement rules; while the crossover operator recognizes and swaps the subsystems in the parents, which possess the same function and similar connection patterns. This approach has been applied to chemical process [204,238] and thermal power plant [239]; however, the major problem lies in the problem-specific replacement rules, which largely limit its extendibility. To cope with this problem, Voll et al. [240] and Wang et al. [241,242] further developed this approach by combing an energy conversion hierarchy (ECH), which allows for generic replacement rules.

4.3. Applications

4.3.1. Superstructure-Based Synthesis

The applications of superstructure-based concepts for various thermal power plants are summarized in Table 3. For these identified references, different types of structural representations were employed: decision tree, graph theory, predefined superstructure, algorithmically-generated superstructure of steam cycles with multiple pressure levels.

Regarding the use of *decision tree*, Hellermeier et al. [213] investigated the design synthesis of feedwater preheater train of thermal power plants via genetic and stochastic optimization techniques. The structural alternatives are represented by (modified) decision trees (as a type of superstructure) to consider the hierarchical parameter dependencies, and a set of rules are defined to find a feasible layout (a tuple of values of decision variables). Once the decision variables are selected, there will be a run through the decision tree with a set of predefined rules to collect and ensure the dependent variables for matching the chosen decision variables. Two solving strategies are tested and compared to cope with structural variables: (1) bi-level hierarchical method with the upper-level algorithms handling structural variables and the lower-level SQP algorithm addressing continuous variables, and (2) single-level evolutionary method with appropriately adapted genetic algorithms handling simultaneously structural and continuous variables. Two case studies with 8 and 26 discrete variables respectively are employed to compare the two solving strategies. It was found that, for both case studies, the one-level evolutionary algorithm performed slightly better with better objective value found and fewer iterations (computational time). The optimal system layouts found by both methods differed from each other. It was also concluded that varying plant layouts had a more prominent effect on the values of objective function than the optimization solving only continuous variables.

Regarding the use of *graph theory*, Grekas and Frangopoulos [200] employed nodes and edges to represent component and streams (connection), and stored the topology of the graphs as digraph of the data structure. The mathematical models of the components are automatically added to the optimization problem by object-oriented programming, once the components are used. With a parametric representation of component usage and connection, a plant layout can be automatically generated with the corresponding mathematical model formed. Then, binary trees for mass and energy balance as well as a digraph for pressure hierarchy are generated to introduce splitting ratios of mass and energy at each node and the pressure of each connection as decision variables to form a feasible process. The optimization is reported to be efficient, less than 5 min for the considered examples.

Table 3. Summary of the applications of superstructure-based synthesis of thermal power plants/steam cycles.

Year	Author	Application	Structural Representation	Platform or Solving Method/Technique	Multi-Level
2000	Hillermeier et al. [213]	Feedwater preheater train	Modified decision tree	Bi-level hierarchical method with SQP; one-level evolutionary optimization	√ / ×
2007	Grekas et al. [200]	Gas-fired combined cycle	Graph theory	Object-oriented programming and application programming interface	×
2010	Ahadi-Oskui et al. [201,202]	Combined-cycle-based cogeneration plant	Directly-coded in GAMS	Outer approximation and branch-and-cut	×
2014	Wang et al. [208,243]	Coal-fired power plants	Graphical flowsheet	Commercial simulator, differential evolution	×
2018	Maréchal and Kalitventzeff [244], Kermani et al. [245], Wallerand et al. [246]	(Organic) Rankine cycle, steam network, heat pump network	Algorithmic generation	AMPL, Integer cut	√

Regarding the use of *predefined superstructure*, it can be represented (i) directly coded via mathematical programming languages [201,202], e.g., AMPL and GAMS, which can easily form a graph with the *set* data type, or (ii) graphically in professional simulators, e.g., [208,243]. The directly-coded superstructure allows simultaneous solving of structural and continuous variables, if the component models and thermodynamic properties of material flows are properly formulated. While the graphical superstructure via simulators is usually integrated into a bi-level solving procedure, where the upper-level iteratively updates the value of decision variables and the lower-level forms a specific structure from the superstructure and employs the simulator to obtain the objective functions. The benefits of such bi-level optimization are efficient solving of nonlinear processes. In [202], a superstructure of gas-fired combined cycle was coded in GAMS and solved with the proposed LaGO solver, which generates a convex relaxation of the MINLP and applies a Branch and Cut algorithm to the convex relaxation. It was concluded that, for the optimal design of combined cycle, the focus should be set on the configuration of the steam cycle with the consideration of process steam extraction, which defined the complexity of the design problem. In [208,243] the superstructure of steam cycle considers up to 10 feedwater preheaters and a secondary steam turbine and was solved with differential evolution, which handles both the structural and continuous variables. The effects of temperature and pressure of main and reheat steam on the plant efficiency were investigated with the optimization of steam-extraction pressure and mass flowrate. Some design guidelines were found to support future plant design. The disadvantages of such *predefined superstructure* are mainly the poor extendibility and the risk of leaving many good alternatives out of consideration; therefore, there are no other applications in this category found for thermal power plants.

For the use of *algorithmically-generated superstructure of steam cycle*, the most representative and applied is the *steam network* routine continuously developed by the group of Industrial Process and Energy Systems Engineering at EPFL, following the original idea published in [244]. The *steam network* is a part of the flagship tool OSMOSE for the optimal conceptual design of industrial processes and energy systems. OSMOSE can flexibly plug-in customized energy technologies to be considered and have been applied to solve various optimization problems with multiple trade-off solutions, e.g., biomass utilization [247,248], energy storage systems [5,249]. The bi-level architecture of OSMOSE is given in Figure 7: The upper (master) level employs evolutionary algorithms to handle the nonlinear variables, with whose values the lower (slave) level prepares the input data with AMPL (AMPL coded superstructure and the corresponding mathematical model for mass and energy balance, utility selection and sizing, heat cascade calculation etc.) and solves the optimization regarding various objective functions (e.g., capital or operational expenses) with CPLEX or Gurobi. The objective values are returned to the upper level for solution comparison and selection. The advantages of OSMOSE platform include (1) easy coupling of professional simulators (e.g., Aspen Plus) to address complex processes, (2) allowing flowsheet decomposition/reuse and easy extension of technology (flowsheet) library, (3) integration of mathematically-formulated heat cascade calculation, (4) integer programming to consider optimal selection of utilities, (5) easy handling of multiple objective functions.

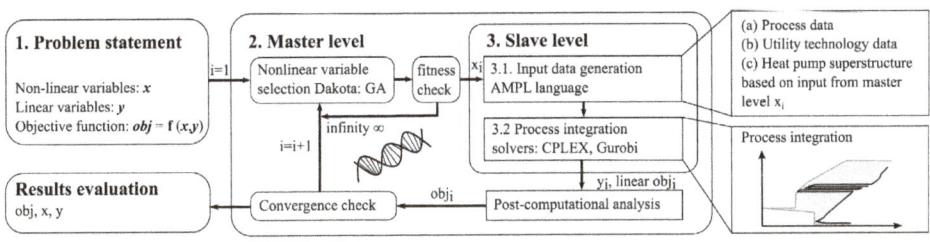

Figure 7. Decomposition method to solve MINLP problems implemented in OSMOSE [246].

For the *steam network* implemented in OSMOSE, the superstructure of steam turbine network is generated based on the predefined pressure levels at the upper level: For any pressure level *i*, a steam turbine is added between the pressure level *i* and any lower pressure level *j* with the calculation of intensive parameters (temperature, pressure and enthalpy) for each pressure level. Sometimes, a steam cycle only allows steam expansion between two neighboring pressure levels, which can be readily considered. At each pressure level, steam can be heated up to a higher temperature before expanding to the lower pressure level, or can be extracted as process steam for other processes or providing heat for the steam cycle itself. With such a bi-level algorithm, the MINLP of steam-network synthesis is converted to MILP at the lower level. Therefore, the decision variables are pressure levels, superheating degree (steam temperature at each pressure levels), subcooling temperature of the condensation level at the upper level, and the use and sizes of steam turbines and steam extractions at the lower level. The synthesis of *steam network* can be solved together with the sequential synthesis of heat exchanger networks, developed in [250], which can yield a specific layout of thermal power plants. Internally, the flexibility and effectiveness of *steam network* have been tested and improved to investigate modern coal-fired power plants with CO_2 capture technologies; however, there have been no published references coping with thermal power plants. The *steam network* can also be employed as a utility to enhance the process integration and has been adapted for organic Rankine cycle [245] and heat pumping network [246]. The computational effort of each run of the lower-level optimization is mainly due to (1) the evaluation of thermodynamic properties and (2) the solving of the MILP problem. Each run of the lower level for a steam cycle with 8 pressure levels would take several seconds and the whole optimization for both levels can take several hours with the time mostly consumed at the lower level.

4.3.2. Superstructure-Free Synthesis

For the superstructure-free synthesis of steam cycles, there are only two methods developed as mentioned before, i.e., the SYNTHSEP method [236,251] and the ECH-based method [241,242], as summarized in Table 4.

Table 4. Summary of the applications of superstructure-free synthesis for thermal power plants/steam cycles.

Year	Author	Application	Platform or Solving Method/Technique	Multi-Level
2007–present	Toffolo, Lazzaretto, and et al. [233,235,236,251]	Thermal power plant	Genetic algorithm and SQP	√
2015–present	Wang et al. [241,242]	Thermal power plant	Energy conversion hierarchy	√

(1) The SYNTHSEP method

The SYNTHSEP method was developed on the basis of HEATSEP method [237], which disaggregates existing energy system configurations into elementary thermodynamic cycles and identifies temperatures that can be varied (decision variables) in the design optimization, as illustrated in Figure 8. The SYNTHSEP method is kind of reversed version of HEATSEP as a bottom-up procedure to generate optimized system configurations by aggregation of elementary thermodynamic cycles. An elementary thermodynamic cycle is composed of four elementary processes, i.e., compression, heating, expansion and cooling. Many elementary cycles can be combined to form a basic system configuration automatically, as illustrated in Figure 9 with 2 elementary cycles sharing one thermodynamic process. Once a basic configuration is generated with the introduced mixers and splitters, thermal cut (as shown in Figure 8 right) can be placed to consider heat integration.

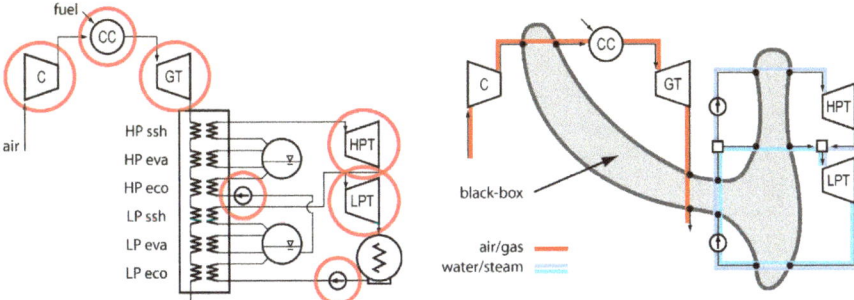

Figure 8. Decomposition of a two-pressure level combined cycle: the basic configuration (left) and the decomposed configuration (right) [235].

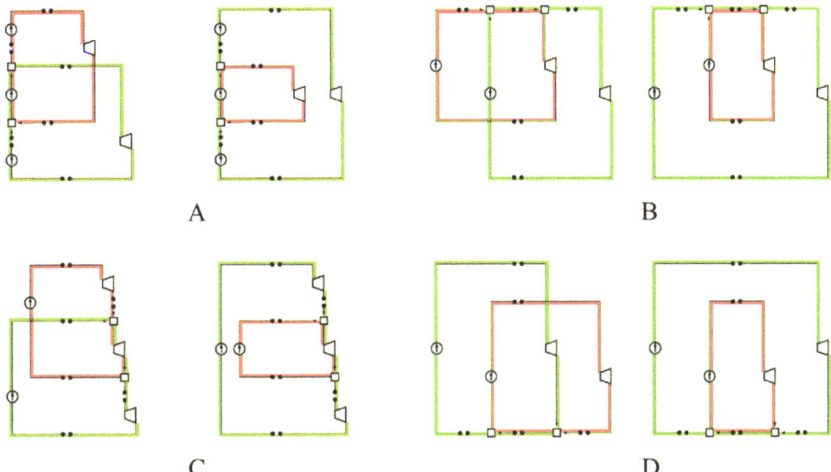

Figure 9. The configurations generated with two elementary cycles sharing one thermodynamic process ((**A–D**) indicating compression, heating, expansion and cooling) [251].

A bi-level hybrid algorithm is employed to find the optimal configuration with design variables: the upper-level evolutionary algorithm searches the configuration space and generates candidate configurations, which is further evaluated by the lower level traditional algorithm with SQP. The lower level searches for the optimal value of the objective function(s) by varying the only mass flow rates of the two elementary cycles of the Basic System Configuration, under the heat transfer feasibility constraint in the associated black-box, as shown in Figure 8. Therefore, each solution ends up with the topology and the design parameters (the optimum values of the temperatures, pressures and mass flow rates at the boundaries of the heat transfer black-box). However, to form a complete flowsheet, heat exchanger network has to be designed according to the techniques suggested by pinch analysis.

The SYNTHSEP method has been applied to various optimization problems for (organic) Rankine cycles and steam cycles [235,236]. It has been illustrated that different configurations can be efficiently generated and optimized, e.g., Figure 10 for steam cycles.

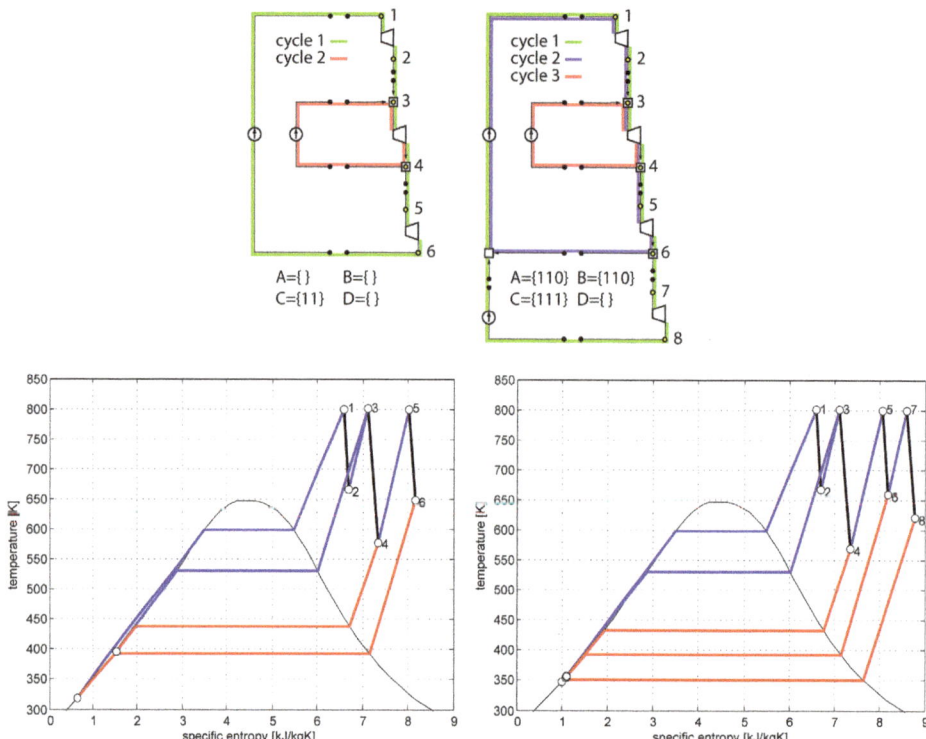

Figure 10. Topologies and T-S diagrams of the optimal steam cycles with 2 and 3 elementary cycles [235].

(2) The ECH-based method

The advantage of the ECH-based method is that it handles heat exchangers simultaneously with the change of pressure levels, which is not handled in the SYNTHSEP method. The algorithmic generation of a specific plant structure is achieved according to the energy conversion hierarchy and six generic replacement rules to change the plant structure. An ECH classifies the energy conversion technologies and links the technologies with the technological functions and replacement rules, as illustrated in Figure 11 for thermal power plants. There are three levels in an ECH: The meta, function and technology levels. Nodes at the meta-level represent the replacement rules. Nodes at the technology level represent specific energy conversion technologies. The connecting nodes on the function level classify energy conversion technologies according to their main functions (solid line) and types of drive (dashed line). With the ECH, it can be clearly shown that which replacement rules are applicable to a considered technology. Six replacement and insertion rules are finalized in the ECH-based method for thermal power plants:

a. Remove one component with all of its interconnections.
b. Remove one component and short-circuit all of its interconnections.
c. Delete one component and insert another component.
d. Delete one component and insert a parallel connection of two other components.
e. Delete one component and insert a serial connection of two other components.
f. Insert one component by replacing the technology-related stream.

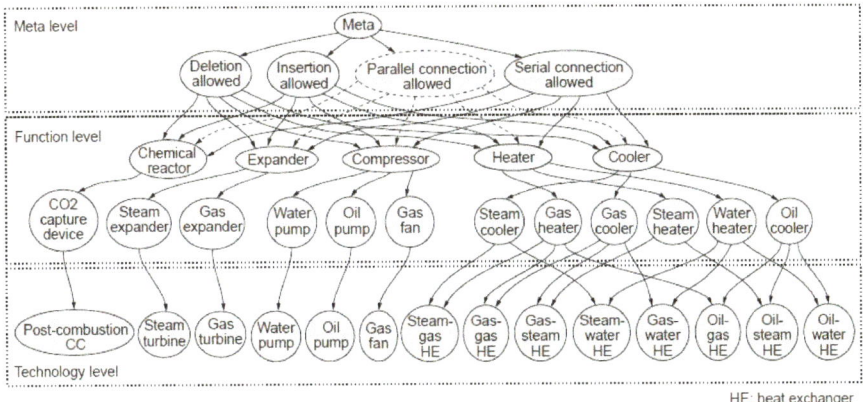

Figure 11. The energy conversion hierarchy for thermal power plants [241].

A bi-level solving procedure is employed: an evolutionary algorithm for the upper level with deterministic optimization for the lower level. The upper-level knowledge-integrated evolutionary algorithm, which is specifically tuned to flexibly integrate the ECH and replacement-insertion rules, generates structural alternatives from given structures, while each alternative generated by the upper level is then optimized deterministically in the lower level, i.e., GAMS used in [241,242].

The ECH-based method has been tested with simple Rankine cycles [241] and employed for solving complex problems [242]. It was found that the ECH-based method is an effective means to explore the design space, with a stable and high rate of the generation of structurally-feasible structures, even for highly complex problems. An example is given in Figure 12: starting from an initial flowsheet (a) with 4 feedwater preheaters and thermal efficiency of 46.49%, an optimal structure shown in (b) is featured with 8 feedwater preheaters, 2 reheaters, 2 de-superheaters, and a secondary turbine supplying steam for one feedwater preheater.

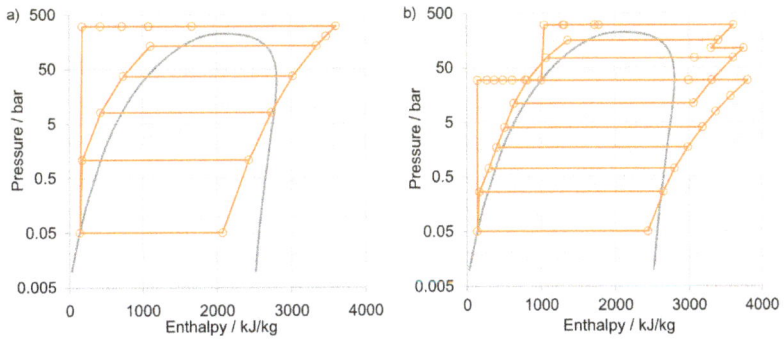

Figure 12. The original structure and one optimal structure found by the ECH-based method [242].

The computational effort is enormous due to the involvement of an evolutionary algorithm. For the complex problems tested in [242], the total time took several days. However, high-quality near-optimal solutions are generated already early in course of searching for the optimal solution. Also, the automatic structural generation without engineers' attendance makes the large computation effort acceptable.

5. Multi-Objective Optimization

In Sections 3 and 4, the optimization of energy systems is discussed with one objective function. However, many energy-system synthesis and design problems are multi-objective in nature, and they have conflicting objectives such as conversion efficiency, profit, cost of the system and environmental impacts [252]. A multi-objective optimization (MOO) problem is formulated as follows:

$$\min_{x} f(x) = (f_1(x), f_2(x), \ldots, f_k(x))^{\mathrm{T}}, \tag{26}$$

$$\text{s.t. } x \in X, \ g(x) \leq 0, \ h(x) = 0, \tag{27}$$

where the vector x ($x \in \mathrm{R}^n$) denoting n independent decision variables in the feasible solution space X. The vector f represents k objective functions $f_k : \mathrm{R}^n \to \mathrm{R}^1$. $g(x)$ and $h(x)$ are respectively inequality and equality constraints in the optimization problem.

Often, an MOO problem has many optimal solutions, known as non-dominated or Pareto-optimal solutions, which represent trade-offs among conflicting objectives. Two solutions are non-dominated to each other if the first solution is better than the second solution in at least one objective, and also the second solution is better than the first solution in at least one other objective. In other words, a solution becomes non-dominated or Pareto solution $\hat{x}$: (1) there is no other feasible solution x which is better in all objective functions, $f(x) \leq f(\hat{x})$, (2) $f_i(x) < f_i(\hat{x})$ for at least one objective function.

There have been many algorithms developed to generate Pareto fronts: classical methods (e.g., weighted sum method [253], ϵ-constraint method [254] and normalized normal constraint method [255]) and metaheuristic methods based on population-based metaheuristics [256], such as genetic algorithm, genetic programming, evolutionary strategy, evolutionary programming and differential evolution [257,258]. These techniques support exploration at the beginning of the search and exploitation towards the end of the search.

This section is organized as follows: In Section 5.1, main-stream multi-objective optimization techniques are introduced with an emphasis on evolutionary algorithms. Then, Section 5.2 summarized the applications to thermal power plants with major contributions.

5.1. Multi-Objective Optimization Techniques

5.1.1. The Weighted Sum Method

MOO problems usually can be converted into single objective optimization (SOO) problems, which can be further solved using deterministic optimization methods such as branch-and-bound or sequential quadratic programming. The simplest technique for converting an MOO problem into an SOO problem is the weighted sum method:

$$\min_{x} f(x) = \sum_{i=1}^{k} w_i \cdot f_i(x), \tag{28}$$

$$\text{s.t. } x \in X, \ g(x) \leq 0, \ h(x) = 0, \tag{29}$$

Each objective f_i is weighted by the positive weighting factors w_i ($\sum_{i=1}^{k} w_i = 1$) to form a super-objective f with inequality and equality constraints (g and h). The weighting factors can be adjusted via the preferences of decision makers or systematically, algorithmically to generate a series of small optimization problems, which can be solved one by one to find the Pareto solutions.

5.1.2. The ϵ-Constraint Method

In the ϵ-constraint method, one primary objective function is chosen, and other objective functions are converted to inequality constraints. It splits the objective function space into many sub-spaces by introducing additional inequality constraints from other objective functions. There could be many sub-spaces depending upon the ϵ values for other objective functions. Hence, ϵ-constraint method transforms a MOO problem into several SOO problems. The optimal solution of each SOO problem gives one Pareto solution. The generic ϵ-constraint method is formulated as follows:

$$\min_{x} f(x) = f_i(x), \forall\, i = 1, 2, \ldots k, \tag{30}$$

$$\text{s.t. } f_j(x) \le \epsilon_{j,p}, \forall j = 1, 2, \ldots k;\ j \ne i;\ p = 1, 2, \ldots n, \tag{31}$$

$$x \in X, g(x) \le 0,\ h(x) = 0. \tag{32}$$

For bi-objective optimization problem, a graphical representation of the ϵ-constraint method is shown in Figure 13. First of all, two anchor solutions ($\hat{\mu}^1$ *and* $\hat{\mu}^2$) are obtained by individually minimizing objective functions 1 and 2. Then, objective function space is divided into many subspaces by introducing bounds on objective function 1, e.g., $\epsilon_1, \epsilon_2, \epsilon_3$ and ϵ_n. The solution of each SOO problem gives one Pareto solution, e.g., s_1 for ϵ_1 and s_2 for ϵ_2. The ϵ-constraint method performs effectively, and the quality of the Pareto front obtained depends on the slope (shape) of the Pareto front and the division of the objective space: The higher the front slope, the denser the division should be to obtain evenly-spread Pareto solutions (Figure 13). Hence, the value of ϵ has to be successively modified for each division of objective function space to find high-quality Pareto front. In case of more than two objectives, the selection of ϵ values becomes difficult to obtain a Pareto front with good spread.

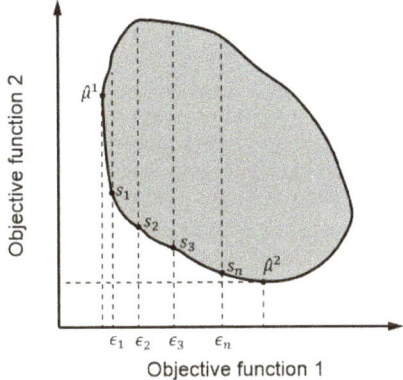

Figure 13. Graphical representation of the ϵ-constraint method for bi-objective optimization problem.

5.1.3. The Normalized Normal Constraint Method

The normalized normal constraint (NNC) method [255] generates evenly-spaced Pareto solutions. NNC introduces normal to Utopia line to divide objective function space (Figure 14) instead of vertical lines in ϵ-constraint method (Figure 13). $\hat{\mu}^1$ and $\hat{\mu}^2$ are the anchor points obtained be successively minimizing objective functions 1 and 2, respectively. After obtaining the anchor points, Utopia line is defined by connecting both anchor points. The Utopia line is divided into several evenly spread points, i.e., μ_1-μ_5. NNC incorporates an additional inequality constraint by adding a normal line to the Utopia line, e.g., NU1 for μ_1 and NU2 for μ_2. In other words, NNC method transforms a MOO problem into several SOO problems. For each SOO problem, the objective function space above the corresponding normal line is the feasible region, and remaining objective function space becomes the infeasible region. The optimal solution of each SOO problem is a Pareto solution, e.g., s_1-s_5. NNC method is able to generate a set of well-distributed Pareto solutions, even those on the non-convex regions of the Pareto front.

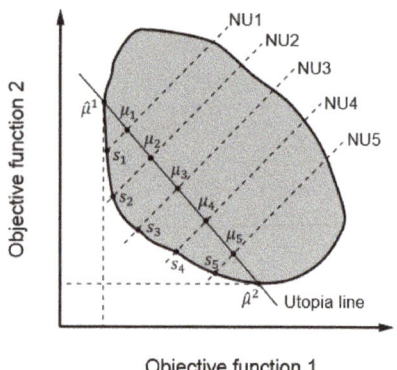

Figure 14. Graphical representation of normalized normal constraint method for bi-objective optimization problem.

5.1.4. Evolutionary Multi-Objective Optimization Algorithms (EMOAs)

Among modern metaheuristics, evolutionary algorithms have a tremendous advantage in solving MOO problems. There are several EMOAs, such as the elitist non-dominated sorting genetic algorithm or NSGA-II [259], S-merit selection EMOA [260], the strength Pareto evolutionary algorithm or SPEA2 [261], the Pareto envelope-based selection algorithm or PESA [262], multi-objective genetic algorithm or MOGA-II [263], multi-objective messy GA or MOMGA [264]. In the following paragraphs, widely used NSGA-II is described in detail.

The NSGA-II algorithm works as described in Figure 15: initialization of population, selection of solution or individual for reproduction operation, crossover and mutation operations, individual selection for the next generation, and the search termination criterion. Initially, a population of NP individuals is randomly generated inside decision variables bounds. Then, two individuals from the current population are selected using binary tournament, and two new individuals are generated by crossover and mutation operations. If there is a decision variable violation for new individual, then that decision variable is randomly generated inside the bounds. In this way, new individuals are generated and then combined with the current/parent population. NSGA-II applies a fast non-dominated sorting of the combined population to assign non-domination ranks to all individuals or solutions. For each solution i in the combined population, the number of solutions that dominate solution i (n_i) are calculated. Solution with $n_i = 0$ are identified as the best Pareto front, $\mathcal{F}_r$. Then, the solutions in the best Pareto front are removed from the combined population, and next Pareto fronts ($\mathcal{F}_{r+1}$, $\mathcal{F}_{r+2}$, etc.) are identified by repeating the procedure.

For constrained MOO problems, feasibility approach is used to rank the solutions in the combined population. If any of the following conditions is true, then solution i is dominating solution j:

- Both solutions are feasible, and $f_i(x) < f_j(x)$ for all objective functions.
- Solution i is feasible and solution j is infeasible.
- Both solutions are infeasible, but solution i has a smaller number of violated constraints (and lesser total absolute constraint violation if both have the same number of violated constraints) compared to solution j.

Solutions from the best Pareto fronts ($\mathcal{F}_r$, $\mathcal{F}_{r+1}$, $\mathcal{F}_{r+2}$, etc.) are selected for the subsequent generation. If all solutions of a Pareto front cannot be selected, then crowding distances (δ) are calculated for each solution, and the least crowded solutions are selected to complete the new solutions for the next generation. NSGA-II calculates crowding distance of solutions for estimating their densities. For each objective function f_m, the solutions in the Pareto front are sorted in a descending order, and solutions with the largest and smallest objective function values are specified with an infinity crowding distances. For each remaining solution i, its crowding distance with respect to the objective function f_m is defined by its two neighbor solutions: $\delta_m(x_i) = f_m(x_{i-1}) - f_m(x_{i+1})$. Therefore, crowding distance of a solution with k objective functions: $\delta(x_i) = \sum_{m=1}^{k} \delta_m(x_i)$. Iterations are repeated for the maximum number of generations (MNG).

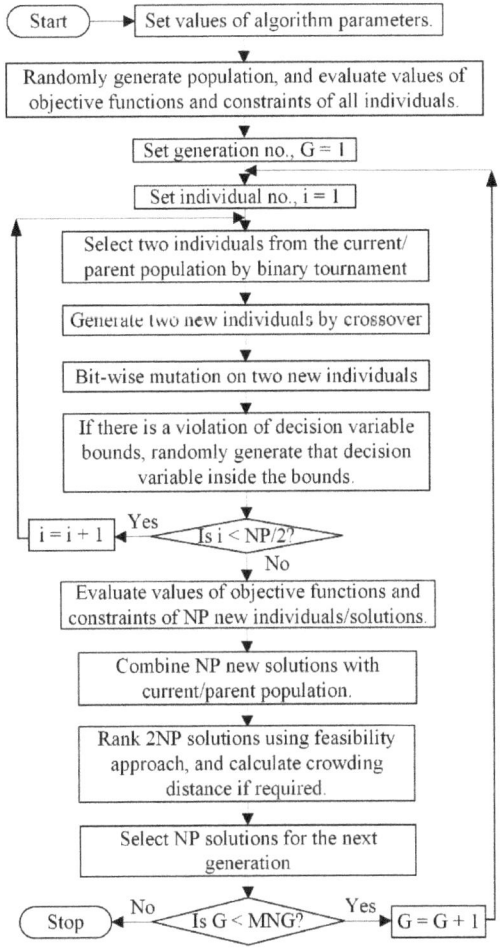

Figure 15. Flow chart of population-based NSGA-II algorithm.

5.2. Applications

There have been many studies on MOO of thermal energy systems, which are mainly focused on the thermodynamic (e.g., energy and exergy efficiency, primary energy-saving ratio, thermal efficiency, electric efficiency, total avoidable exergy destruction rate), economic (e.g., NPV, total cost rate, annual gross profit, annualized cost per unit) and environmental (CO_2 equivalent, pollution damage cost) objective functions. Table 5 summarizes recent MOO studies on the power plants or power-generation systems. The references mostly employ certain multi-objective optimization algorithm to solve specific problems without a significant methodology development. Considering the differences among the considered problems, these references are not be discussed in detail here but are summarized with the following findings:

Table 5. Summary of MOO studies on energy system optimization.

Year	Authors	Applications	Objective Functions	MOO Method
2010	Liu et al. [265]	Methanol/electricity polygeneration plant	NPV, CO$_2$ equivalent	ϵ-constraint
2004	Lazzaretto and Toffolo [266]	Thermal system design	Total cost rate, exergetic efficiency, pollution damage cost	MOEA
2010	Kavvadias and Maroulis [267]	Trigeneration (electricity, heat, cold) generation system	NPV, primary energy savings ratio, emission reduction ratio	MOEA
2012	Fazlollahi et al. [268]	Complex energy system	Total cost, CO$_2$ emission	EMOA and ϵ-constraint method
2014,2016	Wang et al. [242,269]	Thermal power plant	Thermal efficiency and cost of electricity	MOEA
2017	Chen et al. [270]	Nuclear power plant	Primary flow rate, weight	Hybrid NSGA-II
2017	Gimelli et al. [271]	Organic Rankine cycle power plant	Electric efficiency, overall heat exchangers area	MOGA-II
2015	Boyaghchi and Molaie [272]	Combined cycle power plant	Total avoidable exergy destruction rate, CO$_2$ emission	NSGA-II
2016,2017	Yao et al. [273,274]	Combined cooling, heating and power based compressed air energy storage system	Total product unit cost, exergy efficiency	MODE
2011	Avval et al. [275]	Gas turbine power plant	Exergy efficiency, total cost rate, CO$_2$ emission	NSGA-II
2013	Gutierrez-Arriaga et al. [276]	Steam power plant	Annual gross profit, GHG emissions	ϵ-constraint
2011	Hajabdollahi et al. [277]	Heat recovery steam generator	Annualized cost per unit of steam, exergy efficiency	NSGA-II
2006	Li et al. [278]	Combined cycle power plant	Cost of electricity, CO$_2$ emission rate	MOEA
2009	Sayyaadi et al. [279]	Cogeneration system	Exergetic efficiency, cost rate of products, pollution damage cost	MOEA
2016	Gonzalez-Bravo et al. [280]	Power plant and water distribution network	Profit, GHG emission	ϵ-constraint

(1) The trade-off

The trade-off between thermodynamic and economic objective functions usually follows the one illustrated in Figure 16. With an increase in plant efficiency, the fuel cost is reduced but the investment cost increases, which results in a V-shaped profile of total cost with an economic minimum point. The Pareto front starts from the economic minimum point and reaches the maximum efficiency point.

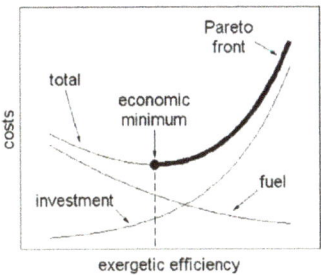

Figure 16. Classical trade-off between plant efficiency and economic objectives [266].

The thermodynamic and environmental objectives usually are not with a trade-off relationship, since increasing efficiency will reduce the fuel consumption, usually leading to lower emission, particularly for fossil-fuel based energy systems. However, trade-off may exist between the economic and environmental objectives following the fuel cost vs. efficiency (emission) as shown in Figure 16: the decrease in cost may possibly increase the pollutant emission, as revealed in [268].

(2)The algorithms and solution quality

As shown in Table 5, the dominating algorithms (over 80%) applied to energy systems are EA, due to the black-box evaluation of nonlinear objective functions and the smoothness of the Pareto fronts. For most cases, a sole evolutionary algorithm is enough to generate high-quality solutions; however, sometimes, the decision-making might be conservative to choose more stable sub-optimal solutions. In such a situation, mathematical programming methods are employed to effectively generate nearby solutions, e.g., by adding integer cut constraints (ICC) [268,276,280].

(3) Computational time

The EA is more effective for searching a Pareto optimal set and nearby solutions, while the ICC combined with the ε constraint is more time-consuming, since it needs to generate most solutions in the feasible space. However, if we only want to obtain a limited number of ordered solutions, then the ICC is powerful and fast [268]. The computational time for EA can be reduced by parallel computation, but not for the ICC method, since generating a new solution with ICC totally depends on previous ones consequently no possibility of using parallel computation [268].

6. Comparison and Perspectives

In the following, a straightforward comparison of the implementation of the identified methodologies to thermal power plants is given first and then future perspectives are provided as a further discussion.

6.1. Comparison of the Identified Methodologies

Considering the current activity and comprehensive impact of the introduced methodology, we highlight the following for thermal power plants: SPECO and advanced exergy-based analysis (analysis method), *steam network* in OSMOSE (superstructure-based synthesis), SYNTHSEP and ECH-based methods (superstructure-free synthesis). The features of these selected methods are compared in Table 6.

Table 6. Comparisons between the methodologies for the analysis and synthesis of thermal power plants.

Terms	Exergy-Based Analysis	Superstructure-Based	Superstructure-Free	
			SYNTHSEP	ECH-based
Name of the selected method	SPECO and advanced analysis	Steam network in OSMOSE	SYNTHSEP	ECH-based
Structure space definition	Specific structure	Superstructure defined by the number of pressure levels	Elementary cycles and number of shared processes	ECH and replacement-insertion rules
Structural generation	Fixed structure	Algorithmic, fixed superstructure	Algorithmic, evolutionary	Algorithmic, evolutionary
Structural evolution algorithm	-	-	EA (mutation, crossover)	EA (only mutation)
Total flowsheet	-	By integrating the synthesis of heat exchanger network		√
Evaluation of objective function	Simulation & Solving a linear equation set	MILP via AMPL with CPLEX solver	MILP by SQP solver	NLP via GAMS by CONOPT3 solver
Num. of meaningless structures?	-	Large	Small	Small
(Near-)optimal solution?	×	√	√	√
Expert knowledge requirement	√	×	×	×
Multi-objective trade-offs	-	√	√	√
Multi-objective selection technique	-	EA	EA	EA
Computational effort needed	Small < 1 s for each simulation <10 s for exergoeconomic analysis	Large (hours)	Large (from hours to days)	Enormous (several days for MOO)
Flexibility and extensibility	-	High	High	High
Target problems	All types	Steam cycle and its integration with all other processes	Thermal cycle	All types after adaption

6.2. Future Perspectives

Three directions of future research on evaluation and optimization-based synthesis of thermal power plants are recommended: real-world designs and retrofits, evaluation and synthesis methodologies, and good modeling practice.

6.2.1. Real-World Optimal Designs and Retrofits

Usually, the grassroots design of energy systems is addressed but not incorporates off-design performances of all involved components. This is reasonable for single-purpose (product), single-source system, as all the employed components are set to operate under partial loads: higher efficiency at the design load would generally lead to higher efficiency at partial loads.

However, when new energy systems are synthesized for multiple products and/or multiple sources, or existing energy systems are expected to be further enhanced by introducing new technologies, the operation-level synthesis must be considered.

For thermal power plants, off-design models for a wide range of technologies, e.g., Stodola ellipse model [149,281] for large steam turbines, Rabek method for feedwater preheater [282], etc. Sound mathematical models to predict off-design performances are to be developed for different components.

A flexible system-level superstructure-free synthesis framework, which is capable of coping with those new challenges when introducing multiple technologies. This approach is expected to extend the range of its application. For thermal power plants, more available technologies can be included in the energy conversion hierarchy, for example, Organic Rankine Cycle [283], supercritical CO_2 cycle [284], CO_2 capture technologies [285,286], solar-thermal utilization technologies [287], energy storage [288,289], etc. Except for the grassroots design, optimal retrofits of thermal power can be solved as well by combining well-developed off-design models of all involved technologies.

The design and retrofits of energy systems must consider more realistic (structural) constraints and objective functions, to ensure that the obtained optimal or near-optimal solutions eventually can be valid for industrial applications. Unfortunately, practical constraints have not been compressively considered yet in most research papers. Specifically to thermal power plants, the steam-extraction pressure for de-aerator should be limited within the range from 5 to 15 bar [213] due to technical reasons; the pressures of steam extractions are not completely continuous but are constrained by the turbine design; the secondary turbine which supplies only one steam extractions is less likely to be implemented in real power plants; the cost functions of all components should be developed with more available industrial data or with the participation of industrial partners. Only when well-established objectives are optimized under reasonable real-world constraints could the number, size and operation mode of each component be well-constrained.

6.2.2. Evaluation and Synthesis Methodologies

Future perspectives on the methodology development concern mainly reasonable estimation approach for endogenous exergy destructions, effective utilization of valuable results from the exergy-based analysis, possible combination of analysis methods with automated synthesis approaches, and further algorithm enhancements of the automatic structural evolutions. A further discussion is given below:

(a) A more rational way of estimating endogenous exergy destructions

There are still some open fundamental problems for calculating endogenous exergy destructions. A drawback of the recently developed calculation method [95] (described in Section 2.3.2) is that, for complex systems, mass flow rates of the streams entering the considered component, whose endogenous exergy destruction is to be calculated, are difficult to determine, since the theoretically-operated components are treated as a reversible black box. In fact, the mass-flow relationships between the streams of the considered component and the black box are hard to establish reasonable, especially when the splitting and mixing of streams are presented in a flowsheet.

Moreover, the fundamentals of the means of handling specific exergies of those streams flowing into the considered component have not been clearly described, particularly for complex systems with, e.g., heat-exchanger and steam-turbine trains.

(b) Adaptive structural evolution strategies

These strategies are expected to help further avoid the generation and evaluation of a number of meaningless structure alternatives. When using genetic algorithms to solve superstructure-based and superstructure-free problems, there are no fundamental differences in mutating which part of the structure. It is almost equivalently assumed that, before each evaluation of the generated structure, the change of any part of the structure would lead to the same effect on the objectives. However, different components, or more clearly, subsystems, have different impacts on the overall system performance. The subsystems, which present larger potentials for improving the objectives, should be given priority to be adjusted. Therefore, proper decomposition methods and adaptive evolution strategies should be developed or coupled to efficiently evolve the structure. For example, decompose the whole system into several subsystems ranked regarding their influences on the performance of the whole system. For another, evolve higher-rank subsystems and estimate their effects on the overall performance of the system. Consequently, frequent adjustments of the parts of the system structure that lead to only limited improvement of overall performance can be, to a large extent, suppressed. In addition, the decomposed subsystems, which are expected to be smaller, easy-to-solve subproblems out of the whole problem, require much fewer computation efforts to be optimized.

(c) Efficient identification of duplicate structures searched

Although many meaningless structural alternatives can be avoided in the superstructure-free approach, duplicate meaningful structures are frequently generated. The duplicate structures may lead to not only huge extra computational time to find the same optimal structure, but also a decrease in the diversity of the preserved solution structures.

Although a large number of meaningless structural alternatives can be avoided in the superstructure-free approach, duplicate meaningful structures are frequently generated. The duplicate structures may lead to not only huge extra computational time to find the same optimal structure, but also a decrease in the diversity of the preserved solution structures. For complex problems, e.g., cost-effective synthesis of thermal power plants, the average evaluation time of an individual structure can be long (over 20 s). The preserve of identical flowsheets can be suppressed in each mutation by discarding those offspring solutions with very similar objective values as the parent solution but generating an offspring solution with a different objective value. However, on the one hand, duplicate structures are distinguished only after the evaluation of the structures, thus it does not help reduce the total computation time; on the other hand, duplicate structural alternatives can still be frequently generated from two independent mutations. Effective algorithms are expected to efficiently identify whether the generated solution has been evaluated or not in the history of the current structural evolution. These algorithms would be quite favorable to further enhance the superstructure-free synthesis approach for synthesizing thermal power plants.

(d) Integration of analysis methods into system synthesis

There is still a large gap between different analysis methods and optimization-based synthesis of thermal systems. It is expected that the coupling of proper analysis methods would further improve the performance of automated synthesis approaches, particularly the optimization of structural alternatives. For example, the thermo-economic functional analysis [50] could formulate the objective functions in explicit relations with decision variables by the insights into the considered energy system. These reformulations of objective functions could lead to proper system decompositions and the removals of surplus intermediate variables and equations. Accordingly, the computational effort for optimizing the same structural alternative can be reduced. However, currently, these approaches are hardly applicable to complex problems and, most importantly, fundamentally, they can only support the optimal synthesis based on a predefined superstructure so far. There are challenges to automatically and properly reformulate the objective functions with respect to different system structures.

For the most widely used accounting methods, e.g., SPECO, few references have been published on employing this analysis information for the parametric optimization; while for structural improvements engineers' expertise usually have to be integrated to judge which parts of the analyzed structure should be modified and how to modify. Thus, for automated optimization-based synthesis, these accounting methods should probably be employed to rank different subsystems for modification.

A far long way is ahead as well for the developing advanced exergy-based analysis to be a supportive method of synthesis approach. There have been no available references yet on how to reasonably use the information from the splitting of exergy destructions and costs for parametric and structural optimization.

6.2.3. Mathematical Modeling Practice

The bad mathematical formulation may evaluate feasible structures as infeasible solutions, good structures as bad solutions as the search procedure may be trapped at local optimums. Good modeling practice, particularly for NLP and MINLP, could help build efficient and sound formulations that can be solved much faster, easier with less possibility of being trapped locally. Various techniques are available for good model formulations, such as setting good initial values and bounds of all variables (including intermediate variables), properly scaling variables and equations, reformulations (piecewise/polynomial/separable approximations or even linearization) of nonlinear formulations, convexification of nonconvex formulations, etc. Note the reformulations of nonlinear or nonconvex equations may lead to the loss of a certain degree of accuracy or even unacceptable optimal solutions, thus they must be carefully developed and checked before replacing the original formulations.

7. Conclusions

System analysis, synthesis and optimization regarding various objective functions are the key leveler to enhance the performances of energy systems. For each topic, we first introduced the fundamentals and methodologies developed in the literature and then summarized their applications to thermal power plants. Considering current status of existing methodologies, we have emphasized and compared the following methods with more details: SPECO and advanced exergy-based analysis for system analysis, *steam network* for superstructure-based synthesis, and SYNTHSEP and ECH-based methods for superstructure-free synthesis.

- For system analysis, the advanced exergy-based analyses aim at paving a step further above traditional exergy analysis to reveal the sources and avoidability of exergy destruction and costs within different components and their interactions. The related methods are still under development and remain several fundamental problems to be addressed, e.g., validation of the splits of exergy dissipation.
- For superstructure-based synthesis, the *steam network*, incorporating algorithmic generation of the steam-cycle superstructure for a predefined number of pressure levels and bi-level hybrid solving algorithm (EA+MILP), is flexible to be employed as standalone thermal power plants or a utility for process integration. However, the method must combine with superstructure-based heat exchanger network for synthesizing complete flowsheets.
- For superstructure-free synthesis, both SYNTHSEP and ECH-based method performs evolutionary structural alternation of given structures based on different concepts. The SYNTHSEP, not for complete flowsheet synthesis, employs elementary cycles and the share of multiple elementary cycles, which limits its use for other processes. The ECH-based method can perform total flowsheet synthesis and can be flexibly extended with well-defined ECH and component models. Both methods employ bi-level decomposition techniques combining EA and mathematical programming.

Future perspectives of methodology development and applications are summarized:

- A straightforward comparison of these chosen methods on a common basis of a benchmarking problem should be made, since the applications given in literature aimed at solving specific optimization problems.
- Three directions of method development and implementation are recommended: real-world designs and retrofits, further methodology development, particularly synthesis methodologies, and good modeling practice.

Author Contributions: L.W. structured the paper, reviewed major references cited and is the main author of the paper. Z.Y., S.S., A.M., T.-E.L. contributed to the writing of relevant sections. F.M. provided valuable comments and discussion. G.T. and Y.Y. provided supervision of the technical contents for L.W.'s doctoral thesis.

Funding: The authors, Z.Y. and Y.Y., received financial support from the National Nature Science Fund of China (51436006).

Acknowledgments: Special thanks to the editors, Andrea Lazzaretto and Andrea Toffolo, for the kind invitation, which encourages L.W. to complete the paper.

Conflicts of Interest: The authors declare no conflict of interest.

Nomenclature

DE	differential evolution
EA	evolutionary algorithm
ECH	energy conversion hierarchy
GDP	generalized disjunctive programming
GRG	generalized reduced gradient
HPT	high-pressure turbine
ICC	Integer cut constraint
IP	integer programming
IPT	intermedia pressure turbine
LP	linear programming
LPT	low-pressure turbine
MILP	mixed integer linear programming
MINLP	mixed integer nonlinear programming
MOO	multi-objective optimization
NLP	nonlinear programming
NNC	normalized normal constraint
SOO	single objective optimization
SPECO	specific exergy costing
SQP	successive quadratic programming
TRR	total revenue requirement
$\dot{E}$	exergy flow, MW
d	design variable
f	objective function
f_m	objective function
$\mathcal{F}$	Pareto front
$\mathcal{F}_r$	the best Pareto front
g	inequality constraint
h	equality constraint; or enthalpy, kJ/kg
$\dot{m}$	mass flow, kg/s
o	operational variable
s	structural variable
T	temperature, °C

$(\nu_1 \cdot \nu_2)$	continuous nonconvex bilinear term
w	weighting factor
x	vector in the feasible solution space
X	solution space
z	variable to replace the bilinear term
Z	investment cost, M\$

Greek letters

Σ	space of all structure alternatives
δ	crowding distance
η	efficiency
λ	size of offspring population
σ	solution structure evolved by mutation
μ	size of parent population; or solution
$\hat{\mu}$	anchor solution
ϵ	objective function

Subscripts and superscripts

A	average
abs	absorption
AV	avoidable
D	destruction
EN	endogenous
EX	exogenous
F	fuel
i, j, k	index
L	loss
N	independent decision variable
P	product
rel	release
UN	unavoidable

References

1. Sieminski, A. *International Energy Outlook*; Energy Information Administration (EIA): Washington, DC, USA, 2014.
2. Iodice, P.; Senatore, A. Atmospheric pollution from point and diffuse sources in a National Interest Priority Site located in Italy. *Energy Environ.* **2016**, *27*, 586–596. [CrossRef]
3. Iodice, P.; Adamo, P.; Capozzi, F.; Di Palma, A.; Senatore, A.; Spagnuolo, V.; Giordano, S. Air pollution monitoring using emission inventories combined with the moss bag approach. *Sci. Total Environ.* **2016**, *541*, 1410–1419. [CrossRef] [PubMed]
4. Qazi, H.W.; Flynn, D. Synergetic frequency response from multiple flexible loads. *Electr. Power Syst. Res.* **2017**, *145*, 185–196. [CrossRef]
5. Wang, L.; Pérez-Fortes, M.; Madi, H.; Diethelm, S.; Herle, J.V.; Maréchal, F. Optimal design of solid-oxide electrolyzer based power-to-methane systems: A comprehensive comparison between steam electrolysis and co-electrolysis. *Appl. Energy* **2018**, *211*, 1060–1079. [CrossRef]
6. Jensen, S.H.; Graves, C.; Mogensen, M.; Wendel, C.; Braun, R.; Hughes, G.; Gao, Z.; Barnett, S.A. Correction: Large-scale electricity storage utilizing reversible solid oxide cells combined with underground storage of CO_2 and CH_4. *Energy Environ. Sci.* **2015**, *8*, 2471–2479. [CrossRef]
7. Fukuda, Y. Development of Advanced Ultra Supercritical Fossil Power Plants in Japan: Materials and High Temperature Corrosion Properties. *Mater. Sci. Forum* **2010**, *696*, 236–241. [CrossRef]
8. International Energy Agency (IEA). *Technology Roadmap: High-Efficiency, Low-Emissions Coal-Fired Power Generation*; International Energy Agency: Paris, France, 2012.
9. Wang, L. Thermo-Economic Evaluation, Optimization and Synthesis of Large-Scale Coal-Fired Power Plants. Ph.D. Thesis, Technical University of Berlin, Berlin/Heidelberg, Germany, 2016.
10. Rukes, B.; Taud, R. Status and perspectives of fossil power generation. *Energy* **2004**, *29*, 1853–1874. [CrossRef]

11. Spliethoff, H. Steam Power Stations for Electricity and Heat Generation. In *Power Generation from Solid Fuels*; Springer: Berlin/Heidelberg, Germany, 2010; pp. 73–219.
12. Silvestri, G.J., Jr.; Westinghouse Electric Corp. Boiler Feedpump Turbine Drive/Feedwater Train Arrangement. U.S. Patent 5,404,724, 11 April 1995.
13. Blum, R.; Kjaer, S.; Bugge, J. Development of a pf fired high efficiency power plant AD700. In Proceedings of the Riso International Energy Conference: "Energy Solutions for Sustainable Development", Roskilde, Denmark, 22–24 May 2007.
14. Stępczyńska, K.; Kowalczyk, Ł.; Dykas, S.; Elsner, W. Calculation of a 900 MW conceptual 700/720°C coal-fired power unit with an auxiliary extraction-backpressure turbine. *J. Power Technol.* **2012**, *92*, 266–273.
15. Eaves, J.; Palmer, F.D.; Wallace, J.; Wilson, S. *The Value of Our Existing Coal Fleet: An Assessment of Measures to Improve Reliability & Efficiency While Reducing Emissions*; The National Coal Council: Washington, DC, USA, 2014.
16. Espatolero, S.; Cortés, C.; Romeo, L.M. Optimization of boiler cold-end and integration with the steam cycle in supercritical units. *Appl. Energy* **2010**, *87*, 1651–1660. [CrossRef]
17. Karthikeyan, M.; Zhonghua, W.; Mujumdar, A.S. Low-rank coal drying technologies–Current status and new developments. *Dry. Technol.* **2009**, *27*, 403–415. [CrossRef]
18. Turchi, C.S.; Langle, N.; Bedilion, R.; Libby, C. Solar-augment potential of US fossil-fired power plants. In Proceedings of the ASME 2011 5th International Conference on Energy Sustainability, Washington, DC, USA, 7–10 August 2011; pp. 641–651.
19. Popov, D. An option for solar thermal repowering of fossil fuel fired power plants. *Sol. Energy* **2011**, *85*, 344–349. [CrossRef]
20. Yang, Y.; Yan, Q.; Zhai, R.; Kouzani, A.; Hu, E. An efficient way to use medium-or-low temperature solar heat for power generation—Integration into conventional power plant. *Appl. Therm. Eng.* **2011**, *31*, 157–162. [CrossRef]
21. Sansaniwal, S.K.; Sharma, V.; Mathur, J. Energy and exergy analyses of various typical solar energy applications: A comprehensive review. *Renew. Sustain. Energy Rev.* **2018**, *82*, 1576–1601. [CrossRef]
22. Tsatsaronis, G. *Thermodynamics and the Destruction of Resources*; Cambridge University Press: New York, NY, USA, 2011; pp. 377–401.
23. Moran, M.J.; Shapiro, H.N.; Boettner, D.D.; Bailey, M.B. *Fundamentals of Engineering Thermodynamics*, 7th ed.; John Wiley Sons: New York, NY, USA, 2010.
24. Iodice, P.; Langella, G.; Amoresano, A.; Senatore, A. Comparative exergetic analysis of solar integration and regeneration in steam power plants. *J. Energy Eng.* **2017**, *143*, 04017042. [CrossRef]
25. Cornelissen, R.L. *Thermodynamics and Sustainable Development: The Use of Exergy Analysis and the Reduction of Irreversibility*; Universiteit Twente: Enschede, The Netherlands, 1997.
26. Tsatsaronis, G. Energy Economics and Management in Industry. In *Energy Economics and Management in Industry*; Pergamon Press: Oxford, UK, 1984; pp. 151–157.
27. Kotas, T.J. *The Exergy Method of Thermal Plant Analysis*; Butterworth Publishers: Stoneham, MA, USA, 1985.
28. Singh, O.K.; Kaushik, S. Energy and exergy analysis and optimization of Kalina cycle coupled with a coal fired steam power plant. *Appl. Therm. Eng.* **2013**, *51*, 787–800. [CrossRef]
29. Sengupta, S.; Datta, A.; Duttagupta, S. Exergy analysis of a coal-based 210 MW thermal power plant. *Int. J. Energy Res.* **2007**, *31*, 14–28. [CrossRef]
30. Aljundi, I.H. Energy and exergy analysis of a steam power plant in Jordan. *Appl. Therm. Eng.* **2009**, *29*, 324–328. [CrossRef]
31. Erdem, H.H.; Akkaya, A.V.; Cetin, B.; Dagdas, A.; Sevilgen, S.H.; Sahin, B.; Teke, I.; Gungor, C.; Atas, S.; et al. Comparative energetic and exergetic performance analyses for coal-fired thermal power plants in Turkey. *Int. J. Therm. Sci.* **2009**, *48*, 2179–2186. [CrossRef]
32. Ray, T.K.; Datta, A.; Gupta, A.; Ganguly, R. Exergy-based performance analysis for proper O&M decisions in a steam power plant. *Energy Convers. Manag.* **2010**, *51*, 1333–1344.
33. Zhao, Z.; Su, S.; Si, N.; Hu, S.; Wang, Y.; Xu, J.; Jiang, L.; Chen, G.; Xiang, J. Exergy analysis of the turbine system in a 1000 MW double reheat ultra-supercritical power plant. *Energy* **2017**, *119*, 540–548. [CrossRef]
34. Tsatsaronis, G. Thermodynamic Optimization of Complex Energy Systems. In *Thermodynamic Optimization of Complex Energy Systems*; Bejan, A., Mamut, E., Eds.; Springer: Amsterdam, The Netherlands, 1999; pp. 93–100.

35. Tsatsaronis, G.; Winhold, M. Exergoeconomic analysis and evaluation of energy-conversion plants—I. Analysis of a coal-fired steam power plant. *Energy* **1985**, *10*, 81–94. [CrossRef]
36. Tsatsaronis, G.; Winhold, M. Exergoeconomic analysis and evaluation of energy-conversion plants—II. A new general methodology. *Energy* **1985**, *10*, 69–80. [CrossRef]
37. Tsatsaronis, G.; Pisa, J. Exergoeconomic evaluation and optimization of energy systems—Application to the CGAM problem. *Energy* **1994**, *19*, 287–321. [CrossRef]
38. Tsatsaronis, G. Thermoeconomic analysis and optimization of energy systems. *Prog. Energy Combust. Sci.* **1993**, *19*, 227–257. [CrossRef]
39. Lazzaretto, A.; Tsatsaronis, G. On the calculation of efficiencies and costs in thermal systems. In Proceedings of the ASME Advanced Energy Systems Division 1999, Nashville, TN, USA, 14–19 November 1999; American Society of Mechanical Engineers: New York, NY, USA, 1999; Volume 39, pp. 421–430.
40. Lazzaretto, A.; Tsatsaronis, G. On the quest for objective equations in exergy costing. In Proceedings of the ASME Advanced Energy Systems Division 1997, Dallas, TX, USA, 16–21 November 1997; American Society of Mechanical Engineers: New York, NY, USA, 1997; Volume 37, pp. 197–210.
41. Valero, A.; Lozano, M.; Muñoz, M. A general theory of exergy saving. III. Energy savings and thermodynamics. In *Computer-Aided Engineering and Energy Systems Second Law Analysis and Modelling*; Gaggioli, R., Ed.; American Society of Mechanical Engineers: New York, NY, USA, 1986; Volume 3, pp. 16–21.
42. Valero, A.; Lozano, M.; Muñoz, M. A general theory of exergy saving. II. On the thermodynamic cost. In *Computer-Aided Engineering and Energy Systems Second Law Analysis and Modelling*; Gaggioli, R., Ed.; American Society of Mechanical Engineers: New York, NY, USA, 1986; Volume 3, pp. 9–15.
43. Valero, A.; Lozano, M.; Muñoz, M. A general theory of exergy saving. I. On the exergetic cost. In *Computer-Aided Engineering and Energy Systems Second Law Analysis and Modelling*; Gaggioli, R., Ed.; American Society of Mechanical Engineers: New York, NY, USA, 1986; Volume 3, pp. 1–8.
44. Valero, A.; Torres, C. Algebraic thermodynamic analysis of energy systems. In *Proceedings of the Approaches to the Design and Optimization of Thermal Systems AES*; American Society of Mechanical Engineers: New York, NY, USA, 1988; Volume 7, pp. 13–23.
45. Valero, A.; Lozano, M.A.; Serra, L.; Torres, C. Application of the exergetic cost theory to the CGAM problem. *Energy* **1994**, *19*, 365–381. [CrossRef]
46. Lozano, M.A.; Valero, A. Theory of the exergetic cost. *Energy* **1993**, *18*, 939–960. [CrossRef]
47. Frangopoulos, C.A. Application of the thermoeconomic functional approach to the CGAM problem. *Energy* **1994**, *19*, 323–342. [CrossRef]
48. Frangopoulos, C.A. Optimal Synthesis and Operation of Thermal Systems by the Thermoeconomic Functional Approach. *ASME J. Eng. Gas Turb. Power* **1992**, *114*, 707–714. [CrossRef]
49. Frangopoulos, C.A. Functional decomposition for optimal design of complex thermal systems. *Energy* **1988**, *13*, 239–244. [CrossRef]
50. Frangopoulos, C.A. *Thermoeconomic Functional Analysis: A Method for Optimal Design or Improvement of Complex Thermal Systems*; Georgia Institute of Technology: Atlanta, GA, USA, 1983.
51. Von Spakovsky, M.R. *A Practical Generalized Analysis Approach to the Optimal Thermoeconomic Design and Improvement of Real-World Thermal Systems*; Georgia Institute of Technology: Atlanta, GA, USA, 1986.
52. Von Spakovsky, M.R.; Evans, R.B. The Design and Performance Optimization of Thermal Systems. *ASME J. Eng. Gas Turb. Power* **1990**, *112*, 86–93. [CrossRef]
53. Von Spakovsky, M.R. Application of engineering functional analysis to the analysis and optimization of the CGAM problem. *Energy* **1994**, *19*, 343–364. [CrossRef]
54. Gaggioli, R.A.; El-Sayed, Y.M. A critical review of second law costing method–II: Calculus procedures. *ASME J. Energy Resour. Technol.* **1989**, *111*, 8–15. [CrossRef]
55. Tsatsaronis, G.; Lin, L.; Pisa, J. Exergy Costing in Exergoeconomics. *ASME J. Energy Resour. Technol.* **1993**, *115*, 9–16. [CrossRef]
56. Lazzaretto, A.; Tsatsaronis, G. SPECO: A systematic and general methodology for calculating efficiencies and costs in thermal systems. *Energy* **2006**, *31*, 1257–1289. [CrossRef]
57. Kanoglu, M.; Ayanoglu, A.; Abusoglu, A. Exergoeconomic assessment of a geothermal assisted high temperature steam electrolysis system. *Energy* **2011**, *36*, 4422–4433. [CrossRef]

58. Kalinci, Y.; Hepbasli, A.; Dincer, I. Exergoeconomic analysis of hydrogen production from plasma gasification of sewage sludge using specific exergy cost method. *Int. J. Hydrogen Energy* **2011**, *36*, 11408–11417. [CrossRef]

59. Al-Sulaiman, F.A.; Dincer, I.; Hamdullahpur, F. Thermoeconomic optimization of three trigeneration systems using organic Rankine cycles: Part I–Formulations. *Energy Convers. Manag.* **2013**, *69*, 199–208. [CrossRef]

60. Alkan, M.A.; Keçebaş, A.; Yamankaradeniz, N. Exergoeconomic analysis of a district heating system for geothermal energy using specific exergy cost method. *Energy* **2013**, *60*, 426–434. [CrossRef]

61. Tsatsaronis, G.; Moran, M.J. Exergy-aided cost minimization. *Energy Convers. Manag.* **1997**, *38*, 1535–1542. [CrossRef]

62. Cziesla, F.; Tsatsaronis, G. Iterative exergoeconomic evaluation and improvement of thermal power plants using fuzzy inference systems. *Energy Convers. Manag.* **2002**, *43*, 1537–1548. [CrossRef]

63. Seyyedi, S.M.; Ajam, H.; Farahat, S. A new approach for optimization of thermal power plant based on the exergoeconomic analysis and structural optimization method: Application to the CGAM problem. *Energy Convers. Manag.* **2010**, *51*, 2202–2211. [CrossRef]

64. Frangopoulos, C.A. Intelligent functional approach: A method for analysis and optimal synthesis-design-operation of complex systems. *Int. J. Energy Environ. Econ.* **1991**, *4*, 267–274.

65. Hua, B.; Yin, Q.; Wu, G. Energy optimization through exergy-economic evaluation. *ASME J. Energy Resour. Technol.* **1989**, *111*, 148–153. [CrossRef]

66. Hua, B.; Chen, Q.L.; Wang, P. A new exergoeconomic approach for analysis and optimization of energy systems. *Energy* **1997**, *22*, 1071–1078. [CrossRef]

67. Valero, A.; Serra, L.; Lozano, M.A. *Structural Theory of Thermoeconomics*; ASME: New York, NY, USA, 1993; Volume 30, pp. 189–198.

68. Torres, C.; Valero, A.; Rangel, V.; Zaleta, A. On the cost formation process of the residues. *Energy* **2008**, *33*, 144–152. [CrossRef]

69. Seyyedi, S.M.; Ajam, H.; Farahat, S. A new criterion for the allocation of residues cost in exergoeconomic analysis of energy systems. *Energy* **2010**, *35*, 3474–3482. [CrossRef]

70. Piacentino, A.; Cardona, E. Scope-Oriented Thermoeconomic analysis of energy systems. Part II: Formation Structure of Optimality for robust design. *Appl. Energy* **2010**, *87*, 957–970. [CrossRef]

71. Piacentino, A.; Cardona, F. Scope-Oriented Thermoeconomic analysis of energy systems. Part I: Looking for a non-postulated cost accounting for the dissipative devices of a vapour compression chiller. Is it feasible? *Appl. Energy* **2010**, *87*, 943–956. [CrossRef]

72. Banerjee, A.; Tierney, M.J.; Thorpe, R.N. Thermoeconomics, cost benefit analysis, and a novel way of dealing with revenue generating dissipative units applied to candidate decentralised energy systems for Indian rural villages. *Energy* **2012**, *43*, 477–488. [CrossRef]

73. Paulus, D.M.; Tsatsaronis, G. Auxiliary equations for the determination of specific exergy revenues. *Energy* **2006**, *31*, 3235–3247. [CrossRef]

74. Cardona, E.; Piacentino, A. A new approach to exergoeconomic analysis and design of variable demand energy systems. *Energy* **2006**, *31*, 490–515. [CrossRef]

75. Kelly, S. *Energy Systems Improvement based on Endogenous and Exogenous Exergy Destruction*; Technische Universit at Berlin: Berlin/Heidelberg, Germany, 2008.

76. Kelly, S.; Tsatsaronis, G.; Morosuk, T. Advanced exergetic analysis: Approaches for splitting the exergy destruction into endogenous and exogenous parts. *Energy* **2009**, *34*, 384–391. [CrossRef]

77. Tsatsaronis, G.; Morosuk, T. A General Exergy-Based Method for Combining a Cost Analysis with an Environmental Impact Analysis: Part I–Theoretical Development. *ASME Conf. Proc.* **2008**, *2008*, 453–462.

78. Gaggioli, R.; Reini, M. Panel I: Connecting 2nd Law Analysis with Economics, Ecology and Energy Policy. *Entropy* **2014**, *16*, 3903–3938. [CrossRef]

79. Rashidi, J.; Yoo, C.K. Exergetic and exergoeconomic studies of two highly efficient power-cooling cogeneration systems based on the Kalina and absorption refrigeration cycles. *Appl. Therm. Eng.* **2017**, *124*, 1023–1037. [CrossRef]

80. Sahin, A.Z.; Al-Sharafi, A.; Yilbas, B.S.; Khaliq, A. Overall performance assessment of a combined cycle power plant: An exergo-economic analysis. *Energy Convers. Manag.* **2016**, *116*, 91–100. [CrossRef]

81. Ahmadzadeh, A.; Salimpour, M.R.; Sedaghat, A. Thermal and exergoeconomic analysis of a novel solar driven combined power and ejector refrigeration (CPER) system. *Int. J. Refrig.* **2017**, *83*, 143–156. [CrossRef]

82. Baghsheikhi, M.; Sayyaadi, H. Real-time exergoeconomic optimization of a steam power plant using a soft computing-fuzzy inference system. *Energy* **2016**, *114*, 868–884. [CrossRef]

83. Wang, L.; Yang, Y.; Dong, C.; Yang, Z.; Xu, G.; Wu, L. Exergoeconomic Evaluation of a Modern Ultra-Supercritical Power Plant. *Energies* **2012**, *5*, 3381–3397. [CrossRef]

84. Cziesla, F.; Tsatsaronis, G.; Gao, Z. Avoidable thermodynamic inefficiencies and costs in an externally fired combined cycle power plant. *Energy* **2006**, *31*, 1472–1489. [CrossRef]

85. Yang, Y.; Wang, L.; Dong, C.; Xu, G.; Morosuk, T.; Tsatsaronis, G. Comprehensive exergy-based evaluation and parametric study of a coal-fired ultra-supercritical power plant. *Appl. Energy* **2013**, *112*, 1087–1099. [CrossRef]

86. Tsatsaronis, G.; Park, M.-H. On avoidable and unavoidable exergy destructions and investment costs in thermal systems. *Energy Convers. Manag.* **2002**, *43*, 1259–1270. [CrossRef]

87. Morosuk, T.; Tsatsaronis, G. Advanced exergy analysis for chemically reacting systems—Application to a simple open gas-turbine system. *Int. J. Thermophys.* **2009**, *12*, 105–111.

88. Morosuk, T.; Tsatsaronis, G. Advanced exergetic evaluation of refrigeration machines using different working fluids. *Energy* **2009**, *34*, 2248–2258. [CrossRef]

89. Tsatsaronis, G.; Morosuk, T. Advanced exergetic analysis of a novel system for generating electricity and vaporizing liquefied natural gas. *Energy* **2010**, *35*, 820–829. [CrossRef]

90. Petrakopoulou, F.; Tsatsaronis, G.; Morosuk, T.; Carassai, A. Advanced Exergoeconomic Analysis Applied to a Complex Energy Conversion System. *ASME J. Eng. Gas Turb. Power* **2011**, *134*, 243–250.

91. Morosuk, T.; Tsatsaronis, G. Exergoeconomic evaluation of refrigeration machines based on avoidable endogenous and exogenous costs. In Proceedings of the 20th International Conference on Efficiency, Cost, Optimization, Simulation and Environmental Impact of Energy Systems, Padova, Italy, 25–28 June 2007; pp. 1459–1467.

92. Morosuk, T.; Tsatsaronis, G. A new approach to the exergy analysis of absorption refrigeration machines. *Energy* **2008**, *33*, 890–907. [CrossRef]

93. Tsatsaronis, G.; Kelly, S.O.; Morosuk, T.V. Endogenous and exogenous exergy destruction in thermal systems. In Proceedings of the ASME 2006 International Mechanical Engineering Congress and Exposition, Chicago, IL, USA, 5–10 November 2006; pp. 311–317.

94. Tsatsaronis, G.; Morosuk, T.; Kelly, S. Approaches for Splitting the Exergy Destruction into Endogenous and Exogenous Parts. In Proceedings of the 5th Workshop Advances in Energy Studies, Porto Venere, Italy, 12–16 September 2006; pp. 12–16.

95. Penkuhn, M.; Tsatsaronis, G. A decomposition method for the evaluation of component interactions in energy conversion systems for application to advanced exergy-based analyses. *Energy* **2017**, *133*, 388–403. [CrossRef]

96. Sorgenfrei, M. *Analysis of IGCC-Based Plants with Carbon Capture for an Efficient and Flexible Electric Power Generation*; Technical University of Berlin: Berlin/Heidelberg, Germany, 2016.

97. Kaushik, S.C.; Reddy, V.S.; Tyagi, S.K. Energy and exergy analyses of thermal power plants: A review. *Renew. Sustain. Energy Rev.* **2011**, *15*, 1857–1872. [CrossRef]

98. Zhu, Y.; Zhai, R.; Peng, H.; Yang, Y. Exergy destruction analysis of solar tower aided coal-fired power generation system using exergy and advanced exergetic methods. *Appl. Therm. Eng.* **2016**, *108*, 339–346. [CrossRef]

99. Ertesvåg, I.S.; Kvamsdal, H.M.; Bolland, O. Exergy analysis of a gas-turbine combined-cycle power plant with precombustion CO_2 capture. *Energy* **2005**, *30*, 5–39. [CrossRef]

100. Reddy, V.S.; Kaushik, S.C.; Tyagi, S.K. Exergetic analysis and performance evaluation of parabolic trough concentrating solar thermal power plant (PTCSTPP). *Energy* **2012**, *39*, 258–273. [CrossRef]

101. Xu, C.; Wang, Z.; Li, X.; Sun, F. Energy and exergy analysis of solar power tower plants. *Appl. Therm. Eng.* **2011**, *31*, 3904–3913. [CrossRef]

102. Hofmann, M.; Tsatsaronis, G. Comparative exergoeconomic assessment of coal-fired power plants – Binary Rankine cycle versus conventional steam cycle. *Energy* **2018**, *142*, 168–179. [CrossRef]

103. Petrakopoulou, F. *Comparative Evaluation of Power Plants with CO2 Capture: Thermodynamic, Economic and Environmental Performance*; Technical University of Berlin: Berlin/Heidelberg, Germany, 2011.

104. Petrakopoulou, F.; Boyano, A.; Cabrera, M.; Tsatsaronis, G. Exergoeconomic and exergoenvironmental analyses of a combined cycle power plant with chemical looping technology. *Int. J. Greenh. Gas Control* **2011**, *5*, 475–482. [CrossRef]

105. Petrakopoulou, F.; Tsatsaronis, G.; Morosuk, T. Advanced Exergoenvironmental Analysis of a Near-Zero Emission Power Plant with Chemical Looping Combustion. *Environ. Sci. Technol.* **2012**, *46*, 3001–3007. [CrossRef] [PubMed]

106. Petrakopoulou, F.; Tsatsaronis, G.; Morosuk, T.; Carassai, A. Conventional and advanced exergetic analyses applied to a combined cycle power plant. *Energy* **2012**, *41*, 146–152. [CrossRef]

107. Wang, L.; Yang, Y.; Morosuk, T.; Tsatsaronis, G. Advanced Thermodynamic Analysis and Evaluation of a Supercritical Power Plant. *Energies* **2012**, *5*, 1850–1863. [CrossRef]

108. Wang, L.; Fu, P.; Wang, N.; Morosuk, T.; Yang, Y.; Tsatsaronis, G. Malfunction diagnosis of thermal power plants based on advanced exergy analysis: The case with multiple malfunctions occurring simultaneously. *Energy Convers. Manag.* **2017**, *148*, 1453–1467. [CrossRef]

109. Fu, P.; Wang, N.; Wang, L.; Morosuk, T.; Yang, Y.; Tsatsaronis, G. Performance degradation diagnosis of thermal power plants: A method based on advanced exergy analysis. *Energy Convers. Manag.* **2016**, *130*, 219–229. [CrossRef]

110. Açıkkalp, E.; Aras, H.; Hepbasli, A. Advanced exergy analysis of an electricity-generating facility using natural gas. *Energy Convers. Manag.* **2014**, *82*, 146–153. [CrossRef]

111. Gökgedik, H.; Yürüsoy, M.; Keçebaş, A. Improvement potential of a real geothermal power plant using advanced exergy analysis. *Energy* **2016**, *112*, 254–263. [CrossRef]

112. Petrakopoulou, F.; Tsatsaronis, G.; Morosuk, T. Evaluation of a power plant with chemical looping combustion using an advanced exergoeconomic analysis. *Sustain. Energy Technol. Assess.* **2013**, *3*, 9–16. [CrossRef]

113. Wang, L.; Yang, Y.; Dong, C.; Xu, G. Improvement and primary application of theory of fuel specific consumption. *Zhongguo Dianji Gongcheng Xuebao (Proc. Chin. Soc. Electr. Eng.)* **2012**, *32*, 16–21.

114. Wang, L.; Wu, L.; Xu, G.; Dong, C.; Yang, Y. Calculation and analysis of energy consumption interactions in thermal systems of large-scale coal-fired steam power generation units. *Zhongguo Dianji Gongcheng Xuebao (Proc. Chin. Soc. Electr. Eng.)* **2012**, *32*, 9–14.

115. Manesh, M.K.; Navid, P.; Marigorta, A.B.; Amidpour, M.; Hamedi, M. New procedure for optimal design and evaluation of cogeneration system based on advanced exergoeconomic and exergoenvironmental analyses. *Energy* **2013**, *59*, 314–333. [CrossRef]

116. Tsatsaronis, G.; Morosuk, T. *Understanding the Formation of Costs and Environmental Impacts Using Exergy-Based Methods. Energy Security and Development*; Springer: Berlin/Heidelberg, Germany, 2015; pp. 271–291.

117. Bolatturk, A.; Coskun, A.; Geredelioglu, C. Thermodynamic and exergoeconomic analysis of Çayırhan thermal power plant. *Energy Convers. Manag.* **2015**, *101*, 371–378. [CrossRef]

118. Frangopoulos, C.A. Methods of energy systems optimization. In Summer school. In Proceedings of the Summer School: Optimization of Energy Systems and Processes, Gliwice, Poland, 24–27 June 2003.

119. Glover, F.; Kochenberger, G.A. *Handbook of Metaheuristics*; Springer: Berlin/Heidelberg, Germany, 2003.

120. Uhlenbruck, S.; Lucas, K. Exergy-Aided Cost Optimization Using Evolutionary Algorithm. *Int. J. Appl. Thermodyn.* **2000**, *3*, 121–127.

121. Nocedal, J.; Wright, S. *Numerical Optimization*, 2rd ed.; Springer: New York, NY, USA, 2006.

122. Biegler, L.T.; Grossmann, I.E. Retrospective on optimization. *Comput. Chem. Eng.* **2004**, *28*, 1169–1192. [CrossRef]

123. Grossmann, I.E.; Biegler, L.T. Part II. Future perspective on optimization. *Comput. Chem. Eng.* **2004**, *28*, 1193–1218. [CrossRef]

124. Floudas, C.A.; Gounaris, C.E. A review of recent advances in global optimization. *J. Glob. Optim.* **2009**, *45*, 3–38. [CrossRef]

125. More, J.; Wild, S. Benchmarking Derivative-Free Optimization Algorithms. *SIAM J. Optim.* **2009**, *20*, 172–191. [CrossRef]

126. Rios, L.M.; Sahinidis, N.V. Derivative-free optimization: A review of algorithms and comparison of software implementations. *J. Glob. Optim.* **2013**, *56*, 1247–1293. [CrossRef]

127. Drud, A. *CONOPT Documentation*; ARKI Consulting and Development A/S: Bagsvaerd, Denmark, 2004.

128. Nocedal, J.; Wright, S. *Conjugate Gradient Methods*; Springer: Berlin/Heidelberg, Germany, 2006.

129. Nesterov, Y.; Nemirovskii, A.; Ye, Y. *Interior-Point Polynomial Algorithms in Convex Programming*; SIAM: Philadelphia, PA, USA, 1994.

130. Matthias, K. *On the Complexity of Nonlinear Mixed-Integer Optimization*; Mixed Integer Nonlinear Programming; Springer: Berlin/Heidelberg, Germany, 2012; pp. 533–557.

131. Bonami, P.; Kilincc, M.; Linderoth, J. *Algorithms and Software for Convex Mixed Integer Nonlinear Programs*; Mixed Integer Nonlinear Programming; Springer: Berlin/Heidelberg, Germany, 2012; pp. 1–39.

132. Grossmann, I.E.; Guillén-Gosálbez, G. Scope for the application of mathematical programming techniques in the synthesis and planning of sustainable processes. *Comput. Chem. Eng.* **2010**, *34*, 1365–1376. [CrossRef]

133. Grossmann, I.; Ruiz, J. Generalized Disjunctive Programming: A Framework for Formulation and Alternative Algorithms for MINLP Optimization. In *The IMA Volumes in Mathematics and Its Applications*; Lee, J., Leyffer, S., Eds.; Springer: New York, NY, USA, 2012; pp. 93–115.

134. Grossmann, I.E.; Trespalacios, F. Systematic modeling of discrete-continuous optimization models through generalized disjunctive programming. *AIChE J.* **2013**, *59*, 3276–3295. [CrossRef]

135. Vecchietti, A.; Grossmann, I.E. LOGMIP: A disjunctive 0–1 non-linear optimizer for process system models. *Comput. Chem. Eng.* **1999**, *23*, 555–565. [CrossRef]

136. Voll, P. *Automated Optimization-Based Synthesis of Distributed Energy Supply Systems*; RWTH Aachen Univeristy: Aachen, Germany, 2014.

137. Beyer, H.G.; Schwefel, H.P. Evolution strategies—A comprehensive introduction. *Nat. Comput.* **2002**, *1*, 3–52. [CrossRef]

138. Deb, K.; Pratap, A.; Agarwal, S.; Meyarivan, T. A fast and elitist multiobjective genetic algorithm: NSGA-II. *IEEE Trans. Evol. Comput.* **2002**, *6*, 182–197. [CrossRef]

139. Storn, R.; Price, K. Differential evolution–A simple and efficient heuristic for global optimization over continuous spaces. *J. Glob. Optim.* **1997**, *11*, 341–359. [CrossRef]

140. Wagner, W.; Cooper, J.R.; Dittmann, A.; Kijima, J.; Kretzschmar, H.J.; Kruse, A. The IAPWS Industrial Formulation 1997 for the Thermodynamic Properties of Water and Steam. *ASME J. Eng. Gas Turb. Power* **2000**, *122*, 150–184. [CrossRef]

141. Savola, T. *Modeling Biomass-Fuelled Small-Scale CHP Plants for Process Synthesis and Optimization*; Helsinki University of Technology: Espoo, Finland, 2007.

142. Angira, R.; Babu, B.V. Optimization of process synthesis and design problems: A modified differential evolution approach. *Chem. Eng. Sci.* **2006**, *61*, 4707–4721. [CrossRef]

143. Luo, X.; Zhang, B.; Chen, Y.; Mo, S. Modeling and optimization of a utility system containing multiple extractions steam turbines. *Energy* **2011**, *36*, 3501–3512. [CrossRef]

144. Manassaldi, J.I.; Mussati, S.F.; Scenna, N.A.J. Optimal synthesis and design of Heat Recovery Steam Generation (HRSG) via mathematical programming. *Energy* **2011**, *36*, 475–485. [CrossRef]

145. Judes, M. *MINLP Optimization of Design and Steady-State Operation of Power Plants Considering Several Operating Points*; Technical University of Berlin: Berlin/Heidelberg, Germany, 2009.

146. TILMedia Suite. Available online: https://www.tlk-thermo.com/index.php/en/tilmedia-suite (accessed on 26 December 2018).

147. Freesteam. Available online: http://freesteam.sourceforge.net/ (accessed on 26 December 2018).

148. Pacherneggm, S.J. *A Closer Look at the Willans-Line*; SAE: Troy, MI, USA, 1969.

149. Mavromatis, S.P.; Kokossis, A.C. Hardware composites: A new conceptual tool for the analysis and optimisation of steam turbine networks in chemical process industries: Part II: Application to operation and design. *Chem. Eng. Sci.* **1998**, *53*, 1435–1461. [CrossRef]

150. Cooke, D.H. *Modeling of Off-Design Multi-Stage Turbine Pressures by Stodola's Ellipse*; Energy Incoporated (PEPSE) User's Group Meeting: Richmond, VA, USA, 1983.

151. Paterson, W.R. A replacement for the logarithmic mean. *Chem. Eng. Sci.* **1984**, *39*, 1635–1636. [CrossRef]

152. Drud, A.S. CONOPT—A large-scale GRG code. *ORSA J. Comput.* **1994**, *6*, 207–216. [CrossRef]

153. Fair Isaac Corporation (FICO). *XPRESS Optimization Suite: MIP Formulations and Linearizations*; Fair Isaac Corporation: San Jose, CA, USA, 2009.

154. Tomlin, J.A. *Special Ordered Sets and an Application to Gas Supply Operations Planning*; Mathematical Programming; Springer: Berlin/Heidelberg, Germany, 1988; Volume 42, pp. 69–84.

155. Tsoukalas, A.; Mitsos, A. Multivariate mccormick relaxations. *J. Glob. Optim.* **2014**, *59*, 633–662. [CrossRef]

156. Haywood, R.W. A generalized analysis of the regenerative steam cycle for a finite number of heaters. *Proc. Inst. Mech. Eng.* **1949**, *161*, 157–164. [CrossRef]

157. Weir, C.D. Optimization of heater enthalpy rises in feed-heating trains. *Proc. Inst. Mech. Eng.* **1960**, *174*, 769–796. [CrossRef]

158. Espatolero, S.; Romeo, L.M.; Cortés, C. Efficiency improvement strategies for the feedwater heaters network designing in supercritical coal-fired power plants. *Appl. Therm. Eng.* **2014**, *73*, 449–460. [CrossRef]

159. Uche, J.; Serra, L.; Valero, A. Thermoeconomic optimization of a dual-purpose power and desalination plant. *Desalination* **2001**, *136*, 147–158. [CrossRef]

160. Xiong, J.; Zhao, H.; Zhang, C.; Zheng, C.; Luh, P.B. Thermoeconomic operation optimization of a coal-fired power plant. *Energy* **2012**, *42*, 486–496. [CrossRef]

161. Conradie, A.; Buys, J.; Kröger, D. Performance optimization of dry-cooling systems for power plants through SQP methods. *Appl. Therm. Eng.* **1998**, *18*, 25–45. [CrossRef]

162. Li, X.; Wang, N.; Wang, L.; Yang, Y.; Maréchal, F. Identification of optimal operating strategy of direct air-cooling condenser for Rankine cycle based power plants. *Appl. Energy* **2018**, *209*, 153–166. [CrossRef]

163. Suresh, M.V.J.J.; Reddy, K.S.; Kolar, A.K. ANN-GA based optimization of a high ash coal-fired supercritical power plant. *Appl. Energy* **2011**, *88*, 4867–4873. [CrossRef]

164. Hajabdollahi, F.; Hajabdollahi, Z.; Hajabdollahi, H. Soft computing based multi-objective optimization of steam cycle power plant using NSGA-II and ANN. *Appl. Soft Comput.* **2012**, *12*, 3648–3655. [CrossRef]

165. Montes, M.J.; Abánades, A.; Martínez-Val, J.M.; Valdés, M. Solar multiple optimization for a solar-only thermal power plant, using oil as heat transfer fluid in the parabolic trough collectors. *Sol. Energy* **2009**, *83*, 2165–2176. [CrossRef]

166. Desai, N.B.; Bandyopadhyay, S. Optimization of concentrating solar thermal power plant based on parabolic trough collector. *J. Clean. Prod.* **2015**, *89*, 262–271. [CrossRef]

167. Biegler, L.T.; Grossmann, I.E.; Westerberg, A.W.; Kravanja, Z. *Systematic Methods of Chemical Process Design*; Prentice Hall PTR: Upper Saddle River, NJ, USA, 1997.

168. Westerberg, A.W. A retrospective on design and process synthesis. *Comput. Chem. Eng.* **2004**, *28*, 447–458. [CrossRef]

169. Frangopoulos, C.A.; von Spakovsky, M.R.; Sciubba, E. A Brief Review of Methods for the Design and Synthesis Optimization of Energy Systems. *Int. J. Thermodyn.* **2002**, *5*, 151–160.

170. Kaibel, G.; Schoenmakers, H. Process synthesis and design in industrial practice. *Comput. Aid. Chem. Eng.* **2002**, *10*, 9–22.

171. Li, X.; Kraslawski, A. Conceptual process synthesis: Past and current trends. *Chem. Eng. Process. Process Intensif.* **2004**, *43*, 583–594. [CrossRef]

172. Barnicki, S.D.; Siirola, J.J. Process synthesis prospective. *Comput. Chem. Eng.* **2004**, *28*, 441–446. [CrossRef]

173. Douglas, J.M. A hierarchical decision procedure for process synthesis. *AIChE J.* **1985**, *31*, 353–362. [CrossRef]

174. Jaksland, C.A.; Gani, R.; Lien, K.M. Separation process design and synthesis based on thermodynamic insights. *Chem. Eng. Sci.* **1995**, *50*, 511–530. [CrossRef]

175. Kravanja, Z.; Grossmann, I.E. Multilevel-hierarchical MINLP synthesis of process flowsheets. *Comput. Chem. Eng.* **1997**, *21*, S421–S426. [CrossRef]

176. Douglas, J.M. *Conceptual Design of Chemical Processes*; McGraw-Hill: New York, NY, USA, 1988.

177. Douglas, J.M. Synthesis of Separation System Flowsheets. *AIChE J.* **1995**, *41*, 2522–2536. [CrossRef]

178. Linnhoff, B. Pinch analysis: A state-of-the-art overview: Techno-economic analysis. *Chem. Eng. Res. Des.* **1993**, *71*, 503–522.

179. Matsuda, K.; Hirochi, Y.; Tatsumi, H.; Shire, T. Applying heat integration total site based pinch technology to a large industrial area in Japan to further improve performance of highly efficient process plants. *Energy* **2009**, *34*, 1687–1692. [CrossRef]

180. Siirola, J.J.; Rudd, D.F. Computer-Aided Synthesis of Chemical Process Designs: From Reaction Path Data to the Process Task Network. *Ind. Eng. Chem. Fund.* **1971**, *10*, 353–362. [CrossRef]

181. Mahalec, V.; Motard, R.L. Procedures for the initial design of chemical processing systems. *Comput. Chem. Eng.* **1977**, *1*, 57–68. [CrossRef]

182. Kirkwood, R.L.; Locke, M.H.; Douglas, J.M. A prototype expert system for synthesizing chemical process flowsheets. *Comput. Chem. Eng.* **1988**, *12*, 329–343. [CrossRef]

183. Sciubba, E. Toward Automatic Process Simulators: Part II–An Expert System for Process Synthesis. *ASME J. Eng. Gas Turb. Power* **1998**, *120*, 9–16. [CrossRef]

184. Manolas, D.A.; Efthimeros, G.A.; Tsahalis, D.T. Development of an Expert System Shell Based on Genetic Algorithms for the Selection of the Energy Best Available Technologies and their Optimal Operating Conditions for the Process Industry. *Expert Syst.* **2001**, *18*, 124–130. [CrossRef]

185. Matelli, J.A.; Bazzo, E.; da Silva, J.C. An expert system prototype for designing natural gas cogeneration plants. *Expert Syst. Appl.* **2009**, *36*, 8375–8384. [CrossRef]

186. Chen, F.; Duic, N.; Manuel Alves, L.; da Graça Carvalho, M. Renewisland–Renewable energy solutions for islands. *Renew. Sustain. Energy Rev.* **2007**, *11*, 1888–1902. [CrossRef]

187. Siirola, J.J. Strategic process synthesis: Advances in the hierarchical approach. *Comput. Chem. Eng.* **1996**, *20* (Suppl. 2), S1637–S1643. [CrossRef]

188. Surma, J.; Braunschweig, B. Case-base retrieval in process engineering: Supporting design by reusing flowsheets. *Eng. Appl. Artif. Intell.* **1996**, *9*, 385–391. [CrossRef]

189. Duran, M.A.; Grossmann, I.E. An outer-approximation algorithm for a class of mixed-integer nonlinear programs. *Math. Program.* **1986**, *36*, 307–339. [CrossRef]

190. Yee, T.F.; Grossmann, I.E. Simultaneous optimization models for heat integration–II. Heat exchanger network synthesis. *Comput. Chem. Eng.* **1990**, *14*, 1165 1184. [CrossRef]

191. Grossmann, I.E.; Caballero, J.A.; Yeomans, H. Advances in mathematical programming for the synthesis of process systems. *Latin Am. Appl. Res.* **2000**, *30*, 263–284.

192. Westerberg, A.W. Process Engineering. In *Perspectives in Chemical Engineering Research and Education*; Colton, C.K., Ed.; Academic Press: Cambridge, MA, USA, 1991; pp. 499–523.

193. Luo, X.; Wen, Q.-Y.; Fieg, G. A hybrid genetic algorithm for synthesis of heat exchanger networks. *Comput. Chem. Eng.* **2009**, *33*, 1169–1181. [CrossRef]

194. Escobar, M.; Trierweiler, J.O.; Grossmann, I.E. A heuristic Lagrangean approach for the synthesis of multiperiod heat exchanger networks. *Appl. Therm. Eng.* **2014**, *63*, 177–191. [CrossRef]

195. Skiborowski, M.; Harwardt, A.; Marquardt, W. Conceptual design of distillation-based hybrid separation processes. *Annu. Rev. Chem. Biomol. Eng.* **2012**, *4*, 45–68. [CrossRef] [PubMed]

196. Ahmetović, E.; Grossmann, I.E. Global superstructure optimization for the design of integrated process water networks. *AIChE J.* **2011**, *57*, 434–457. [CrossRef]

197. Liu, P.; Georgiadis, M.C.; Pistikopoulos, E.N. Advances in energy systems engineering. *Ind. Eng. Chem. Res.* **2010**, *50*, 4915–4926. [CrossRef]

198. Luo, X.; Zhang, B.; Chen, Y.; Mo, S. Operational planning optimization of multiple interconnected steam power plants considering environmental costs. *Energy* **2012**, *37*, 549–561. [CrossRef]

199. Jiang, L.; Lin, R.; Jin, H.; Cai, R.; Liu, Z. Study on thermodynamic characteristic and optimization of steam cycle system in IGCC. *Energy Convers. Manag.* **2002**, *43*, 1339–1348. [CrossRef]

200. Grekas, D.N.; Frangopoulos, C.A. Automatic synthesis of mathematical models using graph theory for optimisation of thermal energy systems. *Energy Convers. Manag.* **2007**, *48*, 2818–2826. [CrossRef]

201. Ahadi-Oskui, T.; Alperin, H.; Nowak, I.; Cziesla, F.; Tsatsaronis, G. A relaxation-based heuristic for the design of cost-effective energy conversion systems. *Energy* **2006**, *31*, 1346–1357. [CrossRef]

202. Ahadi-Oskui, T.; Vigerske, S.; Nowak, I.; Tsatsaronis, G. Optimizing the design of complex energy conversion systems by Branch and Cut. *Comput. Chem. Eng.* **2010**, *34*, 1226–1236. [CrossRef]

203. Friedler, F.; Tarjan, K.; Huang, Y.W.; Fan, L.T. Graph-theoretic approach to process synthesis: Axioms and theorems. *Chem. Eng. Sci.* **1992**, *47*, 1973–1988. [CrossRef]

204. Emmerich, M.; Grötzner, M.; Schütz, M. Design of Graph-Based Evolutionary Algorithms: A Case Study for Chemical Process Networks. *Evol. Comput.* **2001**, *9*, 329–354. [CrossRef] [PubMed]

205. Book, R.V.; Otto, F. *String-Rewriting Systems*; Springer: Berlin/Heidelberg, Germany, 1993.

206. Fraga, E.S. A rewriting grammar for heat exchanger network structure evolution with stream splitting. *Eng. Optim.* **2009**, *41*, 813–831. [CrossRef]

207. Lam, H.L.; Klemeš, J.J.; Kravanja, Z.; Varbanov, P.S. Software tools overview: Process integration, modelling and optimisation for energy saving and pollution reduction. *Asia-Pac. J. Chem. Eng.* **2011**, *6*, 696–712. [CrossRef]

208. Wang, L.; Yang, Y.; Dong, C.; Morosuk, T.; Tsatsaronis, G. Systematic Optimization of the Design of Steam Cycles Using MINLP and Differential Evolution. *ASME J. Energy Resour. Technol.* **2014**, *136*, 031601. [CrossRef]

209. Bertok, B.; Barany, M.; Friedler, F. Generating and Analyzing Mathematical Programming Models of Conceptual Process Design by P-graph Software. *Ind. Eng. Chem. Res.* **2012**, *52*, 166–171. [CrossRef]

210. Hostrup, M.; Gani, R.; Kravanja, Z.; Sorsak, A.; Grossmann, I. Integration of thermodynamic insights and MINLP optimization for the synthesis, design and analysis of process flowsheets. *Comput. Chem. Eng.* **2001**, *25*, 73–83. [CrossRef]

211. Fraga, E.S. The generation and use of partial solutions in process synthesis. *Chem. Eng. Res. Des.* **1998**, *76*, 45–54. [CrossRef]

212. Seferlis, P.; Grievink, J. Optimal Design and Sensitivity Analysis of Reactive Distillation Units Using Collocation Models. *Ind. Eng. Chem. Res.* **2001**, *40*, 1673–1685. [CrossRef]

213. Hillermeier, C.; Hüster, S.; Märker, W.; Sturm, T.F. Optimization of Power Plant Design: Stochastic and Adaptive Solution Concepts. In *Evolutionary Design and Manufacture*; Parmee, I.C., Ed.; Springer: London, UK, 2000; pp. 3–18.

214. Daichendt, M.M.; Grossmann, I.E. Preliminary screening procedure for the MINLP synthesis of process systems–II. Heat exchanger networks. *Comput. Chem. Eng.* **1994**, *18*, 679–709. [CrossRef]

215. Daichendt, M.M.; Grossmann, I.E. Integration of hierarchical decomposition and mathematical programming for the synthesis of process flowsheets. *Comput. Chem. Eng.* **1998**, *22*, 147–175. [CrossRef]

216. Manninen, J.; Zhu, X.X. Level-by-level flowsheet synthesis methodology for thermal system design. *AIChE J.* **2001**, *47*, 142–159. [CrossRef]

217. Yeomans, H.; Grossmann, I.E. A systematic modeling framework of superstructure optimization in process synthesis. *Comput. Chem. Eng.* **1999**, *23*, 709–731. [CrossRef]

218. Chen, K.; Parmee, I.C.; Gane, C.R. Dual mutation strategies for mixed-integer optimisation in power station design. In Proceedings of the 1997 IEEE International Conference on Evolutionary Computation (ICEC '97), Indianapolis, IN, USA, 13–16 April 1997; pp. 385–390.

219. Friedler, F.; Tarjan, K.; Huang, Y.W.; Fan, L.T. Graph-theoretic approach to process synthesis: Polynomial algorithm for maximal structure generation. *Comput. Chem. Eng.* **1993**, *17*, 929–942. [CrossRef]

220. Friedler, F.; Varga, J.B.; Fan, L.T. Decision-mapping: A tool for consistent and complete decisions in process synthesis. *Chem. Eng. Sci.* **1995**, *50*, 1755–1768. [CrossRef]

221. Vance, L.; Cabezas, H.; Heckl, I.; Bertok, B.; Friedler, F. Synthesis of Sustainable Energy Supply Chain by the P-graph Framework. *Ind. Eng. Chem. Res.* **2012**, *52*, 266–274. [CrossRef]

222. Bertok, B.; Kalauz, K.; Sule, Z.; Friedler, F. Combinatorial Algorithm for Synthesizing Redundant Structures to Increase Reliability of Supply Chains: Application to Biodiesel Supply. *Ind. Eng. Chem. Res.* **2012**, *52*, 181–186. [CrossRef]

223. Voll, P.; Klaffke, C.; Hennen, M.; Bardow, A. Automated superstructure-based synthesis and optimization of distributed energy supply systems. *Energy* **2013**, *50*, 374–388. [CrossRef]

224. Martelli, E.; Amaldi, E.; Consonni, S. Numerical optimization of heat recovery steam cycles: Mathematical model, two-stage algorithm and applications. *Comput. Chem. Eng.* **2011**, *35*, 2799–2823. [CrossRef]

225. Bagajewicz, M.J.; Manousiouthakis, V. Mass/heat-exchange network representation of distillation networks. *AIChE J.* **1992**, *38*, 1769–1800. [CrossRef]

226. Papalexandri, K.P.; Pistikopoulos, E.N. Generalized modular representation framework for process synthesis. *AIChE J.* **1996**, *42*, 1010–1032. [CrossRef]

227. Lang, Y.D.; Biegler, L.T. Distributed stream method for tray optimization. *AIChE J.* **2002**, *48*, 582–595. [CrossRef]

228. Stein, O.; Oldenburg, J.; Marquardt, W. Continuous reformulations of discrete-continuous optimization problems. *Comput. Chem. Eng.* **2004**, *28*, 1951–1966. [CrossRef]

229. Urselmann, M.; Barkmann, S.; Sand, G.; Engell, S. A Memetic Algorithm for Global Optimization in Chemical Process Synthesis Problems. *IEEE Trans. Evol. Comput.* **2011**, *15*, 659–683. [CrossRef]

230. Urselmann, M.; Engell, S. Design of memetic algorithms for the efficient optimization of chemical process synthesis problems with structural restrictions. *Comput. Chem. Eng.* **2015**, *72*, 87–108. [CrossRef]

231. Stephanopoulos, G.; Westerberg, A.W. Studies in process synthesis-II: Evolutionary synthesis of optimal process flowsheets. *Chem. Eng. Sci.* **1976**, *31*, 195–204. [CrossRef]

232. Seader, J.D.; Westerberg, A.W. A combined heuristic and evolutionary strategy for synthesis of simple separation sequences. *AIChE J.* **1977**, *23*, 951–954. [CrossRef]

233. Toffolo, A. The synthesis of cost optimal heat exchanger networks with unconstrained topology. *Appl. Therm. Eng.* **2009**, *29*, 3518–3528. [CrossRef]

234. Wright, J.; Zhang, Y.; Angelov, P.; Hanby, V.; Buswell, R. Evolutionary synthesis of HVAC system configurations: Algorithm development. *Hvac R Res.* **2008**, *14*, 33–55. [CrossRef]

235. Toffolo, A. A synthesis/design optimization algorithm for Rankine cycle based energy systems. *Energy* **2014**, *66*, 115–127. [CrossRef]

236. Toffolo, A.; Rech, S.; Lazzaretto, A. Generation of Complex Energy Systems by Combination of Elementary Processes. *J. Energy Resour. Technol.* **2018**, *140*, 112005. [CrossRef]

237. Lazzaretto, A.; Toffolo, A. A method to separate the problem of heat transfer interactions in the synthesis of thermal systems. *Energy* **2008**, *33*, 163–170. [CrossRef]

238. Urselmann, M.; Emmerich, M.T.M.; Till, J.; Sand, G.; Engell, S. Design of problem-specific evolutionary algorithm/mixed-integer programming hybrids: Two-stage stochastic integer programming applied to chemical batch scheduling. *Eng. Optim.* **2007**, *39*, 529–549. [CrossRef]

239. Emmerich, M.T.M. Optimisation of Thermal Power Plant Designs: A Graph-based Adaptive Search Approach. In *Adaptive Computing in Design and Manufacture V*; Parmee, I.C., Ed.; Springer: London, UK, 2002; pp. 87–98.

240. Voll, P.; Lampe, M.; Wrobel, G.; Bardow, A. Superstructure-free synthesis and optimization of distributed industrial energy supply systems. *Energy* **2012**, *45*, 424–435. [CrossRef]

241. Wang, L.; Voll, P.; Lampe, M.; Yang, Y.; Bardow, A. Superstructure-free synthesis and optimization of thermal power plants. *Energy* **2015**, *91*, 700–711. [CrossRef]

242. Wang, L.; Lampe, M.; Voll, P.; Yang, Y.; Bardow, A. Multi-objective superstructure-free synthesis and optimization of thermal power plants. *Energy* **2016**, *116*, 1104–1116. [CrossRef]

243. Wang, L.; Yang, Y.; Dong, C.; Morosuk, T.; Tsatsaronis, G. Parametric optimization of supercritical coal-fired power plants by MINLP and differential evolution. *Energy Convers. Manag.* **2014**, *85*, 828–838. [CrossRef]

244. Maréchal, F.; Kalitventzeff, B. Targeting the optimal integration of steam networks: Mathematical tools and methodology. *Comput. Chem. Eng.* **1999**, *23*, S133–S136. [CrossRef]

245. Kermani, M.; Wallerand, A.S.; Kantor, I.D.; Maréchal, F. Generic superstructure synthesis of organic Rankine cycles for waste heat recovery in industrial processes. *Appl. Energy* **2018**, *212*, 1203–1225. [CrossRef]

246. Wallerand, A.S.; Kermani, M.; Kantor, I.; Maréchal, F. Optimal heat pump integration in industrial processes. *Appl. Energy* **2018**, *219*, 68–92. [CrossRef]

247. Gassner, M.; Maréchal, F. Thermo-economic optimisation of the polygeneration of synthetic natural gas (SNG), power and heat from lignocellulosic biomass by gasification and methanation. *Energy Environ. Sci.* **2012**, *5*, 5768–5789. [CrossRef]

248. Gassner, M.; Maréchal, F. Methodology for the optimal thermo-economic, multi-objective design of thermochemical fuel production from biomass. *Comput. Chem. Eng.* **2009**, *33*, 769–781. [CrossRef]

249. Wang, L.; Düll, J.; Maréchal, F.; Jan van, H. Trade-off designs and comparative exergy evaluation of solid-oxide electrolyzer based power-to-methane plants. *Int. J. Hydrogen Energy* **2018**. [CrossRef]

250. Mian, A.; Martelli, E.; Maréchal, F. Framework for the multiperiod sequential synthesis of heat exchanger networks with selection, design, and scheduling of multiple utilities. *Ind. Eng. Chem. Res.* **2016**, *55*, 168–186. [CrossRef]

251. Lazzaretto, A.; Manente, G.; Toffolo, A. SYNTHSEP: A general methodology for the synthesis of energy system configurations beyond superstructures. *Energy* **2018**, *147*, 924–949. [CrossRef]

252. Rangaiah, G.P.; Sharma, S.; Sreepathi, B.K. Multi-objective optimization for the design and operation of energy efficient chemical processes and power generation. *Curr. Opin. Chem. Eng.* **2015**, *10*, 49–62. [CrossRef]

253. Jubril, A.M. A nonlinear weights selection in weighted sum for convex multiobjective optimization. *Facta Univ.* **2012**, *27*, 357–372.

254. Zhang, W.; Reimann, M. A simple augmented ε-constraint method for multi-objective mathematical integer programming problems. *Eur. J. Oper. Res.* **2014**, *234*, 15–24. [CrossRef]

255. Messac, A.; Ismail-Yahaya, A.; Mattson, C.A. The normalized normal constraint method for generating the Pareto frontier. *Struct. Multidiscip. Optim.* **2003**, *25*, 86–98. [CrossRef]

256. Rangaiah, G.P.; Sharma, S. *Differential Evolution in Chemical Engineering: Developments and Applications*; World Scientific: Singapore, 2017.

257. Deb, K. *Multi-Objective Optimization Using Evolutionary Algorithms*; John Wiley & Sons: Hoboken, NJ, USA, 2001.

258. Coello, C.A.C.; Lamont, G.B.; Van Veldhuizen, D.A. *Evolutionary Algorithms for Solving Multi-Objective Problems*; Springer: Berlin/Heidelberg, Germany, 2007.

259. Sharma, S.; Rangaiah, G. *Hybrid Approach for Multiobjective Optimization and Its Application to Process Engineering Problems*; Applications of Metaheuristics in Process Engineering; Springer: Berlin/Heidelberg, Germany, 2014; pp. 423–444.

260. Beume, N.; Naujoks, B.; Emmerich, M. SMS-EMOA: Multiobjective selection based on dominated hypervolume. *Eur. J. Oper. Res.* **2007**, *181*, 1653–1669. [CrossRef]

261. Zitzler, E.; Laumanns, M.; Thiele, L. *SPEA2: Improving the Strength Pareto Evolutionary Algorithm*; TIK-Report 103; ETH Library: Zurich, Switzerland, 2001.

262. Corne, D.W.; Knowles, J.D.; Oates, M.J. The Pareto envelope-based selection algorithm for multiobjective optimization. In Proceedings of the International Conference on Parallel Problem Solving from Nature, Kraków, Poland, 11–15 September2000; Springer: Berlin/Heidelberg, Germany, 2000; pp. 839–848.

263. Poloni, C.; Giurgevich, A.; Onesti, L.; Pediroda, V. Hybridization of a multi-objective genetic algorithm, a neural network and a classical optimizer for a complex design problem in fluid dynamics. *Comput. Methods Appl. Mech. Eng.* **2000**, *186*, 403–420. [CrossRef]

264. Van Veldhuizen, D.A.; Lamont, G.B. Multiobjective evolutionary algorithms: Analyzing the state-of-the-art. *Evol. Comput.* **2000**, *8*, 125–147. [CrossRef] [PubMed]

265. Liu, P.; Pistikopoulos, E.N.; Li, Z. A multi-objective optimization approach to polygeneration energy systems design. *AIChE J.* **2010**, *56*, 1218–1234. [CrossRef]

266. Lazzaretto, A.; Toffolo, A. Energy, economy and environment as objectives in multi-criterion optimization of thermal systems design. *Energy* **2004**, *29*, 1139–1157. [CrossRef]

267. Kavvadias, K.C.; Maroulis, Z.B. Multi-objective optimization of a trigeneration plant. *Energy Policy* **2010**, *38*, 945–954. [CrossRef]

268. Fazlollahi, S.; Mandel, P.; Becker, G.; Marechal, F. Methods for multi-objective investment and operating optimization of complex energy systems. *Energy* **2012**, *45*, 12–22. [CrossRef]

269. Wang, L.; Yang, Y.; Dong, C.; Morosuk, T.; Tsatsaronis, G. Multi-objective optimization of coal-fired power plants using differential evolution. *Appl. Energy* **2014**, *115*, 254–264. [CrossRef]

270. Chen, L.; Yan, C.; Liao, Y.; Song, F.; Jia, Z. A hybrid non-dominated sorting genetic algorithm and its application on multi-objective optimal design of nuclear power plant. *Ann. Nucl. Energy* **2017**, *100*, 150–159. [CrossRef]

271. Gimelli, A.; Luongo, A.; Muccillo, M. Efficiency and cost optimization of a regenerative Organic Rankine Cycle power plant through the multi-objective approach. *Appl. Therm. Eng.* **2017**, *114*, 601–610. [CrossRef]

272. Boyaghchi, F.A.; Molaie, H. Advanced exergy and environmental analyses and multi objective optimization of a real combined cycle power plant with supplementary firing using evolutionary algorithm. *Energy* **2015**, *93*, 2267–2279. [CrossRef]

273. Yao, E.; Wang, H.; Wang, L.; Xi, G.; Maréchal, F. Multi-objective optimization and exergoeconomic analysis of a combined cooling, heating and power based compressed air energy storage system. *Energy Convers. Manag.* **2017**, *138*, 199–209. [CrossRef]

274. Yao, E.; Wang, H.; Wang, L.; Xi, G.; Maréchal, F. Thermo-economic optimization of a combined cooling, heating and power system based on small-scale compressed air energy storage. *Energy Convers. Manag.* **2016**, *118*, 377–386. [CrossRef]

275. Barzegar Avval, H.; Ahmadi, P.; Ghaffarizadeh, A.; Saidi, M. Thermo-economic-environmental multiobjective optimization of a gas turbine power plant with preheater using evolutionary algorithm. *Int. J. Energy Res.* **2011**, *35*, 389–403. [CrossRef]

276. Gutiérrez-Arriaga, C.G.; Serna-González, M.; Ponce-Ortega, J.M.; El-Halwagi, M.M. Multi-objective optimization of steam power plants for sustainable generation of electricity. *Clean Technol. Environ. Policy* **2013**, *15*, 551–566. [CrossRef]

277. Hajabdollahi, H.; Ahmadi, P.; Dincer, I. An exergy-based multi-objective optimization of a heat recovery steam generator (HRSG) in a combined cycle power plant (CCPP) using evolutionary algorithm. *Int. J. Green Energy* **2011**, *8*, 44–64. [CrossRef]

278. Li, H.; Marechal, F.; Burer, M.; Favrat, D. Multi-objective optimization of an advanced combined cycle power plant including CO_2 separation options. *Energy* **2006**, *31*, 3117–3134. [CrossRef]

279. Sayyaadi, H. Multi-objective approach in thermoenvironomic optimization of a benchmark cogeneration system. *Appl. Energy* **2009**, *86*, 867–879. [CrossRef]

280. González-Bravo, R.; Nápoles-Rivera, F.; Ponce-Ortega, J.M.; El-Halwagi, M.M. Multiobjective Optimization of Dual-Purpose Power Plants and Water Distribution Networks. *ACS Sustain. Chem. Eng.* **2016**, *4*, 6852–6866. [CrossRef]

281. Cooke, D.H. On prediction of off-design multistage turbine pressures by Stodol's ellipse. *ASME J. Eng. Gas Turb. Power* **1985**, *107*, 596–606. [CrossRef]

282. Rabek, G. *Die Ermittlung der Betriebsverhaltnisse von Speisewasservorwarmern bei verschiedenen Belastungen*; Energie und Technik: Winterthur, Swizerland, 1963.

283. Tchanche, B.F.; Lambrinos, G.; Frangoudakis, A.; Papadakis, G. Low-grade heat conversion into power using organic Rankine cycles—A review of various applications. *Renew. Sustain. Energy Rev.* **2011**, *15*, 3963–3979. [CrossRef]

284. Wright, S.A.; Radel, R.F.; Vernon, M.E.; Rochau, G.E.; Pickard, P.S. *Operation and Analysis of a Supercritical CO_2 Brayton Cycle*; Sandia Report, No. SAND2010-0171; Sandia National Laboratories: Livermore, CA, USA, 2010.

285. Harkin, T.; Hoadley, A.; Hooper, B. Using multi-objective optimisation in the design of CO_2 capture systems for retrofit to coal power stations. *Energy* **2012**, *41*, 228–235. [CrossRef]

286. Samanta, A.; Zhao, A.; Shimizu, G.K.H.; Sarkar, P.; Gupta, R. Post-combustion CO_2 capture using solid sorbents: A review. *Ind. Eng. Chem. Res.* **2011**, *51*, 1438–1463. [CrossRef]

287. Jamel, M.S.; Abd Rahman, A.; Shamsuddin, A.H. Advances in the integration of solar thermal energy with conventional and non-conventional power plants. *Renew. Sustain. Energy Rev.* **2013**, *20*, 71–81. [CrossRef]

288. Morandin, M.; Maréchal, F.; Mercangöz, M.; Buchter, F. Conceptual design of a thermo-electrical energy storage system based on heat integration of thermodynamic cycle—Part B: Alternative system configurations. *Energy* **2012**, *45*, 386–396. [CrossRef]

289. Morandin, M.; Maréchal, F.; Mercangöz, M.; Buchter, F. Conceptual design of a thermo-electrical energy storage system based on heat integration of thermodynamic cycles—Part A: Methodology and base case. *Energy* **2012**, *45*, 375–385. [CrossRef]

Article

A Model of Optimal Gas Supply to a Set of Distributed Consumers

Markéta Mikolajková-Alifov [1,*], Frank Pettersson [1], Margareta Björklund-Sänkiaho [2] and Henrik Saxén [1]

[1] Thermal and Flow Engineering Laboratory, Åbo Akademi University, Biskopsgatan 8, 20500 Åbo, Finland; frank.pettersson@abo.fi (F.P.); henrik.saxen@abo.fi (H.S.)
[2] Energy Technology, Åbo Akademi University, Strandgatan 2, 65101 Vasa, Finland; margareta.bjorklund-sankiaho@abo.fi
* Correspondence: mmikolaj@abo.fi

Received: 30 November 2018; Accepted: 16 January 2019; Published: 23 January 2019

Abstract: A better design of gas supply chains may lead to a more efficient use of locally available resources, cost savings, higher energy efficiency and lower impact on the environment. In optimizing the supply chain of liquefied natural gas (LNG), compressed natural gas (CNG) or biogas for smaller regions, the task is to find the best supplier and the most efficient way to transport the gas to the customers to cover their demands, including the design of pipeline networks, truck transportation and storage systems. The analysis also has to consider supporting facilities, such as gasification units, truck loading lines and CNG tanking and filling stations. In this work a mathematical model of a gas supply chain is developed, where gas may be supplied by pipeline, as compressed gas in containers or as LNG by tank trucks, with the goal to find the solution that corresponds to lowest overall costs. In order to efficiently solve the combinatorial optimization problem, it is linearized and tacked by mixed integer linear programming. The resulting model is flexible and can easily be adapted to tackle local supply chain problems with multiple gas sources and distributed consumers of very different energy demands. The model is illustrated by applying it on a local gas distribution problem in western Finland. The dependence of the optimal supply chain on the conditions is demonstrated by a sensitivity analysis, which reveals how the model can be used to evaluate different aspects of the resulting supply chains.

Keywords: gas supply chain; optimization; distributed energy; liquefied natural gas (LNG); compressed natural gas (CNG)

---

1. Introduction

The sustainability of using of natural gas has been widely debated in recent years. Some consider this energy source an environmentally friendlier substitute of other fossil fuels such as coal or oil, but some disagree with this concept. McJeon et al. [1] argue that the role of natural gas as "bridge fuel" is disputable, as its abundancy can lead to even higher energy use, and since there are other low-carbon options available on the market, such as nuclear and renewable sources. Many authors, including Brandt et al. [2] and Levi [3], have also challenged the idea of natural gas as a bridge fuel because of the methane emissions due to leakage during the production, processing and transmission of the gas, and its high greenhouse gas factor. Hausfather and Zhang et al. [4,5] have, on the other hand, stressed that if leakages are minimized in the production and supply chain, natural gas is clearly favorable over coal. Even though the opinions about the advantages of natural gas as an energy source clearly differ, the growing popularity makes it justified to focus attention on its use: the yearly global growth of 1.4% in the natural gas consumption makes it the fastest growing fossil fuel in the world [6]. In Europe, the natural gas consumption is also steadily rising and in 2017 it was about 5200, TWh (530 bcm), which is clearly larger than the European production (2350 TWh, 240 bcm) [7]. The consumption in OECD Europe (all European members of the Organisation for Economic Co-operation and Development) for 2020 is predicted to be in the 5600–6300 TWh (576–646 bcm) range [8]. Naturally, the price influences the popularity and the future use of natural gas. Stern et al. [9] claims that in order for it to be a successful bridging fuel, the price for the high income markets should be below 8 $/MMBtu (about 0.28 $/m^3) and below 6 $/MMBtu (about 0.21 $/m^3) for the low income markets. The transportation to the customer and the supply security are also important issues. Today, natural gas is delivered to Europe mainly by pipeline from the Russian Federation and Norway [7].

Natural gas supplied to Europe is mostly used by the energy sector, households and industry, followed by services and agriculture [10], but due to the shift to using renewable sources in the future, no large growth in the gas consumption is expected in these areas [11]. However, natural gas as a transportation fuel is gaining popularity due to its lower emissions and due to the lower dependence on oil imports [12]. As a transportation fuel it is mostly used in the form of liquefied natural gas (LNG) and compressed natural gas (CNG). Liquefaction of natural gas reduces its volume to 1/600 of the original and is achieved by cooling it at atmospheric pressure to about −160 °C. LNG is from the point of view of SOx, PM and NOx emissions a better fuel than diesel, especially in long-haul freight transport [13]. LNG can be transported over long distances by ship and unloaded in LNG terminals for further distribution in complex supply chains [14]. The biggest suppliers of LNG to Europe in 2017 were Qatar and Algeria, with annual supplies of about 230 TWh and 140 TWh, respectively [7]. A terminal stores large amounts of LNG in specially designed storages, from where it can be delivered onwards in tanks to refueling stations or directly to the users, or after regasification sent out in a pipeline [15]. For heavy-duty vehicles, it is more favorable to use LNG than CNG due to higher energy density and lower pressures, which pose lower demands on strength, size and weight of the tanks [16]. By proper design and operation of LNG refueling stations, the generation of boil-off gas and thus the negative impact on the environment can be kept small [17]. In addition to the road vehicles, LNG can be used to fuel ships. This use is gaining particular popularity in areas with strict regulation of NOx and sulfur emissions [18].

Even though the pressure in a CNG tank is high (typically 250 bar), its energy density is less than half of that of LNG. CNG as an alternative to diesel in transportation became more popular in the USA after 2009 [19]. Its use is especially attractive for low-mileage fleets since the size and weight of the tank do not play a major role in smaller vehicles [16]. Compared with diesel, the noise and the emissions from CNG are lower and after overcoming the technical difficulties such as the lower range, the fuel could become more competitive [20]. In Europe, Italy has a long tradition of using natural gas vehicles [12]. In order to promote further use of natural gas as a fuel for vehicles, the number of refueling stations must grow and the locations have to be selected appropriately. Frick et al. [21] have presented a method for optimizing the locations of compressed natural gas refueling stations in Switzerland.

Biogas produced from biowaste is a valuable methane source. As a natural fuel, it contains a high amount of carbon dioxide, water and sulfur and its use is therefore limited. If upgraded, biogas can reach the same quality as natural gas and can be injected into a natural gas pipeline in the same way as regasified LNG. Synthetic natural gas (SNG), a product of biomass gasification, can also be distributed to customers in the same way.

Optimization of a distribution network supplying gas to customers is not a straightforward task. The expected demand and local availability of different gas sources, such as natural gas, LNG, CNG or biogas, have to be taken into account. Transportation of the fuel is also a crucial element of the planning of gas distribution networks. A pipeline can transport natural gas from the source or regasification site to the customers, both over longer and shorter distances, but since the pressure drop is non-linear with respect to the distance, the optimization of gas pipeline networks is a complex problem. Ríos-Mercado and Borraz-Sánchez [22] provide a thorough review of various problems in the optimization of natural gas distribution and propose possible optimization strategies. Recent approaches to solve gas pipeline optimization problems include a scenario decomposition approach by Schweiger et al. [23], who presented a mixed integer nonlinear programming (MINLP) formulation of the extension of a gas pipeline network. Liang and Hui [24] suggested a convexification of the gas distribution problem in an existing pipeline with multiple demand and supply points in order to minimize the energy demand of the transmission. Mikolajková et al. [25] linearized the non-linear problem and solved the optimal network design and delivery problem by mixed integer linear programming (MILP). Due to the complexity of the problem, the solutions of most pipeline network optimization tasks have been limited to steady state flow. The few optimization studies that consider transient flow have been for pipeline networks of fixed structure: recently, Gugat et al. [26] optimized the transient gas flow in an existing pipe network by MILP. Hante et al. [27] proposed a model for controlling the flow of gas in an existing pipeline network and discussed the problems of selecting appropriate compressors, valves and pipes. Besides pipe length and diameter, elevation differences play a role in gas distribution, particularly in terrains where the pipeline goes through landscapes with large altitude differences. Zhang et al. [28] proposed a model taking into account terrain elevation and other obstacles, optimizing a pipeline connecting production wells.

LNG can be regasified and injected into the same gas pipeline as natural gas from gas wells. Zheng and Pardalos [29] optimized the expansion of the natural gas system with the possible locations of LNG terminals considering the demand/supply uncertainty using a formulation based on Benders decomposition. Since the storage and regasification can account for up to 27% of costs in the LNG value chain [30], also these processes have to be optimized. The place where the regasification unit is installed and the local climate influence the choice of regasification technology. A clear majority of the cases use seawater as heat source in the regasification [31]. In addition to pipeline delivery, LNG from a terminal can be transported by truck to the customers in smaller quantities. Mikolajková et al. [32] used an MILP formulation to optimize the LNG supply from a terminal by truck or after regasification by pipeline to distributed customers.

As for biogas, Hengeveld et al. [33] proposed a pipeline model connecting multiple biogas digesters and an upgrading and injection facility in order to decrease the production costs and energy used for the production of green gas. Hoo et al. [34] studied the injection of upgraded biogas from landfill gas into an existing natural gas pipeline and evaluated scenarios where it is economically and physically viable. Mian et al. [35] developed a multi-objective optimization model of SNG production through gasification of algae feedstock. In the future, the Power-to-Gas (P2G) concept, which uses excess of electricity from renewable resources to produce hydrogen, possibly subsequently converted to methane, may store the gas in a gas pipeline [36].

In summary, recent research activities reflect the importance but also the complexity of optimizing local gas distribution networks with many supply options to find the most cost- and energy-efficient options of gas supply. However, most researchers have focused on optimizing one or a few aspects of the supply chain. The problem of optimizing the gas supply to a region, considering the options of supplying LNG from a terminal by truck, or after regasification by pipeline or by truck as CNG, together with using possible local biogas sources, has not been addressed before. The present paper presents the development of a static model of such a complex local gas supply chain, where the goal is to find the combination of supply technologies that minimizes the total cost of gas delivered. The paper is organized as follows: Section 2 presents the MILP model, its main assumptions and constraints and the resulting cost function to be minimized. Section 3 introduces the parameters for a local problem and a case, where the model is applied to minimize the supply-chain cost for a gas distribution problem in western Finland for a region with 23 consumers. In order to study the dependence of the solution to changes in the costs and market conditions, Section 4 presents results of a sensitivity analysis. Section 5 summarizes the findings and ends with conclusions of the work.

2. Model Description

The model outlined in this section considers several options of simultaneous gas supply to a set of distributed consumers in a region, from a set of sources, including a local and a distant LNG terminal and a biogas plant. The options of using LNG from the local terminal are to regasify it and distribute the gas by pipeline, to deliver the regasified gas in compressed form by containers, or to deliver LNG by tank trucks to the consumers. Biogas sources can be used on the site or injected into a pipeline. To complement the local gas if the local source is too limited or expensive, a supplementary gas source is needed, which is here taken to be LNG from a distant terminal delivered by tank trucks. The objective is to find the optimal supply chain satisfying the demands of gas of all the customers in the region, considering investment and fuel costs as well as operation costs. The costs include investments in pipes of different lengths and diameters, compressor stations, local LNG storage tanks, regasification units, CNG tanking lines and filling units, as well as operation costs, including truck transportation and gas compression. The objective of the optimization can be to design a virgin supply network, or to upgrade or adapt an existing gas supply infrastructure to new suppliers and customers.

2.1. Model Assumptions

In the design and operation of a gas supply network, many technical, economical and physical constraints should be taken into account. However, for optimizing the supply chain, simplifying assumptions have to be made in order to decrease the complexity of the problem. We here list the main assumptions in the model. The system studied is assumed to be in steady-state, and the gas distributed in the pipeline is an ideal gas. The quality of the gas, i.e., its physical properties and chemical composition, is taken to remain constant during the transportation. The gas is characterized by its higher heating value, H, specific heat capacity, c_p, and molar mass, $\overline{M}$. The biogas injected into the pipeline network is for the case of simplicity taken to be upgraded to the same quality as the natural gas. Therefore, the different gases can be interchanged freely in the supply chain.

The system studied has a number of nodes that represent gas sources and sinks. The supply between nodes $(i, j \in I)$ is optimized over a selected time period. If supplied by pipeline, the gas pressure is elevated by compressors to suitable pressure levels so that the desired quantity of gas can flow from the supplier to the customer nodes and be delivered at desired pressure. Since the pressure drop is moderate in a local pipeline network, it was deemed sufficient to install compressors only at the gas injection nodes. Constraints for the maximum and minimum pressures in the network are imposed in the model. The gas injected into the network is assumed to be cooled to the ambient temperature, T_{amb}.

The equations that express the compression power and the pressure drop in the pipeline are non-linear. To reduce the computational burden in the optimization, the equations were linearized to cast the problem into MILP form. The linearization procedure applied is described in detail in Mikolajková et al. [25].

Truck transportation complements the gas supply by pipe. In case of LNG, the gas is supplied from the (local or distant) LNG terminal to the customers' storages by designated trucks. The storage must have an adequate size so that the demand of the customer, and potentially of its neighbor consumers, is covered for a given time period. Supply from a smaller storage to other nearby customers may be realized by a pipeline sub-network. Furthermore, CNG tanking stations can be built to cover the local demand, where trucks distribute the compressed gas in special containers. In this alternative, each customer has a CNG container and a filling equipment, and when the gas pressure in the container falls below a lower limit the container is exchanged by a full one.

In summary, the main constraints of the model are:

- The mass flows in the system are balanced.
- Fuel in adequate quantity covers the customers' demands.
- Technical and physical constraints are obeyed.
- Customers supplied by LNG truck must have adequate storing facilities.

The problem is written as a cost minimization task under the above constraints with the goal to identify the supply network configuration of pipes and trucks that is most economically viable for supplying gas to the customers.

2.2. Constraints

Gas in sufficient quantity, pressurized to the requested level, should cover the energy demand of the customers, D_i. The demand is satisfied by a gas outflow, O_i, from a pipe supplying regasified LNG or biogas, or by a truck that delivers the fuel as LNG or CNG, or a combination of these. The energy balance at the customer's node is therefore:

$$H \cdot \left(O_i + m_i^{\text{truck}}\right) = D_i \tag{1}$$

where H is the (specific) higher heating value and m_i^{truck} is the mass flow rate of gas delivered by truck to the node.

2.2.1. Pipe Transportation

If a pipe connects node i and node j, a binary variable, $y_{i,j,r}$, is activated, indicating that a pipe of type r has been built between the two nodes. The gas mass flow rate through the pipe, $m_{i,j,r}$, is bound to the pipe existence binary variable. Inflows and outflows have to be in balance in each node since losses are assumed negligible. The gas can be supplied to the network at node i with an inflow (injection) rate, S_i, and consumed with an outflow rate, O_i. Therefore, the mass balance can be written as:

$$\sum_{j \in I \mid j \neq i} m_{j,i} + S_i = \sum_{j \in I \mid j \neq i} m_{i,j} + O_i \tag{2}$$

The local LNG terminal size is limited, which restricts the supply of LNG by truck and by pipeline from it to

$$S_{\text{LNG},i^*} + L_{\text{LNG}} + L_{\text{CNG}} \leq S^{\max} \tag{3}$$

where i^* denotes the node number of the local LNG terminal, while $L_{\text{LNG}} = \sum_i L_{\text{LNG},i}$ and $L_{\text{CNG}} = \sum_i L_{\text{CNG},i}$ are the total flows of LNG and CNG delivered from the terminal.

The gas is compressed only at the injection nodes, and pressure drop equations describe the gas flow in the pipeline. The pressure drop for a pipe of length, $l_{i,j}$, and diameter, $d_{i,j}$, is given by Haaland's approximation of the Colebrook-White equation [37]. The gas density, ρ_i, and the friction factor, ζ_i, at node i are needed for this, yielding:

$$p_j^2 = p_i^2 - p_i \cdot \zeta_i \cdot \frac{l_{i,j}}{d_{i,j}} \cdot \rho_i \cdot \left(\frac{m_{i,j}}{\frac{1}{4} \cdot \rho_i \cdot \pi \cdot d_{i,j}^2} \right)^2 \tag{4}$$

Piecewise linearization for each pipe diameter yields a set of linear equations describing the pressure drop in the pipe. The procedure is described in detail in an earlier paper by the present authors [25].

The pressures of flows arriving at a node must be equal and the pressure at the injection nodes equals the compressor discharge pressure. The temperature after ideal compression of the gas at the injection node in n compression stages, where the gas temperature between the compression steps is reduced to the ambient temperature (T_{amb}), is:

$$\widetilde{T}_i = T_{\text{amb}} \cdot \left(\frac{p_i}{p_{\text{amb}}} \right)^{\frac{Rg}{Mcp\,n}} \forall\, i \in I_{\text{sup}} \tag{5}$$

where R_g is the universal gas constant and $\overline{M}$ is the molar mass of the ideal gas. After piece-wise linearization of Equation (5), a set of linear equations controlled by binary variables are introduced into the system model. The compressor discharge temperature after real compression is obtained by applying an adiabatic efficiency factor, η, yielding:

$$T_i = T_{\text{amb}} + \frac{\tilde{T}_i - T_{\text{amb}}}{\eta} \; \forall \, i \in I_{\text{sup}} \tag{6}$$

The temperature differences between the compressor discharge temperature and the ambient temperature gives the power required at the compressor nodes:

$$P_{\text{comp},i} = c_p \, S_i (T_i - T_{\text{amb}}) \; \forall \, i \in I_{\text{sup}} \tag{7}$$

This equation holds a product of two inseparable continuous variables, which is tackled by bilinear interpolation as described by Mikolajková et al. [25].

If the gas is distributed from an LNG storage by pipe, a gasification unit has to be installed and a binary variable, g_i, is activated, using the constraint:

$$S_i \le g_i \cdot M \tag{8}$$

where M is a sufficiently large number ("big M" formulation).

2.2.2. Truck Supply

Trucks can be used to transport the gas to the customer instead of a pipeline, but as it is highly inefficient to build both a pipeline and a local storage supplied by truck, a binary variable $f_{k,i}$ ($k = $ LNG, CNG, ALT) is introduced for the selection between these alternatives. LNG may still be supplied from a distant LNG terminal, controlled by the binary variable, $f_{\text{ALT},i}$. In such a case, the binary variable for CNG supply, $f_{\text{CNG},i}$, and the pipe binary variable, $y_{i,j,r}$, are deactivated, expressed in additional constraints:

$$y_{i,j,r} + \frac{1}{2} f_{\text{LNG},i} + \frac{1}{2} f_{\text{ALT},i} + f_{\text{CNG},i} \le 1 \tag{9}$$

In this case, the mass flow distributed by truck, m_i^{truck}, is the mass flow of LNG from the local LNG terminal, $L_{\text{LNG},i}$, or alternatively, the LNG delivered by truck from distant terminal, $L_{\text{ALT},i}$. Additionally, as the gas may be supplied as CNG by containers, we have:

$$m_i^{\text{truck}} = L_{\text{ALT},i} + L_{\text{LNG},i} + L_{\text{CNG},i} \tag{10}$$

Each truck type has a maximum supply capacity of fuel, U_k^{truck}, $k = $ LNG, CNG, ALT. The number of truck transports to a node during a day is given by:

$$N_i^k = \frac{24 \, \text{h} \cdot L_{k,i}}{U_k^{\text{truck}}}; \; k = \text{LNG, CNG, ALT} \tag{11}$$

If gas is supplied by truck, specific infrastructure is required. In the local LNG terminal a number of loading lines, s, where the LNG is loaded on the trucks, are needed, but as a line cannot fill more than a maximum number of trucks per day, N^{max}, we have:

$$\sum_i N_i^{\text{LNG}} \le s \cdot N^{\text{max}} \tag{12}$$

In practice, due to limited space at a terminal, an upper limit, s^{max}, for the number of loading lines is also imposed.

For a customer that receives LNG by truck, the existence of a storage is considered in the model by an integer variable, $b_{a,i}$, where a is the type of storage (indicating its size), using the constraint:

$$\sum_a b_{a,i} \geq \frac{1}{2} f_{\text{LNG},i} + \frac{1}{2} f_{\text{ALT},i} \tag{13}$$

Installation of a storage facility allows the customers to balance their demand for gas over a period to be able to consider fluctuations in the demand and to guarantee that gas is available in case of delays in the deliveries. In cases where the node is connected to a pipeline, the storage serves as a source for neighboring customers. The storage capacity has to accommodate the amounts of gas consumed at the node and supplied from the storage to the neighboring customers for a multi-day period, t_{mult}. If we assume that no gas is supplied from the pipeline at nodes that inject gas into the pipeline, we get the condition:

$$\sum_a b_{a,i} \cdot U_a^{\text{stor}} \geq (D_i/H + S_i) \cdot t_{\text{mult}} \tag{14}$$

where U_a^{stor} is the size of storage of type a. Note that:

$$m_i^{\text{truck}} = S_i + D_i/H \tag{15}$$

As for compressed gas, if CNG is supplied to a node, a binary variable, $f_{\text{CNG},i} = 1$ and an investment in tanking infrastructure is made, controlled by a binary variable w. The tanking of a CNG container (at the local terminal) takes a certain time, t_{tank}. Therefore, the CNG tanking stations have a limit on the number of containers that can be filled per day:

$$w \cdot \frac{24\,\text{h}}{t_{\text{tank}}} \geq \sum_i N_i^{\text{CNG}} \tag{16}$$

Since the objective is to minimize the total costs, the containers are installed only when necessary. In order to use the CNG tanking and transportation time efficiently, we assume that there are two more containers in the system in addition to the containers that are placed at the customer nodes.

2.3. Costs and Objective Function

The objective function to be minimized is the sum of the cost of the gas supplied to the customers, the operation costs of the system and the investment cost.

The yearly cost for the LNG supplied from the terminal in different forms considers the flows of gas supplied by pipeline and by truck as LNG or CNG, which, using the notation of Equation (3), is:

$$C_{\text{LNG}} = t_{\text{year}} \cdot (S_{\text{LNG},i^*} + L_{\text{LNG}} + L_{\text{CNG}}) \cdot v^{\text{LNG}} \tag{17}$$

where t_{year} is the yearly operation time and v^{LNG} is the LNG unit cost. The yearly cost of biogas injected is the product of operation time, flow of upgraded biogas supplied, S_{BIO}, and the unit cost:

$$C_{\text{BIO}} = t_{\text{year}} \cdot S_{\text{BIO}} \cdot v^{\text{BIO}} \tag{18}$$

The LNG delivered from the distant terminal by a truck has a unit fuel price "at the gate", v^{ALT}, which gives the yearly alternative fuel cost:

$$C_{\text{ALT}} = t_{\text{year}} \cdot L_{\text{ALT}} \cdot v^{\text{ALT}} \tag{19}$$

with $L_{\text{ALT}} = \sum_i L_{\text{ALT},i}$. Thus, the total cost of fuel per year is:

$$C_{\text{fuel}} = C_{\text{LNG}} + C_{\text{BIO}} + C_{\text{ALT}} \tag{20}$$

The investment costs in the gas distribution infrastructure include the cost of the pipes installed to transport the gas from the LNG port, biogas plant or from the storages to the customers. The cost for a pipe of type r depends on the pipe length and the unit cost, v_r^{pipe}, so the total pipe cost can be expressed as:

$$C_{\text{pipe}} = \sum_i \sum_j \sum_r l_{i,j,r} \cdot y_{i,j,r} \cdot v_r^{\text{pipe}} \mid i \neq j \tag{21}$$

To relate this properly to the annual fuel costs, the investment cost of the pipes installed is discounted with an interest rate u over the K_{pipe} years of lifetime:

$$C_{\text{invest}}^{\text{pipe}} = \frac{C_{\text{pipe}}}{(1+u)^{-K_{\text{pipe}}}} \tag{22}$$

The cost of compression for each compressor, $C_{\text{comp},i}$, is obtained by multiplication of the power demand with the unit price of power, v^{pow}. Different costs arise for the truck supply alternatives, including a cost expressed as the product of the distance the truck has to travel between the supplier and the customer and a unit cost per kilometer, v^{dist}. Furthermore, the time needed for the transportation and the time needed for loading and unloading the gas is considered in the unit cost v^{hour}. Since the truck type for CNG container transportation differs from that of LNG transportation, the cost are truck type specific. The number of LNG truck transports, N_i^{LNG}, CNG trucks transports, N_i^{CNG}, and LNG transports from the remote source, N_i^{ALT} (cf. Equation (11)), are multiplied by their corresponding hourly cost and cost for the distance travelled, yielding the total cost of truck transportation to a customer, $C_{\text{truck},i}$. The yearly operation cost of the system is the sum of the cost of compression and the cost of truck supply:

$$C_{\text{oper}} = t_{\text{year}} \cdot \sum_i \left(C_{\text{comp},i} + C_{\text{truck},i} \right) \tag{23}$$

With the number of loading lines, s, needed for filling the trucks distributing LNG from the local port to the customers given by Equation (12), the cost of the load lines is obtained as:

$$C_{\text{load}} = s \cdot v^{\text{load}} \tag{24}$$

The investments also include the cost of the local LNG storages. The storage cost at a node depends on the storage existence integer, $b_{a,i}$ (Equations (13) and (14)) and the storage unit cost, v_a^{stor}, so:

$$C_{\text{stor},i} = \sum_a b_{a,i} \cdot v_a^{\text{stor}} \tag{25}$$

The gas transported from the LNG storage to the customer by pipe must be regasified in a gasification unit. Each installed gasification unit contributes by an investment cost:

$$C_{\text{gasif},i} = g_i \cdot v^{\text{gasif}} \tag{26}$$

with g_i obtained from Equation (8).

Summarizing, the total investment cost in the LNG infrastructure includes the cost of the tank lines, storages and the gasification units, discounted over the their corresponding investment lifetime:

$$C_{\text{invest}}^{\text{LNG}} = \frac{C_{\text{load}}}{(1+u)^{-K_{\text{load}}}} + \frac{\sum_i C_{\text{stor},i}}{(1+u)^{-K_{\text{stor}}}} + \frac{\sum_i C_{\text{gasif},i}}{(1+u)^{-K_{\text{gasif}}}} \tag{27}$$

As for the investments in the CNG infrastructure, the cost of the CNG tanking station is:

$$C_{\text{tank}} = f_{\text{CNG},i} \cdot v^{\text{tank}} \tag{28}$$

With a container unit cost of v^{cont}, the total cost of the containers becomes:

$$C_{\text{cont}} = \left(\sum_i f_{\text{CNG},i} + 2 \right) \cdot v^{\text{cont}} \tag{29}$$

where two extra containers are added as explained in Section 2.2. The cost for the two more containers "on the way" is added to the cost of the containers that are placed at the customer nodes. Furthermore, a customer that uses CNG needs a filling device (unit cost v^{fill}), yielding an investment cost:

$$C_{\text{fill}} = \sum_i f_{\text{CNG},i} \cdot v^{\text{fill}} \tag{30}$$

The total investment cost in the CNG infrastructure includes the costs of the tanking stations, the containers and the filling stations installed at the customers. These are discounted over the their corresponding investment lifetime:

$$C_{\text{invest}}^{\text{CNG}} = \frac{C_{\text{tank}}}{(1+u)^{-K_{\text{tank}}}} + \frac{C_{\text{cont}}}{(1+u)^{-K_{\text{cont}}}} + \frac{C_{\text{fill}}}{(1+u)^{-K_{\text{fill}}}} \tag{31}$$

Finally, the problem of minimizing the total costs is expressed as:

$$\min \left\{ C_{\text{tot}} = C_{\text{fuel}} + C_{\text{invest}}^{\text{pipe}} + C_{\text{oper}} + C_{\text{invest}}^{\text{LNG}} + C_{\text{invest}}^{\text{CNG}} \right\} \tag{32}$$

which can be tackled by MILP since the objective function and constraints are all linear.

2.4. Computational Solution

AIMMS [38] implementing the solver Gurobi, version 7.5, was used to solve the MILP problem of Equation (32) subject to the constraints listed in Section 2.2. The graphical interface of AIMMS helps to identify and understand the changes in the supply chain since the resulting connections between the nodes can be easily visualized and the results readily interpreted.

3. Case Study

This section illustrates how the model can be applied to find the optimal gas supply chain for a region, where the alternative gas sources outlined in Section 2 are available. Section 3.1 lists some general parameters identified for small-scale gas supply problems while the case study and its specific parameters are treated in Section 3.2. Section 3.3 presents the solution referred to as the Base Case, with which the results of the sensitivity analysis in Section 4 are compared.

3.1. Parameters for the Local Gas Supply Problem

To determine the cost terms in the objective function, in addition to the fuel price information about the unit costs of operation and investment are needed. Usually, it is difficult to find such data, because they may be proprietary information and the costs furthermore depend on the location of the energy system. We here present unit costs estimated by the authors based on public information, rules of thumb or personal information from companies with activity in the gas business [39,40]. In many cases, the authors had to resort to extrapolation from known cases because of the specific characteristics of the system studied. The values are reported in Table 1.

The price of all fuels were set equal, 86.5 €/MWh, corresponding to about 1.2 €/kg. Low-pressure pipes ($p \leq 16$ bar) of four diameters, 0.15 m, 0.25 m, 0.4 m and 0.5 m, were considered, with a minimum delivery pressure of $p_{\min} = 4$ bar. The costs of the pipes were extrapolated from costs of larger pipes provided by Gasum [41]. For the local LNG storages at the consumers, three possible LNG tank sizes ($U_{\text{S1}}^{\text{stor}} = 558$ t, $U_{\text{S2}}^{\text{stor}} = 2325$ t and $U_{\text{S3}}^{\text{stor}} = 4650$ t) were considered. Since combinations of such tanks were allowed at the consumers' sites, a wide spectrum of storage sizes could be realized. The costs of the tanks were grossly estimated based on the reported investment costs of tanks per ton and year reported in [42]. The life length of the pipes and LNG tanks was taken to be 30 years. As for auxiliary equipment, including units for LNG loading and gasification, CNG container, loading and filling stations, the rough estimates of the investment costs reported in Table 1 were applied, with life lengths of 15 or 20 years. The hourly cost of transportation was set higher for LNG tank trucks than for trucks transporting CNG containers, because the former trucks are of special design. For investment in energy infrastructure, it is common to use a low interest rate, so we assume this to be 5% ($u = 0.05$). To be able to compare operation and investment costs, the optimization period was taken to be a full year ($t_{year} = 8760$ h), neglecting the effect of maintenance breaks on the results.

As for terms in the constraints, the gas compression was taken to occur in $n = 6$ steps. In the truck transportation, one hour was added to the travelling time for LNG trucks and 30 min for CNG trucks to account for the extra time needed for the manipulation at the supply and customer nodes, while the time needed for loading the LNG trucks and CNG containers was considered separately ($t_{load} = 4.8$ h, $t_{tank} = 4.8$ h). The capacity of the LNG truck and CNG container was 17 t and 2.88 t, respectively, and the average traveling speed was 60 km/h. Distances between the customers and supply nodes can be approximated with the help of the haversine formula (both for the pipe and road connections) [43]. If available, more accurate road distances can be used instead. Based on an earlier study by the authors, piecewise linearization with five segments was found to yield a very accurate approximation of the non-linear equations in the pressure-drop expression of the pipeline, and the bilinear terms of Equation (7) were found to be approximated well by a 4 × 4 segment interpolation scheme. The reader is referred to [25] for a detailed description of the linearization procedures and the accuracy of the approximation.

Table 1. Unit costs and life length of investments.

Component	Specification (Symbol)	Unit Cost	K (a)
Fuel	LNG (v^{LNG})	86.4 €/MWh	-
	CNG (v^{CNG})	86.4 €/MWh	-
	BIO (v^{BIO})	86.4 €/MWh	-
	ALT (v^{ALT})	86.4 €/MWh	-
Pipe	0.15 m (v_1^{pipe})	328 €/m	30
	0.25 m (v_2^{pipe})	386 €/m	30
	0.40 m (v_3^{pipe})	491 €/m	30
	0.50 m (v_4^{pipe})	578 €/m	30
LNG infrastructure	S1 (v_{S1}^{stor})	1800 k€	30
	S2 (v_{S2}^{stor})	7000 k€	30
	S3 (v_{S3}^{stor})	13,000 k€	30
	LNG loading (v^{load})	450 k€	20
	LNG gasification (v^{gasif})	2000 k€	20
CNG infrastructure	CNG container (v^{cont})	90 k€	15
	CNG tanking (v^{tank})	600 k€	20
	CNG filling (v^{fill})	50 k€	15
Truck transportation	Distance (v^{dist})	2 €/km	-
	Time, LNG (v_{LNG}^{time})	200 €/h	-
	CNG (v_{CNG}^{time})	80 €/h	-

3.2. Background of Case Study

The model was applied to a local gas supply optimization problem in Vasa on the Finnish west coast, where an LNG terminal may be built close to the harbor. This terminal would primarily be used to fuel ships in the Gulf of Bothnia, but the LNG could also be used for local power and heat generation. A study by the authors of the energy use in the region identified 23 potential gas consumers (Table 2), with demands varying from very small to quite high, as indicated in Figure 1, where the LNG terminal is represented by the blue and yellow dots, indicating a potential supply of both LNG and CNG. The region has a biogas production unit (green dot in the figure) that can supply biogas to the system. The highest demand among the consumers (263 MW) is for a combined heat and power (CHP) plant, while the total energy demand of the customers is about 582 MW, which for the heating value of $H = 50 \, \text{MJ/kg}$ corresponds to a gas supply rate of 11.64 kg/s. It should be noted that the demands used in the study are estimates by the authors that were considered potential demands under a future gas-based energy supply scenario to the region. In addition to the producers and consumers seen in Figure 1, there is a small customer (1.4 MW) south of the depicted region, and a distant LNG terminal in Pori (located about 250 km south), which is a potential supplier of alternative gas as LNG delivered by trucks.

Table 2. Node numbers, name and coordinates of locations, as well as their energy demand.

Node	Latitude	Longitude	D_i (MW)
1. LNG terminal	63.08	21.57	10.0
2. Biogas plant	63.13	21.76	0.0
3. CHP plant	63.09	21.55	262.9
4. Waste water treatment	63.11	21.59	0.5
5. Gas station I	63.07	21.67	2.0
6. Engine production	63.10	21.61	23.1
7. Industry I	63.06	21.55	0.7
8. Gas station II	63.14	21.76	1.9
9. Hospital	63.08	21.61	1.3
10. University campus	63.11	21.59	157.8
11. Greenhouses I	63.15	21.64	1.6
12. Vasa airport	63.04	21.76	2.1
13. Vasa port	63.09	21.56	3.2
14. Aquaparc	63.09	21.59	15.8
15. Vasa school	63.08	21.64	10.5
16. Industry II	63.08	21.67	21.0
17. Industry III	63.17	21.59	17.9
18. Industry IV	63.03	21.76	0.7
19. Greenhouses II	63.00	21.62	0.9
20. Industry V	63.10	21.73	0.5
21. Industry VI	63.09	21.75	42.1
22. Laihia	62.98	22.00	1.5
23. Pörtom	62.71	21.61	1.4
24. Kvevlax	63.16	21.82	1.3
25. Replot	63.23	21.41	1.2
26. CNG terminal	63.08	21.57	0.0

In order to reduce the complexity of the problem of finding the optimal supply chain, the possible gas pipeline connections were limited to the ones depicted by lines in Figure 2, where also the nodes are numbered. Their names, geographic locations and energy demands are reported in Table 2. Customers remote from the local LNG terminal were not considered for pipeline distribution and must therefore be supplied by truck. Not more than one tanking line for CNG was allowed at the local terminal. The size of the LNG terminal was estimated to 30,000 m³, with a maximum regasification rate of 15 kg/s. The maximum biogas supply is 3 kg/s. Color coding (blue for local LNG, orange for distant LNG, yellow for CNG and green for biogas) will in the following be used to represent the fuel supplied in the figures representing the optimal solutions under different conditions. This formulation resulted in about 55,000 constraints and 38,000 variables (out of which were 14,000 integer variables). The optimization of each case took 5–30 min on a standard PC.

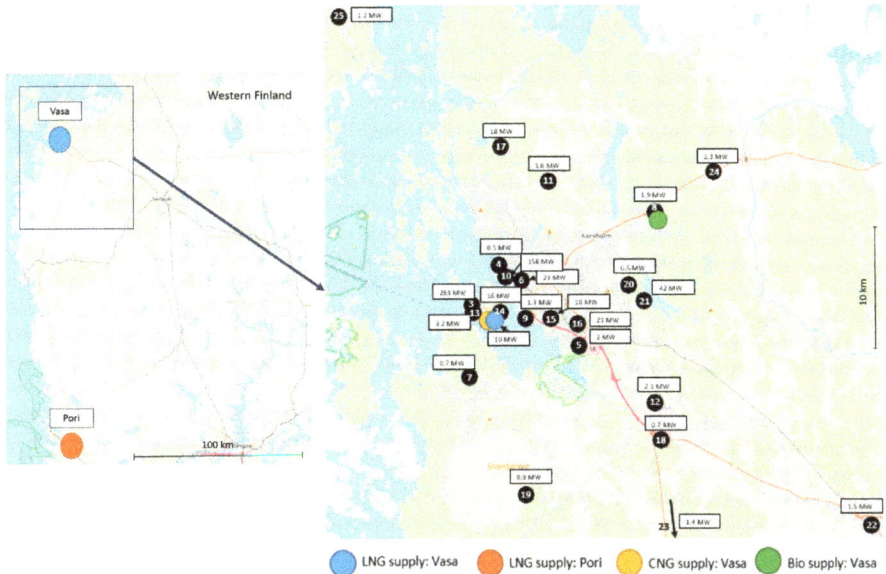

Figure 1. Gas suppliers and consumers. Left: Location of the region studied (Vasa) and the distant LNG terminal (Pori). Right: Consumers in the region with demands reported in boxes (Background map source: © OpenStreetMap contributors).

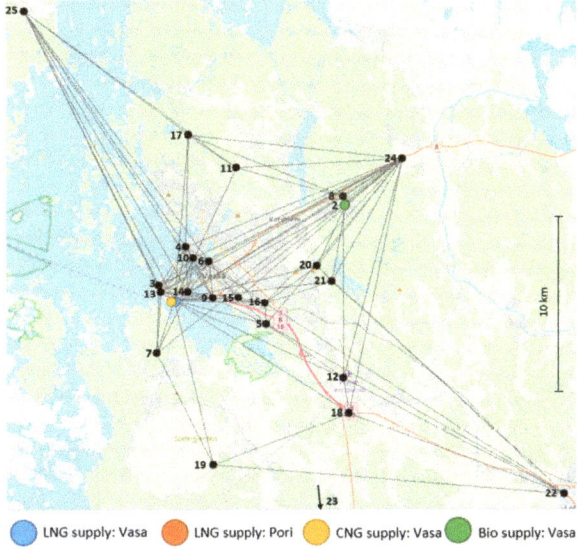

Figure 2. Network scheme with the location of the potential pipe connections (lines), customers (black dots) and suppliers (color dots). Node numbers are reported in red (Background map source: © OpenStreetMap contributors).

3.3. Base Case Solution

The optimal supply chain is a combination of pipeline supply of regasified LNG and upgraded biogas, and truck supply of CNG from the local terminal (Figure 3). There is a separate pipeline from the biogas producer (green dot). The gas is injected at 13 bar at the biogas producer and at 7 bar at the LNG port (node 1) to be supplied to the farthest customers at the required pressure. The total length of the pipeline is 31.6 km. Most pipes have a diameter of 0.15 m, but short sections around the LNG terminal use a diameter of 0.25 m. The discounted cost of the pipeline is 2.45 M€. Seven remote customers are supplied by CNG, requiring a total of nine containers in the system (Table 3). The number of CNG container trucks is limited by the time constraint of the loading line. For instance, the daily delivery to Laihia (node 22) and Industry IV (node 18) is 0.90 and 0.44 containers, respectively. Altogether, 4.74 CNG containers per day are needed (Table 3), i.e., 1713 per year. The local terminal supplies the bigger customers (CHP plant, node 3, and University campus, node 10) as well as the neighboring smaller industrial customers with gas by pipe.

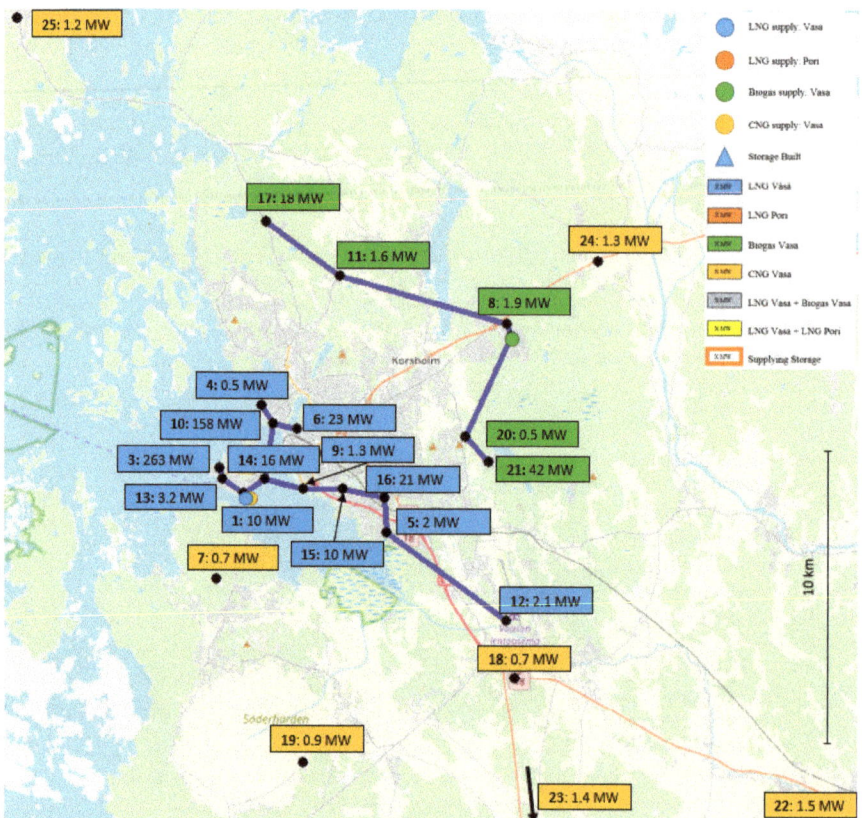

Figure 3. Optimal supply chain for the Base Case. The fuel supplied to the nodes is indicated in the colored boxes (Background map source: © OpenStreetMap contributors).

Table 3. Numbers of daily CNG trucks to the customers in the Base Case.

Node	N^{CNG} (1/d)
7. Industry I	0.44
18. Industry IV	0.44
19. Greenhouses II	0.57
22. Laihia	0.90
23. Pörtom	0.84
24. Kvevlax	0.78
25. Replot	0.72

4. Sensitivity Analysis

This section presents a sensitivity analysis of the model, where some of the values of the Base Case are perturbed, and the effect on the optimal solution of the gas supply chain is analyzed.

4.1. Effect of Gas Price and Investment Costs

The role of the gas price at the local the distant terminals and the investment costs in LNG storages and pipelines on the optimal supply chain is illustrated by four examples. The price or cost levels were set as ±25% compared to the base-case levels, while all other cost parameters were kept unchanged. The first and the third cases have considerably higher gas price in the local terminal compared to the distant terminal, while the opposite holds true for the second and fourth cases. As for investment costs, the cost of storages is low in Cases 1–2, and high in Cases 3–4, while the opposite holds true for the pipe investment costs. Table 4 illustrates the conditions of the four cases and Table 5 summarizes the results (assuming all the productions work during the whole year).

Table 4. Gas price and investment cost changes compared to the Base Case scenario.

Unit Cost Term	Case 1	Case 2	Case 3	Case 4
Local LNG & CNG	+25%	−25%	+25%	−25%
Distant LNG	−25%	+25%	−25%	+25%
Storage investment	−25%	−25%	+25%	+25%
Pipe investment	+25%	+25%	−25%	−25%

Table 5. Main results of optimization of the Base Case and four cases listed in Table 4.

Variables	Unit	Base Case	Case 1	Case 2	Case 3	Case 4
LNG supply, Vasa (pipe+truck)	GWh	4469	0	5029	0	5029
LNG suppy, Pori (truck)	GWh	0	5098	0	5098	0
Biogas supply (pipe)	GWh	560	0	0	0	0
CNG supply (truck)	GWh	69	0	69	0	69
Pipeline length	km	31.6	10.1	16.2	46.9	35.4
Pipeline diameter	m	0.15, 0.25	0.15	0.15, 0.25	0.15	0.15, 0.25
Max. compression pressure	bar	13.0	7.0	8.4	11.6	8.7
LNG storage, S1 units	-	0	17	4	9	0
LNG storage, S2 units	-	0	0	0	0	0
LNG storage, S3 units	-	0	2	0	2	0
LNG storage, total capacity	t	0	14,995	2232	14,322	0
CNG containers	-	9	0	9	0	0
LNG trucks, Vasa	1/a	0	0	866	0	0
LNG trucks, Pori	1/a	0	21,950	0	21,950	0
CNG trucks	1/a	1713	0	1730	0	1730
Total Cost	M€	445.4	370.8	336.5	374.8	335.1

4.1.1. Case 1

Naturally, the largest share of the total cost is the fuel cost, so a change in it affects the optimal supply chain most, which already becomes apparent when the results of Case 1 are studied. As the price of LNG at the local terminal is higher than the price of LNG supplied from the remote terminal, supply from Pori is favored: LNG from Pori is transported to locally built storages in the region and no LNG from the local terminal is used. The lower investment cost of the storages favors their construction over investment in a pipeline. Still, a shorter (10.1 km) pipeline network is constructed to supply gas from the storage in the University campus (node 10, with an own consumption of 158 MW). The discounted cost of this pipeline, which consists of pipes with diameter of 0.15 m, is 1.02 M€. The storages built at the University campus node have a total capacity of 4650 t, enough to supply the node and eight other nodes along the pipeline. The pressure in the pipeline is 7 bar at the injection point, where the LNG is regasified and introduced. This supply of gas requires about 59 trucks per day of LNG to the customers, with the largest amount supplied to the CHP storage (node 3) with storage capacity of 5766 t (26.7 trucks/day) followed by 22.5 trucks/day to the University campus node. Smaller (558 t capacity) storages are built in the nodes not connected by pipeline to cover their own fuel demand. Since the CNG price at the local terminal is high, it is not economically viable to supply containers even to the small remote customers. Therefore, no investments are needed for the LNG and CNG loading lines or the CNG filling stations at the customers. The optimal supply chain of Case 1 is illustrated in Figure 4.

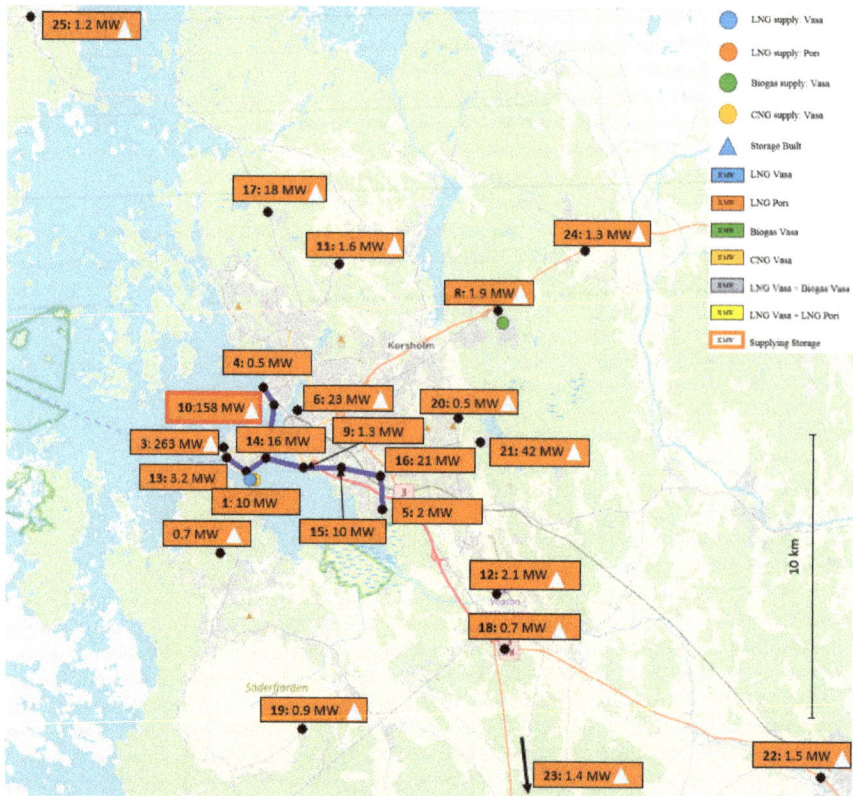

Figure 4. Optimal supply network in Case 1 with storages (triangles) and pipelines (blue lines). The type of fuel used is denoted by color in the rectangles, which also reports the fuel demand. The node number is given by the bold number in the rectangle. The node with the storage supplying regasified LNG into the pipeline is denoted by the red framed rectangle (Background map source: © OpenStreetMap contributors).

4.1.2. Case 2

Case 2 with lower local gas price and higher price at the distant terminal naturally yields an increase in the local fuel supply (Figure 5). LNG is distributed to the local customers from the terminal by pipeline and by trucks. One loading line is used to load LNG on 2.37 trucks daily, and the LNG is stored in distributed storages with total capacity of 2232 t, promoted by the low storage cost. CNG is supplied to the same seven customers as in the base-case solution. Regasified LNG is distributed to customers over a 16.2 km long pipeline from the LNG terminal. The pipeline (diameter 0.25 m) has a maximum pressure of 8.4 bar and a discounted cost of 1.67 M€.

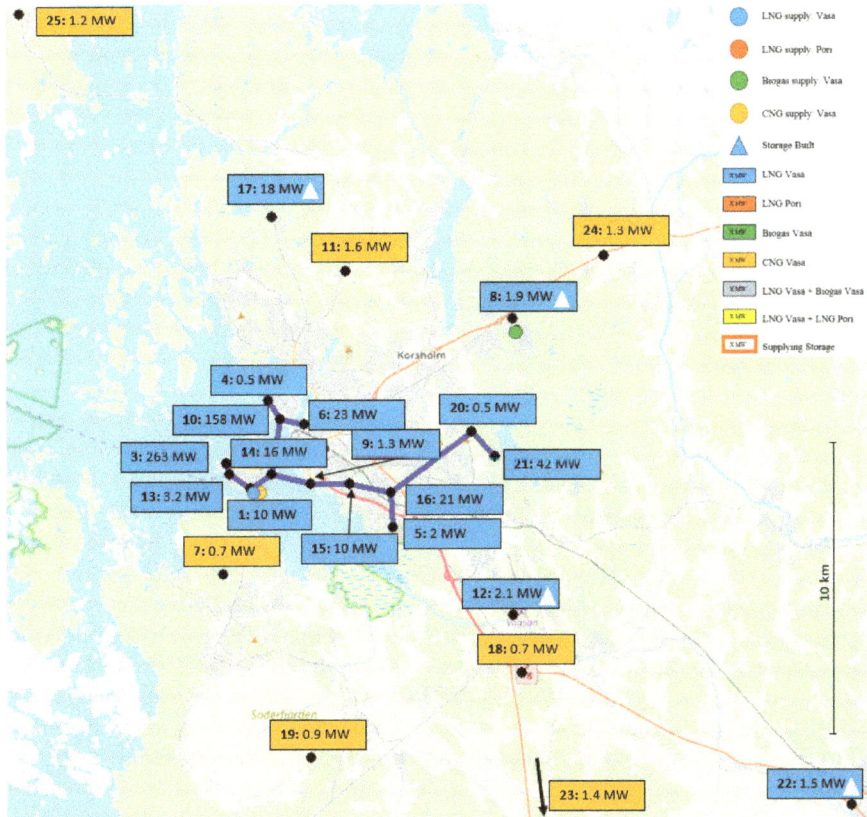

Figure 5. Optimal supply network in Case 2 (Background map source: © OpenStreetMap contributors).

4.1.3. Case 3

In Case 3, the lower alternative LNG price clearly favors the supply from the remote terminal. In combination with lower pipe investment costs, it leads to an extension of the pipeline (Figure 6). The 46.9 km long pipeline (diameter 0.15 m) supplies the consumers along it from storages built in University campus (node 10) with total capacity of 5766 t. The injection pressure is 11.6 bar, and the discounted pipeline cost is 2.66 M€. There are altogether 11 storages at seven customer nodes with a total capacity of 14,322 t. A large amount of LNG is thus supplied from the remote LNG terminal, by 59.2 trucks/day. No biogas is injected into the pipeline. It is interesting to note that this result could be a solution for the case where no LNG terminal exists in Vasa.

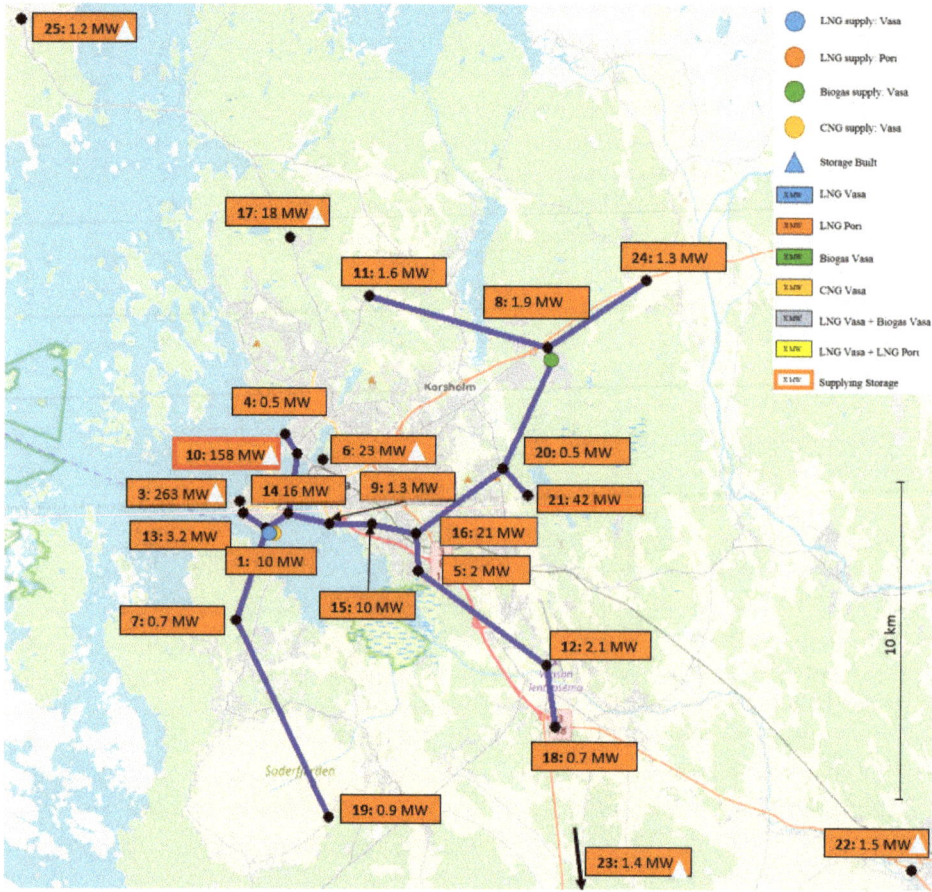

Figure 6. Optimal supply network in Case 3 (Background map source: © OpenStreetMap contributors).

4.1.4. Case 4

Lower local LNG price and pipe investment costs favor the use of local gas and an extended pipeline, as seen in Figure 7 which illustrates the optimal supply chain. The total length of the pipeline network is 35.4 km. The distribution of the regasified LNG is realized by a pipeline system of 0.15 m and 0.25 m diameter pipes of a discounted cost of 2.1 M€. The injection pressure at the LNG terminal is 8.7 bar. The lower price of local LNG also favors the deliveries by CNG containers instead of building long pipelines to the remote nodes or building large LNG storages. The same nodes are served by CNG as in the Base Case and Case 2.

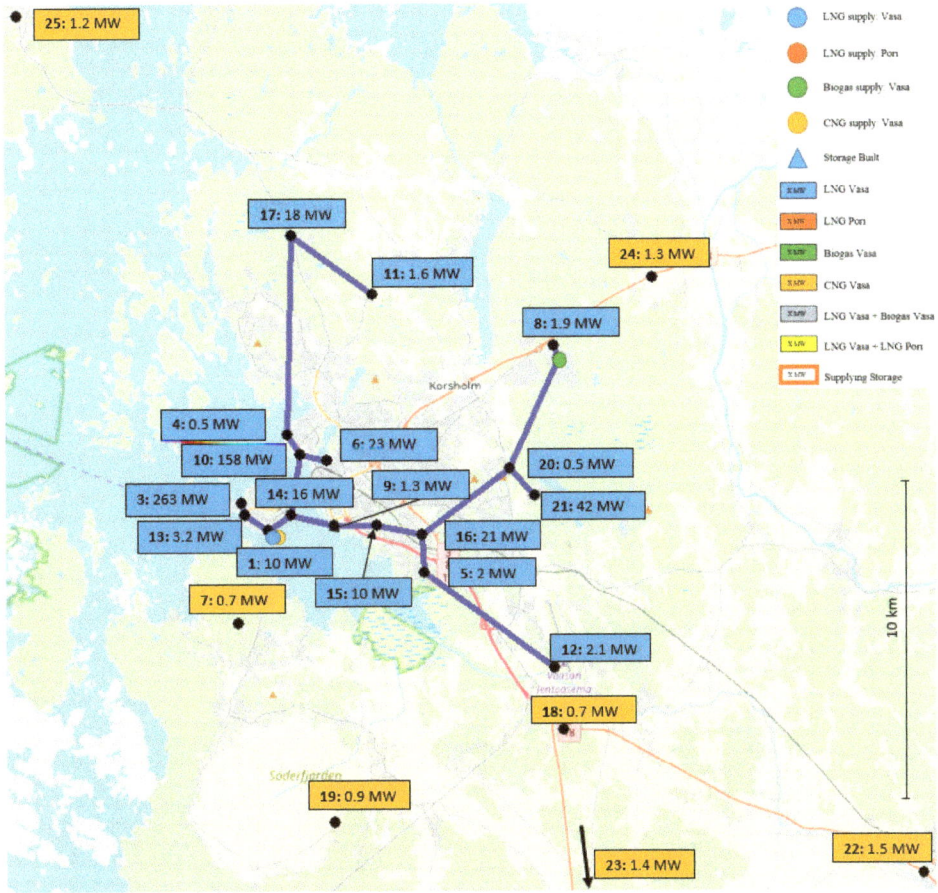

Figure 7. Optimal supply network in Case 4. For a definition of the symbols, see caption of Figure 4 (Background map source: © OpenStreetMap contributors).

In summary, the results of the presented cases emphasize that the optimal supply chain strongly depends on the fuel price and less on the investment cost of the infrastructure needed for the gas distribution.

4.2. Detailed Effect of Alternative Gas Price

To study more accurately the points at which the supply of the locally available gas becomes viable over the supply from the alternative source, the price of alternative gas was gradually increased from 75% to 100% of the nominal value, while the price of the local gas and the investment costs were maintained at their nominal level (100%). Initially, almost the whole fuel demand, is covered by LNG from Pori in the optimal supply chain. The only exception is CNG delivered from the local terminal, which represents 4–5% of the total demand. As expected, the share of alternative gas decreases as its price increases, and the transition occurs in steps. At approximately 82% of the nominal price, the injection of upgraded biogas becomes economically viable and starts complementing the fuel mix of LNG from Pori supplied by trucks. At a low price of the alternative fuel, part of the fuel is delivered to individual storages, and part is redistributed from a large storage to customers along a pipeline. With increasing alternative gas price, the pipeline network first shrinks and then expands again due to the switch of the fuel delivered. As can be seen in Figure 8a, the pipeline network first falls from the initial 29 km total length to 18.7 km and later grows with the increasing price of the alternative fuel from Pori up to 32 km. At an alternative fuel price of about 94% (i.e., approximately 8 cents/kg lower than the price in the local terminal), the supply of LNG from Pori is fully replaced by supply of the local LNG and CNG, and the solution does not change any longer.

(a)

(b)

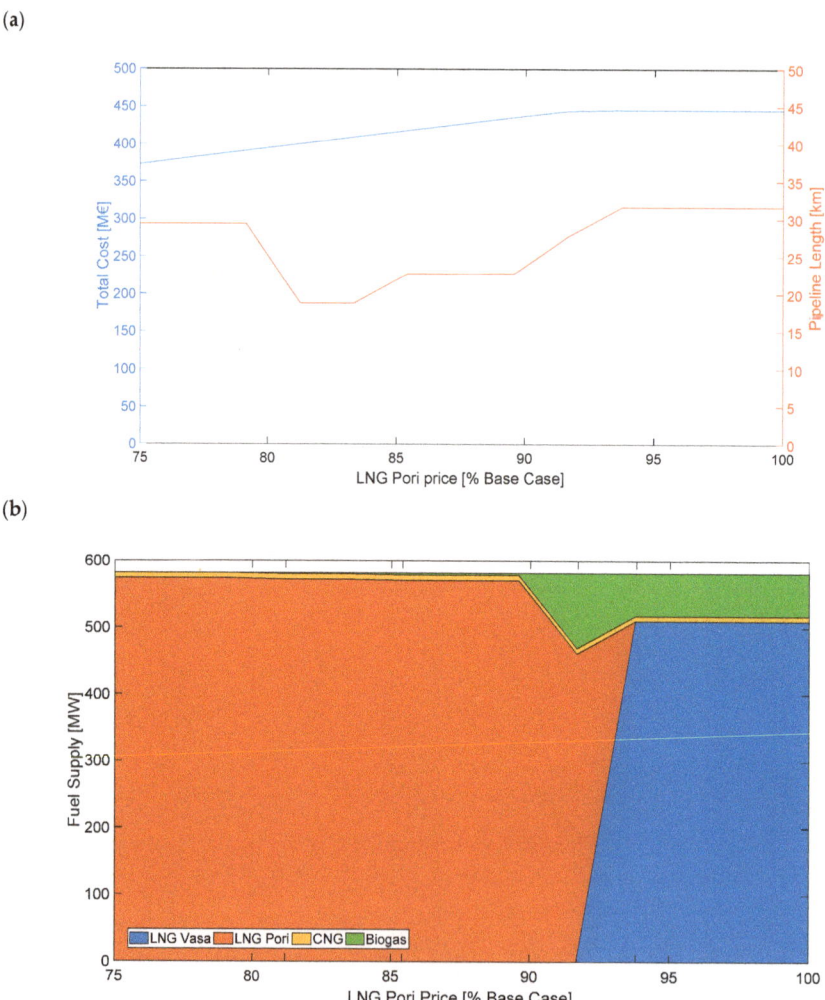

Figure 8. Effect of alternative gas price on the (**a**) total cost and the constructed pipeline length, (**b**) distribution between different gas sources.

A more detailed analysis of the results reveals interesting observations of how the optimal supply chain changes with the price of the alternative fuel. As Figure 8 depicts, there are (at least) five distinct solutions where the price of alternative fuel falls between 78% and 95% of the local fuel price. We therefore study the optimal supply chains at the points where the alternative fuel price is 75.0%, 81.2%, 85.4%, 91.7% and 93.8% of the local fuel price, indicated by the major changes in the fuel type consumption in Figure 8b. Figure 9 depicts the four first solutions, specifying the fuels delivered to the nodes, the position of the storages and the pipeline connections.

For $v^{ALT} = 0.750 \; v^{LNG}$ (Figure 9a) there is one quite large pipeline network, 29.3 km, of pipes of 0.15 m diameter. Its discounted pipe investment cost is 2.23 M€. As indicated in Figure 8b, this pipeline network is entirely supplied by fuel delivered from Pori. The gas is injected at 6.5 bar at node 10, which has a storage capacity of 5208 t (1 × S1 and 1 × S3). There are altogether eight storages in this network with total capacity of 12,648 t. Small and remote customers (at nodes 11, 19, 22–25) obtain CNG delivered in containers.

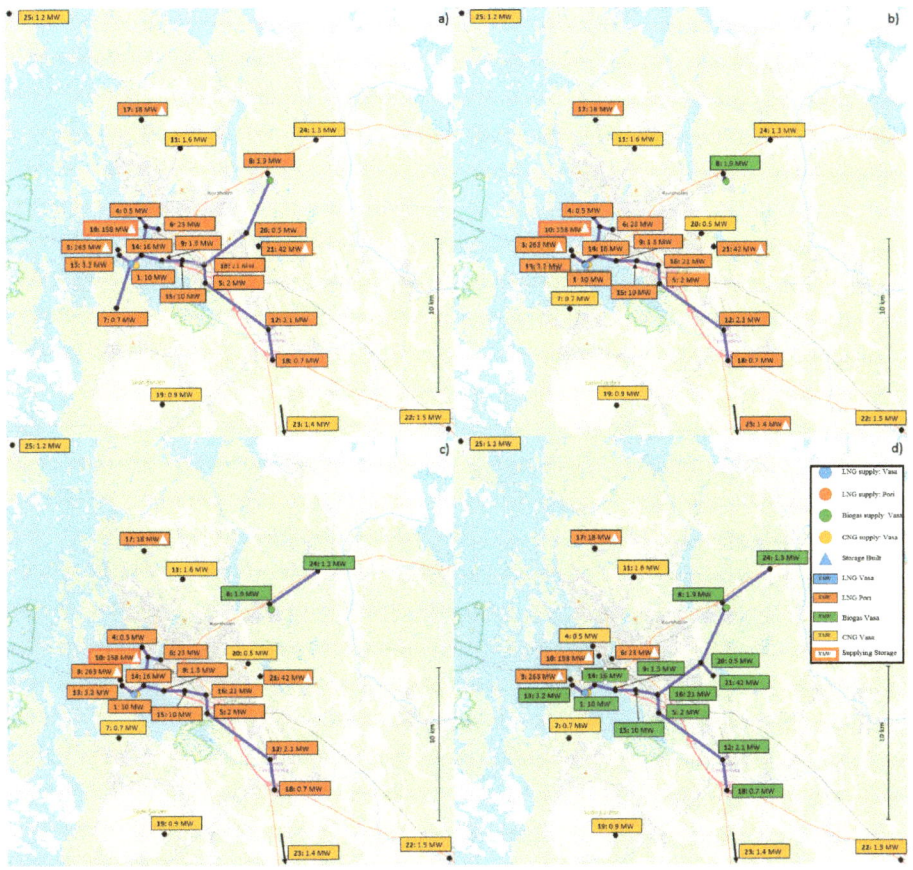

Figure 9. Optimal gas supply chains for an alternative fuel price of (**a**) 75.0%, (**b**) 81.2%, (**c**) 85.4%, and (**d**) 91.7% of the local fuel price (Background map source: © OpenStreetMap contributors).

With increasing price of the alternative fuel ($v^{ALT} = 0.812 \; v^{LNG}$, Figure 9b), the pipeline network becomes shorter (18.7 km), reducing its discounted cost to 1.41 M€. A very short separate pipeline is built between the biogas plant and the neighbor node 8, supplying 1.9 MW. Still, like in the former case, a storage with a capacity of 5208 t in node 10 is required to supply the pipeline network at a slightly lower injection pressure (6.3 bar) by regasified LNG from Pori. Compared to the previous case, the customers in nodes 7 and 20 now obtain CNG instead of the one in the farthest node 23. Therefore, besides the storages in nodes 3, 10, 17 and 21, an LNG storage is built in node 23, and the total storage capacity is 13,206 t (7 × S1 and 2 × S3).

As the alternative fuel price further increases to $v^{ALT} = 0.854 \; v^{LNG}$ (Figure 9c), the local subnetwork distributing upgraded biogas further extends to node 24 supplying 3.2 MW and the total length of the pipeline network is 22.7 km (discounted cost 1.71 M€). The storage in node 10 is still the main supplier of regasified LNG from Pori along the pipeline with an injection pressure of 7.1 bar. The total capacity of the storages has decreased slightly to 12,608 t (6 × S1 and 2 × S3), and CNG is delivered to nodes 7, 11, 19, 20, 22, 23 and 25.

At an alternative fuel price of $v^{ALT} = 0.917\ v^{LNG}$ (Figure 9d), a main pipeline (length 27.7 km, diameter 0.15 m) distributes regasified LNG (from Pori) and upgraded biogas to the customers (discounted cost 2.1 M€). Of the five cases illustrated here, the upgraded biogas consumption is the highest (112 MW), and its injection pressure is 10.8 bar. Still, no local LNG is regasified or supplied by truck. LNG from Pori is only distributed to the storages at the customers with large consumption in nodes 3, 6, 10 and 17. Small customers in nodes 4, 7, 11, 19, 22, 23 and 25 obtain the gas as CNG.

In the last case studied ($v^{ALT} = 0.938\ v^{LNG}$) two separate pipeline subnetworks appear. A regasification unit is built at the port, there are no local storages and it is no longer economically feasible to supply LNG from Pori. The solution is identical to that of the Base Case of Section 3.3 depicted in Figure 3.

Summarizing the findings, the optimal supply is seen to vary considerably with the alternative fuel price, still showing some common subparts. For instance, there is always a pipeline network connecting nodes 13-1-14-9-15-16-5-12, and CNG is always delivered to nodes 11, 19, 22, and 25. Above all, the results demonstrate that the optimization model successfully can tackle problems with numerous options that may become feasible under certain conditions.

4.3. Effect of Gas Demand

The model can also be used to study the optimal supply network when the demand or supply conditions change, which may occur if new customers start using the gas or in case of a sudden disruption in gas supply. To illustrate this behavior, two cases are presented, with results summarized in Table 6, using the base-case settings of the parameters.

Table 6. Results of optimization of the cases with decreased or increased energy demand.

Variables	Unit	Low	High
LNG supply, Vasa (pipe+trucks)	GWh	1278	6544
LNG supply, Pori (trucks)	GWh	0	2312
Biogas supply (pipe)	GWh	1206	1314
CNG supply (truck)	GWh	65	26
Pipeline length	km	23.5	33.8
Pipeline diameter	m	0.15, 0.25	0.15,0.25,0.4
Max. compression pressure	bar	15.8	11.2
LNG storage, S1 unit	-	0	6
LNG storage, S2 unit	-	0	0
LNG storage, S3 unit	-	0	1
LNG storage, total capacity	t	0	7998
CNG containers	1/a	11	4
LNG trucks, Vasa	1/a	0	0
LNG trucks, Pori	1/a	0	117,480
CNG trucks	1/a	1616.9	646.6
Total Cost	M€	223.9	905.2

4.3.1. Low Demand

In the first scenario, the demand in the whole region is decreased to half of the nominal one. With such low demand, there is no a need to supply gas from Pori (Figure 10). The CNG loading line serves remote customers and customers with low demand. No LNG storages are built since all the demand can be covered by CNG or by pipeline. It is more economically viable to build two shorter separate pipelines than a single long one. The shorter pipeline (2.2 km) distributes the regasified LNG to the biggest consumer (CHP plant) and to the consumers closest to the LNG terminal. Smaller farther customers are supplied from the local biogas plant. The two pipelines (total length 23.5 km) consist mostly of pipes with 0.15 m diameter, with a short (0.9 km) section using a pipe diameter of 0.25 m. The regasified LNG has to be compressed to 7 bar while the upgraded biogas is injected at 15.8 bar.

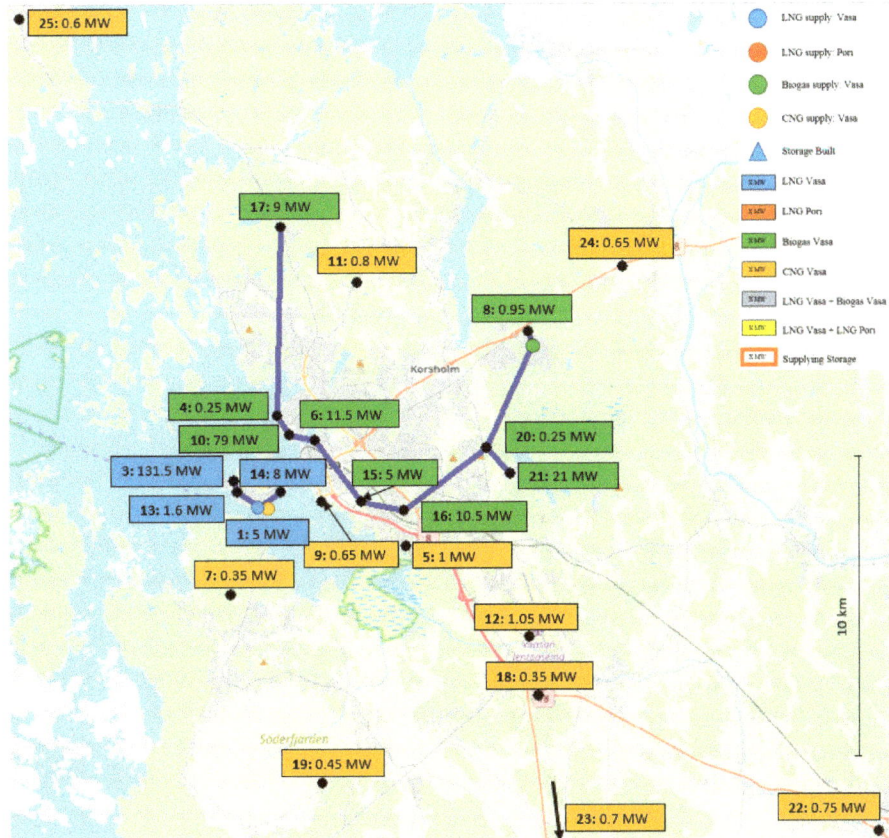

Figure 10. Optimal supply network for the scenario where the demand is half of the nominal one. For a definition of the symbols, see caption of Figure 4 (Background map source: © OpenStreetMap contributors).

4.3.2. High Demand

To study a high demand scenario, the demands were doubled from the nominal level. In contrast to the previous case, it is no longer possible to supply the whole region from local gas resources (Figure 11). The CNG loading capacity is sufficient to supply only two customers, while local LNG is regasified and complemented by upgraded biogas and LNG distributed from the remote LNG terminal to fully satisfy the demand. The LNG from Pori is supplied to customers not connected by the pipeline and to two storages (capacity 5208 t) at Industry VI (node 21). From these storages, regasified LNG is introduced into the pipeline at 10.5 bar pressure. The combination of the regasified LNG from the two terminals covers, together with the injected upgraded biogas (at 11.2 bar), the demand of all customers along the 33.8 km long pipeline. The pipeline consist of pipes of 0.15 m and 0.25 m diameter and a section with a larger (0.4 m) diameter. Since less customer nodes can use CNG, the number of LNG storages increases to seven. The total capacity of all the storages is 7998 t.

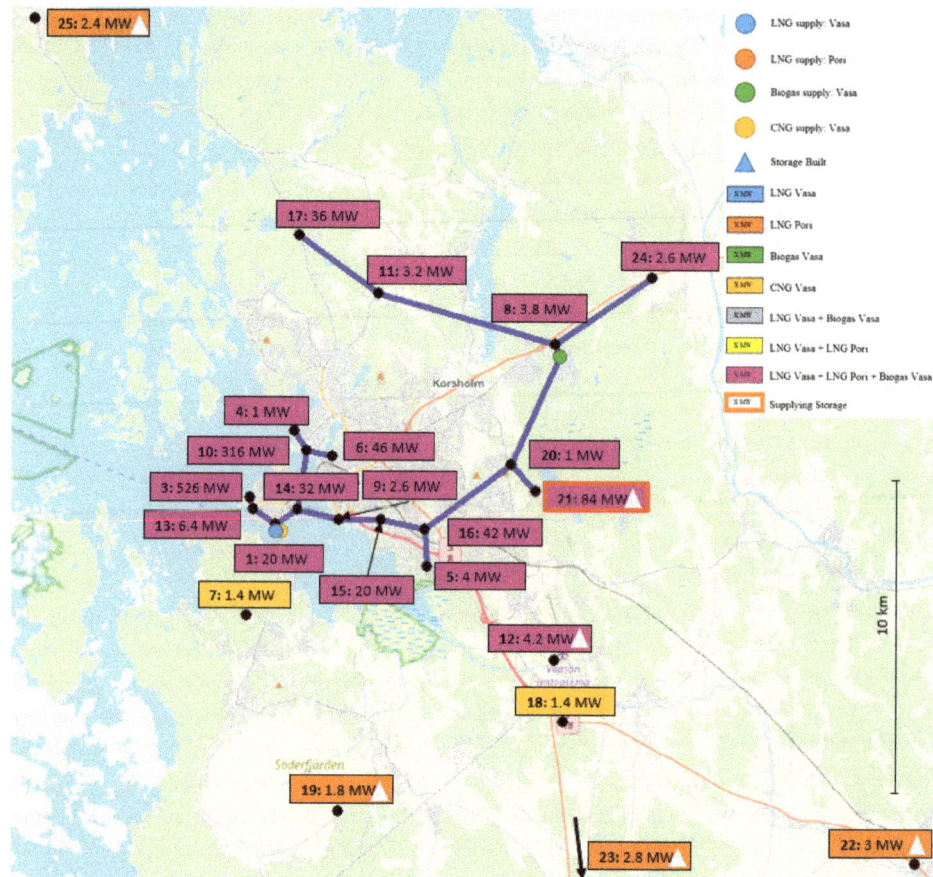

Figure 11. Optimal supply network for the scenario where the demand is double the nominal demand. For a definition of the symbols, see caption of Figure 4 (Background map source: © OpenStreetMap contributors).

The above cases illustrate that changes in the gas demand have a strong effect on the optimal supply chain and gas supply mix. The results of such analysis can give information about the robustness of a selected solution with respect to future changes in the supply or demand.

5. Conclusions

A model for optimization of a gas supply chain, including LNG, CNG and upgraded biogas as potential sources, has been presented in this paper. The model considers gas supply by low-pressure pipeline, and by trucks as LNG or CNG, and considers constraints and costs of the delivery and the investments required to realize the system. After linearization of non-linear expressions in the model, the task of minimizing the overall costs is formulated as a mixed integer linear programming (MILP) problem that can be solved efficiently by state-of-the-art software. Since the model is linear, one can guarantee that the lowest cost is found. The model developed is a flexible tool that can be used to find an appropriate design for new gas supply chains for smaller regions, for what-if analysis to reveal the sensitivity of the solution to changes in the parameters (e.g., constraints or costs) or to study the robustness of a solution to future changes in the gas supply and demand.

The model and its use have been illustrated by studying the gas supply in a region with an emerging gas market. As expected, the fuel price has a major effect on the optimal supply chain, including which fuel sources to use and how to deliver the gas to the customers. The costs of storage and pipes mainly influence the length of the pipeline and the number of storages to be constructed. The simple but illustrative local network with about 25 nodes used in this paper to demonstrate the feasibility of the model may be extended to supply chains encompassing larger regions. Since the solution of the tasks studied in the paper were obtained in 5–30 min on a standard PC, it is expected that systems with up to 50 nodes could be solved in reasonable time (a few days), if, like in the present case, the allowed pipeline connections are limited a priori by excluding clearly infeasible alternatives. Even though the model presented in the paper is presented for a single-period problem (i.e., with fixed demands) it can, following the general procedure outlined in [32], quite easily be extended to multi-period problems. This will, however, increase the complexity of the numerical problem, restricting the size of the problems that can be solved without prohibitive computational burden.

Author Contributions: Funding acquisition, M.M.-A., M.B.-S. and H.S.; Methodology, M.M.-A., F.P. and H.S.; Supervision, F.P. and H.S.; Writing–original draft, M.M.-A.; Writing–review & editing, M.M.-A., F.P. and H.S.

Funding: This research received funding from the AIKO Gas CoE project, Åbo Akademi University and Högskolestiftelsen i Österbotten, Finland. The support is gratefully acknowledged.

Conflicts of Interest: The authors declare no conflicts of interest.

Nomenclature

Binary and Integer Variables

b	integer controlling storages
f	truck supply existence variable
g	gasification existence variable
s	number of tank lines
w	CNG loading line binary variable
y	variable for existing connections

Continuous variables

L	truck supply, kg/s
m	mass flow rate, kg/s
N	number of trucks
O	outflow of natural gas, kg/s
p	pressure, bar
S	supply of natural gas, kg/s
T	temperature, K
$\tilde{T}$	temperature after ideal compression, K

Parameters

c_p	specific heat capacity, kJ/(kg K)
C	cost, €
d	pipe diameter, m
D	energy demand at node, MW
H	heating value, MJ/kg
K	life length of investment, a
l	pipe length, m
M	large positive constant ("big M"), -
$\overline{M}$	average molar mass of natural gas, kg/kmol
n	number of compression steps
O	energy outflow at node, MW

R_g	universal gas constant, J/(mol K)
t	duration of time period, h
u	interest rate, -
U	capacity, kg
v	unit cost, €, €/kWh, €/m or €/kg
Sets	
A	storage type $a \in A$
I	nodes $i \in I$
J	nodes $j \in J$
R	pipe diameter type $r \in R$
Greek	
η	efficiency factor, -
ζ	friction factor, -
ρ	density, kg/m^3
Superscripts	
ALT	alternative fuel
BIO	biogas
CNG	compressed natural gas
dist	distance travelled
k	fuels by truck: LNG, CNG, ALT
load	LNG load line
LNG	liquefied natural gas
max	maximum amount
NG	natural gas
pipe	pipe
pow	power
stor	storage
tank	tanking
time	travelling time
truck	truck transportation
Subscripts	
a	storage type
ALT	alternative fuel
amb	ambient
BIO	biogas
CNG	compressed natural gas
comp	compressor
gasif	gasification cost
i	node
invest	investment cost
j	node
k	fuels by truck: LNG, CNG, ALT
LNG	liquefied natural gas
load	LNG loading line
mult	multi-day
NG	natural gas
oper	operational cost
pipe	pipe investment
r	pipe type
tank	tanking
truck	truck transportation
stor	storage
sup	supply node
year	yearly operation

References

1. McJeon, H.; Edmonds, J.; Bauer, N.; Clarke, L.; Fisher, B.; Flannery, B.P.; Hilaire, J.; Krey, V.; Marangoni, G.; Mi, R.; et al. Limited impact on decadal-scale climate change from increased use of natural gas. *Nature* **2014**, *514*, 482. [CrossRef]

2. Brandt, A.R.; Heath, G.A.; Kort, E.A.; O'sullivan, F.; Pétron, G.; Jordaan, S.M.; Tans, P.; Wilcox, J.; Gopstein, A.M.; Arent, D.; et al. Methane Leaks from North American Natural Gas Systems. *Science* **2014**, *343*, 733–735. [CrossRef] [PubMed]

3. Levi, M. Climate consequences of natural gas as a bridge fuel. *Clim. Chang.* **2013**, *118*, 609–623. [CrossRef]

4. Zhang, X.; Myhrvold, N.P.; Hausfather, Z.; Caldeira, K. Climate benefits of natural gas as a bridge fuel and potential delay of near-zero energy systems. *Appl. Energy* **2016**, *167*, 317–322. [CrossRef]

5. Hausfather, Z. Bounding the climate viability of natural gas as a bridge fuel to displace coal. *Energy Policy* **2015**, *86*, 286–294. [CrossRef]

6. U.S. Energy Information Administration Home Page (EIA). U.S. Energy Information Administration International Energy Outlook. 2017. Available online: https://www.eia.gov/outlooks/ieo/pdf/0484(2017).pdf (accessed on 20 August 2018).

7. *BP Statistical Review of World Energy 2018*, 67th ed.; BP: London, UK, 2018; pp. 26–35. Available online: https://www.bp.com/content/dam/bp/en/corporate/pdf/energy-economics/statistical-review/bp-stats-review-2018-full-report.pdf (accessed on 10 September 2018).

8. Dilaver, Ö.; Dilaver, Z.; Hunt, L.C. What drives natural gas consumption in Europe? Analysis and projections. *J. Nat. Gas Sci. Eng.* **2014**, *19*, 125–136. [CrossRef]

9. Stern, J. *Challenges to the Future of Gas: Unburnable or Unaffordable?* Oxford Institute for Energy Studies: Oxford, UK, 2017.

10. European Environment Agency (EEA). *Final Energy Consumption of Natural Gas by Sector*; European Environment Agency: Copenhagen, Denmark, 2017.

11. International Energy Agency (IEA). *WEO Analysis: Is Natural Gas in Good Shape for the Future?* European Environment Agency: Copenhagen, Denmark, 2018.

12. Yeh, S. An empirical analysis on the adoption of alternative fuel vehicles: The case of natural gas vehicles. *Energy Policy* **2007**, *35*, 5865–5875. [CrossRef]

13. Osorio-Tejada, J.L.; Llera-Sastresa, E.; Scarpellini, S. Liquefied natural gas: Could it be a reliable option for road freight transport in the EU? *Renew. Sustain. Energy Rev.* **2017**, *71*, 785–795. [CrossRef]

14. Bittante, A.; Jokinen, R.; Krooks, J.; Pettersson, F.; Saxén, H. Optimal Design of a Small-Scale LNG Supply Chain Combining Sea and Land Transports. *Ind. Eng. Chem. Res.* **2017**, *56*, 13434–13443. [CrossRef]

15. Bisen, V.S.; Karimi, I.A.; Farooq, S. Dynamic Simulation of a LNG Regasification Terminal and Management of Boil-off Gas. *Comput. Aided Chem. Eng.* **2018**, *44*, 685–690.

16. Arteconi, A.; Polonara, F. LNG as vehicle fuel and the problem of supply: The Italian case study. *Energy Policy* **2013**, *62*, 503–512. [CrossRef]

17. Sharafian, A.; Talebian, H.; Blomerus, P.; Herrera, O.; Mérida, W. A review of liquefied natural gas refueling station designs. *Renew. Sustain. Energy Rev.* **2017**, *69*, 503–513. [CrossRef]

18. Burel, F.; Taccani, R.; Zuliani, N. Improving sustainability of maritime transport through utilization of Liquefied Natural Gas (LNG) for propulsion. *Energy* **2013**, *57*, 412–420. [CrossRef]

19. Xian, H.; Karali, B.; Colson, G.; Wetzstein, M.E. Diesel or compressed natural gas? A real options evaluation of the U.S. natural gas boom on fuel choice for trucking fleets. *Energy* **2015**, *90*, 1342–1348.

20. Engerer, H.; Horn, M. Natural gas vehicles: An option for Europe. *Energy Policy* **2010**, *38*, 1017–1029. [CrossRef]

21. Frick, M.; Axhausen, K.W.; Carle, G.; Wokaun, A. Optimization of the distribution of compressed natural gas (CNG) refueling stations: Swiss case studies. *Transp. Res. Part D Transp. Environ.* **2007**, *12*, 10–22. [CrossRef]

22. Ríos-Mercado, R.Z.; Borraz-Sánchez, C. Optimization problems in natural gas transportation systems: A state-of-the-art review. *Appl. Energy* **2015**, *147*, 536–555. [CrossRef]

23. Schweiger, J.; Liers, F. A decomposition approach for optimal gas network extension with a finite set of demand scenarios. *Optim. Eng.* **2018**, *19*, 297–326. [CrossRef]

24. Liang, Y.; Hui, C.W. Convexification for natural gas transmission networks optimization. *Energy* **2018**, *158*, 1001–1016. [CrossRef]

Energies **2019**, *12*, 351

25. Mikolajková, M.; Saxén, H.; Pettersson, F. Linearization of an MINLP model and its application to gas distribution optimization. *Energy* **2018**, *146*, 156–168. [CrossRef]

26. Gugat, M.; Leugering, G.; Martin, A.; Schmidt, M.; Sirvent, M.; Wintergerst, D. MIP-based instantaneous control of mixed-integer PDE-constrained gas transport problems. *Comput. Optim. Appl.* **2018**, *70*, 267–294. [CrossRef]

27. Hante, F.M.; Leugering, G.; Martin, A.; Schewe, L.; Schmidt, M. Challenges in optimal control problems for gas and fluid flow in networks of pipes and canals: From modeling to industrial applications. In *Industrial Mathematics and Complex Systems*; Springer: Singapore, 2017; pp. 77–122.

28. Zhang, H.; Liang, Y.; Zhang, W.; Wang, B.; Yan, X.; Liao, Q. A unified MILP model for topological structure of production well gathering pipeline network. *J. Pet. Sci. Eng.* **2017**, *152*, 284–293. [CrossRef]

29. Zheng, Q.P.; Pardalos, P.M. Stochastic and Risk Management Models and Solution Algorithm for Natural Gas Transmission Network Expansion and LNG Terminal Location Planning. *J. Optimiz. Theory Appl.* **2010**, *147*, 337–357. [CrossRef]

30. Lee, I.; Park, J.; Moon, I. Key Issues and Challenges on the Liquefied Natural Gas Value Chain: A Review from the Process Systems Engineering Point of View. *Ind. Eng. Chem. Res.* **2018**, *57*, 5805–5818. [CrossRef]

31. Agarwal, R.; Rainey, T.; Rahman, S.; Steinberg, T.; Perrons, R.; Brown, R. LNG Regasification Terminals: The Role of Geography and Meteorology on Technology Choices. *Energies* **2017**, *10*, 2152. [CrossRef]

32. Mikolajková, M.; Saxén, H.; Pettersson, F. Mixed Integer Linear Programming Optimization of Gas Supply to a Local Market. *Ind. Eng. Chem. Res.* **2018**, *57*, 5951–5965. [CrossRef] [PubMed]

33. Hengeveld, E.J.; van Gemert, W.J.T.; Bekkering, J.; Broekhuis, A.A. When does decentralized production of biogas and centralized upgrading and injection into the natural gas grid make sense? *Biomass Bioenergy* **2014**, *67*, 363–371. [CrossRef]

34. Hoo, P.Y.; Hashim, H.; Ho, W.S. Opportunities and challenges: Landfill gas to biomethane injection into natural gas distribution grid through pipeline. *J. Clean. Prod.* **2018**, *175*, 409–419. [CrossRef]

35. Mian, A.; Ensinas, A.V.; Marechal, F. Multi-objective optimization of SNG production from microalgae through hydrothermal gasification. *Comput. Chem. Eng.* **2015**, *76*, 170–183. [CrossRef]

36. Sarić, M.; Dijkstra, J.W.; Haije, W.G. Economic perspectives of Power-to-Gas technologies in bio-methane production. *J. CO2 Util.* **2017**, *20*, 81–90. [CrossRef]

37. Haaland, S.E. Simple and explicit formulas for the friction factor in turbulent pipe flow. *J. Fluids Eng.* **1983**, *105*, 89–90. [CrossRef]

38. Bisschop, J. *AIMMS Optimization Modeling*; Lulu.com: Morrisville, NC, USA, 2014.

39. Gasum Oy Natural Gas Transmission. 2017. Available online: https://www.gasum.com/en/About-gas/Finlands-gas-network/Natural-gas-transmission/ (accessed on 10 September 2018).

40. Wärtsilä Small- and Medium-Scale LNG Terminals. 2018. Available online: https://www.wartsila.com/docs/default-source/Power-Plants-documents/lng/small-and-medium-scale-lng-terminals_wartsila.pdf (accessed on 10 October 2018).

41. Siitonen, S. Natural Gas Pipeline in Finland. Personal communication, 2013.

42. Songhurst, B. LNG Plant Cost Escalation. Available online: https://ora.ox.ac.uk/objects/uuid:40de2a7f-7454-4d7d-977b-972112fb0f9a (accessed on 12 September 2018).

43. Van Brummelen, G. *Heavenly Mathematics: The Forgotten Art of Spherical Trigonometry*; Princeton University Press: Princeton, NJ, USA, 2013.

Article

Identification of Optimal Parameters for a Small-Scale Compressed-Air Energy Storage System Using Real Coded Genetic Algorithm

Thomas Guewouo [1,2], Lingai Luo [2,*], Dominique Tarlet [2] and Mohand Tazerout [3]

[1] Laboratoire de Modélisation et de Simulation Multi Echelle (MSME), CNRS UMR 8208, Université Paris-Est, 5 Bd Descates, F-77454 Marne-la-Vallée CEDEX 2, France; thomas.guewouo@u-pem.fr

[2] Laboratoire de Thermique et Energie de Nantes (LTEN), CNRS UMR 6607, Université de Nantes, La Chantrerie, Rue Christian Pauc, B.P. 50609, F-44306 Nantes CEDEX 3, France; dominique.tarlet@univ-nantes.fr

[3] GEPEA UMR CNRS 6144, BP 406, 37 boulevard de l'Université, 44602 Saint Nazaire, France; mohand.tazerout@imt-atlantique.fr

* Correspondence: lingai.luo@univ-nantes.fr; Tel.: +33-2-4068-3167; Fax: +33-2-4068-3199

Received: 31 October 2018; Accepted: 8 January 2019; Published: 24 January 2019

Abstract: Compressed-Air energy storage (CAES) is a well-established technology for storing the excess of electricity produced by and available on the power grid during off-peak hours. A drawback of the existing technique relates to the need to burn some fuel in the discharge phase. Sometimes, the design parameters used for the simulation of the new technique are randomly chosen, making their actual construction difficult or impossible. That is why, in this paper, a small-scale CAES without fossil fuel is proposed, analyzed, and optimized to identify the set of its optimal design parameters maximizing its performances. The performance of the system is investigated by global exergy efficiency obtained from energy and exergy analyses methods and used as an objective function for the optimization process. A modified Real Coded Genetic Algorithm (RCGA) is used to maximize the global exergy efficiency depending on thirteen design parameters. The results of the optimization indicate that corresponding to the optimum operating point, the consumed compressor electric energy is 103.83 kW h and the electric energy output is 25.82 kW h for the system charging and discharging times of about 8.7 and 2 h, respectively. To this same optimum operating point, a global exergy efficiency of 24.87% is achieved. Moreover, if the heat removed during the compression phase is accounted for in system efficiency evaluation based on the First Law of Thermodynamics, an optimal round-trip efficiency of 79.07% can be achieved. By systematically analyzing the variation of all design parameters during evolution in the optimization process, we conclude that the pneumatic motor mass flow rate can be set as constant and equal to its smallest possible value. Finally, a sensitivity analysis performed with the remaining parameters for the change in the global exergy efficiency shows the impact of each of these parameters.

Keywords: small-scale compressed-air energy storage (SS-CAES) ; energy storage; exergy analysis; optimization; Real Coded Genetic Algorithm (RCGA); Violation Constraint-Handling (VCH)

1. Introduction

The security and reliability of electricity grid need the introduction of a storage system [1]. To these two main objectives of energy storage one can add the reduction of greenhouse gas emissions. Of importance, due to the increased consumption of fossil fuels, the amount of CO_2 emitted increased up to threefold between 1960 and 2008, today reaching more than 32,000 million tons per year. The climate change observed due to these emissions has driven many countries to turn to renewable energy (RE) sources for electricity production to retain global warming within a 2% range [2,3]. However, these sources are strongly related to meteorology and are intermittent [4]. One of the solutions developed to overcome the problem of intermittency is to couple them with an electrical storage system [5,6]. The storage system will be able to play two roles, namely protection and production. By protection it is meant that the system must be able to quickly restore (the response time in the order of a few minutes [7]) the energy stored during the fluctuation of the resource. By production it is meant that the storage system must be able to produce and sustain independently during a sufficiently long period the demands in the absence of the total source.

In the worldwide industries of electricity both mature technologies are used for large scale electricity storage. They are pump hydro (PH) system and compressed-air energy storage (CAES) systems [2,8–12]. As opposed to the hydroelectric pumping stations, the storage systems with compressed air offer flexibility both in size (smaller volumes) and capacity (ranging from several hundred KW to MW). These advantages give CAES the opportunity to be coupled to the power generation system with renewable sources. That is why many researches propose hybrid wind/CAES systems or photovoltaic plant (PV)/CAES systems [13–21]. In periods of low and off-peak energy demand, the CAES system stores electricity in the form of compressed air in a natural or artificial tank. The stored compressed air is released and heated in a combustion chamber burning fossil fuel before being expanded in a turbine connected to a generator for electricity reproduction [11,22,23]. Many studies dealing with the partial or total replacement of combustion chamber exist in the literature [3,24–28]. The heat generated during compression is stored and used to heat air before expansion; such systems are called adiabatic or advanced adiabatic compressed-air energy storage (AA-CAES). Unfortunately, conventional and AA-CAES used natural reservoir (underground caverns, rock formations) for storing compressed air which reduces its penetration potentiality due to the geological restriction [29,30]. Therefore, the attention has been recently focused on the usage of artificial air-tanks. The resulting system is known as micro or small-scale compressed-air energy storage system (SS-CAES). Such system can be used at isolated sites with RE sources or in the residential sector to store electricity during off-peak hours. Generally, in SS-CAES system, fuel combustion is not needed because the compression heat is collected, stored, and re-used to heat the compressed air before being expanded in the turbine or the reciprocating air motor. If the cooling energy in the discharged air is collected, the SS-CAES may act as a tri-generative system, for simultaneous production of cold, heat and electricity [4,9,31–36].

To evaluate the performance of SS-CAES system by means of numerical simulations, numerous thermodynamics models have been developed during recent years. Generally, these models are based on the first law of thermodynamics with imposed design and operating parameters of the analyzed systems [9,32,34]. Unfortunately, energy analysis does not provide the information about the locations of energy degradation in a process and does not quantify the irreversibility in different components of the storage system. Therefore, based on both the first and second laws of thermodynamics, exergy analysis appears to be a powerful tool to overcome the limitations of energy analysis [37].

The purpose of this study is to develop a realistic approach to investigate the performance of SS-CAES system using pressure vessels without fossil fuel. This approach is based on the exergy analysis method. The required equations for modeling different components of the system are presented. These equations are used to build the objective function which is the global exergy efficiency of the storage system. We aim to maximize this objective function depending on thirteen design parameters and seven constraints. These design parameters are respectively: number of compression stage (n), compressor pressure ratio (π), volume of air storage tank (V_t), pressure ratio of high-pressure and low-pressure expansion stages (β_{HP}, β_{LP}), inlet temperature of high-pressure and low-pressure expansion stages (T_{HP}^{in}, T_{LP}^{in}), isentropic efficiency of compressor (η_{IsC}), isentropic efficiency of pneumatic motor ($\eta_{Is,m}$), mechanical efficiency of compressor (η_{mC}), mechanical efficiency of pneumatic motor (η_{mm}), compressor and pneumatic motor mass flow rate ($\dot{m}_C$, $\dot{m}_m$). The ranges of each parameter have been defined in the light of available technology. The optimization is performed using a modified Real Coded Genetic Algorithm (RCGA) in which two crossover methods are randomly selected from one generation to another during evolution. To improve the search efficiency of the RCGA, the Dynamic Random Mutation (DRM) method was used. The coupled modified RCGA-DRM could effectively determine the set of optimal values of influencing parameters that maximizes the global exergy efficiency of the SS-CAES system without fossil fuel used. MATLAB® software is used for all computations (Version 9.1 developed by MathWorks, Inc. whose the headquarters is located in the city of Natick, in the state of Massachusetts in the USA).

2. System Description

The system to optimize is shown in Figure 1. The system operates in two phases: charge and discharge. The first phase is also known as compression phase, which is composed of multistage (CS) reciprocating compressor and intercoolers (HE). In this phase, the compressor is powered by the electrical energy available on grid during off-peak load hours or by the electricity generated by a RE source. Intercoolers are used to recover the compression heat with water as the heat transfer fluid. Hot water is stored in an isolated thermal storage tank (HWt). Cooling of compressed air after each stage has the advantage of reducing the required electrical power of the compressor and increasing the compressed-air storage tank efficiency due to the high density of the cold compressed air. High-pressure cold air from compression and cooling is stored in compressed-air storage tank (CASt). In conventional or diabatic CAES systems, at peak load hours, the stored air is released from the underground cavern, throttled through the regulating valve, and heated by fossil fuel burning in the combustion chamber (CC). However, in an AA-CAES and SS-CAES, the hot water produced during the compression phase can be used to heat air before each stage of expansion through heat exchangers, thus replacing CC. Two stages pneumatics motor (HPe and LPe) coupled to electric generator (G) are used to achieve the expansion process and generate electricity. In Figure 1, the points i for $i \in \{1, 2, ..., 16\}$ denote the states of air transformation during all storage process and $i \in \{17, 18, ..., 26\}$ denote the states of water transformation during all transformation storage process.

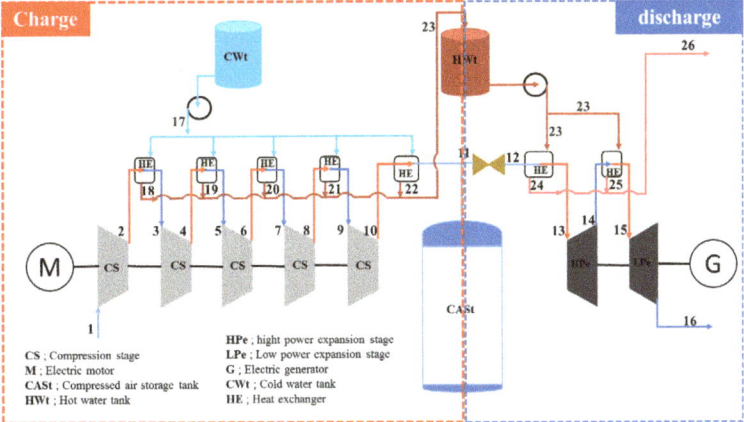

Figure 1. Schematic diagram of the proposed system.

3. Thermodynamic Modeling

To carry on the optimization problem, it is necessary to build the objective function which in this study is the global exergy efficiency. The analytical form of the objective function results from the thermodynamic analysis of each component of the system. Due to the complexity of the system some simplifying but basic assumptions are made following previous works on thermodynamic modeling of energy systems with air as working fluid [38–42]:

- All system components operate under a steady state condition except the CASt for which a dynamic modeling is performed to find the filling and discharge time together with the mean temperature in the tank during the discharge process.
- Air is assumed dry and modeled as an ideal gas.
- Potential and kinetic energy effects are negligible in the energy and exergy balances.
- Pressure drop in the components of system is neglected.
- The reference environment state conditions are $T_0 = 20\,^\circ\text{C}$ and $P_0 = 1.01$ bar which is also used as the system boundary for energy and exergy analyses.
- Isentropic efficiency is assumed constant for compressor and pneumatic motor.
- The mass flow rate of the cooling water is set as constant in every intercooler. Despite the different pressure ratio of expansions stages, we set the mass flow rate of the heating water in each heater as constant and equal to six times that of the cooling water to increase the heat transfer rate and for more simplicity in modeling.
- All the interpolations in thermodynamics tables to find the thermodynamics properties is done using MATLAB built-in function "interp1", with "spline" method. The thermodynamics tables are those of Moran book [43].

3.1. Energy Analysis

In this section, the First Law of Thermodynamics is used for all system components to estimate the temperature of the working fluid (air and water), enthalpy and pressure corresponding to each stage of storage system. These properties depend on the design parameters and finally allow one to evaluate the electrical power consumed by compressor, the electrical power produced by generator coupled to the pneumatic motor, the filling and discharge times, the heat produced during the compression process and the required heat to increase temperature of the compressed air during expansion process.

3.1.1. Compressor

For each stage of compression, the output enthalpy is evaluated as a function of the isentropic efficiency:

$$h_{c,i}^{out} = h_{c,i}^{in} + \frac{h_{c,Is,i}^{out} - h_{c,i}^{in}}{\eta_{IsC}} \tag{1}$$

In Equation (1), $h_{c,i}^{in}$ is the inlet specific enthalpy of the *i*-th stage of compressor estimated at the outlet temperature of intercoolers assumed as constant and equal to 35 °C except for the first stage where it is estimated at ambient temperature T_0, $h_{c,i}^{out}$ is the outlet specific enthalpy of the *i*-th stage of compressor if the compression process is isentropic, and η_{IsC} is the isentropic efficiency of compressor.

The outlet isentropic temperature of each compressor stage can be determined from:

$$S^0(T_{c,Is,i}^{out}) = S^0(T_{c,i}^{in}) + R ln(\pi) \tag{2}$$

Here π is the pressure ratio which is the same for all stages of compression. Since $T_{c,i}^{in}$ is known, $S^0(T_{c,i}^{in})$ would be obtained from Table A-22 in [43], the value of $S^0(T_{c,Is,i}^{out})$ would be calculated with Equation (2), and finally the values of $T_{c,Is,i}^{out}$ and $h_{c,Is,i}^{out}$ would be determined by interpolation. The outlet specific enthalpy of *i*-th stage will be calculated with Equation (1) by knowing $h_{c,Is,i}^{out}$, and outlet temperature will be determined by interpolation.

The outlet pressure $P_{c,i}^{out}$ of the *i*-th compression stage can be calculated by knowing the inlet pressure $P_{c,i}^{in}$ and pressure ratio π as follows:

$$P_{c,i}^{out} = \pi \times P_{c,i}^{in} \tag{3}$$

The electrical power consumed by the compressor can then be calculated as

$$P_{elc,c} = \frac{\dot{m}_c}{\eta_{elc,c}\eta_{m,c}} \sum_{i=0}^{n} \left(h_{c,i}^{out} - h_{c,i}^{in} \right) \tag{4}$$

where $P_{elc,c}$ (often written as $\dot{W}_{elc,c}$ [43]) is the electrical power consumed by the compressor, $\eta_{elc,c}$ is the electric efficiency of the compressor assumed to be constant in this study and equal to 98% (upper value of literature ranges from 90% to 98% [44]), $\eta_{m,c}$ is mechanical efficiency of compressor, $\dot{m}_c$ is the air mass flow rate of compressor and n is the number of compression stages.

3.1.2. Intercoolers

To reduce the work-input required during the compression and prevent the compressor from reaching high temperatures, counter flow air-to-water heat exchangers (called intercoolers and after cooler) cool the compressed air between the stages and after the last stage of the process. By knowing the cooling water mass flow rate $\dot{m}_{cw}$ as well as its inlet temperature, the specific enthalpy of cooling water at intercooler outlet between (i)-th and ($i + 1$)-th stage of compression is computed by an energy balance:

$$h_{cw,i}^{out} = h_{cw,i}^{in} + \frac{\dot{m}_c}{\dot{m}_{cw}} \left(h_{c,i}^{out} - h_{c,i+1}^{in} \right) \tag{5}$$

Since the inlet temperature of cooling water $T_{cw,i}^{in}$ is known, its specific enthalpy $h_{cw,i}^{in}$ is obtained from Table A-22 in [43] and then its outlet temperature $T_{cw,i}^{out}$ can be obtained by interpolation.

The heat stored in the hot water tank (HWt) should be equal to the heat exchanged in the intercoolers (the heat losses are neglected). The total heat transfer rate between air and cooling water in intercoolers during air compression process can be calculated as:

$$\dot{Q}_h = \dot{m}_c \sum_{i=2}^{n+1} \left(h_{c,i}^{out} - h_{c,i+1}^{in} \right) \tag{6}$$

The specific enthalpy of cooling water at the inlet of hot water tank is defined as

$$h_{cw,t}^{in} = \frac{1}{n} \sum_{i=1}^{n} h_{cw,i}^{out} \tag{7}$$

Knowing $h_{cw,t}^{in}$ the final temperature of hot water at the inlet of HWt can be obtained by interpolation from Table A-22 in [43].

3.1.3. Pneumatic Motor

The expansion process can be regarded as an opposite thermodynamic process of compression. The output specific enthalpy of high- and low-power stage is respectively, defined as

$$h_{m,HP}^{out} = h_{m,HP}^{in} - \eta_{Is,m} \left(h_{m,HP}^{in} - h_{m,Is,HP}^{out} \right) \tag{8}$$

$$h_{m,LP}^{out} = h_{m,LP}^{in} - \eta_{Is,m} \left(h_{m,LP}^{in} - h_{m,Is,LP}^{out} \right) \tag{9}$$

In Equations (8) and (9) the outlet isentropic specific enthalpy of each expansion stage $h_{m,Is,HP}^{out}$ and $h_{m,Is,LP}^{out}$ are obtained by interpolation in Table A-22 in [43] knowing $S^0(T_{m,Is,HP}^{out})$ and $S^0(T_{m,Is,LP}^{out})$ given respectively by:

$$S^0(T_{m,Is,HP}^{out}) = S^0(T_{m,HP}^{in}) + Rln\left(\frac{1}{\beta_{HP}}\right) \tag{10}$$

$$S^0(T_{m,Is,LP}^{out}) = S^0(T_{m,LP}^{in}) + Rln\left(\frac{1}{\beta_{LP}}\right) \tag{11}$$

Here β_{HP} and β_{LP} are the pressure ratios of high-pressure and low-pressure expansion stages respectively, $T_{m,HP}^{in}$ and $T_{m,LP}^{in}$ are inlet temperature of air in these expansion stages.

The inlet pressure P_{HP}^{in} of the high-pressure expansion stage can be calculated knowing the pressure ratios β_{HP} and β_{LP} as follows:

$$P_{HP}^{in} = P_0 \beta_{HP} \beta_{LP} \tag{12}$$

The outlet pressure corresponding to each expansion stage is calculated as

$$P_{HP}^{out} = \frac{P_{HP}^{in}}{\beta_{HP}} \tag{13}$$

$$P_{LP}^{out} = \frac{P_{HP}^{out}}{\beta_{LP}} \tag{14}$$

The electric power generated by the electric generator coupled to the pneumatic motor during production phase is evaluated as follows:

$$P_{elc,G} = \dot{m}_m \eta_{elc,G} \eta_{mm} \left[\left(h_{m,HP}^{in} - h_{m,HP}^{out} \right) + \left(h_{m,LP}^{in} - h_{m,LP}^{out} \right) \right] \tag{15}$$

Here $P_{elc,G}$ is the electric power produced by the generator, $\eta_{elc,G}$ is the electric efficiency of generator assumed to be constant in this study and equal to 96% (value of literature range 90% to 98% [44]), η_{mm} is the mechanical efficiency of pneumatic motor, and $\dot{m}_m$ is the air mass flow rate of pneumatic motor which is one of the design parameters.

3.1.4. Heater

To eliminate the use of fossil fuels, to prevent the pneumatic motor from reaching low temperatures and to enhance the power production of pneumatic motor, the CCs usually used in conventional CAES system are replaced by counter flow air-to-water heat exchangers, called heater or air preheater. The hot water produced and stored during compression process is used to warm the air up before each expansion stages. With the assumption that HWt process is adiabatic, the inlet specific enthalpy of hot water h_{hw}^{in}, is known. Then, the specific enthalpy of water at heaters outlet can be respectively expressed as follows:

$$h_{hw,HP}^{out} = h_{hw}^{in} - \frac{\dot{m}_m}{\dot{m}_{hw}} \left(h_{m,HP}^{in} - h_{CASt}^{out} \right) \tag{16}$$

$$h_{hw,LP}^{out} = h_{hw}^{in} - \frac{\dot{m}_m}{\dot{m}_{hw}} \left(h_{m,LP}^{in} - h_{m,HP}^{out} \right) \tag{17}$$

Here, $\dot{m}_{hw}$ is the heating water mass flow rate and h_{CASt}^{out} is the specific enthalpy at the CASt outlet. The temperature of the water at the heater's outlet can then be obtained by interpolation from Table A-2 in [43] knowing enthalpy $h_{hw,HP}^{out}$ and $h_{hw,LP}^{out}$.

The specific enthalpy of cooling water at the expansion train outlet is defined as:

$$h_{hw}^{out} = \frac{1}{2} \left(h_{hw,HP}^{out} + h_{hw,LP}^{out} \right) \tag{18}$$

The thermal power required to warm the air up before it is expanded in the pneumatic motor can be computed as follows:

$$\dot{Q}_{Rh} = \dot{m}_m \left[\left(h_{m,HP}^{in} - h_{CASt}^{out} \right) + \left(h_{m,LP}^{in} - h_{m,HP}^{out} \right) \right] \tag{19}$$

3.1.5. Compressed-Air Storage Tank

This step is crucial because it allows one to determine the charge and discharge time of the CASt as well as the mean temperature of air at the inlet of heater before the high power expansion stage. We consider CASt as one thermodynamic control volume with the total geometric volume V_t. In this study, we assume that the inlet and outlet mass flow rates are constant. The inlet air temperature in the control volume $T_{c,n}^{in}$ is assumed to be equal to that of air exiting the last cooler (35 °C) and the inlet pressure of compressed air $P_{c,n}^{out}$ is assumed to be equal to that of air exiting the last compression stage. The exit air pressure P_{CASt}^{out} is set to the inlet air pressure of the air motor P_{HP}^{in} and its outlet temperature T_{CASt}^{out} is assumed to be the minimum value of temperature inside the control volume. All these assumptions are summarized in Figure 2, where m, P and T represent respectively air mass, air pressure and air temperature inside the tank and $\dot{Q}_{CASt}$ is the thermal power lost through the compressed-air storage tank walls.

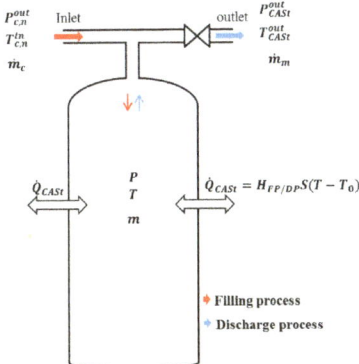

Figure 2. Schematic diagram of compression air storage tank.

Thus, the variations of mass, temperature, and pressure of air in the CASt during filling and discharge process are described by the mass conservation equation, the energy conservation equation and the ideal gas equation of state [43].

$$\begin{cases} \dfrac{dm}{dt} &= \dot{m}^{in} - \dot{m}^{out} \\ \dfrac{dmu}{dt} &= \dot{m}^{in}\left[h^{in} + \tfrac{1}{2}(V^{in})^2 + gZ^{in}\right] - \dot{m}^{out}\left[h^{out} + \tfrac{1}{2}(V^{out})^2 + gZ^{out}\right] - \dot{Q}_{CASt} \\ V_t\dfrac{dP}{dt} &= RT\dfrac{dm}{dt} + Rm\dfrac{dT}{dt} \end{cases} \tag{20}$$

For the filling process where $\dot{m}^{out} = 0$ and $\dot{m}^{in} = \dot{m}_c$ these laws read:

$$\begin{cases} \dfrac{dm}{dt} &= \dot{m}_c \\ \dfrac{dT}{dt} &= \dfrac{1}{m\,(c_p - R)}\left[\dot{m}_c c_p\left(T_{c,n}^{in} - T\right) + \dot{m}_c RT + \dfrac{\dot{m}_c}{2}\left(\dfrac{\dot{m}_c}{\rho\,A_{CASt}^{in}}\right)^2 + \dot{m}_c gH_t - \dot{Q}_{CASt}R\right] \\ \dfrac{dP}{dt} &= \dfrac{\dot{m}_c R}{V_t}T + \dfrac{mR}{V_t}\dfrac{dT}{dt} \end{cases} \tag{21}$$

And for the discharge process where $\dot{m}^{in} = 0$ and $\dot{m}^{out} = -\dot{m}_m$ these laws read:

$$\begin{cases} \dfrac{dm}{dt} = -\dot{m}_m \\[2mm] \dfrac{dT}{dt} = \dfrac{1}{m\,(c_p - R)}\left[-\dot{m}_m c_p T - \dfrac{\dot{m}_c}{2}\left(\dfrac{\dot{m}_m}{\rho\,A_{CASt}^{out}}\right)^2 - \dot{m}_m g H_t - \dot{Q}_{CASt} R \right] \\[2mm] \dfrac{dP}{dt} = -\dfrac{\dot{m}_m R}{V_t}T + \dfrac{mR}{V_t}\dfrac{dT}{dt} \end{cases} \tag{22}$$

With:

$$\begin{cases} c_p(T) &=& A + B\,T + C\,T^2 + D\,T^3 + E\,T^4 \\ A &=& 1.0484 \times 10^3 \\ B &=& -3.837 \times 10^{-1} \\ C &=& 9.4537 \times 10^{-4} \\ D &=& -5.4903 \times 10^{-7} \\ E &=& 7.9298 \times 10^{-11} \end{cases} \tag{23}$$

In this study, the reference state for enthalpy is $0K$ and $h(T = 0K) = 0\ kJ.kg^{-1}.K^{-1}$. In Equations (21) and (22), c_p is the specific heat at constant pressure given by (23) [43] which is evaluated during the numerical resolution at the current time step at the temperature corresponding to the previous step time, A_{CASt}^{in} and A_{CASt}^{out} are respectively inlet and outlet cross section of CASt, H_t is the height of the CASt and g is the acceleration of the gravity. The heat exchange through the tank walls $\dot{Q}_{CASt}$, is modeled through a quasi-steady process assuming a cylindrical geometry with steel shell structure of the CASt. It is calculated as follows:

$$\dot{Q}_{CASt} = H_{FP/DP} S\,(T - T_0) \tag{24}$$

where $H_{FP/DP}$, is the heat transfer coefficient between the CASt wall and the air during filling and discharge process (equal to 40 and 45 $W \cdot m^2 \cdot K^{-1}$, respectively [45]), S is the area of heat transfer between the air and the CASt wall, T is the temperature of air inside the CASt and T_0 is the environment temperature.

Equations (21) and (22) are solved using the fourth-order Runge-Kutta method presented by Press et al. [46]. As initial conditions of air inside the CASt during filling process we use the environmental temperature and the minimal pressure set to inlet pressure $P_{HP}^{in} = P_0 \beta_{HP} \beta_{LP}$ of the high-power expansion stage. The stopping condition relates to the maximum pressure in CASt set to the outlet pressure of last compression stage $P_{c,n}^{out} = P_0 \times \pi^n$. These initial and stopping conditions of air inside the CASt are reversed for the discharge process. At this level of the analysis, the temperature, pressure, and enthalpy of each line of storage system to be optimized are known and for each stage of the process. Therefore, the exergy analysis can be carried out with greater flexibility.

3.2. Exergy Analysis

Exergy is defined as the maximum theoretical work obtainable from an overall system consisting of a system and the environment as the system that comes into equilibrium with the environment [43,47,48]. It can equally be defined as the maximum work that can be obtained from a given form of energy when the reference environment state is defined by the environmental parameters [49]. This second definition is more appropriate to the approach used in this work. To carry out the exergy analysis the exergy rate balance should be applied to each component. In steady state, this exergy rate balance for given component with one inlet and one outlet can be expressed as follows:

$$\dot{Ex}^Q + \dot{Ex}^{in} = \dot{Ex}^W + \dot{Ex}^{out} + \dot{Ex}^D \tag{25}$$

Here $\dot{Ex}^Q$ is the time rate of exergy transfer associated with heat transfer, $\dot{Ex}^W$ is the exergy of the work, $\dot{Ex}^D$ is the time rate of exergy destruction, and $\dot{Ex}^{in}$ and $\dot{Ex}^{out}$ are the time rate of exergy transfer at inlet and outlet of the considered component respectively. These parameters are defined as follows:

$$Ex^Q = \left(1 - \frac{T_0}{T_b}\right)\dot{Q} \tag{26}$$

$$\dot{Ex} = \dot{m}e_x \tag{27}$$

$$e_x = (h - h_0) - T_0(s - s_0) \tag{28}$$

$$\dot{Ex}^W = \dot{W} \tag{29}$$

Here e_x is the specific flow exergy (also known as physical exergy), T_b is the temperature of the boundary where heat transfer ($\dot{Q}$) occurs. Generally, to evaluate $\dot{Ex}^Q$ which is associated with exergy loss for a given component it is necessary to know the heat transfer $\dot{Q}$ across each segment of the boundary and T_b. Although it is sometimes possible to calculate $\dot{Q}$, the temperature of the boundary is more difficult to obtain and requires experimental measurements. Therefore, an alternative approach that often suffices for modeling is to suppose that the boundary is the outer surface of each component where the temperature corresponds to the ambient temperature taken as the temperature of the exergy reference environment. Thus, the heat transfer occurs at T_0 ($T_b = T_0$) and therefore there exists no exergy loss [50]. In this case, the rate of exergy destruction term of Equation (25) accounts for the exergy destruction owing to friction and the irreversibility of heat transfer within the considered component [50]. Then, using Equations (26)–(29), the rate of exergy destruction of all components could be calculated as shown in Table 1.

Table 1. Expression of the rate of exergy destruction for proposed storage system relevant components.

System Component	Exergy Destruction Rate
Air compressor (AC)	$\dot{Ex}_c^D = \sum\limits_{i=1}^{n} \dot{Ex}_{c,i}^{in} - \sum\limits_{i=1}^{n} \dot{Ex}_{c,i}^{out} + P_{elc,c}$
Intercoolers (Int)	$\dot{Ex}_{Int}^D = \sum\limits_{i=1}^{n}\left(\dot{Ex}_{c,i}^{out} - \dot{Ex}_{c,i+1}^{in}\right) - \sum\limits_{i=1}^{n}\left(\dot{Ex}_{cw,i}^{out} - \dot{Ex}_{cw,i}^{in}\right)$
Compressed-air storage tank (CASt)	$\dot{Ex}_{CASt}^D = \dot{Ex}_{c,n}^{out} - \dot{Ex}_{CASt}^{out}$
Heaters (He)	$\dot{Ex}_{He}^D = \sum\limits_{i=HP,LP}\left(\dot{Ex}_{hw,i}^{in} - \dot{Ex}_{hw,i}^{out}\right) - \sum\limits_{i=HP,LP}\left(\dot{Ex}_{m,i}^{in} - \dot{Ex}_{m,i}^{out}\right)$
Pneumatic motor (PM)	$\dot{Ex}_{PM}^D = \sum\limits_{i=HP,LP} \dot{Ex}_{m,i}^{in} - \sum\limits_{i=HP,LP} \dot{Ex}_{m,i}^{out} - P_{elc,G}$

3.3. Storage System Performance Criteria

We use the efficiency as a mean performance parameter of proposed system. We define the global energy efficiency also known as the Round-Trip Efficiency (RTE) and overall exergy efficiency (η_{ex}) as follows. The first one is based on the First Law of Thermodynamics that defines the efficiency of an engineering system by the ratio of energy outputs to inputs. For our proposed system, the energy input is the electricity used by the compressor to produce compressed air. The energy output is the sum of the electrical energy produced by generator and the part of heat recovered during the compression process that has not been used to reheat the air before its expansion. The second and perhaps the most relevant performance criteria of our proposed system is the exergy efficiency. It is an efficiency based on the Second Law of Thermodynamics and it is defined as the ratio of total exergy outputs to exergy inputs [42,48]. Exergy efficiency is also known as the ratio of the product exergy to the fuel exergy. The fuel exergy is defined by the electrical energy consumed by compressor and the product exergy is the difference between the fuel exergy and the sum of exergy destruction in all components of the storage system and total exergy loss associated with the overall considered system [41,51]. To conclude this part, these efficiencies can be express as follows:

$$RTE = \frac{P_{elc,G}t_{DP} + \left(\dot{Q}_h t_{FP} - \dot{Q}_{Rh}t_{DP}\right)}{P_{elc,c}t_{FP}} \tag{30}$$

$$\eta_{ex} = 1 - \frac{\left(\dot{Ex}_c^D + \dot{Ex}_{Int}^D + \dot{Ex}_{c,n}^{out}\right)t_{FP} + \left(\dot{Ex}_{He}^D + \dot{Ex}_{PM}^D - \dot{Ex}_{CASt}^{out}\right)t_{DP} + Ex_{loss}}{P_{elc,c}t_{FP}} \tag{31}$$

Here t_{FP} and t_{DP} are respectively the charge and discharge time, $\dot{Q}_h$ is the total heat transfer rate between air and cooling water in intercoolers during air compression process (see Equation (6)) and $\dot{Q}_{Rh}$ the thermal power required to warm the air up before it is expanded in the pneumatic motor (see Equation (19)).

4. Formulation of Optimization Problem

4.1. Definition of Objective Function

There is a discussion about the definition of global energy efficiency (or RTE) of CAES system. While some, to conform to the First Law of Thermodynamics, define the global energy efficiency by Equation (30), others suggest taking into account the fact that the electrical energy and heat are different energy forms. For the latter, it would be necessary to convert the heat power of the hot water into its electrical equivalent. For this purpose, they assume a virtual thermal power plant that would use the thermal power of water as heat source [31]. The energy efficiency of this virtual power plant given by Equation (32) would allow the deduction of the electrical equivalent of heat from hot water ($E_{elc,eq,hw}$) using Equation (33).

$$\eta_{ref} = \frac{\text{Electrical equivalent of the heat power of the hot water}(E_{elc,eq,hw})}{\left(\dot{Q}_h t_{FP} - \dot{Q}_{Rh}t_{DP}\right)} \tag{32}$$

$$E_{elc,eq,hw} = \left(\dot{Q}_h t_{FP} - \dot{Q}_{Rh}t_{DP}\right)\eta_{ref} \tag{33}$$

Here, η_{ref} is the thermal efficiency of the virtual power plant generally considered equal to that of a reference natural gas power plant (38.2%) [52].

The global energy efficiency of the SS-CAES system would therefore be written as follows:

$$RTE' = \frac{P_{elc,G}t_{DP} + \left(\dot{Q}_h t_{FP} - \dot{Q}_{Rh}t_{DP}\right)\eta_{ref}}{P_{elc,c}t_{FP}} \tag{34}$$

To avoid controversy on good definition of energy efficiency of SS-CAES system, we decided to use as an objective function, the overall exergy efficiency of the system (Equation (31)). This objective function is subject to thirteen design parameters and seven inequality constraints. The mathematical formulation of the optimization problem is given by:

Identify $\vec{X}$ which maximizes $\eta_{ex}(\vec{X})$ subject to seven inequality constraints:

$$\begin{cases} G_i(\vec{X}) \leq 0, \ i = 1, ..., 7 \\ x_j^L \leq x_j \leq x_j^U, \ j = 1, ..., N_{par} \end{cases}$$

Here $\vec{X}$ stands for the solution vector containing the $N_{par} = 13$ design parameters $\vec{X} = \left[x_1, x_2, ..., x_{N_{par}}\right]$ and each of them varies in the range of lower and upper bounds $\left[x_j^L, x_j^U\right]$.

4.2. Constraints

Seven constraints are considered in this optimization procedure:

- The minimum power delivered by the storage system during discharge process may not be smaller than 10^4 W (10 kW) so that they can be implemented in a house of the residential sector.

$$10^4 \leq P_{elc,G} \Rightarrow \frac{10^4}{P_{elc,G}} \leq 1$$

$$\Rightarrow \frac{10^4}{P_{elc,G}} - 1 \leq 0$$

- The proposed system needs to be used to store low cost electricity available during off-peak hours and to store electricity from RE sources. Thus, the charge time must not exceed 43,200 s (12 h) and the discharge time should be greater than 7200 s (2 h).

$$t_{FP} \leq 43{,}200 \Rightarrow \frac{t_{FP}}{43{,}200} \leq 1$$

$$\Rightarrow \frac{t_{FP}}{43{,}200} - 1 \leq 0$$

and

$$7200 \leq t_{DP} \Rightarrow \frac{7200}{t_{DP}} \leq 1$$

$$\Rightarrow \frac{7200}{t_{DP}} - 1 \leq 0$$

- The hot water used to reheat air during discharge process is produced during the compression process. Thus, to eliminate the need for infinitely long heat exchangers, the difference between the hot water temperature and the inlet temperature of air of expansion stages must be larger than 5 K.

$$T_{m,HP}^{in} \leq T_{hw}^{in} - 5 \Rightarrow T_{m,HP}^{in} + 5 \leq T_{hw}^{in}$$

$$\Rightarrow \frac{T_{m,HP}^{in} + 5}{T_{hw}^{in}} \leq 1$$

$$\Rightarrow \frac{T_{m,HP}^{in} + 5}{T_{hw}^{in}} - 1 \leq 0$$

and

$$T_{m,LP}^{in} \leq T_{hw}^{in} - 5 \Rightarrow T_{m,LP}^{in} + 5 \leq T_{hw}^{in}$$

$$\Rightarrow \frac{T_{m,LP}^{in} + 5}{T_{hw}^{in}} \leq 1$$

$$\Rightarrow \frac{T_{m,LP}^{in} + 5}{T_{hw}^{in}} - 1 \leq 0$$

- Due to mechanical constraints and safety problems, the maximum pressure in compressed-air storage tank (CASt) cannot exceed 300×10^5 Pa (300 bar).

$$P_{c,n}^{out} \leq 300 \times 10^5 \Rightarrow \frac{P_{c,n}^{out}}{300 \times 10^5} \leq 1$$

$$\Rightarrow \frac{P_{c,n}^{out}}{300 \times 10^5} - 1 \leq 0$$

- The mass of hot water produced during the compression process must be greater than or equal to that required to warm the air up during the discharge process.

$$2 \times \dot{m}_{hw} \times t_{DP} \leq n \times \dot{m}_{cw} \times t_{FP} \Rightarrow \frac{2 \times \dot{m}_{hw} \times t_{DP}}{n \times \dot{m}_{cw} \times t_{FP}} \leq 1$$

$$\Rightarrow \frac{2 \times \dot{m}_{hw} \times t_{DP}}{n \times \dot{m}_{cw} \times t_{FP}} - 1 \leq 0$$

4.3. Design Parameters

As noted in the introduction section, the design parameters selected for use in this study are: number of compression stages (n), compressor pressure ratio (π), volume of air storage tank (V_t), pressure ratio of high-pressure and low-pressure expansion stages of pneumatic motor (β_{HP}, β_{LP}), inlet temperature of high-pressure and low-pressure expansion stages ($T_{m,HP}^{in}$, $T_{m,LP}^{in}$), compressor and pneumatic motor isentropic efficiency (η_{IsC}, $\eta_{Is,m}$), compressor and pneumatic motor mechanical efficiency (η_{mC}, η_{mm}), compressor and pneumatic motor mass flow rate ($\dot{m}_c$, $\dot{m}_m$). The ranges of each parameter have been specified according to the working specifications of each hardware element and summarized in Table 2.

Table 2. Range of each decision variable.

Decision Variable	Range	Decision Variable	Range
n	2–5	η_{IsC} (%)	70–75
π	2–6.5	$\eta_{Is,m}$ (%)	70–90
V_t (m^3)	0.3–30	η_{mC} (%)	65–75
β_{HP}	6–10	η_{mm} (%)	75–90
β_{LP}	2–6	$\dot{m}_c$ (kg·s^{-1})	0.004–0.0156
$T_{m,HP}^{in}$ (°C)	15–50	$\dot{m}_m$ (kg·s^{-1})	0.066–0.132
$T_{m,LP}^{in}$ (°C)	15–50	–	–

4.4. Modified Real Coded Genetic Algorithm

Developed by John Holland [53], the genetic algorithm (GA) is an optimization and search technique based on the principles of genetics and natural selection. To perform the optimization, GA produces some random numbers for each design variables that form a population of individuals called initial population, where an individual consists of values of the design variables is a potential solution which maximizes the overall exergy efficiency. Based on an analogy with Darwin's laws of natural selection, GA applies to an initial population, the operators of selection, crossover, and mutation to allow it to evolve to a new population that is, the next generation. The type of encoding used to represent these design variables are defined by the type of algorithm, thus when the design variables are continuous (as is the case for this study), it is more logical to represent them by floating-point numbers rather than by binary numbers. This is referred to as RCGA also known as continuous GA [54]. The algorithm has following steps:

- Generate the initial population.
- Evaluate the fitness of each individual of the considered population.
- Select individuals to form the mating pool.
- Select individuals of the mating pool for mating.
- Apply crossover to generate offspring who is individuals of next generation.
- Maintain the diversity in the population by mutation of selected members of the population.
- Terminate the run if the stopping criteria are fulfilled or go back to step 2.

To improve the search efficiency and closer simulate the natural selection which is the fundamental principle of GA, one modification is introduced in the crossover step. In this step, during the evolution process, two crossover operators are randomly selected to generate the children according to the probability of crossover (*Pc*). Thus, for a given generation a random number is generated and compared to the probability of crossover (*Pc*). If this random number is smaller than the probability of crossover, the Simulated Binary crossover (SBX) proposed by Deb and Agrawal [55] is used. Otherwise the Simplex crossover (SPX) developed by Da Ronco and Benini [56] is used. To ensure a good exploration of the search space and avoid convergence towards a local optimum value, a newly developed mutation operator named DRM proposed by Chuang et al. [57] was used. A recent technique of constraint-handling named Violation Constraint-Handling method (VCH) introduced by Chehouri et al. [58] was used in the selection steps during the evolution process. The algorithm configuration of the modified RCGA used in this work is shown in Figure 3.

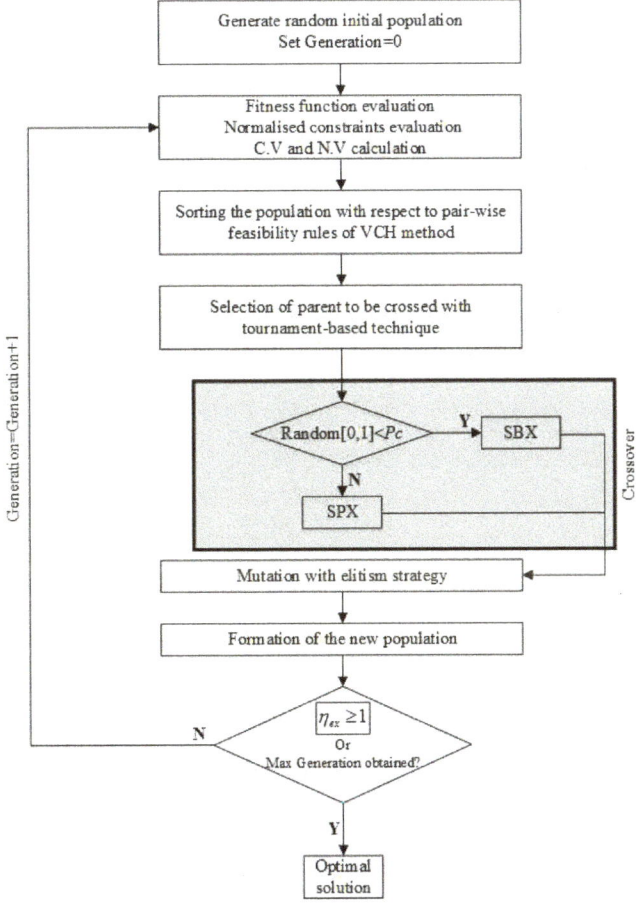

Figure 3. Proposed Real Coded Genetic Algorithm flow chart.

In this algorithm, normalizing an inequality constraint consists of transforming it to ensure that it admits 1 for maximum as presented in Section 4.2. Thus, the normalized constraints used in step 2 in Figure 3, can be expressed as follows:

$$G_1(\vec{X}) = \frac{10^4}{P_{elc,G}} - 1 \tag{35}$$

$$G_2(\vec{X}) = \frac{t_{FP}}{43{,}200} - 1 \tag{36}$$

$$G_3(\vec{X}) = \frac{7200}{t_{DP}} - 1 \tag{37}$$

$$G_4(\vec{X}) = \frac{T_{m,HP}^{in} + 5}{T_{hw}^{in}} - 1 \tag{38}$$

$$G_5(\vec{X}) = \frac{T_{m,LP}^{in} + 5}{T_{hw}^{in}} - 1 \tag{39}$$

$$G_6(\vec{X}) = \frac{P_{c,n}^{out}}{300 \times 10^5} - 1 \tag{40}$$

$$G_7(\vec{X}) = \frac{2 \times \dot{m}_{hw} \times t_{DP}}{n \times \dot{m}_{cw} \times t_{FP}} - 1 \tag{41}$$

Chehouri et al. [58] define Constraint violation factor ($C.V$) and number of violation ($N.V$) that are evaluated here in step 2 for each individual (chromosome) of considered population of possible solutions as follows:

$$C.V = \sum_{l=1}^{7} max(0, G_l) \tag{42}$$

$$N.V = \frac{\text{number of violated constraints}}{7} \tag{43}$$

The pair-wise feasibility rules used in step 3 of Figure 3 separate the population into two families; feasible solutions and unfeasible consisting of individuals that violate at least one of seven constraints. The family of feasible solutions is sorted with respect to their fitness value (exergy efficiency) in descending order. The second family is sorted according to these rules:

- If two considered chromosomes are infeasible, the best is the one with the lowest Number of Violations ($N.V$).
- If both chromosomes have the same ($N.V$), the one with the lowest Constraints Violation ($C.V$) value is the best.

These sorted unfeasible solutions are placed after the sorted feasible solutions and the resulting sorted population is then used in selection of parents to be crossed. The first NKeep individuals of this sorted population are kept to form the mating pool and the rests are discarded and replaced by offspring of parents selected randomly in this mating pool. Since the mating pool is sorted, the tournament selection approach used in this work consists of selecting randomly two chromosomes from the mating pool, the chromosome with the lowest rank becomes a parent. The tournament repeats for every parent needed (twice for SBX and once for SPX). The first individual of mating pool must not be altered by any evolutionary operator (elitism strategy).

5. Validation of the Thermodynamic Model

The storage system we propose in this work has not been constructed in reality. It is inspired from an experimental prototype existing in our lab and schematically illustrated in (Figure 4), due to the poor efficiency 3.4% obtained by experimentally measuring the total electrical energy produced during the discharge phase (0.45 kW h) and consumed by the compressor during the charge phase (13.12 kW h). In this experimental prototype, a three-stage compressor (cylinders) is used to produce compressed air. These compression cylinders (a) are separated by intercoolers (b) and the compressed air produced is stored in six storage tanks (c) having a total volume of 300 L. The maximum pressure in storage tanks is set at 180 bar to maintain the compression ratios of the three cylinders constant throughout the storage phase. The minimum pressure in storage tanks is limited at 16 bar to ensure good regulation of mass flow rate of expander (d). A fan (f) driven by the electric driving motor (M) of the compressor, stirs the ambient air to cool the compressed air passing through the intercoolers. The rated air mass flow rate of the compressor is 14.4 kg h^{-1} for the rated power of driven-motor of 4 kW. The pressure and temperature of air at the inlet and outlet of each compression cylinders are acquired. A pneumatic motor (e) coupled to electric generator (G) is used to achieve the expansion process and generate electricity. The inlet pressure and mass flow rate of pneumatic motor are adjusted by expander (d), its rated air mass flow rate is 97.2 kg h^{-1} and inlet air pressure is 8 bar for a rated output power of 1.2 kW. Because the fan also serves as a flywheel for the drive of the compressor, the ambient air that it stirs also cools the compression cylinders. The measured temperatures cannot fit the model results because of this compression cylinders cooling. That is why, we just use this prototype to validate the model for the filling and the discharge of the tank. The pressure sensor is used for the acquisition of pression inside the storage tanks during charge and discharge process. Figure 5 shows the comparison of the experimental results and the simulation results. The plot in the left panel shows the variation of pressure inside the storage tank with the time during the filling process; the right shows the same physical quantity during the discharge process. A good agreement between the experimental and the simulation results is observed. The rest of the model (compression, expansion) is validated using the data published by Liu et al. [41] where they thermodynamically analyzed a CAES system through an advanced exergetic analysis. We aim to compare the values of the temperature to the different states of their system with those obtained by our model and then to do the same for the specific flow exergy. As shows in Table 3, the differences are quite low (below 2.5%); one may conclude that our model accurately describes both the filling and the discharge process.

Table 3. Comparison between model results and experimental results of filling and discharge time of storage tank.

Parameters	Model Results	Experimental Results	Error (%)
Filling time (s)	14,572	14,874	2.0
Discharge time (s)	1808	1767	2.3

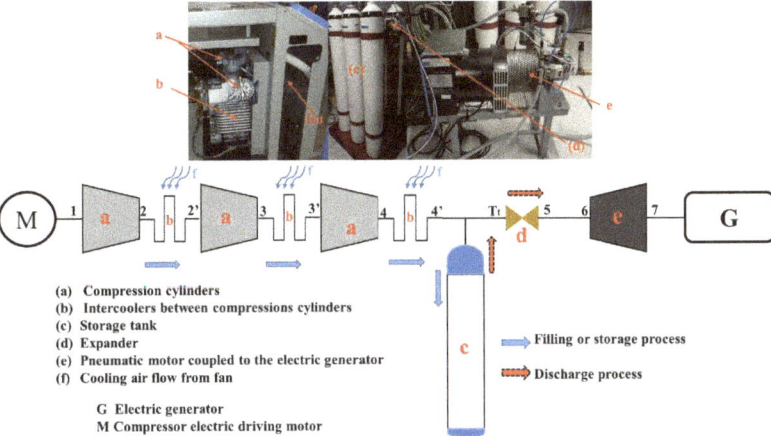

(a) Compression cylinders
(b) Intercoolers between compressions cylinders
(c) Storage tank
(d) Expander
(e) Pneumatic motor coupled to the electric generator
(f) Cooling air flow from fan

G Electric generator
M Compressor electric driving motor

Filling or storage process

Discharge process

Figure 4. Global view and diagram of pilot system.

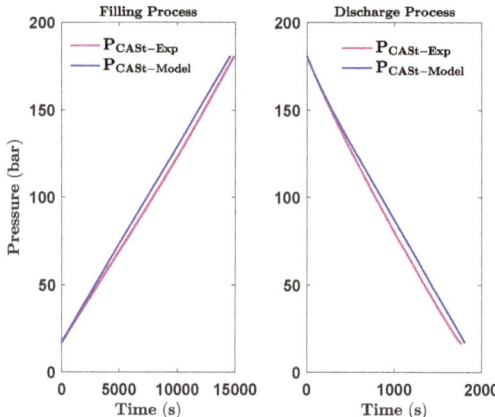

Figure 5. Comparison of model and experimental results for the charge and discharge tests.

The comparison of the temperature and specific flow exergy predicted by our model with previous results presented in Ref. [41] is illustrated in Table 4. We note that the discrepancy is relatively smaller for the compression process than for the expansion process. This can be explained by the fact that we have considered the combustion gases as an ideal gas with the same properties as the air, therefore it could be concluded that the model is accurate.

Table 4. Comparison between obtained results and data published in [41] for compression and expansion process.

State	T (°C)			e_x (kJ·kg^{-1})		
	Present Work	**Reference**	**Error(%)**	**Present Work**	**Reference**	**Error(%)**
1	15.00	15.00	-	0	0	-
2	147.66	148.06	0.27	119.70	128.25	4.31
3	35.00	35	-	110.62	117.27	6.01
4	176.47	177.30	0.47	221.57	241.14	8.83
5	35.00	35	-	183.45	190.86	4.04
6	176.47	177.30	0.47	312.40	314.63	0.71
7	35.00	35	-	274.37	276.37	0.73
8	176.47	177.30	0.47	403.32	406.12	0.70
9	35.00	35.00	-	365.28	363.25	0.56
13	540.00	540.00	-	548.93	552.89	0.72
14	364.68	374.71	2.68	347.52	358.26	3.09
15	957.00	957.00	-	803.68	802.03	0.21
16	461.48	486.06	5.05	203.08	241.37	18.86

6. Results and Discussion

6.1. Optimization Results

In this section, the results of optimization and thermodynamic analysis of optimized system are presented. The simulation code was implemented under MATLAB®. To understand the impact of RCGA parameters to the optimization procedure, we have distinguished six cases with respect to the population size change and change of the number of maximum generation as shown in Table 5; Here *PopSize* is the size of population, *MaxGens* is the maximum number of generations, *Pc* is the probability of crossover and *Pm* is the probability of mutation. In all these cases, the probability of crossover and that of mutation are the same and equal respectively to 60% and 0.5%. The population size in Case 1 is 50 and the maximum number of generations is 100. For more diversity in initial population, the population size is doubled in Case 2 for the same maximum number of generations. In Cases 1 through 5, the population size is held constant at 100 and the evolution time is increased gradually by raising the maximum number of generations from 200 for Case 3 to 1000 for Case 5. Finally, in Case 6, for the purpose of observing the impact of diversity in the initial population on finding an optimal solution, the population size is doubled with the same maximum number of generations as in Case 4 (500).

Table 5. Parameters of modified RCGA for each case.

Parameters	Case 1	Case 2	Case 3	Case 4	Case 5	Case 6
PopSize	50	100	100	100	100	200
MaxGens	100	100	200	500	1000	500
Pc	0.60	0.60	0.60	0.60	0.60	0.60
Pm	0.005	0.005	0.005	0.005	0.005	0.005

The evolutions of the maximum value of the objective function (η_{ex-Max}) in each generation for these six cases have been plotted as show in Figure 6 where *NumGens* is the number of generations.

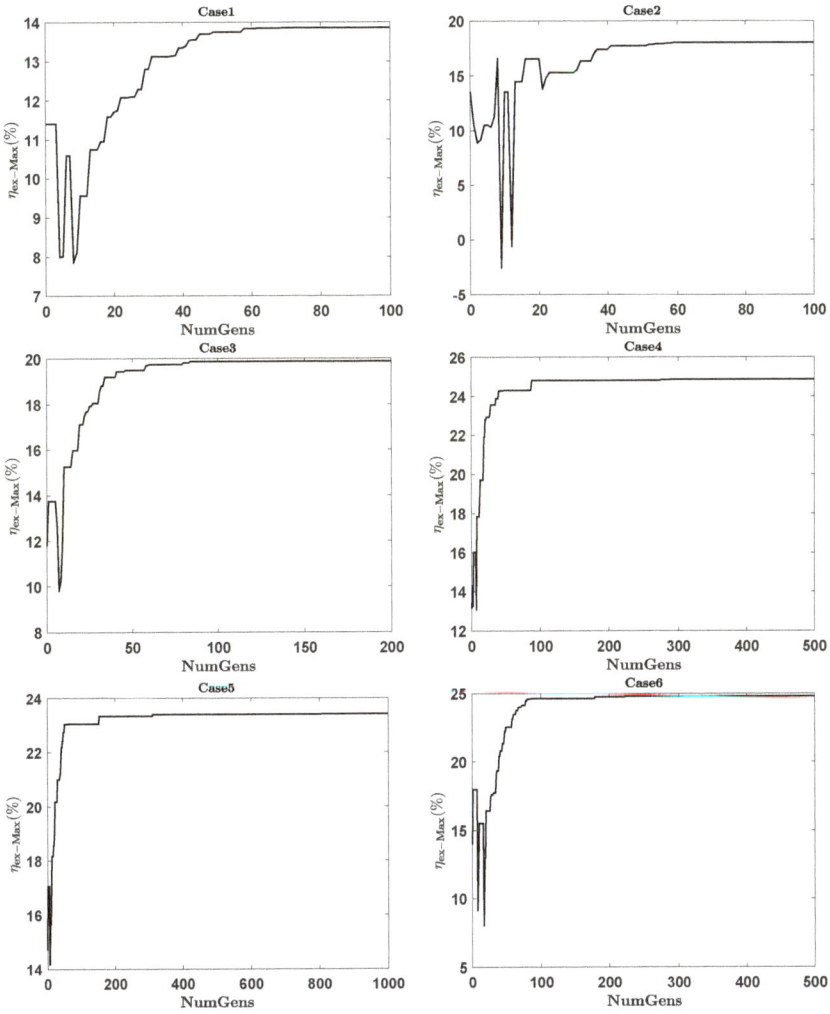

Figure 6. Evolution of the maximum exergy efficiency in each generation.

It can be seen in Case 1 of Figure 6 that, a stable value of 13.85% is reached for minimum generation number of 70, but this value remains stable for at least 30 generations after the plateau is reached. Therefore, we are not sure whether this stable value is the global optimum of the storage system. One possible explanation of this late convergence may be the lack of diversity of the initial population. Therefore, in Case 2, the population size is doubled with the same maximum number of generations. The representative curve of this second case shows that; the population is gradually converging to the stable objective value of 18.02% which is the maximum value of the exergy efficiency reached during all the evolution process. Also, this value is reached for minimum generation number of 72 just 28 generation before the maximum number of predefined generations. Therefore, it is not certain that this maximum exergy efficiency of 18.02% is the optimum solution because the evolution process may not be completed. Therefore, in Case 3, the maximum generation number is doubled with the same population size as in Case 2 to maintain good diversity in the population and a sufficient evolution time to expect a convergence towards the overall optimum. The result of this case shows that the population is converging to a stable value of exergy efficiency. However, it is not sure whether this maximum stable value of 19.88% is the global optimum since, as can be seen, by doubling the evolution times ($NumGens = 200$), the maximum value of exergy efficiency is increased by two percentage points (from 18.02% to 19.88%). That is why, in the fourth case, the evolution time has increased fivefold ($NumGens = 500$) while keeping the same diversity in the initial population as in the third case ($PopSize = 100$).The result of this fourth case shows that the population is converging to a stable value of maximum exergy efficiency that increases by five percentage points to 24.87% in Case 4, compared with 19.88% in case 3. This maximum value of 24.87% remains constant over the last 210 generations during evolution process. It is likely that the algorithm has converged to the global optimum solution. To make sure of that, we wanted to see whether increasing the evolution time could be able to have an impact on the maximum value of the exergy efficiency. Therefore, in the fifth case, the maximum number of generations is doubled ($NumGens = 1000$) with the same population size as in the previous case. As shown in the representative curve of the evolution process of this fifth case, the population converges to a stable value but this stable value of 23.40% is unfortunately lower than that obtained in Case 4. Since an evolution time beyond 500 generations does not have a relevant impact on the maximum value of exergy efficiency, in the last case, the population size used in the fourth case is doubled (for more diversity in the initial population) with the same maximum number of generations. The stable value of maximum exergy efficiency of 24.81% obtained at the end of evolution process was indeed close to that reached in Case 4. Nevertheless, this value remains lower than 24.87% of Case 4 which is certainly the global optimum of this optimization problem.

To return to Case 4, the observation of the final population at the end of the evolution process shows that, all the chromosomes (sets of design parameters) are identical. This means that almost all the chromosomes of the population have converged to the optimum solution. We have then deduced the set of optimal values of influencing parameters that maximizes the global exergy efficiency of the SS-CAES system without fossil fuel used. These optimal parameters are given in Table 6.

Table 6. Optimal parameters from optimization.

Decision Variable	Optimum Value	Decision Variable	Optimum Value
n	3	η_{IsC} (%)	75
π	3.8	$\eta_{Is,m}$ (%)	90
V_t (m^3)	30	η_{mC} (%)	75
β_{HP}	8	η_{mm} (%)	90
β_{LP}	5.2	$\dot{m}_c$ (kg·s^{-1})	0.0156
$T_{m,HP}^{in}$ (°C)	28.71	$\dot{m}_m$ (kg·s^{-1})	0.066
$T_{m,LP}^{in}$ (°C)	28.66	–	–

Using these optimal parameters, the thermodynamic properties of each point in the optimized system (Figure 7) are shown in Table 7. The performance indicator of the optimized system is calculated and shown in Table 8.

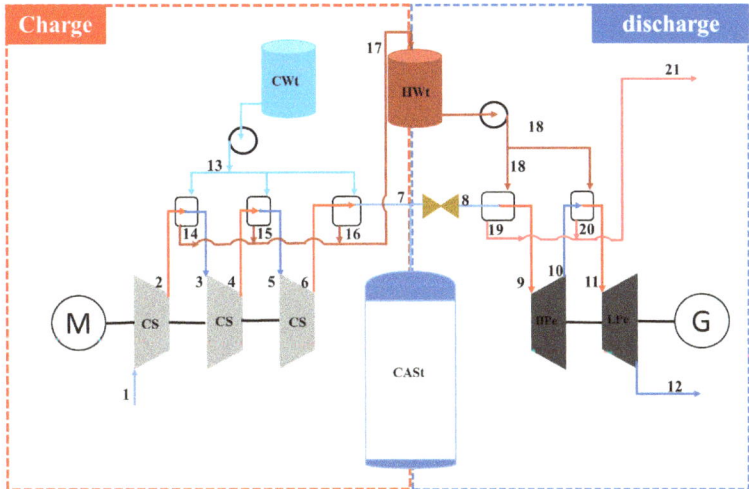

Figure 7. Schematic diagram of optimized system.

Table 7. Thermodynamic properties of each point of the optimized system.

State	Stream	$\dot{m}$ (kg·s^{-1})	T (°C)	P (Bar)	h (kJ·kg^{-1})	s (kJ·kg^{-1}·K^{-1})	e_x (kJ·kg^{-1})
1	Air	0.0156	20	1.01	293.32	6.84	0
2	Air	0.0156	200	3.84	475.55	6.94	152.61
3	Air	0.0156	35	3.84	308.38	6.50	112.69
4	Air	0.0156	223.83	14.58	499.91	6.61	274.56
5	Air	0.0156	35	14.58	308.38	6.12	225.01
6	Air	0.0156	223.83	55.42	499.91	6.22	386.88
7	Air	0.0156	35	55.42	308.38	5.74	337.33
8	Air	0.066	17.57	41.60	290.89	5.76	312.83
9	Air	0.066	28.71	41.60	302.06	5.80	312.95
10	Air	0.066	−94.31	5.20	178.34	5.87	167.87
11	Air	0.066	28.71	5.20	302.06	6.40	138.00
12	Air	0.066	−73.66	1	199.46	6.45	18.40
13	Water	0.0499	20	1.01	83.95	0.30	0
14	Water	0.0499	32.50	1.01	136.19	0.47	1.07
15	Water	0.0499	34.32	1.01	143.81	0.50	1.41
16	Water	0.0499	34.32	1.01	143.81	0.50	1.41
17	Water	0.1497	33.71	1.01	141.27	0.49	1.29
18	Water	0.2981	33.71	1.01	141.27	0.49	1.29
19	Water	0.2981	33.12	1.01	138.80	0.48	1.18
20	Water	0.2981	27.16	1.01	113.89	0.40	0.35
21	Water	0.5963	30.14	1.01	126.35	0.44	0.69

As can be seen in Table 8, under optimal design conditions, the compressors take 8.65 h to fill the CASt while the discharge time of the CASt is equal to 2.02 h. The electric power produced by the generator is 12.77 kW which is well above the electrical energy requirements of an average household (of about 4 individuals). The optimized system also produces 4.64 t of hot water at a temperature of about 30 °C during each cycle of charge/discharge which can be used for heating purposes in a house. The RTE of the optimized system is equal to 79.07% and its exergy efficiency is 24.87%. By calculating the contribution of thermal energy to RTE (CTEtoRTE, Equation (44)), we can see that it represents about 69% of RTE. Unfortunately, the low-temperature hot water containing this thermal energy does not necessarily have the same value as the electrical energy produced during the discharge process. This is one more reason to use the overall exergy efficiency as a performance evaluation criterion of SS-CAES system proposed in this work.

$$CTEtoRTE = \frac{\dot{Q}_h t_{FP} - \dot{Q}_{Rh} t_{DP}}{P_{elc,G} t_{DP} + (\dot{Q}_h t_{FP} - \dot{Q}_{Rh} t_{DP})} \tag{44}$$

Table 8. Results of thermodynamic simulation.

Parameters	Unit	Value
Charge time (t_{FP})	Hour	8.65
Discharge time (t_{DP})	Hour	2.02
$P_{elc,c}$	kW	12.00
$P_{elc,G}$	kW	12.77
$\dot{Q}_h$	kW	8.58
$\dot{Q}_{Rh}$	kW	8.90
m_{hw}	Ton	4.64
RTE	%	79.07
η_{ex}	%	24.87

To identify the locations and magnitudes of storage process inefficiencies, the total exergy destruction of the charge/discharge cycle for each component as well as their exergy destruction ratio are listed in Table 9.

Table 9. Total exergy destruction and exergy destruction ratio of each component for a complete cycle of charge and discharge.

Component	Exergy Destruction (kWh)	Exergy Loss (kWh)	Exergy Destruction Ratio (%)
Compression stages	39.52	...	53.04
Expansion stages	9.50	...	12.75
Intercoolers	17.09	...	22.94
Heaters	4.61	...	6.18
CASt	3.79	...	5.09
Overall system	74.51	3.50	100

As shown in Table 9, the compression stage has the largest exergy destruction, followed by the intercoolers. This can be explained by the fact that, the temperature difference between the inlet and the outlet of each stage is large enough so that the heat transfer to the compression stages walls is no longer negligible compared to the enthalpy change. Another explanation could be given by the advanced exergetic analysis [59,60]. It would certainly indicate that the unavoidable exogenous part of the exergy destruction within these components is important. Indeed, as shown in Table 6, the design of compressor is supposed to be perfect because of the higher values of efficiency (isentropic and mechanical). Thus, the unavoidable endogenous part of the exergy destruction within the compressor is low compared with the unavoidable exogenous part.

6.2. Distribution of Design Parameters

The behavior between lower and upper bounds (dotted lines in Figures 8 and 9) of each optimal design parameters during the evolution is analyzed in this section. The results of this are shown in Figures 8 and 9).

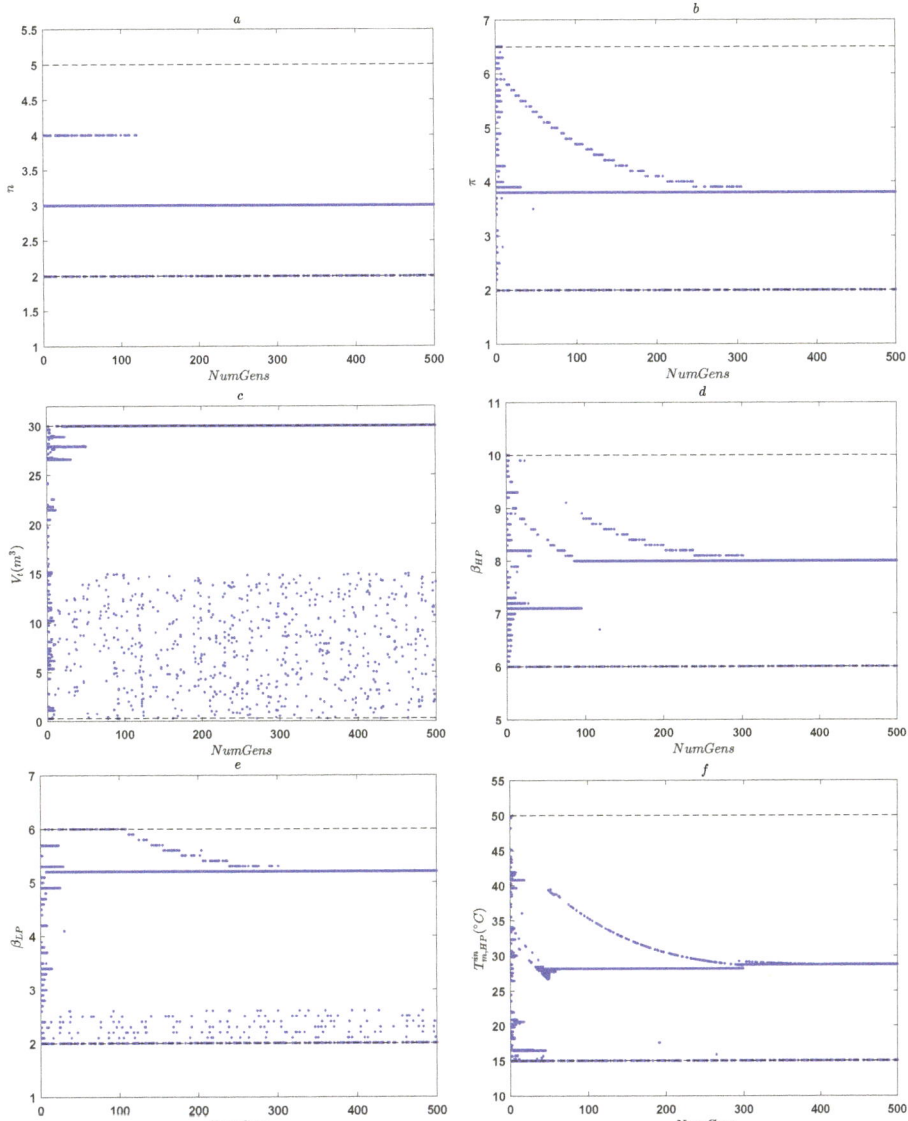

Figure 8. Scattering of optimal design variables during evolution; the case of the number of compression stage (**a**), of the compression pressure ratio (**b**), of the volume of air storage tank (**c**), of the pressure ratio of the high-pressure expansion stage (**d**), of the pressure ratio of the low-pressure expansion stage (**e**) and of the inlet temperature of the high-pressure expansion stage (**f**).

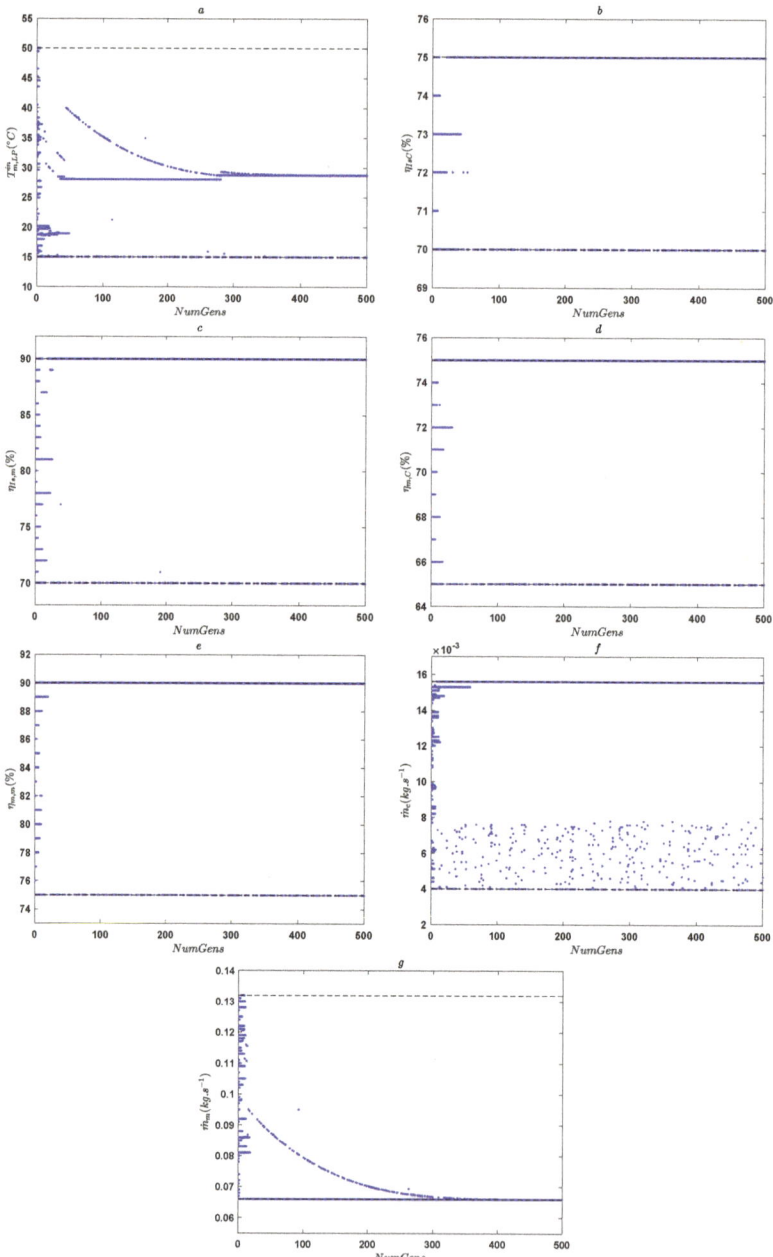

Figure 9. Scattering of optimal design variables during evolution; the inlet temperature of the low-pressure expansion stage (**a**), of the isentropic efficiency of compressor (**b**), of the isentropic efficiency of pneumatic motor (**c**), of the mechanical efficiency of compressor (**d**), of the mechanical efficiency of pneumatic motor (**e**) of the compressor motor mass flow rate (**f**) and of the pneumatic motor mass flow rate (**g**).

According to this distribution, it can be seen in Figure 9g that the pneumatic motor mass flow rate reaches its minimum value. This means that the decreased value of this design variable could improve the global exergy efficiency of system. To reduce the number of variables and thus to speed up the optimization script, the lower born value of the pneumatic motor mass flow rate can be selected and set as a constant.

As also shown in Figure 9b–e concerning respectively the distribution in the population during the evolution process, of the values taken by the isentropic efficiency of compressor, by the isentropic efficiency of pneumatic motor, by the mechanical efficiency of compressor and by the mechanical efficiency of pneumatic motor; during the evolution process, these four design parameters only take almost exclusively their maximum value. This means that, the increased value of these designs variables could improve the global exergy efficiency of system. Similarly, the upper born value of these four designs parameters can be selected and set as a constant.

Other design variables have the scattered distribution which mean that, their variations could have some significant impact on the system performance. To have an idea of these impacts on both global exergy efficiency and number of violated constraints at the optimal point, the sensitivity analyses have been made.

6.3. Effect of Variation of the Design Variables Value on the System Efficiency (Global Exergy Efficiency) and on the Number of Violated Constraints

In this section, we study the effect of variation of the design variables value on the system efficiency and on the number of violated constraints. For this purpose, we vary the value of each design variable in its allowable range specified in Table 2.

6.3.1. Number of Compression Stages and the Compression Ratio Values

Figure 10a shows the decrease of the global exergy efficiency with the increasing number of compression stages at fixed optimal compression ratio. The same trend is observed in Figure 10b for the compression ratio at fixed optimal number of compression stages but, unfortunately, the compression ratio values which provide a maximum exergy efficiency violate some imposed constraints. That is why, as shown in Figure 10c,d, the optimum values of these two design parameters do not violate any constraint.

Thus, for a SS-CAES system, it is not necessary to use more than three compression stages. Furthermore, the compression ratio value of these compression stages should not exceed four.

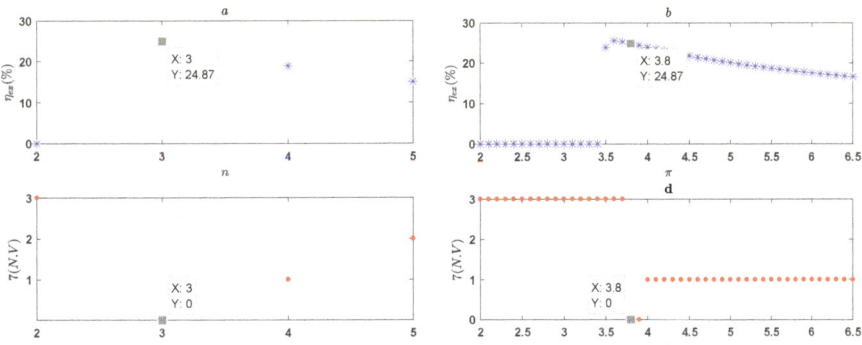

Figure 10. Variation of global exergy efficiency and number of violated constraints with number of compression stages (**a**,**c**) and compressor pressure ratio (**b**,**d**) at optimal point.

6.3.2. Volume of the Air Storage Tank

As can be seen in Figure 11a, the increase in volume of the air storage tank results in an increase of the global exergy efficiency. However, before the optimal value is reached, all other values violate one constraint as shown in Figure 11c. This means that if it had been possible to have a larger tank, the system efficiency would have been greater than the optimal value obtained. However, for a given SS-CAES system, a trade-off must be found between the constraints of space, of charging time, of discharge time and even of the cost of purchasing the storage tanks. Therefore, the optimal value of the air storage tank volume is closely linked to the imposed constraints of the system.

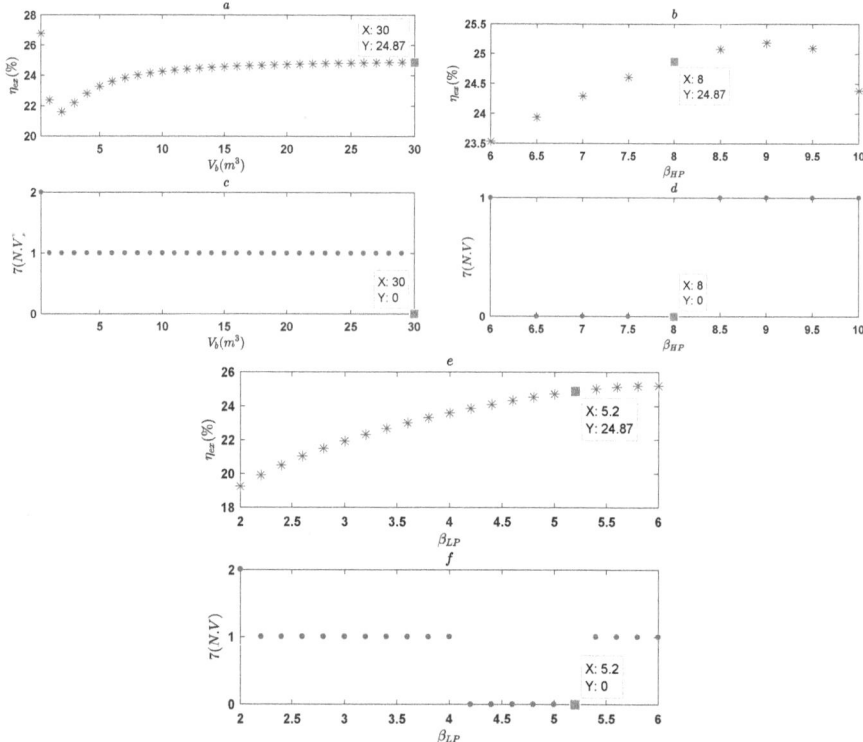

Figure 11. Variation of global exergy efficiency and number of violated constraints with volume of air storage tank (**a,c**), pressure ratio of high-pressure expansion stage (**b,d**) and low-pressure expansion stage (**e,f**) of pneumatic motor at optimal point.

6.3.3. Pressure Ratio Values of High-Pressure and the Low-Pressure Expansion Stages of Pneumatic Motor

As it is shown in Figure 11b,e, the exergy efficiency increases almost linearly with both the pressure ratio of high-pressure and the low-pressure expansion stages of pneumatic motor. However, when the value of pressure ratio of high-pressure expansion stage is greater than 9, the global exergy efficiency begins to decrease (Figure 11b). This can be explained by the low temperature taken by the air at the exit of such expansion stage. In fact, the temperature of hot water produced during compression process is not high enough. Thus, the preheating of air between the expansion stages will no longer be enough to improve the expansion work of the low-pressure stage. Finally, imposed constraints (the minimum power delivered by the storage system during discharge process and the minimum difference between the hot water temperature and the inlet temperature of air of expansion stages) justify the optimal values returned by the optimization algorithm (Figure 11d,f).

For a SS-CAES system using pneumatic motor as expansion system, it is possible to increase its efficiency by using the pressure ratio values of 9 and 6 for the high-pressure and the low-pressure expansion stages, respectively. However, it should be considered in this case to use a water heating system (like solar water heaters) outside of the SS-CAES system. This water heating system would be used to increase the temperature of the hot water produced during the compression process.

6.3.4. Inlet Temperature Values of High-Pressure and the Low-Pressure Expansion Stages of Pneumatic Motor

As shown in Figure 12a,b, the exergy efficiency increases linearly with both the inlet temperature of high-pressure and low-pressure expansion stages of pneumatic motor. Moreover, the optimal values of these two design parameters returned by the GA are the last ones which do not violate any of the imposed constraints (Figure 12c,d). If the temperature of the hot water had been higher, the optimal values of these two variables would have been higher too.

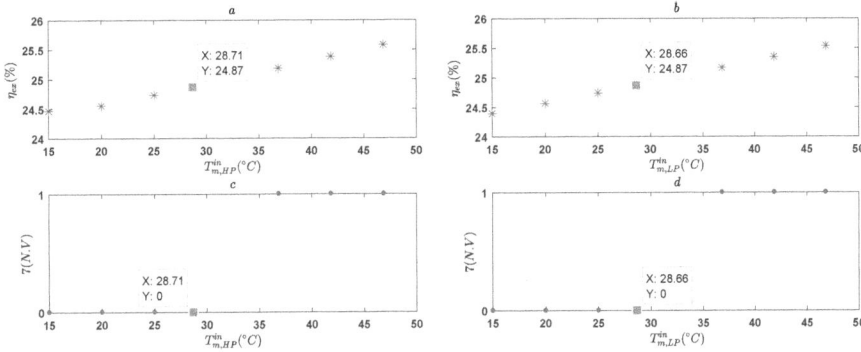

Figure 12. Variation of global exergy efficiency and number of violated constraints with inlet temperature of high-pressure (**a,c**) and low-pressure (**b,d**) expansion stages of pneumatic motor at optimal point.

For a given SS-CAES, only thermomechanical stresses should limit the supply temperature of the expansion stages. Thus, when designing a SS-CAES using pneumatic air motor, consideration should be given, if necessary, to an external water heating system. In that case, during the preheating of the supplying air of the expansion stages during the discharge process, it is possible to move closer to the maximum permissible temperature at the inlet of these stages.

6.3.5. Compressor and Pneumatic Motor Isentropic Efficiency Values, of the Compressor and Pneumatic Motor Mechanical Efficiency Values

As it is shown in Figures 13a,b and 14a,b respectively, the increase in the isentropic and mechanical efficiency of compressor and pneumatic motor leads to a linear increase of the exergy efficiency. Furthermore, it is important to note that this increase is done without any violation of constraints (Figures 13c,d and 14c,d). Therefore, the maximum values of these four design parameters can be selected and set as constants in the cost function for optimization.

We should remember, however, that these results were predictable since for an energy system having turbomachines, the higher is the isentropic efficiency (or mechanical efficiency), the higher will be the system efficiency. Nevertheless, by including these four parameters among the optimization parameters, we are assessing the smooth functioning of the modified RCGA proposed in this work.

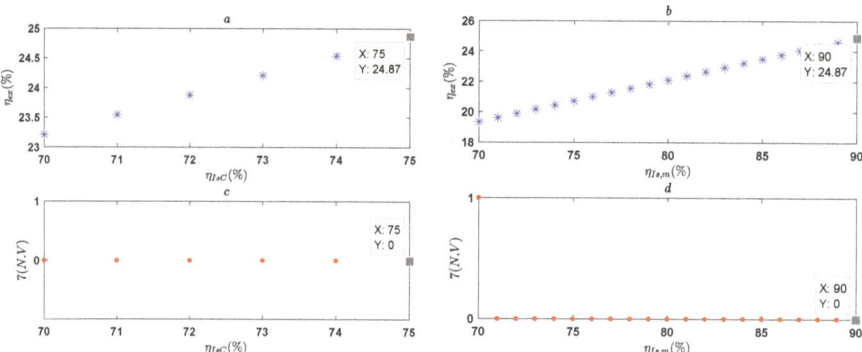

Figure 13. Variation of global exergy efficiency and number of violated constraints with compressor (**a,c**) and pneumatic motor (**b,d**) isentropic efficiency at optimal point.

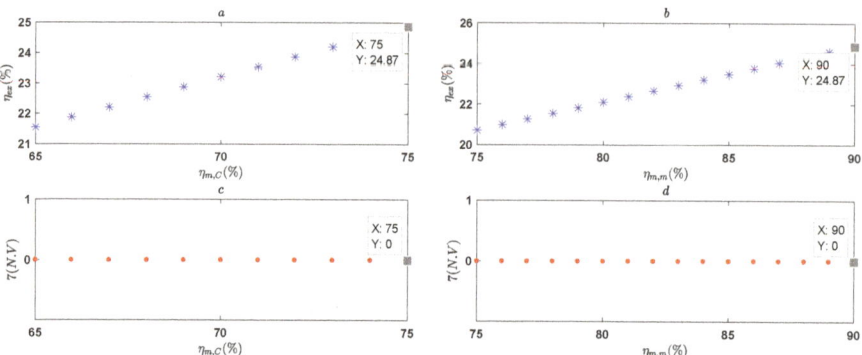

Figure 14. Variation of global exergy efficiency and number of violated constraints with compressor (**a,c**) and pneumatic motor (**b,d**) mechanical efficiency at optimal point.

6.3.6. Compressor and Pneumatic Motor Mass Flow Rate Values

Figure 15a shows that the global exergy efficiency increases with the compressor mass flow rate. In addition, the single value that does not violate any constraint is that returned by the optimization algorithm (Figure 15c). Unfortunately, its scattered distribution (Figure 9f) does not allow one to set it as constant during the optimization procedure. Nevertheless, when designing a SS-CAES system, it would be appropriate to give preference to a compressor with a high mass flow rate.

For the pneumatic motor mass flow rate, its increase leads to a decrease in global exergy efficiency (Figure 15b) and as in the case of compressor mass flow rate, the optimum value of this parameter returned by optimization algorithm is the only one that does not violate any constraint (Figure 15d). In contrast to the scattered distribution of the compressor mass flow rate, a homogeneous convergence towards the minimum value can be noted in Figure 9g. Therefore, this minimum value of the pneumatic motor mass flow rate can be selected and set as a constant in the optimization algorithm. When designing a SS-CAES system, it would be appropriate to give preference to a pneumatic motor with a low mass flow rate.

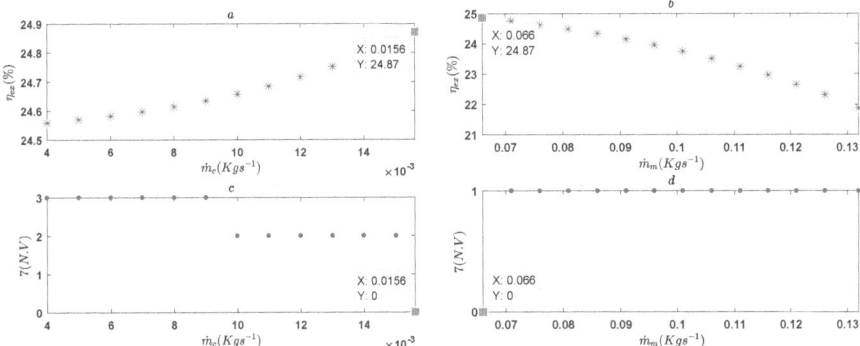

Figure 15. Variation of global exergy efficiency and number of violated constraints with compressor (a,c) and pneumatic motor (b,d) mass flow rate at optimal point.

7. Conclusions

This paper has presented the energy and exergy analyses of a SS-CAES system without fossil fuel used. The resulting thermodynamic model is fed to a modified RCGA to identify the optimal values of thirteen design parameter of the proposed storage system. The modified RCGA that has been clearly presented and tested to verify its stability and robustness. The results of the optimization indicate that for a maximum efficiency of a SS-CAES system using multistage reciprocating compressor and two stages pneumatic air motor:

- The number of compression stage should be less than three.
- The compressor pression ratio of each compression stages should be less than four.
- The maximum value of air storage tank volume allowed by the constraints of spaces, of cost, of charge time and of discharge time should be preferred.
- The pressure ratio together with the temperature of the supplying air of the expansion stages are highly dependent on the temperature of the hot water produced during compression process. However, the use of the maximum technologically acceptable values of these four parameters improve the efficiency of the storage systems.
- A low mass flow of the pneumatic air motor coupled with a high mass flow rate of the compressor improves the efficiency of the storage system.

The presented modified RCGA can be used for optimizing all scale of CAES system under all kinds of constraints if the optimization problem is well formulated.

Author Contributions: Conceptualization, L.L. and T.G.; Methodology, T.G. and L.L.; Software, T.G. and D.T.; Validation, M.T., L.L. and T.G.; Formal Analysis, T.G. and L.L ; Investigation, D.T. and T.G.; Resources, M.T. and L.L.; Data Curation, T.G.; Writing-Original Draft Preparation, T.G.; Writing-Review & Editing, L.L. and T.G.; Visualization, T.G.; Supervision, L.L.; Project Administration, L.L.; Funding Acquisition, L.L. and M.T.

Funding: This research was funded by the European Commission H2020 MSCA programme, for the EU H2020-MSCA-RISE-2016-734340-DEWCOOL-4-CDC project, EU Commission.

Acknowledgments: The authors gratefully acknowledge the financial support of Erasmus Mundus Action 2 CARIBU. Also, the authors want to give their thanks to the support from CPER PREVER project for the experimental prototype.

Conflicts of Interest: The authors declare no conflict of interest.

Notations

Variable	Meaning	Dimension
	Symbols	
A	Cross section	m^2
C_p	Specific heat capacity at constant pressure	$J \cdot kg^{-1} \cdot K^{-1}$
e_x	Specific flow exergy	$kJ \cdot kg^{-1}$
$\dot{Ex}^Q$	Time rate of exergy transfer	kW
g	Acceleration of gravity	$m \cdot s^2$
h	Specific enthalpy	$kJ \cdot kg^{-1}$
H	Heat transfer coefficient	$kW \cdot m^2 \cdot K^{-1}$
m	Mass	kg
$\dot{m}$	Mass flow rate	$kg \cdot s^{-1}$
n	Number of compression stage	$-$
P_0	Pressure of reference environment	bar
P	Pressure	bar
Pc	Probability of crossover	$-$
Pm	Probability of mutation	$-$
P_{elc}	Electric power	kW
$\dot{Q}$	Heat transfer rate	kW
R	Gas constant	$kJ \cdot kg^{-1} \cdot K^{-1}$
s	Specific entropy	$kJ \cdot kg^{-1} \cdot K^{-1}$
S	Area of heat transfer	m^2
T_0	Temperature of reference environment	$°C$
T	Temperature	$°C$
t	Time	s
V_t	volume of air storage tank	kg^3
	Greek symbols	
β_{HP}	high-power pneumatic motor pressure ratio	$-$
β_{LP}	low-power pneumatic motor pressure ratio	$-$
η_{IsC}	compressor isentropic efficiency	$-$
$\eta_{Is,m}$	pneumatic motor isentropic efficiency	$-$
η_{ex}	exergy efficiency	$-$
η_{mC}	compressor mechanical efficiency	$-$
η_{mm}	pneumatic motor mechanical efficiency	$-$
η_{ref}	reference efficiency of a virtual thermal power plant	$-$
π	compressor pressure ratio	$-$

Variable	Meaning
	Subscripts
c	Compressor
cw	cooling water
DP	Discharge Process
FP	Filling Process
G	Generator
h	hot
HP	High power
hw	heating water
i	number of the compression stage
Is	Isentropic
LP	low power
m	motor
n	last compression stage
Rh	Reheat
t	tank

Variable	Meaning
	Superscripts
b	boundary
D	Destruction
in	inlet
j	number of the parameter
L	Lower bound
l	numbers of normalized constraints
Max	Maximum
out	outlet
Q	heat
U	Upper bound
W	Work

Abbreviations

AA-CAES	Advanced Adiabatic Compressed-Air Energy Storage
CAES	Compressed-Air Energy Storage
CASt	Compressed-Air Storage tank
CC	Combustion Chamber
CS	Compression Stage
C.V	Constraint Violation factor
CTEtoRTE	Contribution of Thermal Energy to Round-Trip Efficiency
DRM	Dynamic Random Mutation
G	electric Generator
GA	Genetic Algorithm
HE	Heat Exchanger
HPe/LPe	High- and Low-power expansion stage
HWt	Hot Water tank
M	electric drive Motor
MaxGens	maximum number of generations
N.V	Number of Violation
PopSize	size of population
RCGA	Real Coded Genetic Algorithm
RTE	Round-Trip Efficiency
SBX	Simulated Binary crossover
SPX	Simplex crossover
SS-CAES	Small-Scale Compressed-Air Energy Storage
VCH	Violation Constraint-Handling method

References

1. Zakeri, B.; Syri, S. Electrical energy storage systems: A comparative life cycle cost analysis. *Renew. Sustain. Energy Rev.* **2015**, *42*, 569–596. [CrossRef]
2. Mahlia, T.; Saktisahdan, T.; Jannifar, A.; Hasan, M.; Matseelar, H. A review of available methods and development on energy storage; technology update. *Renew. Sustain. Energy Rev.* **2014**, *33*, 532–545. [CrossRef]
3. Sciacovelli, A.; Li, Y.; Chen, H.; Wu, Y.; Wang, J.; Garvey, S.; Ding, Y. Dynamic simulation of Adiabatic Compressed Air Energy Storage (A-CAES) plant with integrated thermal storage—Link between components performance and plant performance. *Appl. Energy* **2017**, *185*, 16–28. [CrossRef]
4. Chen, L.; Zheng, T.; Mei, S.; Xue, X.; Liu, B.; Lu, Q. Review and prospect of compressed air energy storage system. *J. Mod. Power Syst. Clean Energy* **2016**, *4*, 529–541. [CrossRef]
5. Evans, A.; Strezov, V.; Evans, T.J. Assessment of utility energy storage options for increased renewable energy penetration. *Renew. Sustain. Energy Rev.* **2012**, *16*, 4141–4147. [CrossRef]
6. Yang, C.J.; Jackson, R.B. Opportunities and barriers to pumped-hydro energy storage in the United States. *Renew. Sustain. Energy Rev.* **2011**, *15*, 839–844. [CrossRef]

7. Hedegaard, K.; Meibom, P. Wind power impacts and electricity storage—A time scale perspective. *Renew. Energy* **2012**, *37*, 318–324. [CrossRef]

8. Chen, H.; Cong, T.N.; Yang, W.; Tan, C.; Li, Y.; Ding, Y. Progress in electrical energy storage system: A critical review. *Prog. Nat. Sci.* **2009**, *19*, 291–312. [CrossRef]

9. Jannelli, E.; Minutillo, M.; Lavadera, A.L.; Falcucci, G. A small-scale {CAES} (compressed air energy storage) system for stand-alone renewable energy power plant for a radio base station: A sizing-design methodology. *Energy* **2014**, *78*, 313–322. [CrossRef]

10. Kim, Y.M.; Lee, J.H.; Kim, S.J.; Favrat, D. Potential and Evolution of Compressed Air Energy Storage: Energy and Exergy Analyses. *Entropy* **2012**, *14*, 1501–1521. [CrossRef]

11. Luo, X.; Wang, J.; Dooner, M.; Clarke, J. Overview of current development in electrical energy storage technologies and the application potential in power system operation. *Appl. Energy* **2015**, *137*, 511–536. [CrossRef]

12. Rehman, S.; Al-Hadhrami, L.M.; Alam, M.M. Pumped hydro energy storage system: A technological review. *Renew. Sustain. Energy Rev.* **2015**, *44*, 586–598. [CrossRef]

13. Marano, V.; Rizzo, G.; Tiano, F.A. Application of dynamic programming to the optimal management of a hybrid power plant with wind turbines, photovoltaic panels and compressed air energy storage. *Appl. Energy* **2012**, *97*, 849–859. [CrossRef]

14. Cazzaniga, R.; Cicu, M.; Rosa-Clot, M.; Rosa-Clot, P.; Tina, G.; Ventura, C. Compressed air energy storage integrated with floating photovoltaic plant. *J. Energy Storage* **2017**, *13*, 48–57. [CrossRef]

15. Tong, S.; Cheng, Z.; Cong, F.; Tong, Z.; Zhang, Y. Developing a grid-connected power optimization strategy for the integration of wind power with low-temperature adiabatic compressed air energy storage. *Renew. Energy* **2018**, *125*, 73–86. [CrossRef]

16. Sadreddini, A.; Fani, M.; Aghdam, M.A.; Mohammadi, A. Exergy analysis and optimization of a CCHP system composed of compressed air energy storage system and ORC cycle. *Energy Convers. Manag.* **2018**, *157*, 111–122. [CrossRef]

17. Alami, A.H.; Aokal, K.; Abed, J.; Alhemyari, M. Low pressure, modular compressed air energy storage (CAES) system for wind energy storage applications. *Renew. Energy* **2017**, *106*, 201–211. [CrossRef]

18. Arabkoohsar, A.; Machado, L.; Koury, R. Operation analysis of a photovoltaic plant integrated with a compressed air energy storage system and a city gate station. *Energy* **2016**, *98*, 78–91. [CrossRef]

19. Huang, Y.; Keatley, P.; Chen, H.S.; Zhang, X.J.; Rolfe, A.; Hewitt, N. Techno-economic study of compressed air energy storage systems for the grid integration of wind power. *Int. J. Energy Res.* **2017**, *42*, 559–569. [CrossRef]

20. Zhang, Y.; Yang, K.; Li, X.; Xu, J. Thermodynamic analysis of energy conversion and transfer in hybrid system consisting of wind turbine and advanced adiabatic compressed air energy storage. *Energy* **2014**, *77*, 460–477. [CrossRef]

21. Rabbani, M.; Dincer, I.; Naterer, G. Thermodynamic assessment of a wind turbine based combined cycle. *Energy* **2012**, *44*, 321–328. [CrossRef]

22. Guney, M.S.; Tepe, Y. Classification and assessment of energy storage systems. *Renew. Sustain. Energy Rev.* **2017**, *75*, 1187–1197. [CrossRef]

23. Madlener, R.; Latz, J. Economics of centralized and decentralized compressed air energy storage for enhanced grid integration of wind power. *Appl. Energy* **2013**, *101*, 299–309. [CrossRef]

24. Barbour, E.; Mignard, D.; Ding, Y.; Li, Y. Adiabatic Compressed Air Energy Storage with packed bed thermal energy storage. *Appl. Energy* **2015**, *155*, 804–815. [CrossRef]

25. Grazzini, G.; Milazzo, A. Thermodynamic analysis of CAES/TES systems for renewable energy plants. *Renew. Energy* **2008**, *33*, 1998–2006. [CrossRef]

26. Kere, A.; Goetz, V.; Py, X.; Olives, R.; Sadiki, N. Modeling and integration of a heat storage tank in a compressed air electricity storage process. *Energy Convers. Manag.* **2015**, *103*, 499–510. [CrossRef]

27. Peng, H.; Yang, Y.; Li, R.; Ling, X. Thermodynamic analysis of an improved adiabatic compressed air energy storage system. *Appl. Energy* **2016**, *183*, 1361–1373. [CrossRef]

28. Wolf, D.; Budt, M. LTA-CAES—A low-temperature approach to Adiabatic Compressed Air Energy Storage. *Appl. Energy* **2014**, *125*, 158–164. [CrossRef]

29. Allen, K. CAES: The Underground Portion. *IEEE Trans. Power Appar. Syst.* **1985**, *4*, 809–812. [CrossRef]

30. Raju, M.; Khaitan, S.K. Modeling and simulation of compressed air storage in caverns: A case study of the Huntorf plant. *Appl. Energy* **2012**, *89*, 474–481. [CrossRef]

31. Budt, M.; Wolf, D.; Span, R.; Yan, J. A review on compressed air energy storage: Basic principles, past milestones and recent developments. *Appl. Energy* **2016**, *170*, 250–268. [CrossRef]

32. Facci, A.L.; Sánchez, D.; Jannelli, E.; Ubertini, S. Trigenerative micro compressed air energy storage: Concept and thermodynamic assessment. *Appl. Energy* **2015**, *158*, 243–254. [CrossRef]

33. Li, Y.; Wang, X.; Li, D.; Ding, Y. A trigeneration system based on compressed air and thermal energy storage. *Appl. Energy* **2012**, *99*, 316–323. [CrossRef]

34. Liu, J.L.; Wang, J.H. Thermodynamic analysis of a novel tri-generation system based on compressed air energy storage and pneumatic motor. *Energy* **2015**, *91*, 420–429. [CrossRef]

35. Mei, S.; Wang, J.; Tian, F.; Chen, L.; Xue, X.; Lu, Q.; Zhou, Y.; Zhou, X. Design and engineering implementation of non-supplementary fired compressed air energy storage system: TICC-500. *Sci. China Technol. Sci.* **2015**, *58*, 600–611. [CrossRef]

36. Venkataramani, G.; Parankusam, P.; Ramalingam, V.; Wang, J. A review on compressed air energy storage—A pathway for smart grid and polygeneration. *Renew. Sustain. Energy Rev.* **2016**, *62*, 895–907. [CrossRef]

37. Kanoglu, M.; Dincer, I.; Rosen, M.A. Understanding energy and exergy efficiencies for improved energy management in power plants. *Energy Policy* **2007**, *35*, 3967–3978. [CrossRef]

38. Ahmadi, P.; Dincer, I.; Rosen, M.A. Exergy, exergoeconomic and environmental analyses and evolutionary algorithm based multi-objective optimization of combined cycle power plants. *Energy* **2011**, *36*, 5886–5898. [CrossRef]

39. Ahmadi, P.; Dincer, I. Thermodynamic and exergoenvironmental analyses, and multi-objective optimization of a gas turbine power plant. *Appl. Therm. Eng.* **2011**, *31*, 2529–2540. [CrossRef]

40. Ezzat, M.; Dincer, I. Energy and exergy analyses of a new geothermal–solar energy based system. *Sol. Energy* **2016**, *134*, 95–106. [CrossRef]

41. Liu, H.; He, Q.; Saeed, S.B. Thermodynamic analysis of a compressed air energy storage system through advanced exergetic analysis. *J. Renew. Sustain. Energy* **2016**, *8*, 034101. [CrossRef]

42. Mohammadi, A.; Mehrpooya, M. Exergy analysis and optimization of an integrated micro gas turbine, compressed air energy storage and solar dish collector process. *J. Clean. Prod.* **2016**, *139*, 372–383. [CrossRef]

43. Moran, M.J.; Shapiro, H.N.; Boettner, D.D.; Bailey, M.B. *Fundamentals of Engineering Thermodynamics*; John Wiley & Sons: Hoboken, NJ, USA, 2010.

44. Hartmann, N.; Vöhringer, O.; Kruck, C.; Eltrop, L. Simulation and analysis of different adiabatic Compressed Air Energy Storage plant configurations. *Appl. Energy* **2012**, *93*, 541–548. [CrossRef]

45. Rathore, M.M.; Kapuno, R. *Engineering Heat Transfer*; Jones & Bartlett Publishers: Burlington, MA, USA, 2011.

46. Press, W.H.; Teulolsky, S.A.; Vetterling, W.T.; Flannery, B.P. *Numerical Recipes: The Art of Scientific Computing*; Cambridge University Press: New York, NY, USA, 2007.

47. Cheung, B.C.; Carriveau, R.; Ting, D.S.K. Multi-objective optimization of an underwater compressed air energy storage system using genetic algorithm. *Energy* **2014**, *74*, 396–404. [CrossRef]

48. Dincer, I.; Rosen, M.A. *Exergy: Energy, Environment and Sustainable Development*; Newnes: Oxford, UK, 2012.

49. Kotas, T.J. *The Exergy Method of Thermal Plant Analysis*; Butterworth-Heinemann: London, UK, 1985.

50. Bejan, A.; Tsatsaronis, G.; Moran, M. *Thermal Design and Optimization*; John Wiley & Sons: New York, NY, USA, 1995.

51. Wang, Z.; Xiong, W.; Ting, D.S.K.; Carriveau, R.; Wang, Z. Conventional and advanced exergy analyses of an underwater compressed air energy storage system. *Appl. Energy* **2016**, *180*, 810–822. [CrossRef]

52. Denholm, P.; Kulcinski, G.L. Life cycle energy requirements and greenhouse gas emissions from large scale energy storage systems. *Energy Convers. Manag.* **2004**, *45*, 2153–2172. [CrossRef]

53. Holland, J.H. *Adaptation in Natural and Artificial Systems: An Introductory Analysis with Applications to Biology, Control, and Artificial Intelligence*; University of Michigan Press: Ann Arbor, MI, USA, 1975.

54. Haupt, R.L.; Haupt, S.E. *Practical Genetic Algorithms*; John Wiley & Sons: Hoboken, NJ, USA, 2004.

55. Deb, K.; Agrawal, R.B. Simulated Binary Crossover for Continuous Search Space. *Complex Syst.* **1995**, *9*, 115–148.

56. Da Ronco, C.C.; Benini, E. A Simplex Crossover based evolutionary algorithm including the genetic diversity as objective. *Appl. Soft Comput.* **2013**, *13*, 2104–2123. [CrossRef]

57. Chuang, Y.C.; Chen, C.T.; Hwang, C. A real-coded genetic algorithm with a direction-based crossover operator. *Inf. Sci.* **2015**, *305*, 320–348. [CrossRef]

58. Chehouri, A.; Younes, R.; Perron, J.; Ilinca, A. A Constraint-Handling Technique for Genetic Algorithms using a Violation Factor. *J. Comput. Sci.* **2016**, *12*, 350–362. [CrossRef]

59. Petrakopoulou, F.; Tsatsaronis, G.; Morosuk, T.; Carassai, A. Conventional and advanced exergetic analyses applied to a combined cycle power plant. *Energy* **2012**, *41*, 146–152. [CrossRef]

60. Kelly, S. *Energy Systems Improvement Based on Endogenous and Exogenous Exergy Destruction*; Prozesswissenschaften der Technischen Universität Berlin: Berlin, Germany, 2008.

Article

Opportunities to Optimize the Palm Oil Supply Chain in Sumatra, Indonesia

Fumi Harahap [1,2,*], Sylvain Leduc [2], Sennai Mesfun [2], Dilip Khatiwada [1], Florian Kraxner [2] and Semida Silveira [1]

[1] Energy and Climate Studies Unit, Department of Energy Technology, KTH Royal Institute of Technology, SE-100 44 Stockholm, Sweden; dilip.khatiwada@energy.kth.se (D.K.); semida.silveira@energy.kth.se (S.S.)
[2] International Institute for Applied Systems Analysis (IIASA), A-2361 Laxenburg, Austria; leduc@iiasa.ac.at (S.L.); mesfun@iiasa.ac.at (S.M.); kraxner@iiasa.ac.at (F.K.)
* Correspondence: harahap@kth.se; Tel.: +46-8-790-74-31

Received: 17 December 2018; Accepted: 25 January 2019; Published: 29 January 2019

Abstract: Significant amounts of biomass residues were generated in Indonesia. While untreated, residues emit greenhouse gases during the decomposition process. On the other hand, if efficiently utilized, these residues could be used to produce value-added products. This study investigates opportunities for harnessing the full potential of palm oil residues (i.e., empty fruit bunches, kernel shells, fiber, and mill effluent). As far as we are aware, the study is the first attempt to model the palm oil supply chain in a geographically explicit way while considering regional infrastructures in Sumatra Island, Indonesia. The BeWhere model, a mixed integer linear programming model for energy system optimization, was used to assess the costs and benefits of optimizing the regional palm oil supply chain. Different scenarios were investigated, considering current policies and new practices leading to improved yields in small-scale plantations and power grid connectivity. The study shows that a more efficient palm oil supply chain can pave the way for the country to meet up to 50% of its national bioenergy targets by 2025, and emission reductions of up to 40 MtCO$_2$eq/year. As much as 50% of the electricity demand in Sumatra could be met if residues are efficiently used and grid connections are available. We recommend that system improvements be done in stages. In the short to medium term, improving the smallholder plantation yield is the most optimal way to maximize regional economic gains from the palm oil industry. In the medium to long term, improving electricity grid connection to palm oil mills could bring higher economic value as excess electricity is commercialized.

Keywords: oil palm; palm oil mills; palm oil residues; value-added products; supply chains optimization; spatial analysis; techno-economic analysis

---

1. Introduction

Oil palm is the largest biomass source in Indonesia. The country housed 11 million hectares (Mha) of oil palm plantations and produced 31 million tons (Mt) of crude palm oil (CPO) in 2015 [1]. Oil extraction from palm fruits occurs in palm oil mills. One ton (t) of CPO production results in nearly 5 t of solid biomass waste, including empty fruit bunches (EFB), palm kernel shells (PKS), palm mesocarp fibers (PMF), and palm oil mill effluent (POME), see Figure 1. This implies that, in 2015, Indonesia produced around 155 Mt of palm biomass residue. These residues are the source of significant greenhouse gas (GHG) emissions due to biomass decomposition and, at the same time, result in lost opportunities in terms of economic gains from bio-based products (e.g., fuel for steam boilers, organic fertilizer, or further processed into value-added products, such as briquettes and'pellets) [2,3].

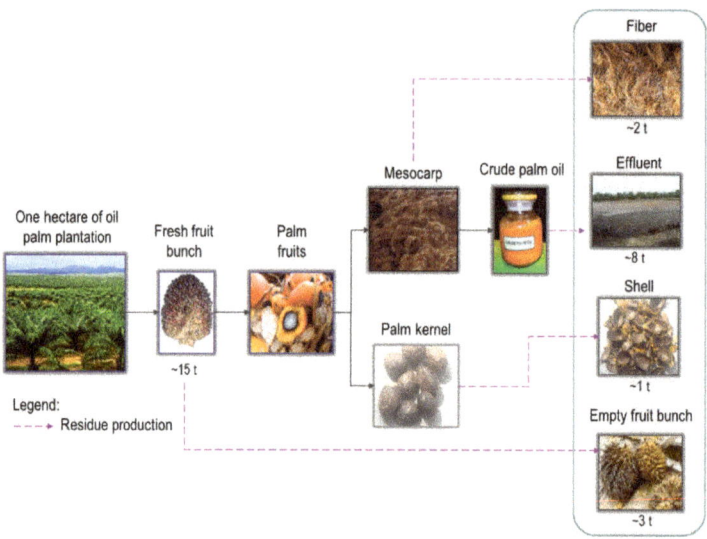

Figure 1. Simplified illustration of palm biomass residue generated from one hectare of oil palm plantation aimed at palm oil production. Note: conversion rate per one-hectare plantation was obtained during fieldwork carried out in a palm oil mill in Sumatra in 2016. Fiber is mesocarp fibers (PMF), effluent is palm oil mill effluent (POME), shell is palm kernel shells (PKS), and empty fruit bunches (EFB).

The Indonesian government has introduced several policy measures to promote the treatment of palm biomass residues as specified in the Indonesian Sustainable Palm Oil or ISPO statute (Regulation 11/2015). This includes methane capture from POME, power generation from solid biomass, and biofertilizer production from POME and EFB. However, the implementation of these policies has evolved slowly [4]. The Indonesian government has also set a national target of 23% renewable energy generation by 2025, which includes 5500 MW_e installed capacity of bioenergy-based plants [5]. This will ultimately help to reduce GHG emissions by 29% by 2030 compared to 2010 levels. Harnessing the full potential of palm biomass residue could contribute to meeting the national targets for renewable energy and emissions reduction, as well as to creating new jobs.

This study examines the utilization of oil palm biomass residues in Indonesia. A cost–benefit analysis was performed to identify optimal ways to fully utilize residues along the palm oil supply chain. We consider the production of the main product (i.e., CPO), byproduct (i.e., palm kernel or PK), as well as the utilization of the palm biomass residues (EFB, PKS, PMF, and POME) for bioenergy and biofertilizer using efficient biomass conversion technologies. We use mathematical modeling through the implementation of mixed integer linear programming (MILP), that has been used in various previous studies of the biomass supply chain [6–9]. Our reference scenario considers the current policy measures as described above. The scenario includes improvements from the conventional palm oil mill as described by Harahap et al. [10]. We investigate other scenarios with enhanced small-scale plantation yields (*Sc-yield*) and improved grid connectivity (*Sc-grid*), as well as with a combination of these measures (*Sc-yield-grid*) to evaluate the economic and environmental impacts that can be obtained along the palm oil supply chain. The geographical focus of the study is Sumatra Island in Indonesia, which has a total land area of 473,481 km². Sumatra accounts for more than 60% of the oil palm plantations and palm oil mills, and close to 70% of total palm oil production in Indonesia [1].

Supply chain optimization considers process engineering to systematically synthesize, design, and analyze biomass processing for multiple purposes [11,12]. A typical biomass supply chain is comprised of biomass harvesting and collection, pretreatment, storage, transport, and biomass conversion to bio-based products [13,14]. Atashbar et al. [15] emphasize that, although the biomass supply chain has been studied extensively, few papers consider the optimization of the whole supply chain. Many studies have been conducted on single parts or a few parts of the logistic system. In addition, very few papers consider the full biomass potential of a single plant. An optimization study is justified in light of the complexity of the palm oil supply chain.

While a large number of studies modeled regions in Malaysia, the world's second largest palm oil producer after Indonesia, only a few have discussed the case of Indonesia. Some studies that analyzed the Malaysian case are worth mentioning. Chiew et al. [16] carried out various scenarios to identify the optimal location of combined heat and power (CHP) plants for treating EFB, aiming at maximizing regional profit. The optimization model took into account biomass availability, transport distances, and the scale and location of CHP plants within the state of Selangor. Also, Foo [6] described the regional bioenergy supply chain for utilizing EFB, minimizing transportation costs to CHP plants for the case of Sabah state in Malaysia. Idris et al. [9] assessed the utilization of EFB, fronds, and trunks for co-firing, using spatial optimization to identify the technology cost. The regional bioenergy supply chain under carbon pricing and trading policies was discussed by Memari et al. [8] to evaluate the impact of using EFB in a CHP plant. Lam et al. [7] and Theo et al. [17] evaluated complex supply chain network designs, considering both the full biomass potential in a single plant and accounting for multiple palm oil mills in Malaysia. Lam et al. [7] proposed a strategy to integrate the solid biomass residues with industrial waste motor oil for fossil fuel substitution, while Theo et al. [17] focused on the utilization of POME and its distribution pathways.

The case of palm biomass supply chain performance in Indonesia was studied by Hadiguna et al. [18]. The supply chain optimization and risk assessment were included in the framework proposed to manage the palm oil supply chain more effectively. The study boundary encompassed the plantation, processing plant, and CPO production up to CPO distribution to the end customer. However, the study focused on operational performance in a single location, and excluded the potential utilization of biomass residues and multiple bio-based products, as well as the upgrading of biomass conversion technologies.

Our work unravels the biomass potential of the mills located in Sumatra Island, and this includes CPO, PK, and residues (i.e., EFB, PKS, PMF, and POME), as well as the biomass supply chain from feedstock collection, biomass processing, bio-products generation, and distribution. As far as we are aware, this study is the first attempt to model the palm oil supply chain in a geographically explicit way on Indonesia's Sumatra Island while considering the regional infrastructures. All datasets were developed for the simulation and had not previously been used in other studies. Our work contributes to the existing literature of the spatial optimization of the palm oil supply chain in Indonesia where half of all global CPO production takes place. Thus, the optimization of the palm oil industry in this region may have both national and global impacts.

Following this introduction, Section 2 describes the structure of the model, the input data, and model assumptions. Section 3 presents the scenarios considered, while Section 4 contains the results and discussions, followed by conclusions and policy recommendations in Section 5.

2. Framework for Analyzing the Palm Biomass Supply Chain

Utilizing palm biomass residues to produce value-added products can bring additional income to a palm oil mill [4]. In this paper, we use the spatially explicit BeWhere model developed at the International Institute for Applied Systems Analysis (IIASA) to optimize the utilization of biomass residue in one palm oil mill, and clustered mills in Sumatra [19]. Leduc [20] and Wetterlund [21] described the core of the BeWhere model. BeWhere is developed in the commercial software GAMS [22], uses a CPLEX solver, and the studied problem is expressed via mixed integer linear programming (MILP). The model is schematically represented with nodes and arcs. Each arc associates to a continuous variable. MILP allows the modeling of discrete (binary) variables. In this study, the binary variables are associated to the plant nodes to select the lowest cost technology for electricity production.

The model chooses the optimal pathways from one set of biomass supply points to a specific plant and, further, to a set of demand points [20]. BeWhere has been used for bioenergy policy evaluation in Europe and elsewhere (e.g., the United States, Brazil, Malaysia, and Vietnam) [23].

BeWhere applications in bioenergy are found in various studies. The integration of biomass with the existing steel industry in Europe, for instance, was studied by Mandova et al. [24]. Natarajan et al. [25] developed the BeWhere model to investigate optimal locations for biodiesel production plants in Finland, and Khatiwada et al. [26] applied the model to optimize ethanol and electricity production in sugarcane biorefineries in Brazil. The case for biogas for transport in Italy was discussed by Patrizio et al. [27], while the optimal biomethane (from palm oil mill effluent) injection into the natural gas grid in Malaysia was investigated by Hoo et al. [28]. Other studies using the BeWhere model can be found in Leduc et al. [23]. For the purpose of the present paper, the model was adjusted with a new algorithm and coding to study the specific case of oil palm in Indonesia.

2.1. BeWhere Model for the Oil Palm Supply Chain

This study defines the objective function to maximize overall profit by minimizing the total cost along the product(s) supply chain. This is formulated as follows:

$$TotProfit = TotIncome - TotCost - TotEnvCost, \tag{1}$$

$$
\begin{aligned}
TotEnvCost = \ & TotEmissions \times CarbonTax + TotPeatPltArea \times EcoLossPeatFire \\
& + TotPltArea \times WaterLoss \times WaterPrice \\
& + TotPltArea \times BiodiversityLoss,
\end{aligned} \tag{2}
$$

where *TotProfit* is the total profit obtained from the system, *TotIncome* is the total revenues generated from selling bio-products, *TotCost* is the supply chain cost, and *TotEnvCost* is the cost of environmental impacts in the studied system.

There are many environmental impacts to consider. *TotEmissions* is the supply chain emissions, *CarbonTax* is the cost of GHG emissions, *TotPeatPltArea* is oil palm plantations on peat land affected by peat fires, and *EcoLossPeatFire* represents economic losses due to peat fires. *TotPltArea* is the area covered by oil palm plantations, *WaterLoss* is the amount of water lost for every one ha of oil palm plantation developed, *WaterPrice* is the price for water, and *BiodiversityLoss* is biodiversity losses from oil palm plantation development.

The profit is maximized by considering product sales, feedstock production costs, feedstock transportation costs from supply points to mills, the investment and production costs of bio-products, and electricity transmission line costs. The supply chain emissions include emissions from feedstock production, transport emissions, emissions from plant operations, and emissions from biomass processing. The cost of GHG emissions is internalized in the model (i.e., in the form of CO_2eq tax). The external cost of biodiversity losses, water supply disruption, and peat fires were quantified based on the affected areas multiplied by factors for monetizing the impacts. Further details on this are available in Section 2.2.6.

The model structure for this regional case (i.e., multiple supply sides and palm oil mills) and multiple residue utilization pathways is illustrated in Figure 2, which depicts the BeWhere model structure for palm oil in Indonesia. The analysis was performed for a period of one year (base year of 2015). This study examines a supply chain optimization problem that consists of fresh fruit bunches (FFB) that originate from small-scale and large-scale plantations (i.e., government and private companies). FFBs are transported from plantations to palm oil mills by truck and using the existing road network. In the palm oil mill, FFB is processed to generate various intermediate products, which were quantified using the conversion rate. The intermediate products encompass CPO and PK, and residues from CPO production, that is, EFB, PKS, PMF, and POME. There are various conversion technologies in the mill for processing biomass into final bio-products, including both energy products (i.e., bioelectricity) and non-energy products (i.e., CPO, PK, PKS, PMF, EFB, and biofertilizer). The bio-products were estimated by multiplying the amount of intermediate products with the respective technological conversion rate. For intermediate products that do not require an upgrading process, and where input is similar to output, an artificial technology with efficiency equal to 1 was created to follow the model structure. We applied material balances from FFB to intermediate products, and from intermediate products to final bio-products.

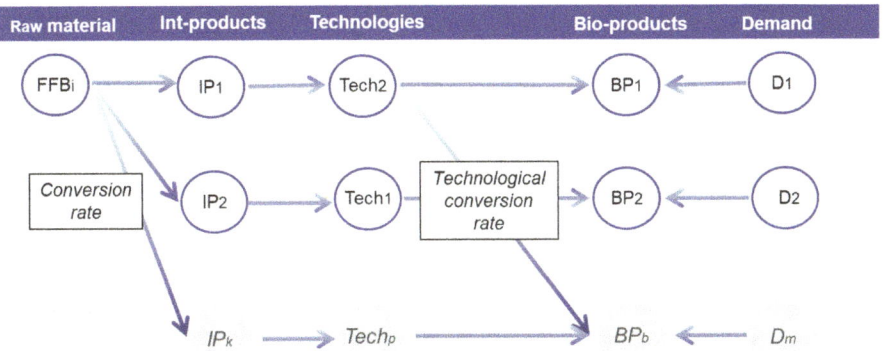

Figure 2. BeWhere model structure for palm oil in Indonesia.

The requirements for residue treatment (i.e., methane capture from POME, power generation from solid biomass, and biofertilizer production from EFB and POME), set by the Indonesian government, were taken into account. Low-efficiency CHP is the technology used to treat PKS and PMF, while high-efficiency CHP can also burn EFB. A preprocessing of the EFB is needed to reduce the moisture content before burning it in the boiler [29]. Alternatives considered for CHP plants consist of 1 MW low-efficiency CHP, and 4 MW and 9 MW high-efficiency CHP configurations. To reduce GHG emissions and increase bioelectricity generation, a 1 MW and 2 MW anaerobic digester biogas plant, that captures methane from POME, was incorporated in the model. The electricity produced is used for FFB processing in the mill, and excess electricity is fed into the electricity grid. Co-composting of the EFB and the POME was introduced in the model for the production of biofertilizer. The optimization model chose the most optimal biomass and biogas combustion system for each mill. Due to a variety of technical restrictions associated with the utilization of residue and upgrading to bio-products, relationships had to be defined within the BeWhere model. Details of the possible combinations of intermediate products, technologies, and bio-products are available in the Supplementary Material (Figure S1).

2.2. Input Data

This section describes the parameter input for the model comprising spatial data, technical data, investment costs and commodity prices, and emissions. All data are for 2015. All costs and prices were adjusted for inflation corresponding to 2015 prices and using the Indonesian GDP deflator [30].

2.2.1. Feedstock Production and Availability

A 25 km × 25 km grid size was chosen to match the resolution of a previous work on forest fires in Indonesia [31], the results of which we intend to merge with the results of this work to serve as input for a harmonized risk assessment study in the future. The grid size gives a total amount of 656 grid points in Sumatra Island, which is a relevant number for the model to be resolved in a reasonable amount of time without losing accuracy in the results. The availability of feedstock (i.e., FFB) represented at grid level was derived from the district statistical data obtained from the Indonesian Ministry of Agriculture (MoA) [1]. Notice that the statistical data do not provide geographical information on the plantation. Still, at the time of writing, this was the best dataset available to perform the regional analysis. The spatial data from the Ministry was not publicly available, and plantation maps were provided by Global Forest Watch (GFW) [32], which only included the concession area allocated by the government for large-scale oil palm plantations, as opposed to the actual area planted. In addition, the latter excludes the small-scale plantations.

The availability of FFB was calculated based on the total plantation area and the average plantation yield per district. Thus, each grid point contains information about the raw material available from small-scale and large-scale plantations, the plantation area, and the average yield. The physical characteristics of small-scale and large-scale plantations considered in the analysis are listed in Table 1. The spatial representation of FFB availability can be found in the Supplementary Material (Figure S2).

Table 1. Physical characteristics of small-scale and large-scale plantations in Sumatra, Indonesia.

Item	Unit	Small-Scale Plantation	Large-Scale Plantation
Total Fresh Fruit Bunches (FFB) production in Sumatra, 2015	Mt/y	41.9 [1]	54.5 [1]
Total plantation area in Sumatra, 2015	Mha	3.33 [1]	3.24 [1]
Average productivity, 2015	t/ha	13.57 [1]	18.82 [1]
Feedstock production cost [a]	USD/t	98.14	75.52
$N_{fertilizer}$ consumption [b]	kg/ha	92.94 [33]	138.72 [33]

Notes: [a] The production cost comprises the cost of fertilizers, field maintenance, and harvesting, as well as general expenses. The feedstock production cost per year (y) per t_{FFB} is 1332 USD/ha for small-scale plantations [34] and 1421 USD/ha/y for large-scale plantations [35], multiplied with the respective average productivity. Notice that the higher cost for large-scale plantations is related to higher fertilizer use. [b] The amount of nitrogen fertilizer applied was calculated based on an equation by Khasanah [33] for the relation between nitrogen fertilizer applied and the plantation yield in Indonesia: $y = 1.1386x^2 - 28.157x + 265.36$, where y is the amount of nitrogen fertilizer applied and x is the plantation yield.

2.2.2. Processing FFB in Palm Oil Mills

The FFB are transported from the plantation area to the palm oil processing plant (palm oil mills). Processing FFB in palm oil mills generates various products. Here, these products are defined as intermediate products, consisting of the main product (i.e., CPO), the byproduct (i.e., PK), and biomass residues (PKS, PMF, EFB, and POME). The quantity of each intermediate product was derived using conversion rates, which are available in the Supplementary Material (Table S1).

According to the MoA, there were 636 mills across Indonesia in 2015, of which, 415 were located in Sumatra. This number of mills is not consistent with the GFW database [32,36], according to which there were 410 mills in Indonesia and 339 in Sumatra in 2015. In fact, Indonesia does not have any official database of palm oil mills specifying their geographical locations and processing capacity (mill data is at district level, and not fully compiled at the national level). We used various sources from the MoA, ISPO, the Roundtable on Sustainable Palm Oil (RSPO), and a database of palm oil suppliers, to compile a database of mills in Sumatra. For the purpose of our analysis, we considered 415 mills in Sumatra, which is in line with the numbers provided by the MoA and Hambali et al. [37]. The geographic location of palm oil mills and the annual FFB processed is provided in the Supplementary Material (Figure S3). All mills are assumed to operate for 4800 h per year (h/y). Previous studies used operating hours ranging between 3600 [28] and 6000 h/y [7]. In reality, every mill has different operating hours. This assumption causes a 16% discrepancy between the total CPO production considered in this study (i.e., 18 Mt_{CPO}/y) and the MoA's national statistic [1] (i.e., 22 Mt_{CPO}). This also means that our study does not overestimate the added value generated from the industry.

A mill requires 22 kWh_{el}/t_{FFB} and 500 kg_{steam}/t_{FFB} for operation [4]. The mill costs consist of annualized capital costs and operation and maintenance (O&M) costs (5% of annual capital cost [38], equal to 1.96 USD/t_{FFB}). The annuity factor was 8.45% using an interest rate of 6.83% [30] and a project lifetime of 25 years.

2.2.3. Palm Oil Biomass Utilization and Biomass Conversion Technologies

Our assessment considers a strategic level of decision-making that involves the selection of biomass conversion technologies. Technology selection plays a key role in the biomass supply chain, as it influences the choice of biomass, capital, and operational costs of the supply chain and related environmental impacts [39]. In this study, the equipment of the new technology is installed in the existing facility, assuming the availability of land.

CHP plants (low- and high-efficiency systems), biogas plants with a covered lagoon, and co-composting are the biomass technologies considered in this study. The technologies were chosen in line with the government policy promoting the utilization of palm biomass residues, as specified in ISPO Regulation 11/2015. The CHP system consists of steam boilers, back pressure turbines, and electrical networks. The biogas plant consists of biodigester (i.e., continuous stirred tank reactor), scrubber, gas engine, boiler, and flare. In this study, we predefined the installed capacity for the CHP and biogas systems, since only systems below 10 MW are considered for special tariffs from the government (Regulation 12/2017). The low-efficiency CHP system has 1 MW capacity, which only meets the energy demand of the palm oil mill. Alternatively, in places with higher demand, high-efficiency CHP with 4 MW or 9 MW capacity are considered. The biogas system has 1 MW or 2 MW installed capacity, which are typical capacity sizes for some mills in Indonesia [40]. Excess electricity is fed into the national grid.

The technological conversion rate was used to quantify the amount of bio-products generated from treating the residue using a specific technology. The cost of technology consisted of O&M costs and annualized capital costs. Similar to the palm oil mill, an annuity factor of 8.45% was used.

Notice that the production of electricity is driven by electricity demand at the mill and in the district. A palm oil mill requires 22 kWh_{el}/t_{FFB} [4]. Sumatra's electricity demand is 34 TWh/y, as further described in Section 2.2.5. The production of other bio-based products is influenced by their market values. The residues that are not treated at the mill site are sold to the market at a specific price. Indonesia has been exporting PKS to Japan to be used as feedstock for biomass power plants [41].

Details about the technological conversion rates, the technology costs, and the prices of bio-products considered in this study are provided in the Supplementary Material (Tables S2–S4).

2.2.4. FFB Transport to Palm Oil Mills

The prompt transport of FFB to mills is crucial to maintain the quality of palm oil with a low level of free fatty acids and other impurities, and to allow high oil extraction [42]. Fatty acids can quickly build up, which means that FFB must be processed within 24 h of harvest [36]. Transporting FFB can take from 30 min to more than three hours. Hence, we restricted the travel time from the plantation side to the mill to four hours [43]. The average truck speed is 50 km/h. The network analysis performed using ArcGIS software identified 272,241 possible route combinations from plantation to mill. The restriction of travel time reduced the possible routes to 26,590. The original data on the road network were obtained from DIVA-GIS [44], which was modified to improve resolution prior to running the network analysis using the ArcGIS software. Information about Sumatra's road network is available in the Supplementary Material (Figure S4). The transport cost considered the variable cost (0.2 USD/t_{FFB}/km), and fixed cost (0.5 USD/t_{FFB}) [7], as well as the return trip.

2.2.5. District Electricity Demand and Transmission Lines

In this work, electricity was delivered to high voltage transmission lines, that is, 150, 275, and 500 kV. The location of the distribution transformers is detailed in the Supplementary Material (Figure S5). For modeling purposes, each distribution transformer contains information on aggregated district electricity demand as presented in the Supplementary Material (Figure S6). Electricity demand was calculated by district population in 2015 [45] and average electricity consumption per capita in Indonesia (812 kWh/y [46]). The transmission cost considered the amount of excess electricity, the annualized capital cost of transmission lines (58 USD/MW/km/y), the distance from the mill to the nearest power grid, and the annualized capital cost for connection (17.5 USD/MW/y) [47]. The annuity factor is 0.06, assuming a 40-year life span and a 5% interest rate.

2.2.6. Environmental Impacts

The palm oil industry generates significant positive economic and social impacts (e.g., job creation) for the region. However, the expansion of palm oil plantations has also led to non-sustainable land use practices in past years, particularly deforestation [48]. Thus, the negative environmental impacts from supply chain emissions, haze and peat fires, water supply disruption, and biodiversity losses were quantified in the analysis.

Supply chain emissions account for direct GHG emissions associated with emissions from feedstock production, transport emissions, emissions from plant operations, and emissions from biomass processing. The emissions factors used in this study are summarized in the Supplementary Material (Table S5). Direct and indirect land use change effects are not in the scope of the analysis. The main GHGs considered in the biomass combustion system are carbon dioxide (CO_2), methane (CH_4), and nitrous oxide (N_2O) [49], which are converted to CO_2 equivalent (i.e., CO_2eq) by using a global warming potential (GWP) of 1, 25, and 298, respectively [50]. Currently, Indonesia does not apply carbon taxes. Yet, the inclusion of a carbon tax to compensate for CO_2 emissions could play an important role in substituting carbon-intensive fossil-based electricity in the future [26]. The carbon-pricing gap varies widely, both across countries and across sectors within countries. Mexico, Poland, and Ukraine, for example, adopted a carbon tax below 1 USD/tCO_2eq, while Sweden imposed 139 USD/tCO_2eq, the highest in the world [51]. In this study, we used a carbon tax of 25 USD/tCO_2eq. In the optimization model, carbon tax is applied to the total GHG emissions.

The conversion factors for monetizing the other environmental impacts (i.e., emissions from peat fires, disruption in water supply and biodiversity losses) were taken from Agustira et al. [52]. The economic losses from peat fires were based on 41.73 USD/ha/year [52]. We estimate that 39,000 ha are affected [31]. The water supply disruptions were calculated by multiplying the total number of oil palm plantations with the amount of water lost for every one ha of oil palm plantation developed (500 m^3/ha/year) and water price (0.402 USD/m^3) [52]. The estimation of biodiversity losses took into account the total number of oil palm plantations and 30 USD/ha to represent biodiversity losses from oil palm plantation development [52].

The estimation of emissions reductions considers the substitution of products with higher fossil fuel consumption with bio-products, such as substituting electricity in the Sumatra grid (grid emission factor is 0.855 tCO_2/MWh [53]) with biomass electricity, and replacing chemical fertilizers with biofertilizers. In one hectare of plantation, 1 $t_{biofertilizer}$ replaces 0.25 $t_{chemical-fertilizer}$, based on the ratio of nutrient values [54]. The methane avoidance from POME treatment was also included in the calculation of emissions reduction.

2.2.7. Model Assumptions and Limitations

Here, we summarize the key assumptions made. The estimation of FFB availability was based on the aggregated regional values from national statistics instead of on the actual planted area. The location of palm oil mills was fixed. The capacity of the mill in FFB processed per year assumes 4800 operating hours per year. The transportation of FFB from the supply site to the mill was limited to a maximum of four hours travel time to meet the 24 hour requirement from FFB harvesting until processing. The model selected the optimal routes to transport FFB to palm oil mills. The assessment was performed for one year (base year 2015), which means that the model optimizes the production of CPO and PK, as well as the utilization of biomass residue (PKS, PMF, EFB, and POME) for processing them into bio-products or selling them to the market. All bio-products were sold at the gate (mill) except bioelectricity, which was delivered to the distribution transformer via transmission lines. In the reference scenario, only palm oil mills that were located within 10 km of the distribution transformer could supply electricity to the grid. The study assumed that there was sufficient need for heat generated from the CHP plant for internal use in the mill.

3. Scenario Development

The base case or reference scenario (*Sc-ref*) in this study incorporated the official policy for utilizing biomass residue in Indonesia (i.e., methane capture from POME, power generation from solid biomass, and biofertilizer production from EFB). Depending on the parameter inputs and spatial location, the model selected the most cost-effective biomass conversion technologies for each mill. The alternative technologies and possible combinations of biomass input are presented in in the Supplementary Material (Figure S1).

Apart from the *Sc-ref*, we consider a scenario with higher yields in small-scale plantations (*Sc-yield*). Smallholder plantations are often associated with low yield due to poor agricultural practices, bad quality of seed, and insufficient fertilizer application [55]. Over decades, the yield of smallholder plantations in Indonesia has been improving. Several studies attribute this to cooperation between smallholders and companies [34]. The government views this as an opportunity to improve the livelihoods of farmers and is continuing with a program for improving smallholder yield, while also providing financial incentives channeled through the biodiesel fund program [2]. The *Sc-yield* scenario proposes an improvement of smallholder yields to levels similar to those found in large-scale plantations, that is, 18.82 t_{FFB}/ha. Hence, there is an increase in the FFB availability in this scenario.

Harnessing the full potential of palm biomass for electricity generation requires sufficient grid connection infrastructure to deliver bioelectricity. The lack of electricity grid connections in palm oil mills has curtailed increases in bioelectricity production from palm biomass. We developed a scenario (*Sc-grid*) with enhanced grid connectivity that allows increased bioelectricity delivery. With this scenario, we evaluate the opportunities offered by palm biomass in terms of meeting national bioenergy targets, as well as the overall cost along the supply chain studied. To deliver excess electricity to external consumers, the *Sc-ref* scenario included mills located within a 10 km radius from the power grid. *Sc-grid* considers the connection of all mills (415) to the distribution transformers and, therefore, their increased access to biomass electricity. However, this was constrained by electricity demand at the district level. Notice that the model selected the nearest point from the mill.

Finally, the third scenario (*Sc-yield-grid*) was developed to assess the impact of both higher yields from small-scale plantations and improved grid electricity connection for the mills. Table 2 presents the different parameters of the scenarios analyzed.

Table 2. Different parameters of the scenarios analyzed for the palm biomass supply chain in Sumatra, Indonesia.

Scenario	Parameters Related to Small-Scale Plantations				Number of Mills Connected to Power Grid	Average Annual Mill Operating Hours (h/y)	Scenario Description
	Plantation Yield	$N_{fertilizer}$ Consumption	Emissions	Raw Material Production Cost *			
	t_{FFB}/ha	kg/ha	tCO_2eq/t_{FFB}	USD/t_{FFB}			
Sc-ref	13.57	92.94	0.089	98.14	65	4800	Reference scenario incorporating government policy to foster the utilization of palm biomass residue
Sc-yield	18.82	138.72	0.096	81.93	65	5000	Improved yield in small-scale plantations, which also affects fertilizer consumption and associated emissions
Sc-grid	13.57	92.94	0.089	98.14	415 (all mills)	4800	All palm oil mills are connected to the nearest distribution transformers
Sc-yield-grid	18.82	138.72	0.096	81.93	415 (all mills)	5000	Combines the improvement specified in Sc-yield and Sc-grid

Note: * The production cost consists of the cost of fertilizers, field maintenance and harvesting, as well as general expenses. The feedstock production cost per t_{FFB} for small-scale plantations is based on 1332 USD/ha/y for the reference value [34] and 1542 USD/ha/y for higher yields (using a similar fertilizer cost as for large-scale plantations). The cost per hectare is multiplied with the respective average productivity.

Sensitivity Analysis

A sensitivity analysis was carried out to investigate the importance of different parameters, as well as their interactions on total cost and total profit. We performed a 2^k factorial design of experiments in two levels, with all input factors also set at two levels. We evaluated the impact of five parameters as shown in Table 3. The 2^5 factorial designs resulted in 32 model runs. The analysis was carried out using R code following the steps described in [56] by first observing the response (i.e., total cost and total profit) for each factor using a box plot, identifying the significant effects with a theoretical model, and confirming the most significant effects using normal probability plots.

Table 3. Upper and lower values of variables for the sensitivity analyses.

Factor	Description	Unit	−1	Ref.	+1
A	Mill operating hours	h/y	4000	4800	4900
B	Palm oil extraction rate	t_{CPO}/t_{FFB}	0.18	0.2	0.25
C	Raw material production cost of large-scale plantations	USD/t_{FFB}	65	75.52	98
D	Capital cost of 1 MW Combined Heat and Power (CHP) system	USD/MWh	17	23.65	30
E	Transport cost	USD/t_{FFB}/km	0.1	0.2	0.3

Previous studies used operating hours ranging between 3600 [28] and 6000 h/y [7]. The reference value of 4800 h/y was chosen, with 4000 h/y as the lower level and 4900 h/y as the upper level. Notice that the model is constrained by the availability of raw material, implying that there should be sufficient biomass for processing in the mill. The second parameter for sensitivity analysis is the FFB to CPO conversion rate or oil extraction rate. CPO is the main product of the industry, which means that the revenue generated is driven by production. Depending on the varieties of the oil palm (e.g., dura, tenera, psifera), the oil extraction rate is between 16% and 30% [55]. The reference value of the oil extraction rate is 0.2 t_{CPO}/t_{FFB}. Here, the upper value considered is 0.25 t_{CPO}/t_{FFB} or a 25% oil extraction rate, and the lower value is 0.18 t_{CPO}/t_{FFB} or an 18% oil extraction rate. The cost of raw material has been cited in some studies as the key parameter affecting the production cost of bio-based products (i.e., biofuels) [57–59]. We investigate the change in raw material production cost of large-scale plantations, since large-scale plantations account for more than half of the total FFB production in Indonesia. In this sensitivity test, the raw material production cost of small-scale plantations is not changed. We assume the upper level of the cost to be similar to the cost of small-scale plantations (98 USD/t_{FFB}) and set the lower level of the cost at 65 USD/t_{FFB}. Apart from the raw material cost, Petterson [60] and Solikhah [61] indicated that the cost for transporting feedstock is significant in relation to the total cost of feedstock. We assume a range of 50% from the reference value to present the upper and lower levels. The capital cost of biomass conversion technology is an important parameter that influences the decision of a mill to adopt the technology. A 1 MW CHP plant is the minimum improvement for a mill to utilize biomass for bioenergy production, which can then be consumed internally. A reference value of 23.65 USD/MWh was chosen with the lower and upper levels set at 17 USD/MWh and 30 USD/MWh, respectively.

The parameters used for the sensitivity analysis are presented in Table 3, together with their extreme values (−1 or low level and +1 or high level).

4. Results and Discussion

4.1. Technology Selection and Quantity of Bio-Products Generated

The model chooses the most cost-effective technology at each mill depending on the scenario setup. Figures 3 and 4 present the total bio-products generated in all mills. For power production, PMF is the preferred feedstock for the CHP system, due to its high heating value compared to PKS and EFB. The results show that *Sc-ref* and *Sc-grid* provided 3 TWh/y and 17 TWh/y of excess electricity to the grid, respectively. Not surprisingly, when all mills are connected to the power grid (*Sc-grid* and *Sc-yield-grid*), there is higher potential to generate excess electricity. While *Sc-ref* has the potential to meet 8% of electricity demand in Sumatra, *Sc-grid* could cover up to 50% of the electricity demand on the island (i.e., 34 TWh/y). This is quite significant and means that oil palm biomass could play a major role in the development of electrification in Sumatra.

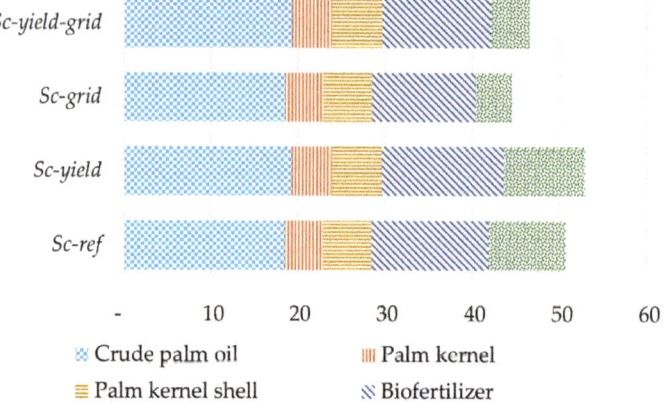

Figure 3. Total bio-products generated in palm oil mills (Mt/y). Legend: *Sc-ref*: Incorporating the government policy to foster the utilization of palm biomass residue; *Sc-yield*: Improving the yield of small-scale plantations; *Sc-grid*: Improving bioelectricity delivery; *Sc-yield-grid*: Combination of *Sc-ref* and *Sc-grid*.

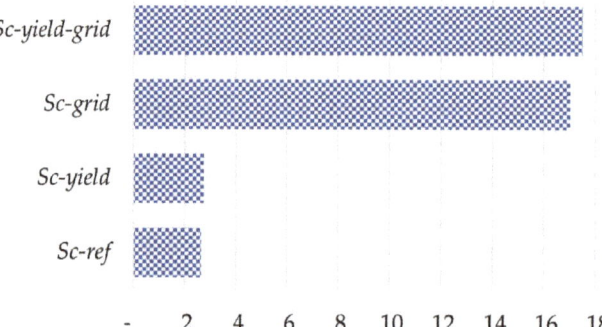

Figure 4. Total excess power to the grid (TWh/y). Legend: *Sc-ref*: Incorporating the government policy to foster the utilization of palm biomass residue; *Sc-yield*: Improving the yield of small-scale plantations; *Sc-grid*: Improving bioelectricity delivery; *Sc-yield-grid*: Combination of *Sc-ref* and *Sc-grid*.

In terms of technology selection for electricity production, most mills in *Sc-ref* installed 1 MW low-efficiency CHP plants for treating PMF, generating just enough electricity to run the mill operations. Out of 415 mills, 345 chose a 1 MW CHP system, 30 mills upgraded to a 4 MW CHP system, and the remaining 40 mills upgraded to 9 MW CHP systems. In the *Sc-yield-grid* scenario, 129 mills upgraded to 4 MW and 266 upgraded to 9 MW high-efficiency CHP systems. Figure 5 provides a schematic representation of the total installed capacity of CHP systems in the districts of Sumatra.

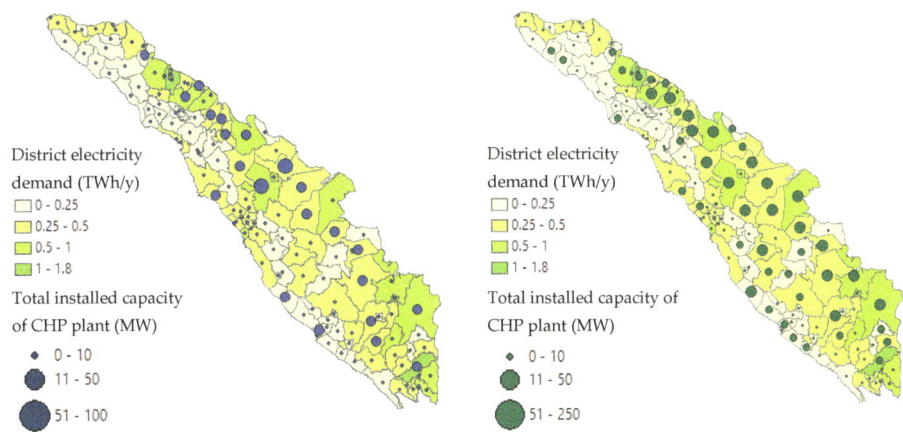

Figure 5. Total installed capacity of CHP plants per district, *Sc-ref* (**left**) and *Sc-yield-grid* (**right**).

The results demonstrate that *Sc-ref* has the potential to install a capacity equivalent to 670 MW$_e$ to use the palm biomass residues available, thus contributing to meeting 12% of the national bioenergy target (i.e., 5500 MW$_e$ [5]) and 38% of the target for Sumatra (i.e., 1755 MW$_e$ [5]) by 2025. *Sc-ref* requires 250 million USD/y in new investments for upgrading the biomass technologies and building transmission lines. About four-fold of installed capacity of bioenergy plants can be achieved in *Sc-yield-grid*—equivalent to 2800 MW$_e$. This means that 50% of the national bioenergy target and more than 100% of the target for Sumatra can be met by 2025 if the full potential of palm biomass residues in Sumatra are harnessed. Investments in the order of 760 million USD/y (i.e., upgrading the equipment and transmission lines) are needed to implement this scenario.

Only 19 plants installed 1 MW biogas plants (anaerobic digester) to treat POME in *Sc-ref*, and none in *Sc-grid* and *Sc-yield-grid*. No mills installed a 2 MW biogas plant. This occurs due to the high capital cost of the technology. Economically, it is more attractive to produce biofertilizer than bioelectricity from POME. This suggests that a cluster model with POME collection from several palm oil mills can achieve economies of scale. The concept of industrial symbiosis can be explored to facilitate mills' cooperation [62]. As described by Martin et al. [63], industrial symbiosis can improve regional sustainable development by enhancing the environmental performance of the industries and the socioeconomic status of the region. Such a model could be the way forward for making biogas plants economically more attractive.

4.2. Total Costs and Benefits

Figures 6 and 7 show the costs, income, and profit from the palm oil industry in Sumatra. In all scenarios, the highest costs relate to FFB production and feedstock transportation to the mills. *Sc-yield-grid* shows the highest total cost of 12.3 billion USD/y but, at the same time, provides the highest profit for the palm oil industry (4 billion USD/y). *Sc-ref* and *Sc-yield* offer profits of 2.5 billion USD/y and 3.6 billion USD/y, respectively. However, *Sc-yield-grid* requires new investments of 760 million USD/y for upgrading the biomass technologies and building new transmission lines, whereas *Sc-yield* could achieve higher profits than *Sc-ref* with three times lower investment than *Sc-yield-grid*. This suggests priorities and cost-efficient measures to consider when planning an upgrading program, depending on resource availability for new investments.

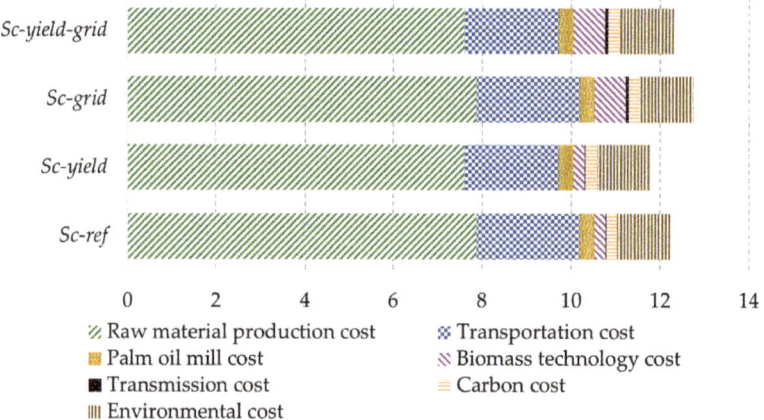

Figure 6. Total costs of a more efficient palm oil supply chain (billion USD/y). Legend: *Sc-ref*: Incorporating the government policy to foster the utilization of palm biomass residue; *Sc-yield*: Improving the yield of small-scale plantations; *Sc-grid*: Improving bioelectricity delivery; *Sc-yield-grid*: Combination of *Sc-ref* and *Sc-grid*.

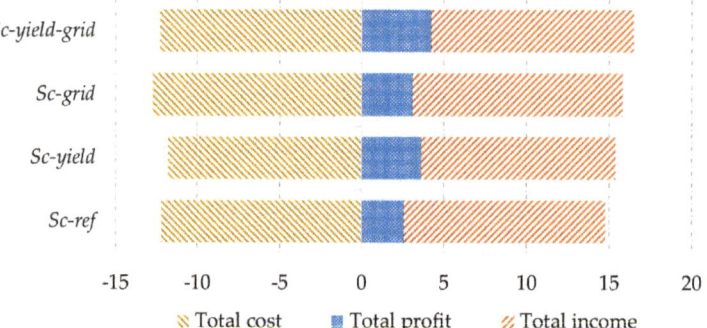

Figure 7. Total costs, income, and profits of a more efficient palm oil supply chain (billion USD/y). Legend: *Sc-ref*: Incorporating the government policy to foster the utilization of palm biomass residue; *Sc-yield*: Improving the yield of small-scale plantations; *Sc-grid*: Improving bioelectricity delivery; *Sc-yield-grid*: Combination of *Sc-ref* and *Sc-grid*.

4.3. Total Emissions, Emissions Reduction, and Technology Abatement Cost

The total emissions of each scenario include emissions from raw material production and transportation, the use of diesel in the palm oil mill, and the emissions from biomass processing. The emissions reduction is based on the emissions saving potential derived from product substitution with bio-products and methane avoidance from POME treatment. For example, electricity from biomass replaces the electricity in the Sumatra grid, and biofertilizer is applied in plantations instead of chemical fertilizers. Figures 8 and 9 show GHG emissions at activity level and emissions reduction from product substitution and methane avoidance. Net emissions savings between 17 and 30 MtCO$_2$eq/y can be achieved in a more efficient system.

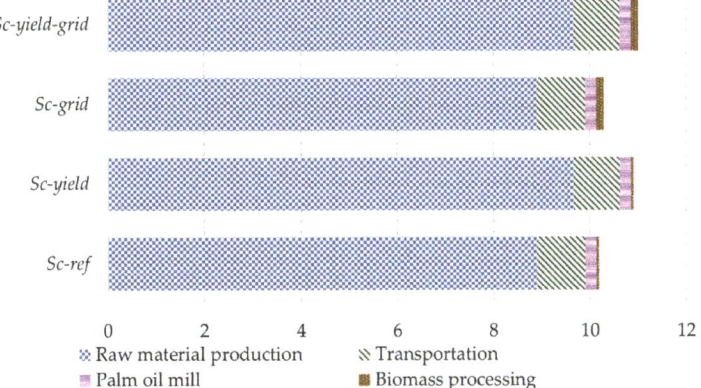

Figure 8. GHG emissions from a more efficient palm oil supply chain (MtCO$_2$eq/y). Legend: *Sc-ref*: Incorporating the government policy to foster the utilization of palm biomass residue; *Sc-yield*: Improving the yield of small-scale plantations; *Sc-grid*: Improving bioelectricity delivery; *Sc-yield-grid*: Combination of *Sc-ref* and *Sc-grid*.

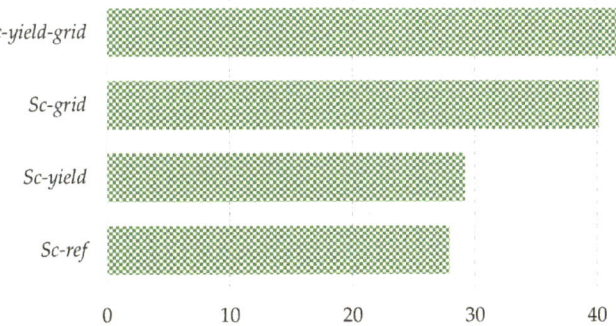

Figure 9. Emissions reduction from a more efficient palm oil supply chain (MtCO$_2$eq/y). Legend: *Sc-ref*: Incorporating the government policy to foster the utilization of palm biomass residue; *Sc-yield*: Improving the yield of small -scale plantations; *Sc-grid*: Improving bioelectricity delivery; *Sc-yield-grid*: Combination of *Sc-ref* and *Sc-grid*.

According to the Nationally Determined Contribution (NDC), Indonesia aims to reduce GHG emissions by 834 MtCO$_2$eq/y by 2030 (including emissions from sectors: energy, waste, industrial processes, agriculture, and forestry) from business-as-usual (BAU) levels in 2010 [64]. For the energy and waste sectors, the GHG emissions reduction targets for 2030 are 314 MtCO$_2$eq/y and 11 MtCO$_2$eq/y, respectively [64]. *Sc-grid* and *Sc-yield-grid* can potentially reduce emissions between 28 and 40 MtCO$_2$eq/y. Thus, a more efficient utilization of oil palm biomass residues can contribute 3% to 5% of the national GHG emissions reduction target for 2030. Bioenergy production from palm biomass residue in Sumatra can serve 1.3% to 5% of the target for the energy sector (excluding methane avoidance from POME management). Additional bioenergy will reduce Sumatra's dependence on highly fossil-based electricity, which is currently more than 55% crude oil and coal-based [65]. In addition, emissions are significantly reduced through methane avoidance—up to 22 MtCO$_2$eq/y. This means that efforts to manage POME in Sumatra can meet the country's emissions reduction target from the waste sector for the unconditional mitigation scenario (i.e., 11 MtCO$_2$eq/y) and nearly all for the conditional mitigation scenario (i.e., 26 MtCO2eq/y) by 2030 [64]. This indicates that Indonesia can set a higher target to reduce emissions from the waste sector.

We analyzed the technology abatement cost, which is the cost to reduce one tCO2eq. Each bar in Figures 10 and 11 represents the abatement cost of each analyzed mill. It suggests which mills can be prioritized to achieve the most in terms of emissions reduction. The average abatement cost of all mills in *Sc-ref* (i.e., 8.5 USD/tCO2eq) is 50% lower than in *Sc-yield-grid* (i.e., 18.5 USD/tCO2eq). However, the total emissions reduction in *Sc-yield-grid* is 1.5 times higher than in *Sc-ref*.

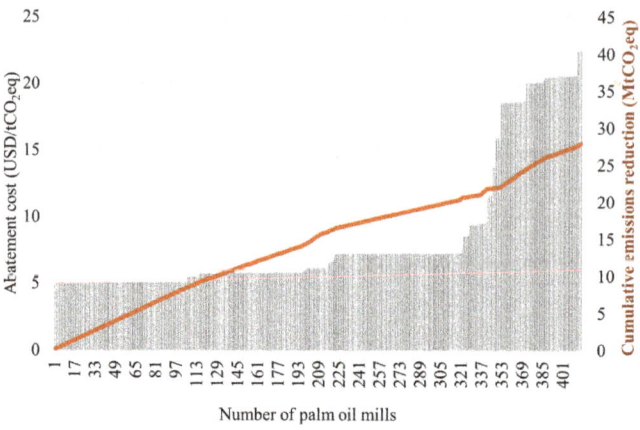

Figure 10. Technology abatement cost (bar chart, primary Y-axis) and cumulative emissions reduction (line, secondary Y-axis) of *Sc-ref*.

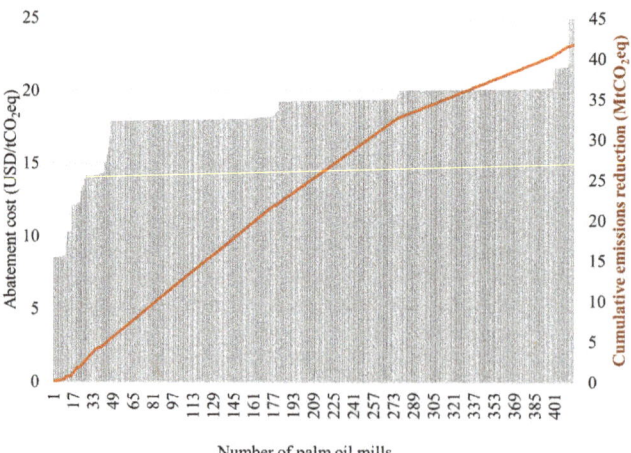

Figure 11. Technology abatement cost (bar chart, primary Y-axis) and cumulative emissions reduction (line, secondary Y-axis) of Sc-yield-grid.

4.4. Effects of Significant Parameters on Cost and Profit

The sensitivity and interactions of the parameters were analyzed using a 2^5 factorial design of experiments, in which the influence on the costs of five parameters as well as their interactions with each other, were studied. This study was carried out in 32 runs with different combinations between the parameters. For each run, the total cost and profit were determined with the parameter changes listed in Table 3. The full factor combinations and the results from the factorial designs of the 32 runs are provided in Appendix A.

The normal probability plots of the effects on total cost and total profit are shown in Figure 12. The effects considered to be significant are labeled. The significant effects deviate substantially from the straight line (i.e., normal distribution line). Non-significant effects should effectively follow an approximately normal distribution with the same location and scale. The mill operating hours (A) appear to have a larger effect on total cost, followed by the raw material production cost (C) and transport cost (E). The interaction between mill operating hours and transport cost (AE) also has a significant effect on total cost. The magnitude of the effect estimates show that the palm oil extraction rate (B) is the most important factor in total profit. The oil extraction rate (B) is not influential on the total cost because, in this study, the cost function is determined by raw material availability (i.e., FFB) instead of CPO. The interaction of raw material availability with operating hours (AB) is relevant because a higher number of operating hours leads to higher amounts of FFB processed which, in turn, leads to a higher oil extraction rate and CPO production, the latter being the main source of income in the mill. Other single factors that have a significant effect on total profit are raw material production cost (C) and transport cost (E). The interactions that appear to be significant in terms of total profit include those between (i) the mill operating hours and the oil extraction rate (AB) and (ii) between mill operating hours and transport cost (AE).

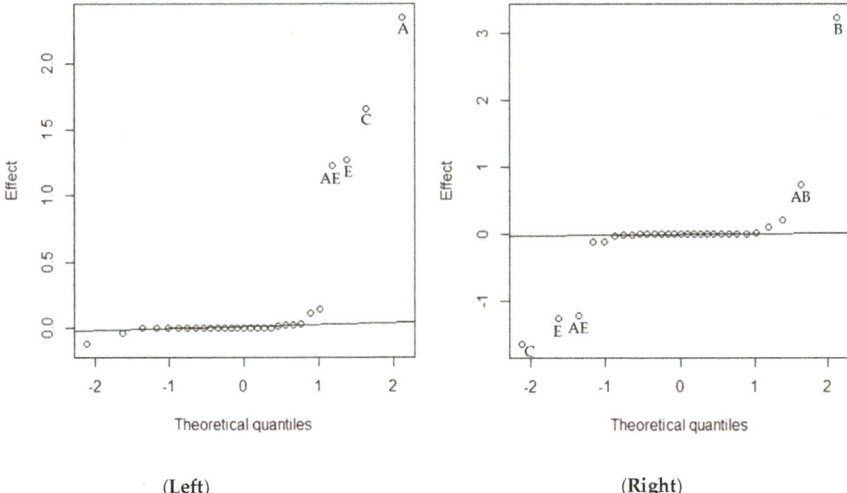

(**Left**)   (**Right**)

Figure 12. Normal probability plots of the effects on total cost (**left**) and total profit (**right**). Legend: A: Mill operating hours; B: Palm oil extraction rate; C: Raw material production cost of large-scale plantations; D: Capital cost of 1 MW CHP system; E: Transport cost.

The analysis reveals that, when mill operating hours, raw material cost, and transport cost are at the upper levels, and the efficiency of the mill is at the lower level, the costs are higher than the revenues, and the system runs at a loss (Run 22 and 30 in Appendix A).

Plots of the low and high levels of the main effects, and their interactions with total cost and total profit, are shown in Figures 13 and 14, respectively. The highest cost of 14 billion USD/y is achieved when mill operating hours and transport cost (AE) are at their highest level. The lowest cost is 9 billion USD/y, and is obtained when both variables are at their lowest level. This can be explained by the importance of each factor, that is, mill operating hours (A) and transport cost (E). The highest profit of 5.3 billion USD/y is achieved when the palm oil extraction rate (B) is at the upper level, while the lowest profit of 1.1 billion USD/y is reached when mill operating hours and transport cost (AE) are at their lowest levels.

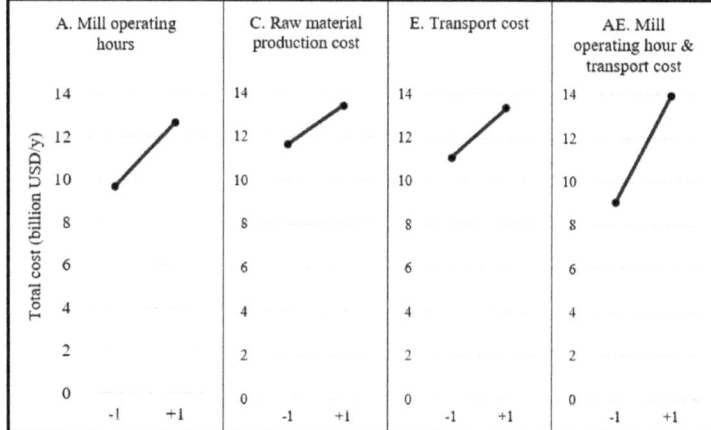

Figure 13. Plots of the main effects and significant two-way interactions on total cost.

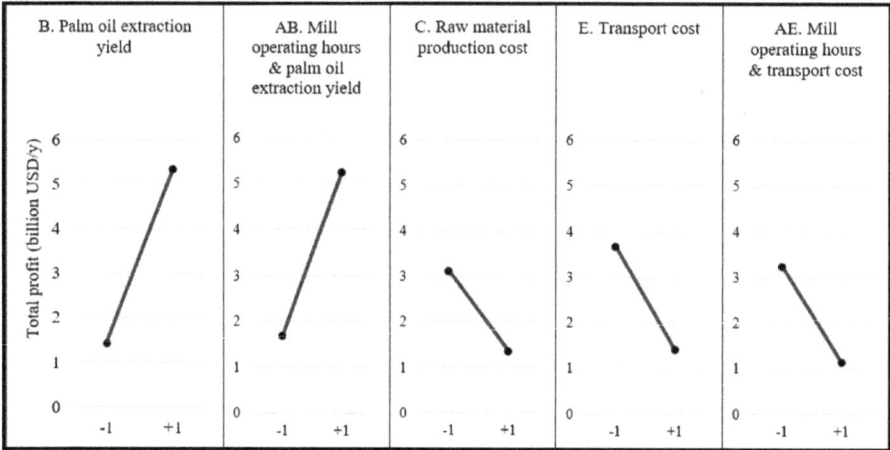

Figure 14. Plots of the main effects and significant two-way interactions on total profit.

5. Conclusions: Pathways to Enhance the Utilization of Palm Biomass Residue in Indonesia

This work scrutinizes the most important agro-industry in Indonesia—palm oil. The results show that it is economically feasible to fully utilize solid and liquid biomass residues, and improve regional economic gains from the palm oil industry in Sumatra. The results can be used to assist in strategic decision-making around the planning and operation of the palm oil supply chain in terms of the cost competitiveness of oil palm biomass utilization, emissions reduction, biomass technology selection, and the prioritization of grid connectivity. The implementation of the results will depend on the availability of financial resources for new investments. Meanwhile, better databases, policy monitoring processes, and more stringent consequences for non-compliance are needed to fully take advantage of current policies, and achieve climate and sustainable development targets.

Data availability poses challenges for the spatial analysis and monitoring of the palm oil industry. A reliable palm oil database is crucial for future research in the oil palm industry, so that relevant policy recommendations can be provided, and the sustainability of palm oil production can be improved.

An actual mapping of oil palm plantations encompassing both small-scale and large-scale plantations, as well as their geographical location and characteristics, is needed for a more robust estimation of FFB availability, monitoring of production, and planning for optimal investments. A complete database of palm oil mills, their geographical location, and operational capacity is also necessary to perform future research on the industry, and plan possible integration with other industries. A plan to improve the database can be carried out together with the definition of monitoring processes for policy implementation. The MoA could collaborate with the regional government in improving the database.

The study demonstrates that *Sc-yield* (i.e., improved yield of small-scale plantations) could obtain 1 billion USD/y more profit than *Sc-ref*. Both *Sc-yield* and *Sc-grid-yield* result in significant benefits in terms of resource efficiency and the reduction of GHG emissions. However, *Sc-grid-yield* requires 70% more investments for upgrading biomass conversion technologies and building transmission lines compared to *Sc-yield*. This suggests that, in the short to medium term, improving smallholder plantation yields is the optimal way to maximize regional economic gains from the palm oil industry. This means that it is worth continuing the financial incentive that the government currently provides to improve small-scale plantation yields. In addition, the use of biofertilizer for substituting chemical fertilizer can contribute to significant emissions reductions. Besides, the potential leaching of biofertilizer into freshwater ecosystems is significantly less harmful than that of conventional fertilizers, which are mixed formulations [66]. These benefits are difficult to materialize without policy support for fertilizer substitution, which could start with a partial substitution of chemical fertilizers in the plantations. More importantly, the effort will significantly reduce the amount of methane that would have been released from untreated POME.

Sc-ref shows that harnessing the full potential of palm biomass residue in Sumatra can contribute to meeting 12% of the national bioenergy target (i.e., 5500 MW_e of bioenergy installed capacity) or 38% of the target for Sumatra (i.e., 1755 MW_e of bioenergy installed capacity), while also delivering 28 $MtCO_2eq/y$ of GHG emissions reductions. Up to 50% of the national target of the bioenergy installed capacity, and emissions reductions of up to 40 $MtCO_2eq/y$, can be achieved with improved grid connectivity. Therefore, in the medium to long term, improving grid electricity connections to palm oil mills will make it attractive to invest in larger electricity capacity at mills, and supply excess electricity to the grid. This measure could be part of an integrated program to increase electricity access in Sumatra, while also upgrading the palm oil industry. Although not analyzed in this study, the high investment for improved grid connectivity provides an opportunity to explore small grid options for mill clusters to meet the electricity demand of surrounding areas. This option can be aligned with the cluster model for biogas plants as described in Section 4.1. Furthermore, it provides an opportunity to develop industrial symbiosis in Sumatra by enhancing interaction and cooperation among mill owners, leading to regional sustainable development.

While the main business of a typical palm oil mill is to process FFB and sell CPO and its byproducts (i.e., PK), it is important to sensitize palm oil mill owners to the opportunities at hand in terms of incorporating energy production in their business models. To realize the existing potential, palm oil mills will have to acquire new expertise or collaborate with new players to develop independent power production capacity. A review of the potential interaction between industrial policy for waste management and energy policy would be valuable to guide policymakers. Likewise, forging partnerships between mill owners and independent power producers can help trigger investments.

In this study, we investigated optimal ways to fully harness oil palm biomass residue for generating both energy and non-energy products. Ultimately, the results show that there are alternatives for decoupling the palm oil industry's growth from plantation expansion, while curtailing deforestation and promoting sustainable development. Existing policy frameworks are working in the right direction, but the various bottlenecks identified here need to be addressed to realize the full bioenergy potential of this industry in Indonesia.

Supplementary Materials: The following are available online at https://www.mdpi.com/1996-1073/12/3/420/s1, Figure S1: List of raw materials, intermediate products (IP), technologies (Tech), bio-products (BP) and associated demands optimized in the BeWhere model. Figure S2: Representation of FFB availability in Sumatra, grid of 25 km × 25 km (left: small-scale plantations, right: large-scale plantations). Figure S3: Mills capacity in grid of 25 km × 25 km. Figure S4: The road network of Sumatra. Figure S5: Geographical location of distribution transformers. Figure S6: District's electricity demand. Table S1: Conversion rate from FFB to intermediate products. Table S2: Technological conversion rate for processing the biomass residue into bio-products. Table S3: Annualized cost of biomass-based technology. Table S4: Prices of bio-products used in the analysis. Table S5: Emissions factor applied in the model.

Author Contributions: "Conceptualization, F.H.; Methodology, F.H., S.L. and S.M.; Software, F.H., S.L. and S.M.; Validation, F.H. and S.L.; Formal Analysis, F.H.; Investigation, F.H.; Data Curation, F.H.; Writing—Original Draft Preparation, F.H.; Writing—Review & Editing, F.H., S.L., S.S. and D.K.; Visualization, F.H.; Supervision, S.L., S.S., F.K.; Funding Acquisition, S.S, F.K.", please turn to the CRediT taxonomy for an explanation of terminology used.

Funding: This research was developed during the Young Scientists Summer Program at the International Institute for Systems Analysis (IIASA), with financial support from Ferrero Trading Lux S.A. The first author's research project is financed by the Swedish Energy Agency [T6473].

Acknowledgments: We would like to thank the Ministry of Agriculture of Indonesia for the data provided. This work was supported by the IIASA Tropical Futures Initiative (TFI) (www.iiasa.ac.at/tropics) and the RESTORE+ project (www.restoreplus.org), which is part of the International Climate Initiative (IKI), supported by the Federal Ministry for the Environment, Nature Conservation and Nuclear Safety (BMU) based on a decision adopted by the German Bundestag.

Conflicts of Interest: The authors declare no conflict of interest. The funders had no role in the design of the study; in the collection, analyses, or interpretation of data; in the writing of the manuscript, or in the decision to publish the results.

Appendix A

Table A1. Results from the factorial design.

Run	A. Mill Annual Operating Hours (h/y)	B. Palm Oil Extraction Rate (t_{CPO}/t_{FFB})	C. Raw Material Production Cost of Large-Scale Plantations (USD/t_{FFB})	D. Capital Cost 1 MW CHP Plant (USD/MWh)	E. Transport Cost ($USD/t_{FFB}/km$)	Total Cost (billion USD)	Total Profit (billion USD)
1	4000	0.18	65	17	0.1	8.50	2.89
2	4900	0.18	65	17	0.1	10.86	3.10
3	4000	0.25	65	17	0.1	8.50	6.14
4	4900	0.25	65	17	0.1	10.86	7.08
5	4000	0.18	98	17	0.1	10.16	1.23
6	4900	0.18	98	17	0.1	12.62	1.32
7	4000	0.25	98	17	0.1	10.16	4.48
8	4900	0.25	98	17	0.1	12.65	5.29
9	4000	0.18	65	30	0.1	8.54	2.85
10	4900	0.18	65	30	0.1	10.89	3.07
11	4000	0.25	65	30	0.1	8.54	6.10
12	4900	0.25	65	30	0.1	10.89	7.05
13	4000	0.18	98	30	0.1	10.19	1.20
14	4900	0.18	98	30	0.1	12.66	1.29
15	4000	0.25	98	30	0.1	10.19	4.45
16	4900	0.25	98	30	0.1	12.68	5.26
17	4000	0.18	65	17	0.3	9.78	1.61
18	4900	0.18	65	17	0.3	13.35	0.59
19	4000	0.25	65	17	0.3	9.78	4.86
20	4900	0.25	65	17	0.3	13.37	4.56
21	4000	0.18	98	17	0.3	11.31	0.07
22	4900	0.18	98	17	0.3	15.14	(1.19)
23	4000	0.25	98	17	0.3	11.33	3.31
24	4900	0.25	98	17	0.3	15.16	2.78
25	4000	0.18	65	30	0.3	9.81	1.58
26	4900	0.18	65	30	0.3	13.39	0.56
27	4000	0.25	65	30	0.3	9.81	4.83
28	4900	0.25	65	30	0.3	13.40	4.53
29	4000	0.18	98	30	0.3	11.34	0.03
30	4900	0.18	98	30	0.3	15.17	(1.22)
31	4000	0.25	98	30	0.3	11.36	3.28
32	4900	0.25	98	30	0.3	15.19	2.75

References

1. Ministry of Agriculture. *Tree Crop Estate Statistics of Indonesia 2015–2017*; Ministry of Agriculture: Jakarta, Indonesia, 2017.
2. Salema, A.A.; Ani, F.N. Pyrolysis of oil palm empty fruit bunch biomass pellets using multimode microwave irradiation. *Bioresour. Technol.* **2012**, *125*, 102–107. [CrossRef]
3. Kasivisvanathan, H.; Ng, R.T.L.; Tay, D.H.S.; Ng, D.K.S. Fuzzy optimisation for retrofitting a palm oil mill into a sustainable palm oil-based integrated biorefinery. *Chem. Eng. J.* **2012**, *200*, 694–709. [CrossRef]
4. Harahap, F. An Evaluation of Biodiesel Policies—The Case of Palm Oil Agro-Industry in Indonesia. Licentiate Thesis, KTH Royal Institute of Technology, Stockholm, Sweden, 2018.
5. Ministry of Energy and Mineral Resources. *Indonesia Energy Roadmap 2017–2025*; Ministry of Agriculture: Jakarta, Indonesia, 2017; p. 3804242.
6. Foo, D.C.Y.; Tan, R.R.; Lam, H.L.; Kamal, M.; Aziz, A.; Rí, J.; Kleme, J. Robust models for the synthesis of flexible palm oil-based regional bioenergy supply chain. *Energy* **2013**, *55*, 68–73. [CrossRef]
7. Lam, H.L.; Ng, W.P.Q.; Ng, R.T.L.; Ng, E.H.; Aziz, M.K.A.; Ng, D.K.S. Green strategy for sustainable waste-to-energy supply chain. *Energy* **2013**, *57*, 4–16. [CrossRef]
8. Memari, A.; Ahmad, R.; Rahim, A.R.A.; Jokar, M.R.A. An optimization study of a palm oil-based regional bio-energy supply chain under carbon pricing and trading policies. *Clean. Technol. Environ. Policy* **2018**, *1*, 113–125. [CrossRef]
9. Idris, M.N.M.; Hashim, H.; Razak, N.H. Spatial optimisation of oil palm biomass co-firing for emissions reduction in coal-fired power plant. *J. Clean. Prod.* **2018**, *172*, 3428–3447. [CrossRef]
10. Harahap, F.; Silveira, S.; Khatiwada, D. Cost competitiveness of palm oil biodiesel production in Indonesia. *Energy* **2019**, *170*, 62–72. [CrossRef]
11. Ng, D.K.; Ng, R.T. Applications of process system engineering in palm-based biomass processing industry. *Curr. Opin. Chem. Eng.* **2013**, *2*, 448–454. [CrossRef]
12. Grossmann, I.E. Challenges in the new millennium: Product discovery and design, enterprise and supply chain optimization, global life cycle assessment. *Comput. Chem. Eng.* **2004**, *29*, 29–39. [CrossRef]
13. Mafakheri, F.; Nasiri, F. Modeling of biomass-to-energy supply chain operations: Applications, challenges and research directions. *Energy Policy* **2014**, *67*, 116–126. [CrossRef]
14. Sahoo, K.; Mani, S.; Das, L.; Bettinger, P. GIS-based assessment of sustainable crop residues for optimal siting of biogas plants. *Biomass Bioenergy* **2018**, *110*, 63–74. [CrossRef]
15. Atashbar, N.Z.; Labadie, N.; Prins, C. Modeling and optimization of biomass supply chains: A review and a critical look. *IFAC-PapersOnLine* **2016**, *49*, 604–615. [CrossRef]
16. Chiew, Y.L.; Iwata, T.; Shimada, S. System analysis for effective use of palm oil waste as energy resources. *Biomass Bioenergy* **2011**, *35*, 2925–2935. [CrossRef]
17. Theo, W.L.; Lim, J.S.; Ho, W.S.; Hashim, H.; Lee, C.T.; Muis, Z.A. Optimisation of oil palm biomass and palm oil mill effluent (POME) utilisation pathway for palm oil mill cluster with consideration of BioCNG distribution network. *Energy* **2017**, *121*, 865–883. [CrossRef]
18. Hadiguna, R.A.; Tjahjono, B. A framework for managing sustainable palm oil supply chain operations: A case of Indonesia. *Prod. Plan. Control* **2017**, *28*, 1093–1106. [CrossRef]
19. IIASA. "BeWhere". Available online: www.iiasa.ac.at/bewhere%0A (accessed on 2 September 2018).
20. Leduc, S. Development of an Optimization Model for the Location of Biofuel Production Plants. Ph.D. Thesis, Luleå University of Technology, Luleå, Sweden, 2009.
21. Wetterlund, E. *Interim Report Optimal Localization of Biofuel Production on a European Scale*; IIASA Publ.: Laxenburg, Austria, 2010; p. 50.
22. Rosenthal, R. *GAMS-A User's Guide*; GAMS Development Corporation: Washington, DC, USA, 2017.
23. Leduc, S.; Wetterlund, E.; Dotzauer, E.; Schmidt, J.; Natarajan, K.; Khatiwada, D. Policies and Modeling of Energy Systems for Reaching European Bioenergy Targets. In *Handbook of Clean Energy Systems*; John Wiley & Sons, Ltd.: Hoboken, NJ, USA, 2015.
24. Mandova, H.; Leduc, S.; Wang, C.; Wetterlund, E.; Patrizio, P.; Gale, W.; Kraxner, F. Possibilities for CO_2 emission reduction using biomass in European integrated steel plants. *Biomass Bioenergy* **2018**, *115*, 231–243. [CrossRef]
25. Natarajan, K.; Leduc, S.; Pelkonen, P.; Tomppo, E.; Dotzauer, E. Optimal locations for second generation Fischer Tropsch biodiesel production in Finland. *Renew. Energy* **2014**, *62*, 319–330. [CrossRef]

26. Khatiwada, D.; Leduc, S.; Silveira, S.; McCallum, I. Optimizing ethanol and bioelectricity production in sugarcane biorefineries in Brazil. *Renew. Energy* **2016**, *85*, 371–386. [CrossRef]

27. Patrizio, P.; Leduc, S.; Chinese, D.; Dotzauer, E.; Kraxner, F. Biomethane as transport fuel - A comparison with other biogas utilization pathways in northern Italy. *Appl. Energy* **2015**, *157*, 25–34. [CrossRef]

28. Hoo, P.Y.; Patrizio, P.; Leduc, S.; Hashim, H.; Kraxner, F.; Tan, S.T.; Ho, W.S. Optimal Biomethane Injection into Natural Gas Grid - Biogas from Palm Oil Mill Effluent (POME) in Malaysia. *Energy Procedia* **2017**, *105*, 562–569. [CrossRef]

29. Garcia-Nunez, J.A.; Rodriguez, D.T.; Fontanilla, C.A.; Ramirez, N.E.; Lora, E.E.S.; Frear, C.S.; Stockle, C.; Amonette, J.; Garcia-Perez, M. Evaluation of alternatives for the evolution of palm oil mills into biorefineries. *Biomass Bioenergy* **2016**, *95*, 310–329. [CrossRef]

30. Bank Indonesia. Data BI Rate—Bank Sentral Republik Indonesia. 2016. Available online: http://www.bi.go.id/en/moneter/bi-rate/data/Default.aspx (accessed on 24 October 2017).

31. Krasovskii, A.; Khabarov, N.; Pirker, J.; Kraxner, F.; Yowargana, P.; Schepaschenko, D.; Obersteiner, M. Modeling burned areas in Indonesia: The FLAM approach. *Forests* **2018**, *9*, 437. [CrossRef]

32. Global Forest Watch. Global Forest Watch Open Data Portal. Available online: http://data.globalforestwatch.org/datasets/palm-oil-mills (accessed on 23 October 2017).

33. Khasanah, N.; van Noordwijk, M.; Ekadinata, A.; Dewi, S.; Rahayu, S.; Ningsih, H.; Setiawan, A.; Dwiyanti, E.; Octaviani, R. The Carbon Footprint of Indonesian Palm Oil Production. Available online: http://old.icraf.org/sea/Publications/files/policybrief/PB0047-12.pdf (accessed on 24 January 2019).

34. Molenaar, J.W.; Persch-Orth, M.; Lord, S.; Taylor, C.; Harms, J. Diagnostic Study on Indonesian Oil Palm Smallholders. Available online: http://www.aidenvironment.org/media/uploads/documents/201309_IFC2013_Diagnostic_Study_on_Indonesian_Palm_Oil_Smallholders.pdf (accessed on 20 September 2017).

35. Pardamean, M. *Mengelola Kebun Dan Pabrik Kelapa Sawit Secara Profesional*; Penebar Swadaya: Jakarta, Indonesia, 2014.

36. Dowell, L.; Rosenbarger, A.; December, S.L. Palm Oil Mill Data: A Step Towards Transparency. 2015. Available online: https://www.wri.org/blog/2015/12/palm-oil-mill-data-step-towards-transparency (accessed on 15 June 2018).

37. Hambali, E.; Rivai, M. The Potential of Palm Oil Waste Biomass in Indonesia in 2020 and 2030. *IOP Conf. Ser. Earth Environ. Sci.* **2017**, *65*, 012050. [CrossRef]

38. IRENA. *Renewable Energy Prospects: Indonesia, a REmap Analysis*; IRENA: Abu Dhabi, UAE, 2017.

39. McKendry, P. Energy Production from Biomass (Part 2): Conversion Technologies. *Bioresour. Technol.* **2002**, *83*, 47–54. [CrossRef]

40. Rahayu, A.S.; Karsiwulan, D.; Trisnawati, H.I.; Mulyasari, S.; Rahardjo, S.; Hokermin, S.; Paraminta, V. Handbook POME-to-Biogas Project Development in Indonesia. Available online: https://www.winrock.org/wp-content/uploads/2016/05/CIRCLE-Handbook-2nd-Edition-EN-25-Aug-2015-MASTER-rev02-final-new02-edited.pdf (accessed on 24 January 2019).

41. BioEnergy Consult. Biomass Market in Japan: Perspectives. 2018. Available online: https://www.bioenergyconsult.com/biomass-market-japan/ (accessed on 25 November 2018).

42. Corley, R.; Tinker, P.B. *The Oil Palm*; Wiley: Backwell, UK, 2016.

43. Golden Agri-Resources. Plantation to Mill in 24 h. 2017. Available online: https://goldenagri.com.sg/plantation-mill-24-hours/ (accessed on 14 October 2018).

44. DIVA-GIS. GIS Data. Available online: http://www.diva-gis.org/gdata (accessed on 7 September 2018).

45. BPS. BPS Database. Available online: http://www.bps.go.id/ (accessed on 7 September 2018).

46. World Bank. Country Profile Indonesia: World Development Indicators. 2016. Available online: http://databank.worldbank.org/data/Views/Reports/ReportWidgetCustom.aspx?Report_Name=CountryProfile&Id=b450fd57&tbar=y&dd=y&inf=n&zm=n&country=IDN (accessed on 9 February 2018).

47. Mesfun, S.; Leduc, S.; Patrizio, P.; Wetterlund, E.; Mendoza-Ponce, A.; Lammens, T.; Staritsky, I.; Elbersen, B.; Lundgren, J.; Kraxner, F. Spatio-temporal assessment of integrating intermittent electricity in the EU and Western Balkans power sector under ambitious CO_2 emission policies. *Energy* **2018**, *164*, 676–693. [CrossRef]

48. Harahap, F.; Silveira, S.; Khatiwada, D. Land allocation to meet sectoral goals in Indonesia—An analysis of policy coherence. *Land Use Policy* **2017**, *61*, 451–465. [CrossRef]

49. Sher, F.; Pans, M.A.; Afilaka, D.T.; Sun, C.; Liu, H. Experimental investigation of woody and non-woody biomass combustion in a bubbling fluidised bed combustor focusing on gaseous emissions and temperature profiles. *Energy* **2017**, *141*, 2069–2080. [CrossRef]
50. USEPA. *EPA Climate Leadership—Emission Factors November 2015*; USEPA: Washington, DC, USA, 2015.
51. World Bank. *State and Trends of Carbon Pricing 2018*; World Bank: Washington, DC, USA, 2018.
52. Agustira, M.A.; Rañola, R.F., Jr.; Sajise, A.J.U. Economic Impacts of Smallholder Oil Palm (Elaeis guineensis Jacq.) Plantations on Peatlands in Indonesia. *J. Econ. Manag. Agric. Dev.* **2016**, *1*, 105–123.
53. Institute for Global Environmental Strategies (IGES). List of Grid Emission Factors, version 10.3. 2018. Available online: https://pub.iges.or.jp/pub/iges-list-grid-emission-factors (accessed on 25 November 2018).
54. UNFCCC. *PDD Co-Composting of EFB and POME—MG BioGreen Sdn.Bhd*; UNFCCC: Bonn, Germany, 2006.
55. Woittiez, L.S.; van Wijk, M.T.; Slingerland, M.; van Noordwijk, M.; Giller, K.E. Yield gaps in oil palm: A quantitative review of contributing factors. *Eur. J. Agron.* **2017**, *83*, 57–77. [CrossRef]
56. NIST/SEMATECH. e-Handbook of Statistical Methods. 2013. Available online: http://www.itl.nist.gov/div898/handbook/ (accessed on 13 October 2018).
57. Ong, H.C.; Mahlia, T.M.I.; Masjuki, H.H.; Honnery, D. Life cycle cost and sensitivity analysis of palm biodiesel production. *Fuel* **2012**, *98*, 131–139. [CrossRef]
58. Posada, J.A.; Rincón, L.E.; Cardona, C.A. Design and analysis of biorefineries based on raw glycerol: Addressing the glycerol problem. *Bioresour. Technol.* **2012**, *111*, 282–293. [CrossRef]
59. Moncada, J.; Tamayo, J.; Cardona, C.A. Evolution from biofuels to integrated biorefineries: Techno-economic and environmental assessment of oil palm in Colombia. *J. Clean. Prod.* **2014**, *81*, 51–59. [CrossRef]
60. Pettersson, K.; Wetterlund, E.; Athanassiadis, D.; Lundmark, R.; Ehn, C.; Lundgren, J.; Berglin, N. Integration of next-generation biofuel production in the Swedish forest industry—A geographically explicit approach. *Appl. Energy* **2015**, *154*, 317–332. [CrossRef]
61. Solikhah, M.D.; Kismanto, A.; Raksodewanto, A.; Peryoga, Y. Profitability and sustainability of small—Medium scale palm biodiesel plant. *AIP Confer. Proc.* **2017**, *1855*, 070005.
62. Ng, R.T.L.; Hassim, M.H.; Ng, D.K.S.; Tan, R.R.; El-Halwagi, M.M. Multi-objective design of industrial symbiosis in palm oil industry. *Comput. Aided Chem. Eng.* **2014**, *34*, 579–584.
63. Martin, M.; Harris, S. Prospecting the sustainability implications of an emerging industrial symbiosis network. *Resour. Conserv. Recycl.* **2018**, *138*, 246–256. [CrossRef]
64. GoI. *Nationally Determined Contribution Republic of Indonesia*; GoI: Jakarta, Indonesia, 2016.
65. PLN. *RUPTL PLN 2016–2025*; PLN: Jakarta, Indonesia, 2016.
66. Malusà, E.; Pinzari, F.; Canfora, L. Effi cacy of Biofertilizers: Challenges to Improve Crop Production. In *Microbial Inoculants in Sustainable Agricultural Productivity: Vol. 2: Functional Applications*; Springer: New Delhi, India, 2016; Volume 2016, pp. 1–308.

Article

Weptos Wave Energy Converters to Cover the Energy Needs of a Small Island

Lucia Margheritini * and Jens Peter Kofoed

Civil Engineering Department, Aalborg University, Thomas Manns Vej 23, 9220 Aalborg Ø, Demark; jpk@civil.aau.dk
* Correspondence: lm@civil.aau.dk; Tel.: +45-9940-8512

Received: 25 November 2018; Accepted: 21 January 2019; Published: 29 January 2019

Abstract: This paper presents the details of a study performed to investigate the feasibility of a wave energy system made up of a number of Weptos wave energy converters (WECs) and sets of batteries, to provide the full energy demands of a small island in Denmark. Two different configurations with 2 and 4 Weptos machines respectively with a combined installed power of 750 kW (and a capacity factor of 0.2) are presented. One full year simulation, based a detailed hourly analysis of the power consumption and wave energy resource assessment in the surrounding sea, is used to demonstrate that both configurations, supplemented by a 3 MWh battery bank and a backup generator, can provide the energy needs of the island. The proposed configurations are selected on the basis of a forecast optimization of price estimates for the individual elements of the solutions. The simulations show that Weptos WECs actually deliver 50% more than average consumption over the year, but due to the imbalance between consumption and production, this is not enough to cover all situations, which necessitates a backup generator that must cover 5–7% of consumption, in situations where there are too few waves and the battery bank is empty.

Keywords: wave energy; battery storage; price estimation; hourly distribution; electricity production; electricity demand

---

1. Introduction

The temporal variability of wave energy has effects on the electricity supply to the grid. The comparison of the hourly variability of the resource, together with the hourly distributions of demands and production, provides the first indications of the dimensioning and design of an effective system. Seasonal and annual variability is increasing, even more so under the effects of climate change [1], and should also be taken into consideration in accordance with the expected lifetime of the power plant; doing so allows for the prediction of underproduction, downtime and possibly strategic storage systems. Transmission and distribution networks can also be affected by irregular power production, and relative fluctuations must be considered further.

The design of an effective system requires three main reliable sets of data: hourly distribution of electricity consumption, a long record of validated metocean data and analytics for the specific location and validated power production of the adopted wave energy technology. These last two are not easy to obtain; the wave energy resource is unevenly mapped around the globe, and very few wave energy technologies have reached the pre-commercial stage [2], so limited validated data exist relative to power production in real sea. Indeed, while the variability of the resource has been under discussion for some years [3], there are only few studies analyzing the performance of wave energy systems on an hourly basis and, at the same time, considering annual variability [4].

The application of wave energy in small islands could be ideal; the location is surrounded by the resource, and consumption is often off the grid. There may however be occasional need to rely on diesel generators with consequent high costs (that include transportation of the fuel) [5].

The study presented in this paper is part of a broader investigation by the Danish Ministry of Defense, the Advisory Department/Building and Energy Division, aimed at evaluating new energy supply plans for a Danish small island in the Baltic Sea. Aalborg University (AAU) has been asked to contribute to it with wave energy expertise. We report here there results of this analysis. The aim of this paper is to describe the design of a wave energy system for a small island of circa 80 inhabitants and significant touristic activity, creating a very uneven and unfavorable demand on the grid. The island and its commitments to sustainability make it an interesting case for the application of wave energy. The island is Natura 2000 Under the EU Birds and Habitats Directive, while its wetlands, coastline and marine environment are protected under the HELCOM and Rasmar convention. All energy and heat supply on the islands is today via 3 oil tanks shipped to location. Electricity is produced by 3 generators powered by diesel fuel. Heating is done via 2 oil boilers and heat recovery on flue gas from generators and 1 oil fired boiler. Additionally, seasonal tourism (June–September) brings circa 80,000 tourist per year. Despite the tourists being mostly daily visitors, some do spend more time on the island, and tourism has a significant impact on the energy consumption and variability of demand. The island is sheltered on the east side by a bigger island, so that a suitable location for the wave energy installation is on the NNE-facing side, with a bathymetry that exceeds 80 m water depth.

2. Weptos Wave Energy Technology

Weptos is an A-shaped floating structure that absorbs wave energy through multiple wave absorbing bodies, i.e., the rotors (Figure 1). It has been through rigorous research and development, which has brought it to the forefront of the wave energy sector [6]. Throughout the development of the Weptos, the design of the full-scale device has been optimized, which has significantly increased its power production, while reducing its weight and cost. This resulted in favorable forecasts of its cost-of-energy. The A-shaped structure has the particularity that it can adjust the angle between the two legs, from 13° up to 120° for the optimization of power absorption as well as the reduction of structural loads in extreme conditions. Furthermore, it provides a natural power smoothening effect, as the rotors on a leg interfere with a wave successively, limiting peak loads on the power take off (PTO), thereby resulting in unusually high load factors on its generators. Depending on the location of installation and the wave climate, the rotors (as well as the whole structure) and power take off are scaled for the specific site conditions. Two sets of 10 rotors are part of the Weptos device.

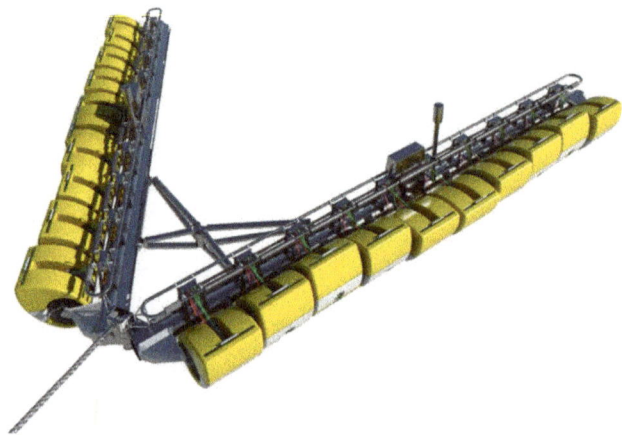

Figure 1. Weptos WEC.

The shape of these rotors was inspired by the Salter duck; their high efficiency was proven already in the 70's [7]. The PTO consists of a fully mechanical transmission, resulting in very high efficiency. The power transmission from the rotors to the axle flows at both upwards or downwards strokes of the rotors. Their weight is optimized, and they feed power to a common uni-directional rotating axle on each leg. The transmission between the rotor and power transmission axle gears the rotational speed up and the torque down. Generator houses are built into the structure and located in the middle of each leg. The mechanical power is then transformed into electrical power by the generator. Depending on the wave conditions during operation, the generator loading is regulated by inverter control settings. The machine is moored to the seabed through a single anchor leg mooring system with rigid members and a buoy which also holds a mechanism allowing Weptos to serve as a weather vane. The machine and its development is described in further detail in [8,9]. The device has been through preliminary real sea testing on a moderate scale during the summer/autumn of 2017, where a 6 kW Weptos prototype called OFFSHORE #1 was tested in Lillebælt (Denmark), near the island Brandsø. These preliminary tests indicated that the performance observed in laboratory testing could be achieved [10]. The Technology Readiness Level for the device is currently considered to be at level 6 [11].

3. Method

In order to design the Weptos WECs for a specific location, the wave climate must be described. The wave climate is defined as the distribution of wave characteristics over a long period of time (several years). The main wave parameters are the significant wave height, Hs, and energy period, Te. A long record of these two statistical parameters is collected in scatter diagrams; each cell of the scatter diagram is identified by a range of Hs and Te, the sea states, and the relative probability of occurrence. For each cell, it is then possible calculate the corresponding wave power and the probability of occurrence of the specific sea state [12]. The scatter diagram is therefore site specific, and describes the wave climate at a specific location. The power matrix or the efficiency curve are the characteristics of the WEC. They can be used to present the performance of a specific device in terms of efficiency, i.e., how much energy is produced for a given sea state. Depending on the stage of development, these are obtained from laboratory tests, numerical simulations or real sea trials; the more advanced the stage of development, the more reliable the power matrix or efficiency curve. In this study, we utilize the efficiency curve of the Weptos. This was defined through laboratory tests and validated during sea trials [8,9], and describes the efficiency of the WEC (in terms of converting the power in the waves into mechanical energy at the unidirectionally rotating axle of the power take-off) as a function of the wave period, based on the active width of the rotors. When applying the efficiency curve to the real sea scale of the WEC, the wave period must be scaled according to Froude scaling, using the assumed scale of the rotor diameter.

For the specific application dealt with in this paper—power supply for a small island grid—the balance of production and consumption is of paramount importance. Hence, it is not sufficient to look at power production based on scatter diagram combined with power matrix/efficiency curve. It is necessary to look at the short-term power balance. At this preliminary stage, it was decided to use one-hour averages as the basis for analysis. Thus, time series of the wave parameters, as well as power consumption on the grid with a one-hour resolution are needed. Using the wave parameters (and, therefore, the wave power flux), combined with the efficiency curve (for an assumed scaling ratio, and with a specific installed generator capacity), the time series of the power production from the WEC at one-hour intervals can be calculated. By matching the power production time series with the power consumption time series, the balance can be made and the difference is fed into or extracted from the battery; as such, the overflow or undersupply can be calculated accordingly.

For this study, one year (1 September 2016 to 31 August 2017) of hourly data for both waves and electricity consumption is available. The designs take into account only hourly and seasonal variability. For this preliminary study, the magnitude of the annual variability of the resource is calculated on a longer record of wave data obtained from hourly wind data (period 1 January 1979–31 December 2016) by calculating the standard deviation of the average available wave power.

In the following sections, the analytics of the wave energy resource is described. It must be noted that while we provide the scatter diagram at location for characterization purposes, the power production is calculated on an hourly basis (and, therefore, with a higher degree of resolution than the scatter diagram) for the period 1 September 2016 to 31 August 2017, compared to the power consumption provided directly by the Utilities, taking into account the storage capacity of the batteries, as well as backup power, when needed.

3.1. One Year Wave Data

One year of wave data was obtained from the Danish Hydraulic Institute (DHI) MetOcean data portal (2016) in order to make a detailed hourly analysis of energy generation and consumption. The wave data set covers the period 1 January 2016–31 December 2016 at the location. The data is provided as time series of the significant wave heights [Hm_0], peak periods [T_p] and related wave direction, one each hour for the year 2016. The wave rose in Figure 2 shows that most of the waves come from WSW direction (circa 45% of the wave events) while the rest mainly come from ENE and E directions.

By grouping the results based on significant wave heights [Hm_0] and peak periods [T_p] into bins using the median value for the indicated value, we can then identify the most common wave conditions (Table 1), i.e., Hm_0 = 0.75 m and T_p = 4.5 s. 15.3% of all calculated wave conditions that fall into this category. Other very common conditions are Hm_0 = 0.25 m and T_p = 3.5 s (12.3% of the time) and Hm_0 = 1.25 m and T_p = 5.5 s (12.2%). Therefore, for the wave conditions in the scatter diagram, the theoretical wave power, calculated with Equation (1), is reported in Table 2. Preliminarily, the most suitable location for the installation of a Weptos WEC to supply power to the island is considered to be circa 2.5 km NNE of the Island in a relatively shallow water location (small plateau of 40 m water depth, surrounded by deeper water > 80 m). A coarse estimation of the deep water limit h/L < 1/2, where h is the water depth and L is the wave length, results in circa 80/170 for H = 4.25 m and T = 9.5 s. It should be noted that for T_p less than 6 s, there is a less than 2% effect of the water depth limit, but as T_p increases, the effect also increases, so at T_p of 11 s, the deep water formulation underestimates the P_{wave} by up to 15%. However, the bulk of the energy contents is found for T_p below 7 s, where the underestimation is at around 5%. This is the difference between using the general expression for wave power density vs. deep water approach [13]. It is also worth considering that sea state data is not derived from a model where this locally reduced water depth is present, and thus, using the model sea states and applying it to the lower water depth will not be correct either, as wave transformation will probably also take place, which is not correctly modelled. Ultimately, since the wave model is not including the finer seabed features, such as the local plateau, it is most accurate to use the general water depth in the area (>80 m), which means that the deep water assumption is fair, and the bulk energy error is below 2%. In any case, since the deep water formulation underestimates the resource slightly, the approach is considered conservative from a production standpoint.

$$P_{wave} = \frac{\rho g^2 H_{m0}^2 T_e}{64\pi} \tag{1}$$

with a relation of T_p/T_e = 1.15, assuming a JONSWAP spectrum with a gamma of 3.3, as in [12] and with ρ of 1023 kg/m³.

To understand where most of the power to be exploited is located among the different wave conditions, it is important to multiply the probability of occurrence to the wave power for the specific location (Table 3). The location under study has a wave climate of 4.38 kW/m (based on wave data from 2016).

Table 1. Probability of occurrence (%) for different wave conditions.

Hm0 \ Tp	1.5	2.5	3.5	4.5	5.5	6.5	7.5	8.5	9.5	10.5	11.5	Total (%)
0.25	0.022769	3.210383	12.31785	7.570583	2.982696	1.104281	0.136612	0	0.011384	0.011384	0	27.367942
0.75	0	0	4.610656	15.25501	7.821038	2.470401	1.161202	0.671676	0.364299	0.045537	0	32.399818
1.25	0	0	0	1.54827	12.21539	3.927596	1.035974	0.432605	0.147996	0	0	19.307832
1.75	0	0	0	0	1.525501	7.61612	0.9449	0.432605	0.07969	0.056922	0	10.655738
2.25	0	0	0	0	0	2.436248	3.449454	0.500911	0	0.011384	0.022769	6.420765
2.75	0	0	0	0	0	0.182149	1.423042	0.808288	0	0.034153	0.011384	2.459016
3.25	0	0	0	0	0	0	0.102459	0.512295	0.136612	0.022769	0	0.774135
3.75	0	0	0	0	0	0	0	0.056922	0.170765	0	0	0.227687
4.25	0	0	0	0	0	0	0	0.022769	0.136612	0.045537	0	0.204918
4.75	0	0	0	0	0	0	0	0	0.011384	0.170765	0.034153	0.182149
	0.022769	3.210383	16.92851	24.37386	24.54463	17.73679	8.253643	3.438069	1.058743	0.398452	0.034153	100

Table 2. Theoretical power for each wave condition (kW/m).

Hm0 \ Tp	1.5	2.5	3.5	4.5	5.5	6.5	7.5	8.5	9.5	10.5	11.5	Total (kW/m)
0.25	0.03924	0.06541	0.09157	0.11773	0.14390	0.17006	0.19622	0.22239	0.24855	0.27471	0.30088	
0.75	0.35320	0.58867	0.82414	1.05961	1.29507	1.53054	1.76601	2.00148	2.23695	2.47241	2.70788	
1.25	0.98112	1.63519	2.28927	2.94335	3.59743	4.25151	4.90558	5.55966	6.21374	6.86782	7.52189	
1.75	1.92299	3.20498	4.48697	5.76897	7.05096	8.33295	9.61494	10.89694	12.17893	13.46092	14.74291	
2.25	3.17882	5.29803	7.41724	9.53645	11.65567	13.77488	15.89409	18.01330	20.13251	22.25173	24.37094	
2.75	4.74860	7.91434	11.08008	14.24581	17.41155	20.57729	23.74302	26.90876	30.07450	33.24023	36.40597	
3.25	6.63235	11.05391	15.47548	19.89705	24.31861	28.74018	33.16174	37.58331	42.00488	46.42644	50.84801	
3.75	8.83005	14.71675	20.60345	26.49015	32.37685	38.26355	44.15025	50.03695	55.92365	61.81035	67.69705	
4.25	11.34171	18.90285	26.46399	34.02513	41.58627	49.14741	56.70855	64.26969	71.83082	79.39196	86.95310	
4.75	14.16733	23.61221	33.05709	42.50198	51.94686	61.39174	70.83663	80.28151	89.72639	99.17128	138.61616	2487.98115

Table 3. Realistic wave power for each sea state considering the relative probability of occurrence = Pwave*Prob (kW/m).

Hm0 \ Tp	1.5	2.5	3.5	4.5	5.5	6.5	7.5	8.5	9.5	10.5	11.5	Total (kW/m)
0.25	0.00001	0.00210	0.01128	0.00891	0.00429	0.00188	0.00027	0.00000	0.00003	0.00003	0.00000	
0.75	0.00000	0.00000	0.03800	0.16164	0.10129	0.03781	0.02051	0.01344	0.00815	0.00113	0.00000	
1.25	0.00000	0.00000	0.00000	0.04557	0.43944	0.16698	0.05082	0.02405	0.00920	0.00000	0.00000	
1.75	0.00000	0.00000	0.00000	0.00000	0.10756	0.63465	0.09085	0.04714	0.00971	0.00766	0.00000	
2.25	0.00000	0.00000	0.00000	0.00000	0.00000	0.33559	0.54826	0.09023	0.00000	0.00253	0.00555	
2.75	0.00000	0.00000	0.00000	0.00000	0.00000	0.03748	0.33787	0.21750	0.00000	0.01135	0.00414	
3.25	0.00000	0.00000	0.00000	0.00000	0.00000	0.00000	0.03398	0.19254	0.05738	0.01057	0.00000	
3.75	0.00000	0.00000	0.00000	0.00000	0.00000	0.00000	0.00000	0.02848	0.09550	0.00000	0.00000	
4.25	0.00000	0.00000	0.00000	0.00000	0.00000	0.00000	0.00000	0.01463	0.09813	0.03615	0.00000	
4.75	0.00000	0.00000	0.00000	0.00000	0.00000	0.00000	0.00000	0.00000	0.01021	0.16935	0.00000	4.38184

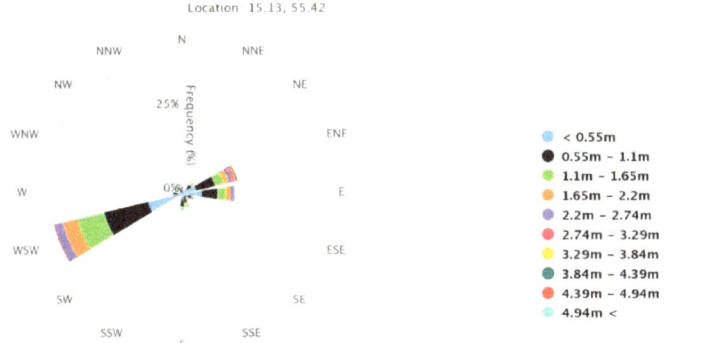

Figure 2. Wave rose at 55.42N, 15.13E for the period 01.01.1016 to 31.12.2016.

3.2. Annual Variability

The year-by-year variability on the wave energy resource should be taken into consideration as the energy content in waves can vary considerably from one year to the next, while the detailed power production and consumption balance is analyzed for one specific year. For this study, considered to be a preliminary assessment, we use the long-term set of data to make a preliminary assessment (from hindcast period 1 January 1979–31 December 2016) [14].

The set of historical wind data covers the period 1 January 1979–31 December 2016 (one wind speed [m/s] measurement each hour, for a total of 333.122 data points). The data has been retrieved as open access from DHI MetOcean data portal (Global, Wind Parameters at 10 m, Climate Forecast System Reanalysis (CFSR), National Centers for Environmental Prediction (NCEP) National Oceanic and Atmospheric Administration (NOAA), 2017). The representative location is 55.42N, 15.13E, circa 10 Km NNW from the island. The wind rose at location for the selected years is presented in Figure 3. We can see that the dominant winds come from WSW, covering circa 12% of all wind data. W, WSW, SW directions together cover 33.7% of all wind while 46.4% is covered by WNW, W, WSW, SW, SSW directions only. The fetch and wind speed are directly related to the wave condition, and therefore, from each wind speed it is possible to calculate a corresponding significant wave height and peak wave period (Hm0 and Tp) using the SPM 1984 Wave Hindcast Model [15] if wind duration, direction, fetch and water depth along the wind directions are known (Figure 4).

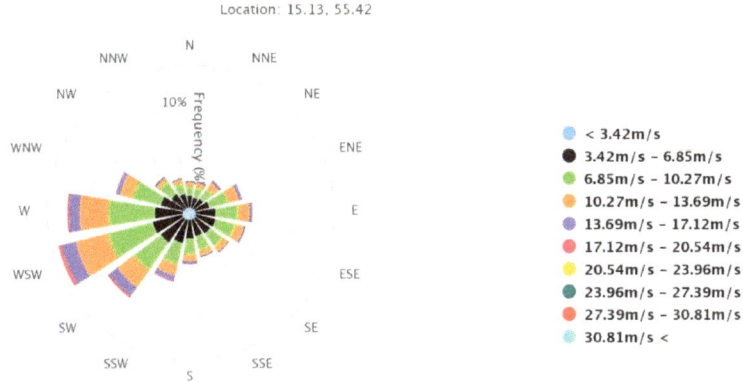

Figure 3. Wind rose near the Island, for the period 01.01.1979 to 31.12.2016.

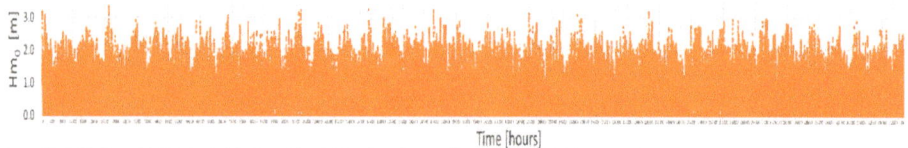

Figure 4. Wave heights in meters from hindcast data, period 01.01.1979 to 31.12.2016. Used only to assess year-by-year variability.

After the calculation of the wave power for each wave condition from 1979 to 2016 with Equation (1), the average wave power for each year was calculated. The mean and standard deviation of P_{wave} year-by-year is given in the Table 4. The mean P_{wave} is here calculated as arithmetic mean of all P_{wave} hourly values. The standard deviation over the whole population of data, from 1979 to 2016 is 4.89. The standard deviation of the annual mean values over all 38 years is 0.41. It can be expected that the variation in the power produced by a wave energy device will roughly follow the variation in the available P_{wave}.

Table 4. Mean and standard deviation Pwave for each year of the data set.

Year	1979	1980	1981	1982	1983	1984	1985	1986	1987	1988	1989	1990	
Average energy [kW/m]	2.90	3.11	2.91	2.09	3.14	3.18	2.87	3.17	3.08	2.70	2.74	3.08	
STDeviation	5.21	4.57	4.55	4.55	5.03	5.07	4.99	5.26	5.83	4.43	4.72	4.97	
Year	**1991**	**1992**	**1993**	**1994**	**1995**	**1996**	**1997**	**1998**	**1999**	**2000**	**2001**	**2002**	**2003**
Average energy [kW/m]	2.68	2.82	3.82	3.18	3.28	3.80	2.62	3.41	2.62	2.47	2.51	3.86	2.60
STDeviation	4.30	4.47	5.89	4.77	5.38	6.24	3.98	4.94	4.31	4.07	4.28	5.48	4.04
Year	**2004**	**2005**	**2006**	**2007**	**2008**	**2009**	**2010**	**2011**	**2012**	**2013**	**2014**	**2015**	**2016**
Average energy [kW/m]	2.86	3.01	2.23	3.08	3.10	2.89	3.91	2.69	2.40	3.01	2.98	2.96	2.88
STDeviation	4.33	5.20	3.78	5.08	4.87	4.53	7.15	4.16	3.82	5.08	4.55	4.56	4.76

4. Results

The most suitable location for the installation of a Weptos WEC to supply power to the island is considered to be circa 2.5 km NNE of the Island in a relative shallow water location (small plateau of 40 m water depth, surrounded by deeper water > 80 m). The location is outside the protected areas around the island, the water depth is less than the surroundings (therefore reducing the cost of moorings), and it is still exposed to the Western wave and wind conditions responsible for most of the favorable wave climate. The wave resource is 4.38 kW/m (in 2016). As a basis for discussion, two different cases will be presented: Four Weptos WECs and Two Weptos WECs, respectively. A coarse cost-of-energy optimization, together with fixed target load factor of 0.20 have been used as the basis in order to find a good balance between installed capacity, scale of the machine, battery capacity etc. At the end, the battery pack size was kept equal to 3000 kWh in both cases and the total installed generator capacity was set to 750 kW (sum across all machines). The load factor has been chosen so to keep results on the conservative side [16].

It is assumed that the potential cost increase due to more and smaller generators, in the case of a park with 4 Weptos compared with the one with 2, is balanced by the reduced cost of the structure, due to higher serial production gains, and therefore, the CAPEX (excluding installation) is assumed to be the same for both cases (same cost per installed capacity, overall). It is expected that properly placing more devices [17,18] can have a smoothing effect in the power output, but only in the short term.

It is therefore necessary to quantify the variability, estimate the underproduction and overproduction times and, consequently, design a system that integrates storage in an optimal way.

The basis for the coarse cost-of-energy optimization resulting in the suggested configuration is given below:

- Cost of WEPTOS: 37,500 DKK/kW generator capacity.
- Cost of battery packs: 3000 DKK/kWh storage capacity.
- Cost of backup generation: 4 DKK/kWh production.

It must be noted that these numbers can only be considered ball park estimates, and are only applicable to this coarse optimization exercise.

4.1. System with 4 Weptos WECs

For the wave conditions at the location, a rotor width of 3 m is found to be suitable for the 4 machines' configuration. With 10 rotors per side, the total active width of each side is 30 m (60 m per WEC). Given the power consumption of the island, it was estimated that 4 Weptos WECs with the above given characteristics will be necessary. Each machine is equipped with a 187.5 kW generator capacity. The peak efficiency of the device is then expected for wave periods around 2.9–3.4 s (Figure 5). In this case, the efficiency (ratio between mechanical power available to the generators P_{mech} and P_{wave} times the active width of the absorbers) is above 50%.

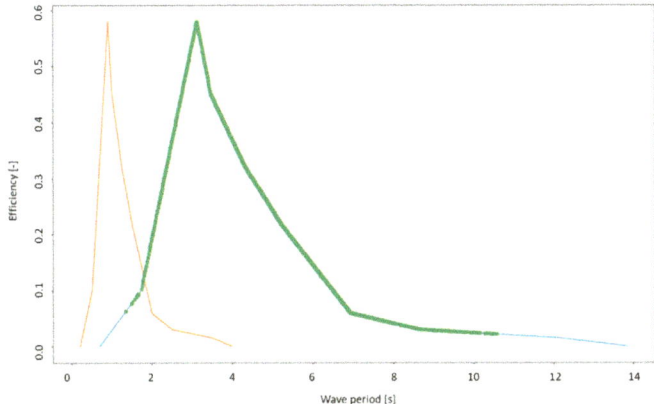

Figure 5. Efficiency of a Weptos with 3 m rotors diameter over different wave periods at location (green) and in the laboratory (yellow).

Figure 6 presents time series of the Power consumption, Power production and Power per meter of wave crest (P_{cons}, P_{prod} and P_{wave}) for the year 2016 (based on hourly data). On the *x*-axis, the hours range from 1 September 2016 to 31 August 2017, where hour 1 is the 1st hour of the 1 September 2016. It can be seen that the highest P_{wave} occurs during the first 6 months, which are the winter months from September to February. On the other hand, the power consumption on the island does not follow the same seasonal trend, as it has its highest peak in the summer months, particularly July, probably due to tourists visiting in this period. When analyzing the power balance, defined here as the power production–power consumption (Figure 7), it can be seen that it is negative for long periods in the months of July (low wave energy resource and high power consumption), as well as in March. To handle this, batteries of 3 MWh and a backup generator system have been considered (Figure 8). When the battery level goes above 3 MWh, the remaining power must be dissipated or not harvested (through PTO detuning), while for the energy below 0 kWh, the backup generator must be used.

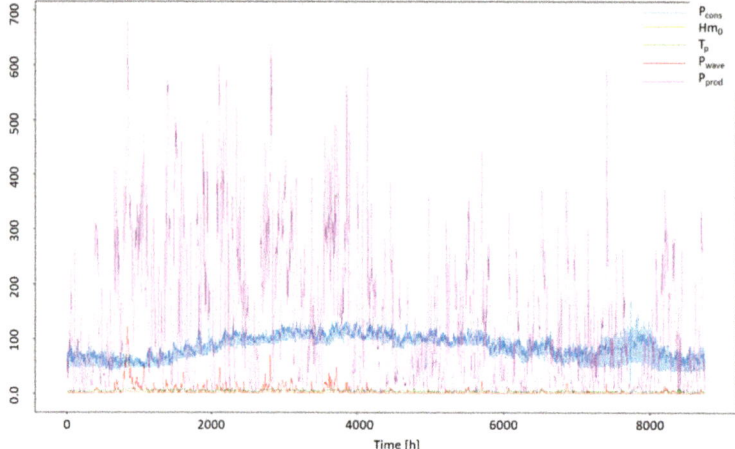

Figure 6. Power consumption, Power production and Power per meter of wave crest (Pcons, Pprod and Pwave) in kW for 2016, 4 Weptos configuration.

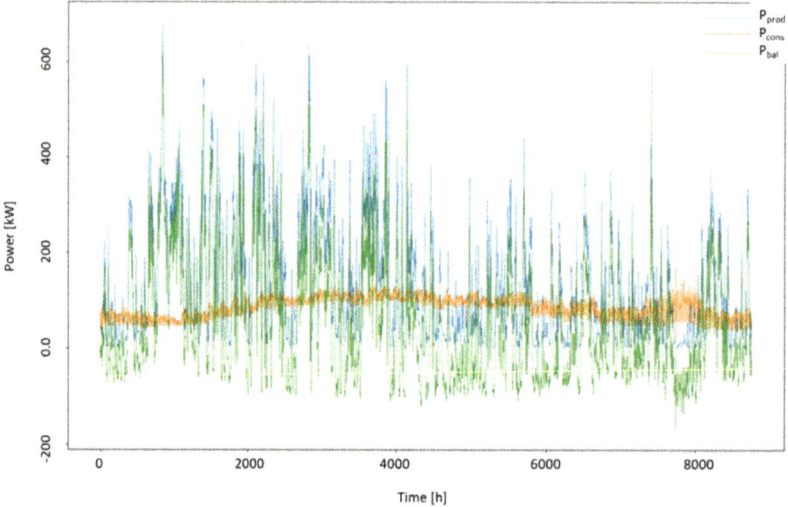

Figure 7. Power production, Power consumption and difference between the two (Power balance), in KW for 2016, 4 Weptos configuration.

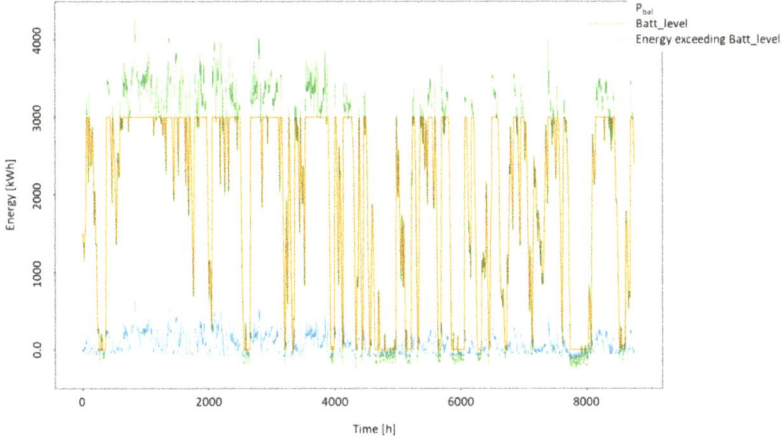

Figure 8. 2 Weptos configuration. Power balance during 8760 hours from 1 September 2016 to 31 August 2017. Pbal = Pprod − Pcons (blue), Batt_level: battery stored energy (orange), Energy Exceeding Batt_level (green): the energy that cannot be stored in the batteries, either because batteries are already full or because there is not enough energy to be stored, as it is all used for the demand.

The main results for the proposed system are presented in Table 5. The average power production in the considered year is 157 kW, while the average consumption is 83.5 kW. The overall efficiency (wave to wire) of the Weptos is 13%. To these figures corresponds an electricity consumption of 731 MWh with a production of 1339 MWh; therefore, the electricity balance = electricity produced − electricity consumed, i.e., 608 MWh/y. With 3 MWh batteries, it is possible to store the electricity surplus for times when the resource is scarce. Nevertheless, the battery capacity is exceeded by 671 MWh, which is also the amount of energy not harvested (50.1%). The deficit in supply of 64 MWh/y needs to be provided by backup generators.

Table 5. Summary of results for 4 Weptos configuration.

Key Parameters	Estimated Figures
Number of Weptos	4
Installed capacity per WEC	187 kW
Total active width of all 4 WECs	240 m
Power production, average	153 kW
Power consumption, average	84 kW
Electricity production	1339 MWh/y
Electricity consumption	731 MWh/y
Electricity balance	608 MWh/y
Battery_over production	671 MWh/y
Battery_under production	−64 MWh/y
Surplus	50.1%
Deficit	4.8%
Load Factor	0.20
Overall efficiency	0.13

4.2. System with 2 Weptos Devices

While the most common wave periods at the location are around 4.5 s, most of the energy is around periods of 6.5 s. The efficiency of the machine can therefore be improved by having larger rotors that will perform better for longer periods. We consider that a rotor width of 3.96 m is suitable. With 10 rotors per side, the total active width of each side of the machine would result in roughly 39.6 m (79.2 m per WEC). Given the power consumption of the island, it was estimated that 2 Weptos

WEC of the above characteristics would be necessary; each machine is equipped with 375 kW generator capacity. The efficiency of the machine is then expected to be maximal for wave periods equal to 3.5–4.0 s (Figure 9). Comparing with Figure 5, a shift to longer periods is seen, to better match the energy distribution of the site. The figures related to power consumption and production are here reported in Figures 10–12 and Table 6. In terms of energy production, the 2 Weptos produce, on average, 148 kW, compared to 153 kW of the 4 WECs configuration in the previous section. Generally, only small variations are found; the most notable is probably that there is a larger need for supplementary energy production in the case with 2 WECs compared with the case with 4 WECs, namely 4.8% vs. 7.2%. These results are not completely unexpected, as we do have the same total installed capacity and load factors in the two cases.

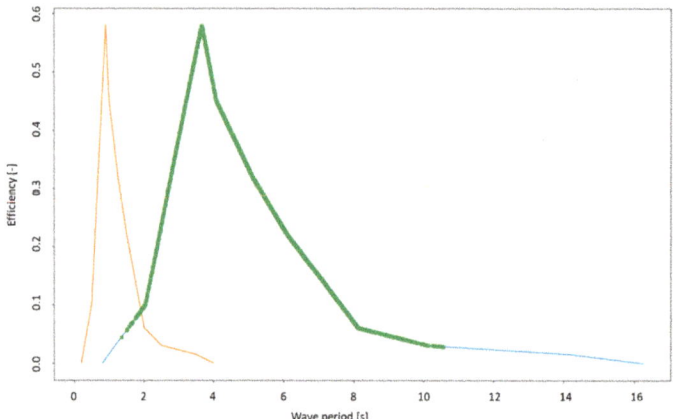

Figure 9. Efficiency of a Weptos with 3.96 m rotors diameter over different wave periods (green) and in the laboratory (yellow).

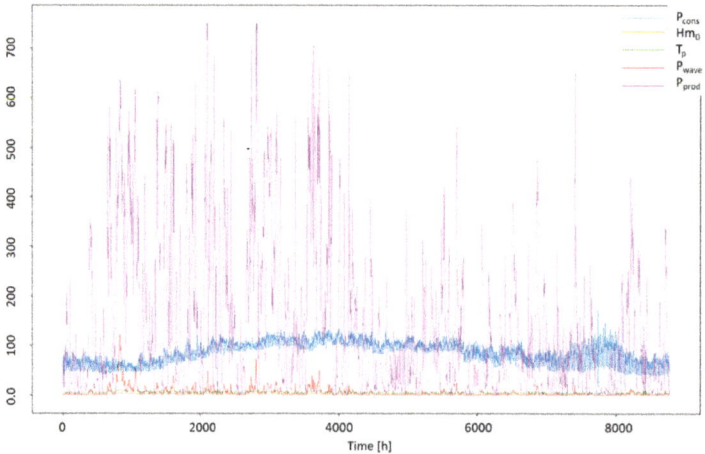

Figure 10. Power consumption, Power production and Power per meter of wave crest (Pcons, Pprod and Pwave) in kW for 2016, 2 Weptos configuration.

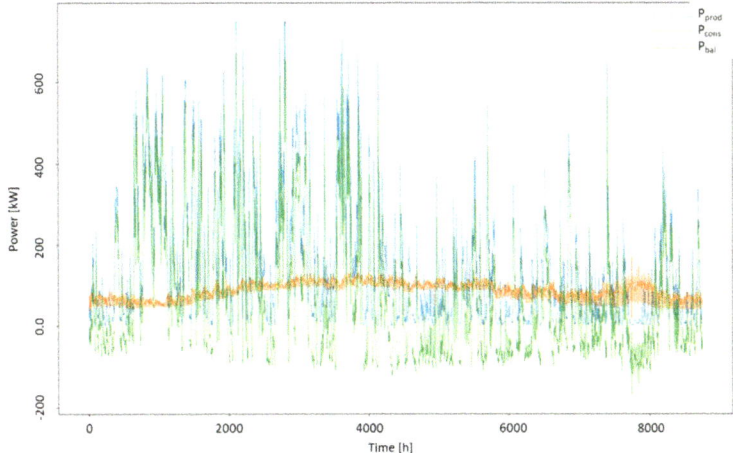

Figure 11. Power production, Power consumption and difference between the two (Power balance), in KW for 2016, 2 Weptos configuration.

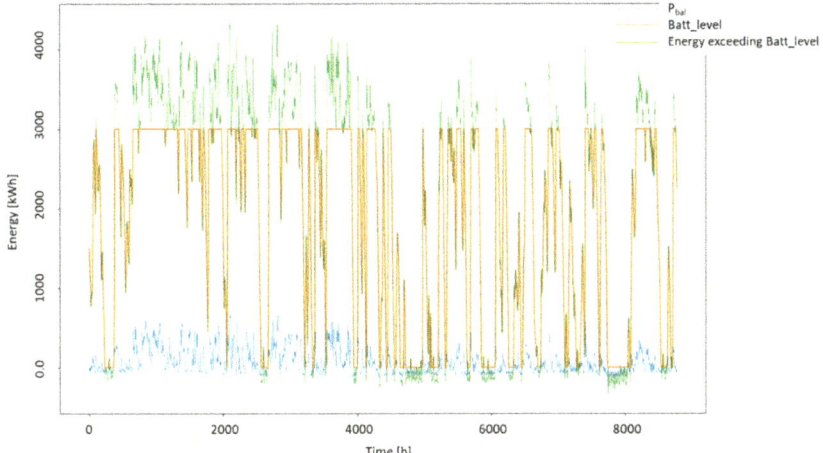

Figure 12. 4 Weptos configuration. Power balance during 8760 hours from 1 September 2016 to 31 August 2017. Pbal = Pprod − Pcons (blue), Batt_level: battery stored energy (orange), Energy Exceeding Batt_level (green): the energy that cannot be stored in the batteries, either because batteries are already full or because there is not enough energy to be stored, as it is all used for the demand.

Table 6. Summary of results for 2 Weptos configuration.

Key Parameters	Estimated Figures
Number of Weptos	2
Installed capacity per WEC	375 kW
Total active width of both WECs	158.4 m
Power production, average	148 kW
Power consumption, average	84 kW
Electricity production	1295 MWh/y
Electricity consumption	731 MWh/y
Electricity balance	564 MWh/y

Table 6. *Cont.*

Key Parameters	Estimated Figures
Battery over production	657 MWh/y
Battery under production	−93 kWh/y
Surplus	50.7%
Deficit	7.2%
Load Factor	0.20
Overall efficiency	0.19

5. Application

In the above, two different cases for deployment of Weptos WECs to cover the energy needs of the island have been presented: Case #1 utilizes 4 smaller Weptos WECs (individual rotor widths of 3.0 m), while Case #2 utilize 2 larger ones (individual rotor widths of 3.96 m).

At this preliminary stage, it is expected that the cost of the machines will be similar between the two cases. In terms of energy production, it was also shown that only small variations are found; the most notable is probably that there is a larger need for supplementary energy production in Case #2 compared to Case #1, namely 4.8% vs. 7.2%. However, on the other hand, only installing and operating 2 instead of 4 machines is expected to be simpler and cheaper, although having 4 machines provides more redundancy, and thereby, robustness, in case of failures and corresponding downtime of individual machines.

The Weptos mooring system is of the slack type and designed to allow >360° rotations. For Case #1 placed at a water depth of 30–40 m, it is expected that each machine will have a watch circle with a radius of roughly 200 m around the mooring point. For placing 4 machines, two different options are considered:

- Placing the 4 machines on a straight line (placed in N–S). In this case, the machines will occupy a rectangle of roughly 400 × 1600 m (area 0.64 km²).
- Placing the 4 machines in a staggered grid formation, i.e., in a diamond with a small angle of 60° (baseline placed in N–S). In this case, the machines will occupy a diamond shaped area with side lengths of rough 800 m (0.55 km²).

Comparing these two options, the first will give the least array/shadowing effects, while the second will result in lower cabling costs.

For Case #2, the watch circle will have a radius of roughly 250 m, and the obvious layout will simply be putting the two machines in a straight line (placed N–S). In this case the machines will occupy a rectangle of roughly 500 × 1000 m (0.5 m²).

Comparing case #1 and #2, it can be seen that #2 is favorable in terms of area usage and cabling lengths.

In this study, losses in the generators, inverters and transmission have been neglected. However, overall losses in generators and inverters are expected to be less than 10%. Also, no array effects have been considered, as these are expected to be minimal for reasonable array layouts. Regarding the installation, operation and maintenance (O&M) of the machines, the following could be considered. Installation on site would be done in two phases: first, the mooring system (anchor, tether, buoy and hawser), including grid connection, would be deployed probably using a barge and a floating crane, and second, the machine itself would be floated to the site and attached to the mooring system. The attachment to the mooring system is designed for ease of connection, meaning hook-on/hook-off operations should be feasible at sea without specialized vessels. i.e., a standard tug should be sufficient for installation and O&M operations, where it is needed to take the machines to port. However, most O&M operations are expected to be carried out on site using a smaller vessel (e.g., a pilot boat, large Rigid-hulled inflatable boat or some other work boat), and the WEC is accessible by boat, even in more severe weather conditions, as the berthing onto the machine (inside the V shape) is protected by the wave absorbing rotors. Thus, it is only expected that it will be necessary to take the machine to port in case of unforeseen needs for repairs. An estimate for needed O&M, once the machines have been properly commissioned, could comprise yearly inspections and repairs on site, and the need to tow it to port once every 10 years.

6. Discussion

The study wants to make a contribution to the investigation of the performance of wave energy systems on an hourly basis based on the energy demands of a small island. For this reason, a detailed analysis of the resource is presented, while the consumption was provided on an hourly basis by the Utilities. The wave data includes a "short" record (1 September 2016 to 31 August 2017) of hourly wave data measurements that have been directly compared to consumption and a long, less precise wave data record, obtained by hourly wind measurements (1 January 1979–31 December 2016) that has been used for the estimations on annual variability of the resource only. Additionally, in the analysis of the previous sections, only the power variation in terms of one-hour averages has been considered. However, power fluctuations within a sea state of one-hour duration must be anticipated. Below, the expected level of fluctuations are illustrated in Figure 13 in terms of normalized instantaneous power (average values of power from a single machine over 2–3 s) as a function of time.

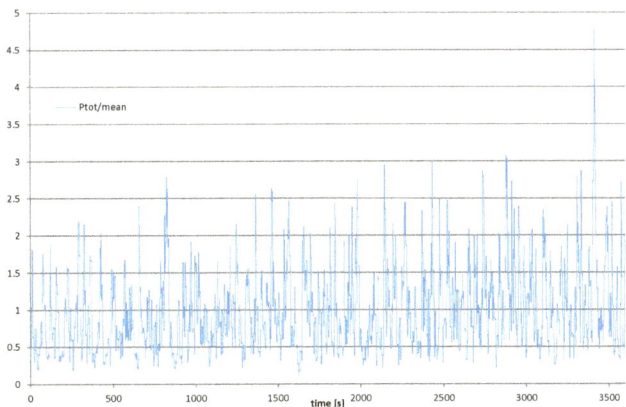

Figure 13. Illustration of short-term variability of produced power given in terms of normalized power (instantaneous power divided by the one-hour average) as function of time. Top: A full hour record, bottom: A zoom of the last 10 min where largest peak occurs.

The data is normalized by the mean power over the considered duration. The coefficient of variation (standard deviation divided by mean) is here 58%. The max-to-mean ratio is 4.8, but this value is heavily dominated by a single instance. Generally, the max-to-mean ratio is in the order of 3, and discarding (power shaving) peaks above this level will have hardly any influence on the mean power. In the more energetic power production conditions, i.e., where 3 times the one-hour average exceeds the installed capacity of the machine, a decrease in production compared to the power production data given in previous sections is to be expected. However, as these conditions typically occur where excess power is produced, the net result on yearly average power production is expected to be minimal. Furthermore, the fluctuations within a one-hour period are expected to be absorbed and smoothed by the battery storage system.

Ideally, the final design should take into account all the power fluctuations mentioned above, as they influence differently the system components.

7. Conclusions

All in all, a few different configurations and layouts for a wave energy-based solution to supply the Island with primarily renewable energy has been provided. It has been demonstrated that for a selected full year (1 September 2016 to 31 August 2017), Weptos machines with a total installed generator capacity of 750 kW (distributed on 2 to 4 machines) and a load factor of 0.20, supplemented by a 3 MWh battery storage and a backup generator (e.g., biodiesel or fuel cell based) can supply the demand on an hour-by-hour basis. The suggested configurations have been based on coarse optimizations based on the rough capital expenditure (CAPEX) estimates of the individual elements of the solution.

In fact, the Weptos WECs can deliver in excess of 50% more energy than what is demanded over the year. However, due to the misalignment of demand and supply, this is not sufficient to cover all instances, and a backup generator is needed to supply 5–7% of the demand, covering the cases where there are too few waves and the battery storage has already been depleted. This backup generator can use traditional combustion, but it is considered appealing to use e.g., a fuel cell solution instead, where (parts of) the excess energy production from the Weptos WECs could be used for hydrogen production. Even though the round trip efficiency (electricity–hydrogen–electricity) is rather low (probably around 50%), the excess energy production should be sufficient (by far) to cover the deficit. However, the economics of such a solution has not been considered here.

The suggested cases will occupy an area of 0.5–0.64 km^2 (depending on the specific choice of layout) at circa 40 m of water depth at a location roughly 2.5 km NNE of the island.

Some considerations regarding both the year-by-year, as well as intra-hour, variations in power have been given. While the intra-hour variations are not expected to impact the feasibility of the solution in a significant way, year-by-year variation can potentially have a very significant impact. It is strongly recommended that a more detailed analysis be done, covering a longer time period than one year, should the wave energy based scenario be selected for further investigations. It is also noted that only very coarse cost estimates, neglecting cabling, installation, operation, maintenance and decommissioning, have been used in this preliminary analysis. It is recommended that a much more thorough cost analysis, and additional optimization loops, mush be carried out before final design. Nevertheless, it can be concluded that the suggested solutions using Weptos WECs in a wave energy-based solution for a renewable energy supply for this island is indeed feasible.

Author Contributions: Conceptualization, L.M. and J.P.K.; Methodology, L.M., J.P.K.; Software, J.P.K.; Validation, L.M.; Formal Analysis, L.M. wave data analysis. J.P.K. power production analysis; Investigation, L.M. and J.P.K.; Data Curation, J.P.K.; Writing-Original Draft Preparation, L.M.; Writing-Review & Editing, L.M. and J.P.K.; Visualization, L.M. and J.P.K.; Supervision, J.P.K.; Project Administration, J.P.K.

Funding: This research was funded by Strunge Jensen A/S and the Advisory Department/Building and Energy Division of the Danish Minister of Defense.

Acknowledgments: Strunge Jensen A/S and the Advisory Department/Building and Energy Division of the Danish Minister of Defense for historic data and encouragement and enthusiasm.

Conflicts of Interest: The authors declare no conflicts of interest.

References

1. Fairley, I.; Smith, H.C.M.; Robertson, B.; Abusara, M.; Masters, I. Spatio-temporal variation in wave power and implications for electricity supply. *Renew. Energy* **2017**, *114*, 154–165. [CrossRef]
2. Alessi, A.; Bannon, E.; Bould, D.; De Marchi, E.; Frigaard, P.B.; Soares, C.G.; Todalshaug, J.H.; Heward, M.; Hofmann, M.; Holmes, B.; et al. *Workshop on Identification of Future Emerging Technologies in the Ocean Energy Sector*; Joint Research Centre (European Commission): Petten, The Netherlands, 2018. [CrossRef]
3. Mackay, E.B.; Bahaj, A.S.; Challenor, P.G. Uncertainty in wave energy resource assessment. Part 2: Variability and predictability. *Renew. Energy* **2010**, *35*, 1809–1819. [CrossRef]
4. Hernandez-Torres, D.; Bridier, L.; David, M.; Lauret, P.; Ardiale, T. Technico-economical analysis of a hybrid wave power-air compression storage system. *Renew. Energy* **2015**, *74*, 708–717. [CrossRef]

5. Fadaeenejad, M.; Shamsipour, R.; Rokni, S.D.; Gomes, C. New Approaches in Harnessing Wave Energy: With Special Attention to Small Islands. *Renew. Sustain. Energy Rev.* **2013**, *29*, 345–354. [CrossRef]

6. Magagna, D.; Margheritini, L.; Moro, A.; Schild, P. Consideration on future emerging technologies in the Ocean Energy Sector. In Proceedings of the 3rd International Conference on Renewable Energies Offshore, Lisbon, Portugal, 8–10 October 2018.

7. Salter, S.H. Wave power. *Nature* **1974**, *249*, 720–724. [CrossRef]

8. Pecher, A.; Kofoed, J.P.; Larsen, T. Design Specifications for the Hanstholm WEPTOS Wave Energy Converter. *Energies* **2012**, *5*, 1001–1017. [CrossRef]

9. Pecher, A.; Kofoed, J.P.; Larsen, T.; Marchalot, T. Experimental Study of the WEPTOS Wave Energy Converter. In Proceedings of the 31th International Conference on Ocean, Offshore and Arctic Engineering (OMAE), Rio de Janeiro, Brazil, 1–6 July 2012.

10. Kofoed, J.P.; Tetu, A.; Ferri, F.; Margheritini, L.; Sonalier, N.; Larsen, T. Real Sea Testing of a Small Scale Weptos WEC Prototype. In Proceedings of the 37th International Conference on Ocean, Offshore & Arctic Engineering, Madrid, Spain, 17–22 June 2018.

11. De Rose, A.; Buna, M.; Strazza, C.; Olivieri, N.; Stevens, T.; Peeters, L.; Tawil-Jamault, D. *Technology Readiness Level: Guidance Principles for Renewable Energy Technologies*; European Commission: Petten, The Netherlands, 2017. [CrossRef]

12. Pecher, A.; Kofoed, J.P. Volume 7 Ocean Engineering & Oceanography. In *Handbook of Ocean Wave Energy*; Springer: Berlin/Heidelberg, Germany, 2017; pp. 43–77. [CrossRef]

13. Pecher, A.; Kofoed, J.P. *Handbook of Ocean Wave Energy*; Springer: Berlin/Heidelberg, Germany, 2017; pp. 221–260. [CrossRef]

14. Margheritini, L.; Kofoed, J.P. *Wave Energy for Grøn Christiansø*; Civil Engineering Department Report; DCE Contract Report No. 192; Aalborg University: Aalborg, Denmark, 2018.

15. Hurdle, D.P.; Stive, R.J.H. Revision of SPM 1984 wave hindcast model to avoid inconsistencies in engineering applications. *Coast. Eng.* **1989**, *12*, 339–351. [CrossRef]

16. Rusu, E.; Onea, F. A review of the technologies for wave energy extraction. *Clean Energy* **2018**, *2*, 10–19. [CrossRef]

17. Salter, S.H. World progress in wave energy. *Int. J. Ambient Energy* **1988**, *10*, 3–24. [CrossRef]

18. Sjolte, J.; Tjensvoll, G.; Molinas, M. Power Collection from Wave Energy Farms. *Appl. Sci.* **2013**, *3*, 420–436. [CrossRef]

Article

Development of Complex Energy Systems with Absorption Technology by Combining Elementary Processes

Kosuke Seki [1,*], Keisuke Takeshita [2] and Yoshiharu Amano [1,3]

1   Department of Applied Mechanics, Waseda University, Tokyo 162-0044, Japan; yoshiha@waseda.jp
2   Waseda Research Institute for Science and Engineering, Tokyo 162-0044, Japan;
    take@power.mech.waseda.ac.jp
3   Advanced Collaborative Research Organization for Smart Society, Tokyo 162-0044, Japan
*   Correspondence: k_seki@power.mech.waseda.ac.jp; Tel.: +81-3-3203-4337

Received: 27 December 2018; Accepted: 2 February 2019; Published: 4 February 2019

Abstract: Optimal design of energy systems ultimately aims to develop a methodology to realize an energy system that utilizes available resources to generate maximum product with minimum components. For this aim, several researches attempt to decide the optimal system configuration as a problem of decomposing each energy system into primitive process elements. Then, they search the optimal combination sequentially from the minimum number of constituent elements. This paper proposes a bottom-up procedure to define and explore configurations by combining elementary processes for energy systems with absorption technology, which is widely applied as a heat driven technology and important for improving system's energy efficiency and utilizing alternative energy resources. Two examples of application are presented to show the capability of the proposed methodology to find basic configurations that can generate the maximum product. The demonstration shows that the existing absorption systems, which would be calculated based on the experience of designers, could be derived by performing optimization with the synthesis methodology automatically under the simplified/idealized operating conditions. The proposed bottom-up methodology is significant for realizing an optimized absorption system. With this methodology, engineers will be able to predict all possible configurations and identify a simple yet feasible optimal system configuration.

Keywords: synthesis/design optimization; cycle configuration; absorption technology; absorption refrigerator

1. Introduction

Recently, in the design of smart cities, etc., it is necessary to derive an optimum system capable of supplying energy satisfying the required specifications while effectively utilizing resources such as renewable or unused waste heat. Optimization for energy systems is performed at three stages: synthesis (configuration), implying the definition of set of components and their interconnections; design (component characteristics), implying the definition of technical specifications of each component and the properties of the working fluids at nominal load; and, operation, implying the definition of operating properties of the working fluid under specified conditions [1]. In particular, the fundamental research on the synthesis/design optimization methodology of energy conversion systems is extremely difficult because there are too many parameters to be considered. Most of the optimization problems are solved using a superstructure prepared by the designer in advance and a pruning strategy for the search process leading to the definition of the optimal configuration [2–4]. In addition to this, research groups have also developed superstructure-free methodologies [5–8] that start from an existing system configuration and add/remove parts of it using evolutionary algorithms to define new design alternatives. However, these approaches have some disadvantages: first, the definition of superstructure is still based on the designer's experience or the starting solution and limits the search space; and, second, the optimal solution sometimes would be too complex and infeasible.

In general, the more complex a system configuration, the more ideal efficiency improvement and product increase can be expected according to the given thermodynamic conditions. However, from the economic point of view, a strategy to design a system that can generate the utmost utility while being a minimum component is necessary. Therefore, the ultimate goal in the optimal design of energy systems is to construct the methodology to realize an energy system that utilizes available resources to maximize the utility with minimum components. For the goal, it is appropriate to decide the optimal system configuration as a problem of decomposing the energy system into primitive process elements and searching the optimal combination sequentially from the minimum number of constituent elements, which is a bottom-up approach. Toffolo and Lazzaretto proposed a general criterion, named SYNTHSEP methodology [9–12], to generate a complex energy conversion system by combining elementary cycles based on the original idea that the elementary thermodynamic cycle is fundamental to the construction of any energy system configuration. The pioneering SYNTHSEP methodology borrows ideas from the HEATSEP methodology [13–16], in which designers focus on a set of fundamental thermodynamic processes (compression, heating, expansion, cooling) in the flowsheet—the so-called "basic configuration" of the system. In this basic configuration, heat transfer devices between system components are replaced with "thermal cuts". All heat transfer processes required for varying the specific enthalpy of working fluid are assumed to occur inside a "black box" of unknown configuration. Design of a heat exchangers network is left to a later process. The SYNTHSEP methodology aims to represent and explore the search space of synthesis/design optimization problems. The basic configurations of energy conversion systems are defined as the combination of the elementary cycles, which are defined as the consequence of four processes (compression, heating, expansion and cooling), obtained by sharing some fundamental processes. With this bottom-up methodology, engineers will be able to predict all possible configurations in advance and identify a simple yet feasible optimal system configuration. The original idea that the basic configuration is defined as a set of elementary cycles can be applied to any types of energy systems, however, the concrete procedure to assemble elementary cycles and to codify and apply the idea to optimization of a system is constructed only for power generation system operating with one pure working fluid so far.

On the other hand, absorption technology is widely applied as a heat driven technology and it is important for improving systems' energy efficiency and utilizing alternative energy resources such as solar power and low-temperature waste heat. The absorption technology-aid system is an energy system operating with mixture and including some absorption and generation processes and the benefits include using significantly less electricity. An absorption power and cooling system can be mentioned as a representative one. A mixed refrigerant cycle for power production and cooling was proposed by Goswami [17]; it combines a Rankine cycle and a refrigeration cycle using absorption technologies. Xu et. al. performed a parametric study for the proposed absorption power and cooling system and proved that the system could be optimized to produce maximum power, refrigeration capacity, and system efficiency [18]. Martin and Goswami also carried out a theoretical and experimental study of the Goswami cycle [19]. The absorption power and cooling system, which attempts to generate both heating/refrigeration capacity and a net power output, sometimes requires the superheating and sub-cooling capabilities of working fluids. Fontalvo et al. performed an exergy analysis for the power and cooling system operating with an ammonia-water mixture and demonstrated the importance of superheater for exergy efficiency of the overall system [20]. The absorption refrigerator, which runs on a heat source instead of electric power, is also a widely used absorption technology. Herold et. al. suggested expressing an absorption refrigeration system as the combination of power generation and refrigeration cycles [21]. The main difference between an absorption refrigerator/heat pump and an absorption power and cooling system is that the former does not require a net power output to function. Examples of the absorption technologies include systems utilizing solar, geothermal, and biomass energy, and engine waste heat recovery [22–25]. Many researchers have attempted to perform optimization on the absorption refrigeration system, e.g., the entropy generation minimization methodology was introduced for designing optimal real devices [26] and Myat et al. demonstrated that minimizing the entropy generated in an absorption refrigeration system leads to the maximization of its coefficient of performance (COP) [27]. Moreover, in order to improve the COP of a system driven by low temperatures, several studies have attempted to change the cycle configurations. Such modified cycles include absorber heat exchanger (AHX) cycle [28] and generator absorber heat exchanger cycle [21], which uses absorption heat to increase generator temperature. However, these studies on the optimization of cycle configuration are only parametric or heuristic, whereas the design solutions are still dependent on experience of designers. To derive the optimal design solution from all possible system candidates, a general procedure to search cycle configurations of the absorption refrigeration system is necessary. Ziegler and Alefeld proposed a procedure to define the cycle configurations for absorption refrigeration system [29,30]. An absorption refrigeration system configuration is defined as the combination of elementary refrigeration cycles and elementary heat transformer cycles. Based on this methodology, Inoue presented a simplified analysis to decide optimal the number of generation processes in an absorption refrigerator for a specific heat-source temperature [31]. However, these methodologies could not guarantee that the absorption refrigerator is the optimal solution among all the possible absorption technology-aided energy systems under specific environmental conditions.

Considering these limitations of previous approaches, this study aims to represent and explore the search space of synthesis/design optimization problems for absorption technology-aided energy systems. The objective of this study is to extend the SYNTHSEP methodology to an absorption technology-aid energy system, which can be a type of power and cooling cycle and operating with different kinds of working fluids and the mixture, and to develop a bottom-up methodology for defining and exploring all possible configurations of the absorption system. This paper, at first, explain the details of the proposed synthesis methodology for absorption system. After that, examples of application are presented to show the capability of the proposed bottom-up methodology to find basic configurations that can generate the maximum product.

2. Bottom-Up Synthesis Methodology

The new methodology aims to generate the basic configuration of the candidate solutions of a synthesis/design optimization problem for an absorption technology-aid system. The proposed methodology derives many ideas from SYNTHSEP to explore the basic configurations, which is defined as a set of elementary thermodynamic cycles [9]. This methodology should be organized and codified so that an optimization algorithm can implement the organized rules to generate new candidate configurations in the optimization problem.

2.1. Main Idea

This research proposes a general procedure to define and explore the basic configuration for energy systems with absorption technologies. The basic configuration of energy conversion systems, including the absorption system, could be expressed as the consequence of four fundamental processes: compression (A), heating (B), expansion (C), and cooling (D) [11]. The basic configurations are defined by combining elementary cycles, which are simply expressed as the consequence of the four processes, sharing some of fundamental processes. A process shared by two or more elementary cycles is called "shared process", while a process operating with a separated working fluid in an elementary cycle is called "non-shared process" in this paper. There are two patterns in the order of the four processes in the elementary cycle: compression, heating, expansion, and cooling in the elementary Rankine cycle;

and compression, cooling, expansion, and heating in the elementary Refrigeration cycle as shown in Figure 1. Complex Rankine cycle configurations are defined as a set of the elementary Rankine cycles [11]. On the other hand, the basic configuration of an absorption system, including a type of combined power and cooling cycle, would sometimes be defined by combining both the elementary Rankine cycles and the elementary Refrigeration cycles.

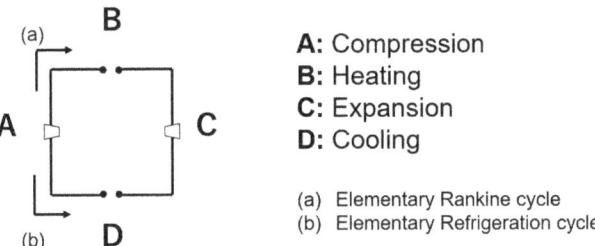

A: Compression
B: Heating
C: Expansion
D: Cooling

(a) Elementary Rankine cycle
(b) Elementary Refrigeration cycle

Figure 1. Basic configuration of elementary thermodynamic cycle.

During the assembly of the elementary cycles, mixers and splitters must be introduced to mark the nodes at which the working fluid in the cycles enters a shared process or leaves from a shared process. For absorption system operating with different kinds of working fluid, the working fluid in shared process is a mixture of absorbent solution and refrigerant. Two different working fluids flow into the mixers, and the resulting mixture in liquid state flows out; therefore, the absorption heat must be dissipated to the external environment. Hence, absorbers are located in mixing points and used as mixers. And the absorption process in this study includes not only the mixing and cooling processes in absorbers but also a cooling process to sub-cool the mixed solution at the outlet of absorbers. Therefore, absorption processes are indicated as a sort of placeholders, which are used to leave the temperature/enthalpy of working fluids at the inlet/outlet of the absorption process free to vary. Additionally, to separate the mixture into its constituent working fluids at the splitters, heat is transferred from the external environment to the mixture; therefore, the generators are located in splitting points and used as splitters. Generation process in this study includes not only the heating and splitting process in generators but also a heating process to superheat the refrigerant at the outlet of the generator and a cooling process to sub-cool the solution at the outlet of generators. Therefore, generation processes are indicated as a sort of placeholders, which are used to leave the temperature/enthalpy of working fluids at the inlet/outlet of the generation process free to vary.

Next, in the region in which cycle processes occur, the intensive design parameters are limited by maximum pressure (temperature), minimum pressure (temperature), minimum specific entropy, maximum specific entropy, maximum mass fraction, and minimum mass fraction. The "mass fraction" means the ratio of the mass of a substance to the total mass of mixture according to each working fluid. These values are set according to operational, technological or environmental constraints The extensive design parameter, indicating mass flow rate, is optimized to find a solution that generate maximum products under the condition of a constant heat source capacity and constraint about the heat transfer feasibility within the undefined heat transfer section.

There are four types of non-shared processes operating with refrigerant: compression, expansion, heating, and cooling, in which the mass fraction and the mass flow rate of refrigerant does not change. In shared processes operating with a mixed solution and non-shared processes operating with an absorbent solution, in this simplified study, it is assumed that the solutions are in their saturated liquid state at all operating nodes and it is assumed that there are only two kinds of processes: compression and expansion. Then, the addition following assumptions are used:

- A solution in its compressed liquid state right after a compression process is heated to be in its saturated liquid state, and this process is considered as a part of compression process in this study.

- A solution right before an expansion process is cooled until the specific entropy of the solution equals that in its saturated liquid state at the outlet of the expansion process, and this process is considered as a part of expansion process in this study.
- Solution heat dissipated is recovered and utilized to heat solutions at other nodes, and the difference of the heat exchange rate from/to a solution is considered as a part of absorption/generation heat.

In these processes, the mass fraction and the mass flow rate of solution do not change. In absorption/generation processes, the pressure is regarded as a constant value, whereas the mass fraction and the mass flow rate of the working fluids are variable and calculated based on the mass balance equations in the mixing/splitting processes.

Figure 2 represents examples of the defined basic configuration for the absorption system consisting of two elementary cycles, in which each elementary cycle shares a part of compression(left)/expansion(right) process. It should be noted that the pressure of separated fluids at node 1/3 could sometimes be equal to the pressure at node 2/4, where the basic configuration does not have a non-shared compression(left)/expansion(right) process between node 1/3 and node 2/4.

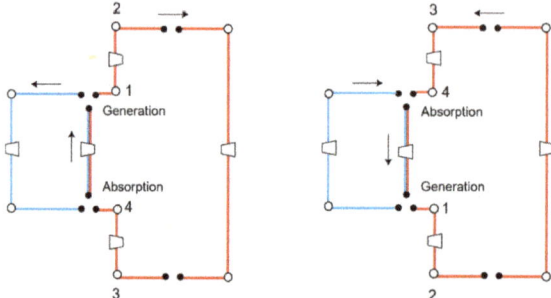

Figure 2. Examples of the basic configuration for the absorption system that is obtained by assembling two elementary cycles, in which each elementary cycle shares a part of compression(left)/expansion(right) process. (red line—refrigerant, blue line—absorbent solution).

There are various possible basic configurations classified according to not only the number of elementary cycles but also the number of absorption/generation processes. For example, a basic configuration made of two elementary cycles with one absorption/generation process and that with two absorption/generation processes are different in view of thermodynamics, where the input/output power in compression/expansion processes would be different from each other because of the difference of pressure range of shared compression/expansion processes operating with mixture. Therefore, it is proposed to search the configurations from those consisting of a few elementary thermodynamic cycles and a few absorption/generation processes.

2.2. Codification Method

The representation of a solution to the synthesis/design optimization problem is based on the decision variables, which should include information on the construction of cycle topologies and design parameters of the target system [9]. In addition, the assembling procedure must follow logical rules to make the candidate configurations feasible solutions. In fact, information on topology sometimes conflicts with information on design parameters. For instance, a case may occur where the pressure values at two nodes show the same value when calculated using topology codes and different value when calculated using design parameter codes. To avoid this situation, the general procedure involves instructions on which information designers should prioritize and which information they must ignore.

2.2.1. The Codification of Topology

The goal of this section is to show the codification of the topology of the basic configurations for the absorption system. The code of the topology has to include the information about the shared elementary cycles and processes, the pattern of which is called "sharing pattern" in this paper. The sharing pattern must be defined in each phase between an absorption process and a generation process, because the main idea of this research proposes to search the configurations consisting of a fixed number of absorption/generation processes, which have the function to change the sharing pattern of each phase. Figure 3 shows an example of the order of the phases with each sharing pattern and the absorption/generation processes. Then, there is no constraint about the order of the absorption/generation processes and engineers should consider all possible orders. The number of the sharing patterns equals the sum of number of absorption processes and generation processes in a thermodynamically closed system.

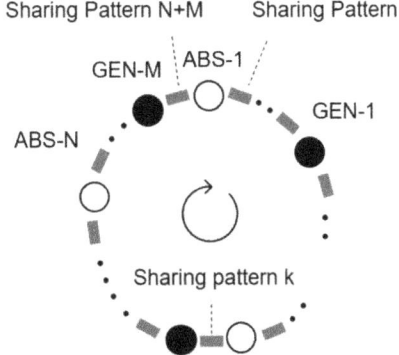

Figure 3. Order of the phases with each sharing pattern and the absorption/generation processes.

The codification features two types of lists of shared processes, one for each type of process (A: compression, C: expansion). Each list represents the elementary cycles shared and the working fluid mixed in each sharing pattern k, the numbering of which can be decided arbitrarily by the designer. List items are binary digits, and their number equals the number of aggregated elementary cycles. A list may be "1" at points that are linked in the shared process, represented by a vector. Moreover, a list may even be empty in case all the possible processes of that particular type occur separately in all elementary cycles. In other mathematical terms, this part of the codification for the sharing pattern k in the elementary Cycle i can be expressed as an organized collection of binary variables b:

$$A_k = \left\{ b_{A_{k,1}}, b_{A_{k,2}}, \ldots, b_{A_{k,i}}, \ldots, b_{A_{k,I}} \right\}, k \in \{1,2\ldots(N+M)\},\ i \in \{1,2\ldots I\} \quad (1)$$

$$C_k = \left\{ b_{C_{k,1}}, b_{C_{k,2}}, \ldots, b_{C_{k,i}}, \ldots, b_{C_{k,I}} \right\}, k \in \{1,2\ldots(N+M)\},\ i \in \{1,2\ldots I\} \quad (2)$$

Example 1: The following are four lists of shared processes in a basic configuration made of two elementary cycles and one absorption/generation process:
$A_1 = \{\, 1\,1\,\}, C_1 = \{\}, A_2 = \{\}, C_2 = \{\}$ indicate that (Figure 4):

- $A_1 = \{\, 1\,1\,\}$; Cycle 1 and Cycle 2 share the compression process in sharing pattern 1.
- $C_1 = \{\}$; The expansion process of each elementary cycle in sharing pattern 1 is isolated.
- $A_2 = \{\}$; Cycle 1 and Cycle 2 does not share the compression process in sharing pattern 2.
- $C_2 = \{\}$; The expansion process of each elementary cycle in sharing pattern 2 is isolated.

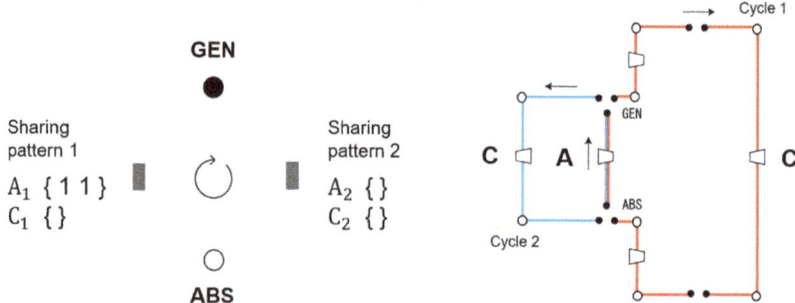

Figure 4. An example of topology definition of a basic configuration, including two elementary cycles and one absorption/generation process. (red line—refrigerant, blue line—absorbent solution).

As for the basic configuration made of one absorption/generation process, the order of these processes is determined as shown in Figure 4.

Example 2: The following are six lists of shared processes in a basic configuration made of three elementary cycles, one absorption process, and two generation processes:
$A_1 = \{\}$, $C_1 = \{111\}$, $A_2 = \{\}$, $C_2 = \{011\}$, $A_3 = \{\}$, $C_3 = \{\}$ indicate that (Figure 5):

- $A_1 = \{\}$; The compression process of each elementary cycle in sharing pattern 1 is isolated.
- $C_1 = \{111\}$; Cycle 1, Cycle 2 and Cycle 3 share the expansion process in sharing pattern 1.
- $A_2 = \{\}$; The compression process of each elementary cycle in sharing pattern 2 is isolated.
- $C_2 = \{011\}$; Cycle 2 and Cycle 3 share the expansion process in sharing pattern 2.
- $A_3 = \{\}$; The compression process of each elementary cycle in sharing pattern 3 is isolated.
- $C_3 = \{\}$; The expansion process of each elementary cycle in sharing pattern 3 is isolated.

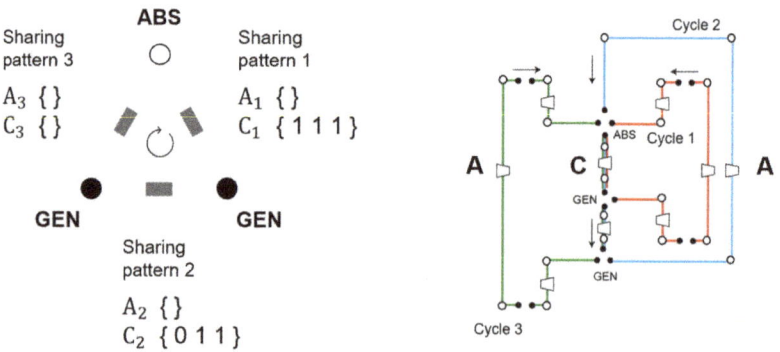

Figure 5. An example of topology definition of a basic configuration for the absorption system including three elementary cycles and one absorption process and two generation processes. (red line, green line—refrigerant, blue line—absorbent solution).

As for the basic configuration made of one absorption process and two generation processes, the order of these processes is determined as shown in Figure 5.

2.2.2. The Codification of Design Parameters

The goal of this section is to show the codification of the intensive design parameters of the basic configurations for the absorption system. The design parameters in non-shared processes, indicating as "sep" in superscripts, are defined with seven real variables per working fluid of each elementary cycle (pressure at two nodes: the inlet and outlet of the compression process, specific enthalpy at the four nodes: the inlet and outlet of the compression process, the inlet and outlet of the expansion process, and mass fraction at any node), resulting in three matrixes or matrices such as:

$$P_{ij}^{sep} \; i \in \{1, 2 \ldots I\}, \, j \in \{1, 2\} \tag{3}$$

$$s_{ij}^{sep} \; i \in \{1, 2 \ldots I\}, \, j \in \{1, 2, 3, 4\} \tag{4}$$

$$z_i^{sep} \; i \in \{1, 2 \ldots I\} \tag{5}$$

with i: the number indicating each elementary cycle and j: the number indicating each node in elementary cycles; node 1: the inlet of the compression process, node 2: the outlet of the compression process, node 3: the inlet of the expansion process, node 4: the outlet of the expansion process. Note that the specific entropy of working fluids at node 2 could be calculated based on the information about the specific entropy at node 1, the pressure at nodes 1 and 2, and the equipment performance characteristics in the compression process, in the basic configuration with specific topology the elementary cycle does not share the compression process. In the same manner, the specific entropy of working fluids at node 4 could be calculated from the specific entropy at node 3 in the basic configuration with specific topology the elementary cycle does not share the expansion process. Moreover, in non-shared processes operating with absorbent solution, the design parameters about the specific entropy does not have to be defined, because that of the solution in its saturated liquid state could be calculated based on the information about the other intensive design parameters.

In shared processes, it is assumed that the mixture is in its saturated liquid state and the mass fraction of mixture could be calculated based on the information about the mass balance equations at absorption/generation processes, so the codification of the design parameters in shared processes includes one real variable, namely pressure at each absorption/generation process, resulting in two vectors:

$$P_{ABS_n}, P_{GEN_m} \; n \in \{1, 2, \ldots N\}, \, m \in \{1, 2, \ldots M\} \tag{6}$$

where n represents the number indicating absorption processes, and m is the number indicating generation processes.

2.2.3. Interaction between Two Parts of the Codification: Topology and Design Parameters

The two parts of the codification are actually interdependent as shown in [9]. Considering the decision variables of the design parameters and topology independently leads to the generation of a large number of infeasible basic configurations. Moreover, information on topology sometimes conflicts with information on design parameters, and some information on one part of the codification has to be either corrected or ignored. The goal of this section is to construct a rule to consider the interaction between the two parts of the codification.

To assemble some elementary cycles into a basic configuration, the absorption and generation processes must be introduced at the inlet or the outlet of each shared process. In the absorption process, two different working fluids flow into the mixer and are mixed, after which the absorption heat is dissipated to external environment and the mixed working fluid enters its liquid phase. Then, there is no constraint on the operating properties of the separated working fluids right before the absorption process. On the other hand, during the generation process, the mixture working fluid is heated until some quantity of the working fluid becomes a vapor while the rest becomes liquid, after which the two working fluids are entirely separated. Therefore, the decision variables about design parameters indicating the operating properties of the separated working fluids right after generation must follow the constraint imposed by this real situation, where the operating properties of one fluid is in liquid phase while those of the other are in vapor phase.

In addition, the pressure exerted by the working fluids at the inlet of the absorption and generation processes and that of the working fluids at the outlet of these processes must be equal. Therefore, the pressure exerted by the mixture at the inlet of each shared compression process, in which some quantity of working fluids in the compression process of each elementary cycle is shared, is equal to the pressure exerted by the separated working fluid at the same point. In addition, this pressure must be larger than or equal to the pressure exerted by the separated working fluids at the inlet of the compression process. The pressure exerted by the mixture at the outlet of each shared compression process must be smaller or equal to than the pressure exerted by the separated working fluids at the outlet of the compression process. In the same manner, for the shared expansion process, the decision variables about design parameters must follow the constraint that the pressure values of the mixture are in the pressure range of the separated working fluids.

Information on topology may limit the pressure range of a working fluid in some elementary cycles as shown in Figure 6. In these basic configurations with specific topologies, the three working fluids of each elementary cycle are shared in the compression(left)/expansion(right) process right before a generation process and the two working fluids of elementary Cycles 2 and 3 are shared in the expansion(left)/compression(right) process right after the generation process; there are no separated compression/expansion processes after the generation process. Therefore, the decision variables about design parameters must meet the constraint that the pressure values of the two separated working fluids at the outlet of the compression(left)/expansion(right) process in the elementary Cycles 2 and 3 are equal to the pressure at the generation process. In other basic configurations with specific topologies, some working fluids of each elementary cycle are shared in the compression/expansion process right before an absorption process, and in the expansion/compression process right after the absorption process. For the basic configurations, in the same manner, the decision variables about the design parameters must follow a rigorous logical constraint about pressures of the working fluids.

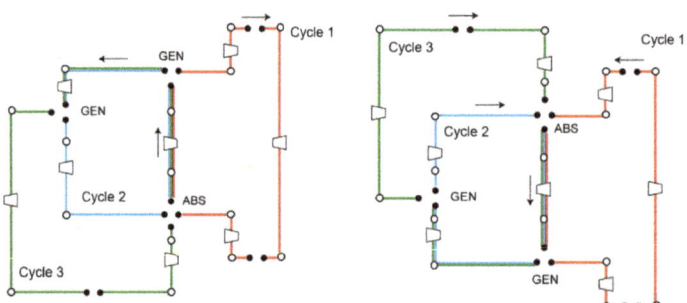

Figure 6. Examples of topologies that limit pressure range of working fluids in non-shared processes. (red line, green line—refrigerant, blue line—absorbent solution). $A_1 = \{111\}$, $C_1 = \{\}$, $A_2 = \{\}$, $C_2 = \{011\}$, $A_3 = \{\}$, $C_3 = \{\}$ (**left**); $A_1 = \{\}$, $C_1 = \{111\}$, $A_2 = \{011\}$, $C_2 = \{\}$, $A_3 = \{\}$, $C_3 = \{\}$ (**right**).

The design parameters of the working fluids in a non-shared process are defined by 7 real variables per working fluid of each elementary cycle. However, some basic configurations with specific topology would require additional decision variables for the specific entropy of certain separated working fluids, as can be seen Figure 7.

In these basic configurations, there are some non-shared processes between the shared compression/expansion processes. The specific entropy at the inlet/outlet of these non-shared processes should be considered in the optimization problem as additional decision variables because the properties cannot be determined based on information on other design parameters and topologies.

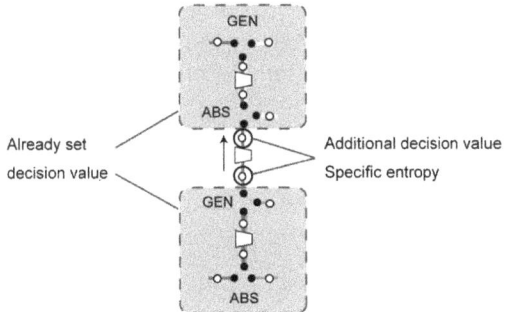

Figure 7. An example of topologies that require definition of several additional design parameters. (red line—refrigerant, blue line, yellow line—absorbent solution).

2.2.4. All Possible Topologies

The goal of this section is to clarify the number of all possible basic configurations for the absorption systems. The number is lower than that of all binary variable patterns in Equations (1) and (2), because some constraints must be considered to define the basic configurations:

(1) Elementary cycles share either "compression" or "expansion" in each sharing pattern and could not share both. Therefore, all binaries for either shared process have to be zero in each sharing pattern.

$$b_{A_{k,i}} = 0 \text{ or } b_{C_{k,i}} = 0 \ \forall i \in \{1,2..I\} \forall k \in \{1,2.., (N+M)\} \tag{7}$$

(2) In the shared process, two or more elementary cycles are shared. In the non-shared process, no elementary cycle is shared. Therefore, there cannot be a pattern in which the sum of all binaries for a shared process is one.

$$\sum_i b_{A_{k,i}} + \sum_i b_{C_{k,i}} \neq 1 \tag{8}$$

(3) The number of elementary cycles shared after absorption is larger than that before the absorption. If an absorption process is located between the phase with sharing pattern k_{ABS} and the phase with sharing pattern $k_{ABS} + 1$, then the following equation can be established:

$$\sum_i b_{A_{k_{ABS},i}} + \sum_i b_{C_{k_{ABS},i}} < \sum_i b_{A_{k_{ABS}+1,i}} + \sum_i b_{C_{k_{ABS}+1,i}} \tag{9}$$

If an absorption process is located between the phase with sharing pattern $N + M$ and the phase with sharing pattern 1, then the following equation can be established:

$$\sum_i b_{A_{N+M,i}} + \sum_i b_{C_{N+M,i}} < \sum_i b_{A_{1,i}} + \sum_i b_{C_{1,i}} \tag{10}$$

(4) The number of elementary cycles shared after the generation process is smaller than that shared before it. If a generation process is located between the phase with sharing pattern k_{GEN} and the phase with sharing pattern $k_{GEN} + 1$, then the following equation can be established:

$$\sum_i b_{A_{k_{GEN},i}} + \sum_i b_{C_{k_{GEN},i}} > \sum_i b_{A_{k_{GEN}+1,i}} + \sum_i b_{C_{k_{GEN}+1,i}} \tag{11}$$

If a generation process is located between the phase with sharing pattern $N + M$ and the phase with sharing pattern 1, then the following equation can be established:

$$\sum_i b_{A_{N+M,i}} + \sum_i b_{C_{N+M,i}} > \sum_i b_{A_{1,i}} + \sum_i b_{C_{1,i}} \tag{12}$$

Example 1: For basic configurations made of two elementary cycles, one absorption process, and one generation process, there could be two possible topologies for the absorption systems that meet the constraints.

$$\begin{aligned}
A_1 &= \{\,1\,1\,\}, \; C_1 = \{\,\}, \; A_2 = \{\,\}, \; C_2 = \{\,\} \\
A_1 &= \{\,\}, \; C_1 = \{\,1\,1\,\}, \; A_2 = \{\,\}, \; C_2 = \{\,\}
\end{aligned} \tag{13}$$

Example 2: For basic configurations made of three elementary cycles, one absorption process, and two generation processes, there could be twelve possible topologies for the absorption systems that meet the constraints.

$$\begin{aligned}
A_1 &= \{\,1\,1\,1\,\}, \; C_1 = \{\,\}, \; A_2 = \{\,1\,1\,0\,\}, \; C_2 = \{\,\}, \; A_3 = \{\,\}, \; C_3 = \{\,\} \\
A_1 &= \{\,1\,1\,1\,\}, \; C_1 = \{\,\}, \; A_2 = \{\,0\,1\,1\,\}, \; C_2 = \{\,\}, \; A_3 = \{\,\}, \; C_3 = \{\,\} \\
A_1 &= \{\,1\,1\,1\,\}, \; C_1 = \{\,\}, \; A_2 = \{\,1\,0\,1\,\}, \; C_2 = \{\,\}, \; A_3 = \{\,\}, \; C_3 = \{\,\} \\
A_1 &= \{\,1\,1\,1\,\}, \; C_1 = \{\,\}, \; A_2 = \{\,\}, \; C_2 = \{\,1\,1\,0\,\}, \; A_3 = \{\,\}, \; C_3 = \{\,\} \\
A_1 &= \{\,1\,1\,1\,\}, \; C_1 = \{\,\}, \; A_2 = \{\,\}, \; C_2 = \{\,0\,1\,1\,\}, \; A_3 = \{\,\}, \; C_3 = \{\,\} \\
A_1 &= \{\,1\,1\,1\,\}, \; C_1 = \{\,\}, \; A_2 = \{\,\}, \; C_2 = \{\,1\,0\,1\,\}, \; A_3 = \{\,\}, \; C_3 = \{\,\} \\
A_1 &= \{\,\}, \; C_1 = \{\,1\,1\,1\,\}, \; A_2 = \{\,1\,1\,0\,\}, \; C_2 = \{\,\}, \; A_3 = \{\,\}, \; C_3 = \{\,\} \\
A_1 &= \{\,\}, \; C_1 = \{\,1\,1\,1\,\}, \; A_2 = \{\,0\,1\,1\,\}, \; C_2 = \{\,\}, \; A_3 = \{\,\}, \; C_3 = \{\,\} \\
A_1 &= \{\,\}, \; C_1 = \{\,1\,1\,1\,\}, \; A_2 = \{\,1\,0\,1\,\}, \; C_2 = \{\,\}, \; A_3 = \{\,\}, \; C_3 = \{\,\} \\
A_1 &= \{\,\}, \; C_1 = \{\,1\,1\,1\,\}, \; A_2 = \{\,\}, \; C_2 = \{\,1\,1\,0\,\}, \; A_3 = \{\,\}, \; C_3 = \{\,\} \\
A_1 &= \{\,\}, \; C_1 = \{\,1\,1\,1\,\}, \; A_2 = \{\,\}, \; C_2 = \{\,0\,1\,1\,\}, \; A_3 = \{\,\}, \; C_3 = \{\,\} \\
A_1 &= \{\,\}, \; C_1 = \{\,1\,1\,1\,\}, \; A_2 = \{\,\}, \; C_2 = \{\,1\,0\,1\,\}, \; A_3 = \{\,\}, \; C_3 = \{\,\}
\end{aligned} \tag{14}$$

3. Demonstration for Synthesis/Design Optimization

This section presents examples of application to show the capability of the proposed bottom-up methodology to find optimal basic configurations for absorption technology-aid systems. Two case studies for optimization are performed to represent that the existing absorption refrigerator is rightfully chosen as the optimal solution from all possible absorption systems under the specific operating conditions. Besides, in earlier studies, it is predicted that the optimal effect number for an absorption refrigerator can be determined based on the heat source temperature under ideal operating conditions [31]. This section also attempts to evaluate and compare the calculated optimal solutions with those derived based on the previous optimal design methodology for the absorption refrigeration system [31].

3.1. Optimization Problem

3.1.1. Target System

This research focuses on an absorption system operating with water-lithium-bromide mixture, in which water is the refrigerant in saturated state, where the saturated liquid and saturated vapor coexist in vapor-liquid equilibrium state or either one exists, and water-lithium-bromide is the solution in saturated liquid state at all operating points.

This system aims to transfer heat from heat source to provide heat to supplied hot water and to remove heat from supplied chilled water. The simplified study assumes that the hot water temperature and the chilled water temperature are constant. The product of the system is either the heating capacity ($\dot{Q}_M$) at the hot water temperature (T_M) or the refrigeration capacity ($\dot{Q}_L$) at the chilled water temperature (T_L), while the system does not require net power output. It is set that the absorption heat in absorption processes is utilized as a part of heating capacity of a system ($\dot{Q}_M$).

To obtain ideal conditions for the operating properties, the following assumptions are made:

(1) There is no heat loss or pressure loss.
(2) In heat exchange process, the difference between the temperatures of the hot and cold side at the pinch point is 0 °C.
(3) In compression/expansion processes, the working fluid undergoes a reversible change and the properties undergo an isentropic change.

The demonstration targets the two types of basic configurations, a configuration made up of two elementary cycles and an absorption/generation process, and a configuration made up of three elementary cycles, an absorption process, and two generation processes, the topologies of which are as indicated in Equations (13) and (14).

3.1.2. Calculation Method

Figure 8 shows a flow chart of the synthesis/design optimization procedure. The optimization problem is solved using both the full search method and linear programming (LP) method, which is used to optimize the operating properties about specific entropy of refrigerants. To express the linear relationship between the parameters and the energy transfer rate from/to a refrigerant, we assume limited condition that a refrigerant is in its saturated state, where the saturated liquid and saturated vapor coexist or either one exists. Of course, if we apply a more general optimization methodology, we can explore a wider range of feasible solutions. The energy transfer rate from/to a refrigerant is calculated as the product of mass flow rate and the difference between specific enthalpy of refrigerant at the inlet and that at the outlet of each process. Then, the specific enthalpy of a refrigerant in saturated state at each node can be calculated based on the information about specific entropy at that node, with the linear relationship between specific entropy and specific enthalpy of the refrigerant in saturated state under the constant pressure (temperature) condition (Clausius—Clapeyron equation) [32]. Therefore, the relationship between the design parameters for specific entropy and the energy transfer rate from/to a refrigerant can be expressed as a linear one.

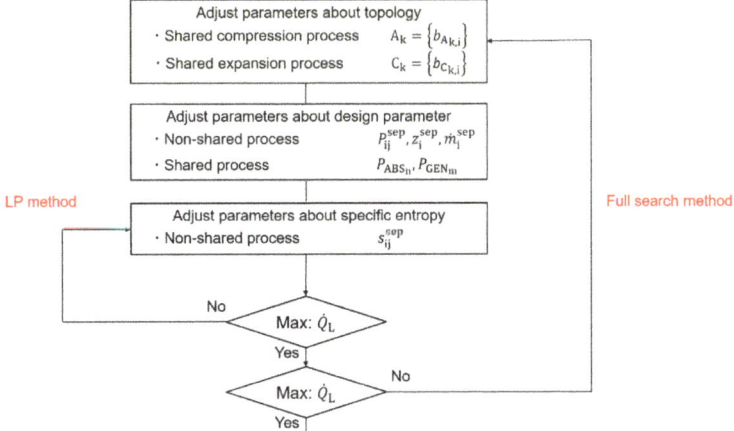

Figure 8. Optimization calculation procedure.

3.1.3. Decision Variables

The optimal topology and design parameters of the basic configuration are calculated by evaluating the following parameters:

<Topology variables>

- Lists of shared compression and expansion processes

$$A_k = \left\{ b_{A_{k,i}} \right\} \tag{15}$$

$$C_k = \left\{ b_{C_{k,i}} \right\} \tag{16}$$

<Design parameter variables>

- Operating properties of working fluid in non-shared process

$$P_{ij}^{sep} j \in \{1, 2\} \tag{17}$$

$$s_{ij}^{sep} j \in \{1, 2, 3, 4\} \tag{18}$$

$$z_i^{sep} \tag{19}$$

- Pressure of mixture at absorption/generation processes

$$P_{ABS_n} \tag{20}$$

$$P_{GEN_m} \tag{21}$$

- Mass flow rate of working fluid in each elementary cycle

$$\dot{m}_i^{sep} \tag{22}$$

3.1.4. Constraints

The constraints include mass balance, energy balance, equipment performance characteristics, and operating conditions at several specific points. This study assumes that water is in the saturated state and water-lithium bromide is in the saturated liquid state in non-shared process, and the mixture of the two is in the saturated liquid state in shared process at all the nodes. In addition, it is assumed that the temperature of hot thermal stream is higher than or equal to the temperature of cold thermal stream in the same heat duty condition in the grand composite curves of the overall system. In the idealized condition, the minimum pinch temperature would be 0 °C. Hence, the constraint is established as follows:

$$\Delta H_h(T) \geq \Delta H_c(T) T \in (T_{amb} < T) \tag{23}$$

- Hot thermal stream

  Heat duty is calculated as the heat transfer rate the working fluid and the heat source supplies to other fluids. This simplified demonstration does not consider the heating capacity of a system ($\dot{Q}_M$) as a part of heat duty in the composite curves but adds a constraint that the temperature of refrigerant in heat exchange process with supplied hot water in each elementary cycle is higher than or equal to the hot water temperature (T_M).

- Cold thermal stream

  Heat duty is calculated as the heat transfer rate the working fluid accepts from the heat source and other working fluids. This demonstration does not consider the refrigeration capacity ($\dot{Q}_L$) as a part of heat duty in the composite curves but adds a constraint that the temperature of refrigerant in heat exchange process with supplied chilled water in each elementary cycle is lower than or equal to the chilled water temperature (T_L).

The constraints also include information on an evaluation indicator as shown in Equation (24). This constraint compares the Carnot factor of a vapor compression refrigeration cycle between the same temperature (T_M and T_L) and the ratio between refrigeration capacity and input power in the absorption system, which is of high value in a system using heat as an energy source instead of an input power source such as the existing absorption systems. The constraints would enable us to get the existing absorption system as a solution from any energy systems by setting the value f_{abs}, which is chosen by the designers based on information about environmental conditions and use applications.

$$\frac{\dot{Q}_L}{\dot{W}_{in}} > f_{abs} \frac{T_L}{T_M - T_L} \tag{24}$$

3.1.5. Objective Function

The objective function is set as the output refrigeration capacity $\dot{Q}_L$ obtained under the fixed heat source conditions. In addition, the COP is defined as the evaluation indicator of system performance. The COP is expressed as the ratio of output refrigeration capacity to the input energy rate of the system, which equals the net heat rate transferred from the external environment in generation processes and can be calculated as in Equation (25), where $\dot{Q}_{GEN,rec}$ is the heat rate recovered from an internal working fluid flowing in other processes.

$$COP = \frac{\dot{Q}_L}{\dot{Q}_{in}} = \frac{\dot{Q}_L}{\dot{Q}_{GEN} - \dot{Q}_{GEN,rec}} \tag{25}$$

3.2. Demonstration

This section aims to perform optimization for an absorption system under specific operating conditions and represent the capability of the proposed methodology to find basic configurations that can generate the maximum refrigeration capacity. The calculated solutions are evaluated and compared with those derived from the previous optimal design methodology [31].

3.2.1. Input Parameters

Input parameters are shown in Table 1, including the heat source condition and operating conditions at a few points. The two cases have a difference in the operating condition—heat source temperature.

Table 1. Input parameters of the operating properties of working fluids in the case studies.

Item	Symbol	Unit	Value Case X	Value Case Y
Mass flow rate of heat source fluid	$\dot{m}_H$	kg/s	10	
Specific heat at constant pressure of heat source	c_{P_H}	kJ/(kg·°C)	4.217	
Heat source temperature at system inlet	$T_{H,in}$	°C	90	130
Heat source temperature at system outlet	$T_{H,out}$	°C	$T_{H,in} - 1$	
Hot water temperature	T_M	°C	38	
Chilled water temperature	T_L	°C	5	
Mass fraction of working fluid in Cycle 1	z_1^{sep}	kg/kg	0	
Mass fraction of working fluid in Cycle 3	z_3^{sep}	kg/kg	0	
Coefficient included in the constraint (24)	f_{abs}	-	100	

3.2.2. Results and Discussion

This section describes the calculated optimal solution based on the defined optimization problem and evaluate the solutions by comparing the calculated solutions with those derived based on the previous optimal design methodology for the absorption refrigerator. In case X, it is known from a literature [31] that the optimal absorption refrigeration system with the highest *COP* is the single-effect one under a constant heat source temperature of 90 °C. The calculated solution in this study seems to be reasonable because the optimal configuration and operating properties in the Dühring chart is very similar to those in an existing single-effect absorption refrigeration system, as shown in Figure 9. In both cycle configurations, the mixture (node 2) is separated into the absorbent solution fluid (node 3) and the refrigerant fluid (node 5) in the generation process. The separated absorbent solution is expanded and flows to the absorption process (node 4), while the refrigerant is used as a working fluid to produce the heating/refrigeration capacity of the system before passing it to the absorption process (nodes 5, 6, 7, 8). The basic configuration has a shared compression process between two elementary cycles (nodes 1, 2), but no non-shared compression process. Therefore, only the mixed solution flows in the compression process to reduce the required input power as in Table 2. Moreover, the operating points in the calculated solution is also similar to that in the solutions in the previous study, as shown Figure 9. As in Figure 9d, the difference of mass fraction of absorbent solution and mixture cannot appear in the Dühring chart of the solution in the previous study because it is assumed that the mass flow rate of absorbent solution is much larger than that of refrigerant, resulting in a little difference of operating points of two systems. Because of the too idealized assumptions in the previous problem, the two solutions have a little difference, however, it could be said that the calculated solution in this demonstration could express the main features of the solution in the previous study. As a result, the *COP* in calculated solution is 0.82, which is near to the *COP* of the solution in the previous study: 0.9.

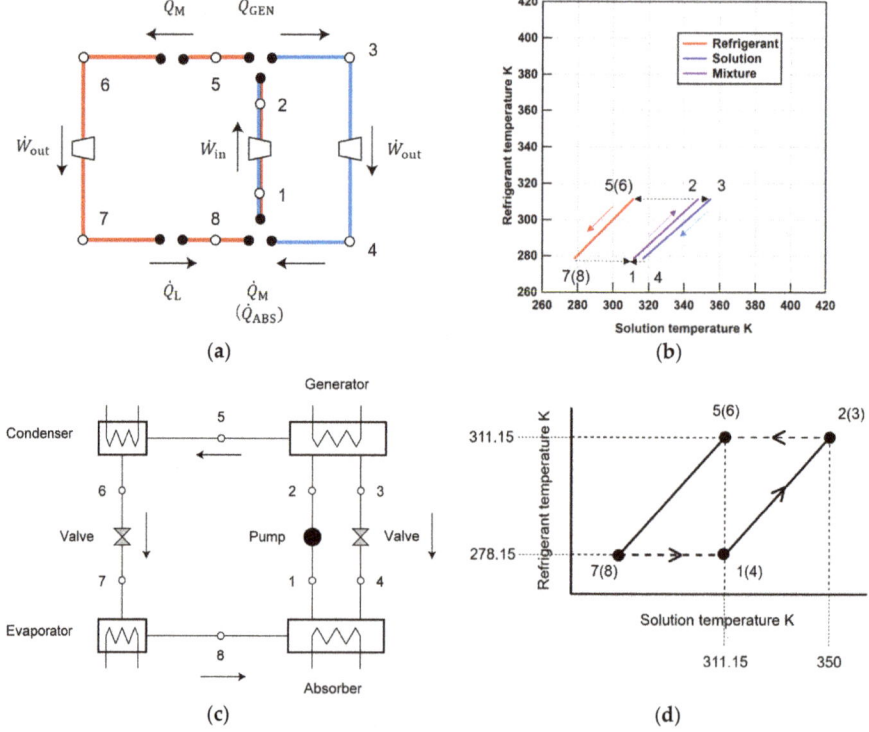

Figure 9. Optimal solution in case X (red line—refrigerant, blue line—absorbent solution): (**a**) Basic configuration calculated; (**b**) Operating properties calculated; (**c**) Cycle configuration of an existing single-effect absorption refrigerator; (**d**) Operating properties of an existing system.

Table 2. Calculated values about the energy exchange rate in case X.

Item	Symbol	Unit	Value
Output power	$\dot{W}_{out}$	W	12.25
Input power	$\dot{W}_{in}$	W	0.76
Output heating capacity	$\dot{Q}_M$	W	7637
Generation heat rate	$\dot{Q}_{GEN}$	W	4201
Heat recovery rate	$\dot{Q}_{GEN,rec}$	W	0
Heat source potential energy	$\dot{Q}_H$	W	4217
Coefficient of performance	COP	-	0.82
Output refrigeration capacity	$\dot{Q}_L$	W	3448

In case Y, it is known from [31] that the optimal absorption refrigeration system with the highest COP is the double-effect one under a constant heat source temperature of 130 °C. The calculated solution in this study seems to be reasonable because the optimal configuration and operating properties in the Dühring chart is very similar to those in an existing double-effect absorption refrigeration system, as shown in Figure 10.

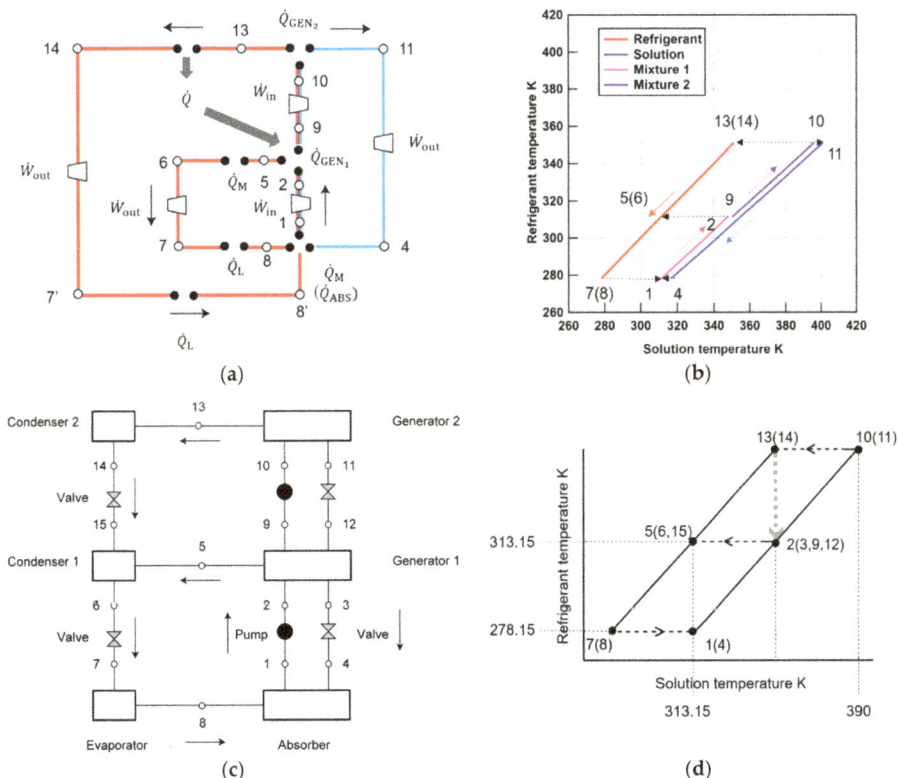

Figure 10. Optimal solution in case Y (red line—refrigerant, blue line—absorbent solution): (**a**) Basic configuration calculated; (**b**) Operating properties calculated; (**c**) Cycle configuration of an existing double-effect absorption refrigerator; (**d**) Operating properties of an existing system.

In both cycle configurations, the dilute mixed solution (node 1) is compressed (node 2) and separated into the concentrated solution fluid (node 9) and the refrigerant fluid (node 5) in generation process 1. The separated refrigerant fluid is used as the working fluid to produce heating/refrigeration capacity of the system before being transported to the absorption process (nodes 5, 6, 7, 8). The separated solution fluid is compressed (node 10) and separated into the absorbent solution (node 11) and the refrigerant (node 13). The separated solution is expanded and transported to the absorption process (node 4), while the refrigerant is cooled (node 13, 14); the heat is reused as a part of generation heat in generation process 1. After this, the refrigerant is expanded (node 7′) and used as the working fluid to produce refrigeration capacity of the system (nodes 7′, 8′). The refrigeration capacity of the system is produced through two mass flows in the calculated solution; however, the refrigeration capacity calculated is the same as that in the configuration in which the two mass flows are shared, because the mass fraction, pressure, and temperature of the two working fluids are the same at the cooling processes and the absorption process. On the other hand, the calculated solution of the optimal basic configuration does not consider heat exchange between the absorbent solution (nodes 4, 11) and the other working fluids in the generation process 1 because of the assumption that the solution fluid is in the saturated liquid state at all operating points in the defined optimization problem. However, the calculated solution could still express the main benefits of an existing absorption system. First, the optimal basic configuration has a shared compression processes between two or three elementary cycles and only the mixture solution flows into the compression process to reduce the required input power as in Table 3. Second, the heat dissipated from the refrigerant right after the generation process at higher pressure (between nodes 13 and 14) is recovered and utilized as the generation heat at a lower pressure in the system as shown in Figure 11. Moreover, the operating points in the calculated solution are also similar to that of the solution in the previous study as shown in Figure 10b,d. It could be said that the calculated solution in this demonstration could express the main features of the solution in the previous study. As a result, the *COP* in calculated solution is 1.54, which is near to the *COP* of the solution in the previous study: 1.7.

Table 3. Calculated values about the energy exchange rate in case Y.

Item	Symbol	Unit	Value
Output power	$\dot{W}_{out}$	W	59.51
Input power	$\dot{W}_{in}$	W	3.27
Output heating capacity	$\dot{Q}_M$	W	8989
Generation heat rate	$\dot{Q}_{GEN}$	W	6666
Heat recovery rate	$\dot{Q}_{GEN,rec}$	W	3471
Heat source potential energy	$\dot{Q}_H$	W	4217
Coefficient of performance	COP	-	1.54
Output refrigeration capacity	$\dot{Q}_L$	W	4908

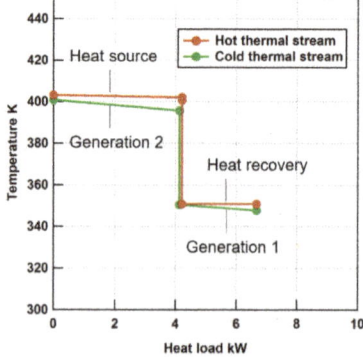

Figure 11. Composite curves of the optimal solutions in case Y.

4. Conclusions

Our study aims to propose a general bottom-up methodology to develop basic configurations for energy system with absorption technology by extending the pioneering SYNTHSEP methodology. The methodology proposes to define a basic configuration for the absorption system as a set of elementary Rankine/Refrigeration cycles and absorption/generation processes and develops the codification method so that an optimization algorithm can implement the organized rules to generate new candidate basic configurations for the absorption system in the optimization problem. Two examples of application are presented to show the capability of the proposed methodology to find basic configurations that can generate the maximum product. The demonstration shows that the existing absorption systems, which would be calculated based on the experience of energy conversion system designers, could be derived by performing the synthesis/design optimization automatically using the proposed synthesis methodology under the simplified/idealized operating conditions. The proposed bottom-up methodology is significant for realizing the synthesis/design optimization for the absorption system, because it can allow engineers to predict all possible configurations in advance and identify a simple and feasible optimal system configuration. The development of the bottom-up synthesis methodology for more types of energy systems may allow us to approach the ultimate goal: the construction of the methodology to derive an optimized system that utilizes available resources to generate the maximum product with minimum components, which can be applied to all types of energy system.

Future work should consider heating/cooling processes operating with absorbent solution or mixed solution in its saturated state or compressed liquid state. It would enable engineers to explore the search space of optimization problem more deeply. Furthermore, certain working fluids would be superheated at certain points in the exiting absorption power and cooling systems, in which the net power is considered as a system product. This methodology should be explored to optimize systems considering superheating and sub-cooling of the working fluids in order to perform the optimization for absorption power and cooling system.

Author Contributions: Original draft preparation K.S.; supervision Y.A.; conceptualization, K.S. and Y.A. and K.T.; methodology, K.S.; writing—review and editing, K.S. and Y.A.; visualization, K.S.; numerical simulations, K.S.

Funding: A part of this work is supported by CREST, the Japan Science and Technology Agency, Grant Number JPMJCR15K5.

Acknowledgments: We would like to express my sincere gratitude to Andrea Lazzaretto and Andrea Toffolo for their valuable comments. Moreover, a part of this work is supported by CREST, headed by Yasuhiro Hayashi, the Japan Science and Technology Agency, Grant Number JPMJCR15K5.

Conflicts of Interest: The authors declare no conflict of interest.

Nomenclature

A	List expressing compression process
b	Binary variable
B	List expressing heating process
C	List expressing expansion process
c_P	Specific heat at constant pressure kJ/(kg·°C)
COP	Coefficient of performance
D	List expressing cooling process
f	Coefficient utilized in the optimization problem
h	Specific enthalpy kJ/kg
I	The number of elementary cycles in a system
$\dot{m}$	Mass flow rate kg/s
M	The number of generation processes in a system
N	The number of absorption processes in a system
P	Pressure MPa

$\dot{Q}$	Heat exchange rate kW
s	Specific entropy kJ/(kg·°C)
T	Temperature °C or K
$\dot{W}$	Power rate kW
z	Mass fraction kg/kg
Subscripts	
abs	Absorption system
ABS	Absorption process
amb	Ambient condition
c	Cold thermal stream
GEN	Generation process
h	Hot thermal stream
H	Heat source
i	Number indicating each elementary cycle
in	Inlet
j	Number indicating operating node
k	Number indicating each sharing pattern
L	Refrigerant capacity
M	Heating capacity
m	Number indicating each generation process
n	Number indicating each absorption process
out	Outlet
rec	Heat recovery
sep	Working fluid in non-shared process

References

1. Frangopoulos, C.A.; Spakovsky, M.; Sciubba, E. A brief review of methods for design and synthesis optimization of energy systems. *Int. J. Appl. Thermodyn.* **2002**, *5*, 151–160.
2. Bertran, M.O.; Frauzem, R.; Sanchez-Arcilla, A.S.; Zhang, L.; Woodley, J.M.; Gani, R. A generic methodology for processing route synthesis and design based on superstructure optimization. *Comput. Chem. Eng.* **2017**, *106*, 892–910. [CrossRef]
3. Cui, C.; Li, X.; Sui, H.; Sun, J. Optimization of coal-based methanol distillation scheme using process superstructure method to maximize energy efficiency. *Energy* **2017**, *119*, 110–120. [CrossRef]
4. Kwon, S.; Won, W.; Kim, J. A superstructure model of an isolated power supply system using renewable energy: Development and application to Jeju Island, Korea. *Renew. Energy* **2016**, *97*, 177–188. [CrossRef]
5. Emmerich, M.; Grötzner, M.; Schütz, M. Design of graph-based evolutionary algorithms: A case study for chemical process networks. *Evol. Comput.* **2001**, *9*, 329–354. [CrossRef] [PubMed]
6. Angelov, P.; Zhang, Y.; Wright, J.; Hanby, V.; Buswell, R. Automatic design synthesis and optimization of component-based systems by evolutionary algorithms. In In Proceedings of the Genetic and Evolutionary Computation Conference, Chicago, IL, USA, 12–16 July 2003; Cantú-Paz, E., Foster, J.A., Deb, K., Davis, L.D., Roy, R., O'Reilly, U.-M., Beyer, H.-G., Standish, R., Kendall, G., Wilson, S., et al., Eds.; Springer: Berlin/Heidelberg, Germany, 2003; p. 213.
7. Urselmann, M.; Emmerich, M.T.; Till, J.; Sand, G.; Engell, S. Design of problem-specific evolutionary algorithm/mixed-integer programming hybrids: Two-stage stochastic integer programming applied to chemical batch scheduling. *Eng. Optimiz.* **2007**, *39*, 529–549. [CrossRef]
8. Wright, J.; Zhang, Y.; Angelov, P.; Hanby, V.; Buswell, R. Evolutionary synthesis of HVAC system configurations: Algorithm development (RP-1049). *HVAC&R Res.* **2008**, *14*, 33–55.
9. Toffolo, A. A synthesis/design optimization algorithm for Rankine cycle based energy systems. *Energy* **2014**, *66*, 115–127. [CrossRef]
10. Toffolo, A.; Rech, S.; Lazzaretto, A. Combination of Elementary Progresses to Form a General Energy System Configuration. In Proceedings of the ASME 2017 International Mechanical Engineering Congress & Exposition IMECE2017, Tampa, FL, USA, 3–9 November 2017.

11. Toffolo, A.; Rech, S.; Lazzaretto, A. Generation of Complex Energy Systems by Combination of Elementary Processes. *J. Energy Resour. Technol.* **2018**, *140*, 1–11. [CrossRef]

12. Lazzaretto, A.; Toffolo, A. A practical tool to generate complex energy system configurations based on the SYNTHSEP methodology. In Proceedings of the ECOS 2018, 31st International Conference of Efficiency, Cost, Optimization, Simulation and Environmental Impact of Energy Systems, Guimarães, Portugal, 17–22 June 2008.

13. Lazzaretto, A.; Segato, F. Thermodynamic optimization of the HAT cycle plant structure—Part I: Optimization of the "basic plant configuration". *J. Eng. Gas Turb. Power* **2000**, *123*, 1–7. [CrossRef]

14. Lazzaretto, A.; Segato, F. Thermodynamic optimization of the HAT cycle plant structure—Part II: Structure of the heat exchanger network. *J. Eng. Gas Turb. Power* **2000**, *123*, 8–16. [CrossRef]

15. Lazzaretto, A.; Toffolo, A. A method to separate the problem of heat transfer interactions in the synthesis of thermal systems. *Energy* **2008**, *33*, 163–170. [CrossRef]

16. Morandin, M.; Toffolo, A.; Lazzaretto, A. Superimposition of elementary thermodynamic cycles and separation of the heat transfer section in energy systems analysis. *J. Energy Resour. Technol.* **2013**, *135*, 021602. [CrossRef]

17. Goswami, D.Y. Solar thermal power technology: Present status and ideas for the future. *Energy Sources* **1998**, *20*, 137–145. [CrossRef]

18. Xu, F.; Goswami, D.Y.; Bhagwat, S.S. A combined power/cooling cycle. *Energy* **2000**, *25*, 233–246. [CrossRef]

19. Martin, C.; Goswami, D.Y. Effectiveness of cooling production with a combined power and cooling thermodynamic cycle. *Appl. Therm. Eng.* **2006**, *26*, 576–582. [CrossRef]

20. Fontalvo, A.; Pinzon, H.; Duarte, J.; Bula, A.; Quiroga, A.G.; Padilla, R.V. Exergy analysis of a combined power and cooling cycle. *Appl. Therm. Eng.* **2013**, *60*, 164–171. [CrossRef]

21. Herold, K.E.; Radermacher, R.; Klein, S.A. *Absorption Chillers and Heat Pumps*, 2nd ed.; CRC Press: Boca Raton, FL, USA, 1996; pp. 7–22 & 235–254.

22. Swartman, R.K.; Ha, V.; Swaminathan, C. Comparison of ammonia-water and ammonia-sodium thiocyanate as the refrigerant-absorbent in a solar refrigeration system. *Sol. Energy* **1975**, *17*, 123–127. [CrossRef]

23. Best, R.; Heard, C.L.; Fernández, H.; Siqueiros, J. Developments in geothermal energy in Mexico-Part five: The commissioning of an ammonia/water absorption cooler operating on low enthalpy geothermal energy. *Heat Recovery Syst.* **1986**, *6*, 209–216. [CrossRef]

24. Siddiqui, M.A. Optimum generator temperatures in four absorption cycles using different sources of energy. *Energy Convers. Manag.* **1993**, *34*, 251–266. [CrossRef]

25. Fernández-Seara, J.; Vales, A.; Vázquez, M. Heat recovery system to power an onboard NH_3-H_2O absorption refrigeration plant in trawler chiller fishing vessels. *Appl. Therm. Eng.* **1998**, *18*, 1189–1205. [CrossRef]

26. Bejan, A. *Advanced Engineering Thermodynamics*, 4th ed.; John Wiley & Sons Inc.: Hoboken, NJ, USA, 2016; pp. 531–599.

27. Myat, A.; Thu, K.; Kim, Y.; Chakraborty, A.; Chun, W.G.; Ng, K.C. A second law analysis and entropy generation minimization of an absorption chiller. *Appl. Therm. Eng.* **2011**, *31*, 2405–2413. [CrossRef]

28. Wu, W.; Wang, B.; Shi, W.; Li, X. An overview of ammonia-based absorption chillers and heat pumps. *Renew. Sustain. Energy Rev.* **2014**, *31*, 681–707. [CrossRef]

29. Ziegler, F.; Alefeld, G. Coefficient of performance of multistage absorption cycles. *Int. J. Refrig.* **1987**, *10*, 285–295. [CrossRef]

30. Alefeld, G.; Radermacher, R. *Heat Conversion System.*; CRC Press: Boca Raton, FL, USA, 1994; pp. 179–206.

31. Inoue, N. Studies on the Characteristics of Absorption Cycles and their Applications. Ph.D. Thesis, Waseda University, Tokyo, Japan, 3 March 2005.

32. Kolaczkiewics, J.; Bauer, E. Clausius-Clapeyron equation analysis of two- dimensional vaporization. *Surf. Sci.* **1985**, *155*, 700–714. [CrossRef]

Article

Stochastic Optimization for Integration of Renewable Energy Technologies in District Energy Systems for Cost-Effective Use

Thomas T. D. Tran [1] and Amanda D. Smith [2,*]

1   Indiana Institute of Technology, 1600 E Washington Blvd, Fort Wayne, IN 46803, USA;
    thomasdtran@indianatech.edu
2   Site-Specific Energy Systems Laboratory, Department of Mechanical Engineering, University of Utah,
    Salt Lake City, UT 84112, USA
*   Correspondence: amanda.d.smith@utah.edu

Received: 31 December 2018; Accepted: 31 January 2019; Published: 7 February 2019

Abstract: Stochastic optimization of a district energy system (DES) is investigated with renewable energy systems integration and uncertainty analysis to meet all three major types of energy consumption: electricity, heating, and cooling. A district of buildings on the campus of the University of Utah is used as a case study for the analysis. The proposed DES incorporates solar photovoltaics (PV) and wind turbines for power generation along with using the existing electrical grid. A combined heat and power (CHP) system provides the DES with power generation and thermal energy for heating. Natural gas boilers supply the remaining heating demand and electricity is used to run all of the cooling equipment. A Monte Carlo study is used to analyze the stochastic power generation from the renewable energy resources in the DES. The optimization of the DES is performed with the Particle Swarm Optimization (PSO) algorithm based on a day-ahead model. The objective of the optimization is to minimize the operating cost of the DES. The results of the study suggest that the proposed DES can achieve operating cost reductions (approximately 10% reduction with respect to the current system). The uncertainty of energy loads and power generation from renewable energy resources heavily affects the operating cost. The statistical approach shows the potential to identify probable operating costs at different time periods, which can be useful for facility managers to evaluate the operating costs of their DES.

Keywords: district energy system; optimization; renewable energy systems; combined heat and power; operating cost; uncertainty

---

1. Introduction

Power generation from distributed energy resources has become increasingly popular [1]. Traditional power generation is often associated with large-scale power plants [2], but distributed power generation tends to occur on smaller scales and is located near the end users. The introduction of distributed generation allows the local energy demand to be less dependent on the grid [3] and may provide additional efficiencies by using local energy resources. Renewable energy systems in recent years have been considered within distributed generation systems [4]. One of the advantages of renewable energy systems is that they can be configured in various system sizes to meet the local energy demand [5].

The options for integrating renewable energy systems into district energy systems (DES) vary depending on the location and the objective of the DES [6]. The integration of renewable energy systems into DES has been actively explored in recent literature. Various renewable energy technologies are utilized in DES, ranging from small scale to large scale [7]. The types of technology are also dependent on the location of the DES [8]. In other words, the availability of local renewable energy resources determines the type of renewable energy systems which can be employed in that area [9]. Solar and wind power are the two most popular types of renewable energy technologies integrated into DES [10]. Furthermore, geothermal, biomass, and hydropower are also attractive options for locations containing the respective energy resources [11]. Other types of distributed energy systems such as combined heat and power (CHP) are also used for DES planning [12].

Power generation of solar and wind energy systems can be interrupted due to the intermittency and uncertainty of solar irradiation [13] and wind speed [14]. Mathematical models have been developed to address the intermittency of renewable power generation in DES [15]. For example, Parisio et al. proposed a Robust Optimization technique to control energy carriers into an energy hub [16]. Evins et al. utilized Mixed-integer Linear Programming to address operational constraints [17]. Mavromatidis et al. introduced a two-stage stochastic programming approach to optimal design of distributed energy systems [18]. Jabbari-Sabet et al. used Particle Swarm Optimization and Unit Commitment to solve for DES operation and management of a 10-bus system in the day-ahead model [19]. Fioriti et al. investigated a hybrid minigrid under load and renewable generation uncertainty [20]. The results from these studies show that stochastic optimization can be used to address the uncertainties associated with DES.

There are a number of ways to incorporate stochasticity in the simulation and optimization of renewable energy systems. In recent literature, the uncertainties of power generation from renewable energy systems have been investigated for meeting the electricity demand [21], heating demand [22], and cooling demand [23]. Najibi et al. investigated stochastic scheduling of renewable energy resources to meet the electricity demand under uncertainties of solar photovoltaics (PV) power generation [24]. Similarly, Nikmehr et al. studied the operating cost optimization of a network of energy hubs to fulfill the electricity demand [25]. Balaman and Selim focused on meeting the heating demand in a heating district system [26]. Lu et al. presented a modeling solution to coordinate dispatch of a multi-energy system with district heating network [27]. Comparatively, Sameti and Haghighat studied optimization methods for a cooling network together with a district heating network [28]. Furthermore, Gang et al. presented an uncertainty-based design optimization for stand-lone district cooling systems [29]. Overall, these DES planning studies are mostly focused on meeting one of the three major energy demands, with electricity as the main focus.

There are limited studies on optimizing the operating cost of DES with consideration of all three major energy demands. Li et al. optimized building cooling heating and power system with consideration of uncertainty of energy demands [30]. Recently, Mavromatidis et al. incorporated uncertainty and global sensitivity analysis to optimize design of an energy hub [31]. There is a need to further investigate all three major energy use types, especially in the presence of uncertainty and global sensitivity analysis. The addition of cooling and heating demand to the analysis is important for the overall operation of the DES. This is because systems such as CHP can utilize thermal energy, which is a by-product of power generation, for fulfilling the heating demand. Furthermore, the uncertainties in power generation can impact the fulfillment of the cooling demand since cooling equipment is run by electrical power. In this work, these uncertainties can be included in the model for optimization without requiring a stochastic programming approach, in which later decision stages depend on uncertainties in earlier decision stages [32].

The novelty of this paper is to establish an optimization framework with uncertainty incorporated in terms of stochastic renewable power generation systems and stochastic energy use (loads) based on a day-ahead model to optimize the operating cost of a modeled DES and potentially reduce dependence on the grid. To address the existing gap in the literature, all three major categories of energy consumption (electricity, heating, and cooling) are considered in the presence of renewable power generation and energy usage uncertainties. These energy needs of the DES are met by a mix of renewable energy systems (solar PV and wind turbines), CHP system, energy storage, natural gas boilers, and the electrical grid.

This study uses real data from a group of existing buildings on the campus of the University of Utah as a case study. These buildings are metered to measure both the electrical and thermal energy consumption. The uncertainties of energy consumption patterns and power generation from the renewable energy resources (i.e., solar irradiation and wind speed) are analyzed based on the Monte Carlo approach. The DES is optimized with the population-based Particle Swarm Optimization (PSO) algorithm. The objective of the optimization is to minimize the operating cost of the DES on the day-ahead model.

Unlike the aforementioned works in the literature, the results from this study will lay the groundwork for the adding renewable energy systems into DES with (1) considering of all three major types of energy consumption (electricity, heating, and cooling) in buildings; (2) including of the uncertainties of energy consumption and power generation from renewable energy resources; (3) incorporating the Monte Carlo statistical approach into the population-based PSO; and (4) generating statistical distributions of the operating cost at different time periods in order to stochastically optimize for the operating cost of the DES.

2. Problem Formulation

2.1. District Energy System Description

The methodology described in this section can be applied to any DES with known electricity, heating, and cooling demands. Energy system sizes in the DES can be adjusted according to the demands of the given DES. The stochastic optimization methodology proposed in this paper is examined by utilizing a group of existing buildings on the campus of the University of Utah as a case study. The buildings are predominantly used as offices and classrooms. As previously mentioned, the chosen buildings are metered for energy usage, disaggregate into electricity, heating, and cooling. The study examines the DES on four different days of the year (20 March, 21 June, 22 September, and 21 December), which occur at the beginning of the four astronomical seasons. During these four days, offices and classrooms are open. Throughout this paper, the existing energy system (utilizing the electrical grid and natural gas) will be referred to as the current energy system. On the other hand, the modeled DES (utilizing a mixed of energy systems) will be referred to as the proposed DES. Furthermore, the non-cooling electricity load will be referred to as simply the electricity load, while the electrical energy required to run cooling equipment will be referred to as the cooling load.

The average energy loads of the buildings comprising the proposed DES are detailed in Table 1. To obtain average energy loads representing the four example days, hourly energy data is taken from 10 preceding days and 10 subsequent days to capture typical load patterns during the time of year around that particular example day. Figure 1 shows the daily energy loads of 21 June and its 20 neighboring days to illustrate the process of obtaining data for the study. Instead of using only the actual energy data for 21 June, incorporating data from neighboring days allows for a representative energy load that includes stochastic variation in the likely energy load on such an example day, which will be discussed further in Section 2.3.1. The data represents the real energy loads in 2017 and is obtained from the university's SkySpark installation, a building analytics platform that collects building data [33]. The mean and standard deviation of each set of hourly energy data are shown in the Appendix A (Tables A1–A3). The average values are plotted in Figures 2–4. The daily electricity load is consistent while the daily heating and cooling loads vary throughout the year. The DES is located at Salt Lake City, UT, which is part of American Society of Heating, Refrigerating and Air-Conditioning Engineers (ASHRAE) climate zone 5 (i.e., cool and dry) [34]. Buildings in this climate zone typically exhibit considerable heating and cooling demands in winter and summer, respectively.

Table 1. Average loads on four example days.

Design Day	Electricity (kWh$_e$)	Heating (kWh$_t$)	Cooling (kWh$_t$)
20 March	15,747	15,243	5940
21 June	15,669	12,675	6897
22 September	14,981	14,981	4749
21 December	13,641	24,693	62

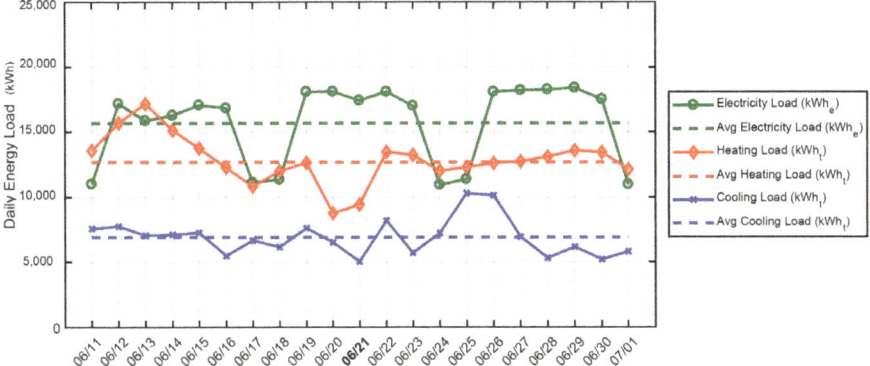

Figure 1. Daily energy loads on 21 June and its neighboring days.

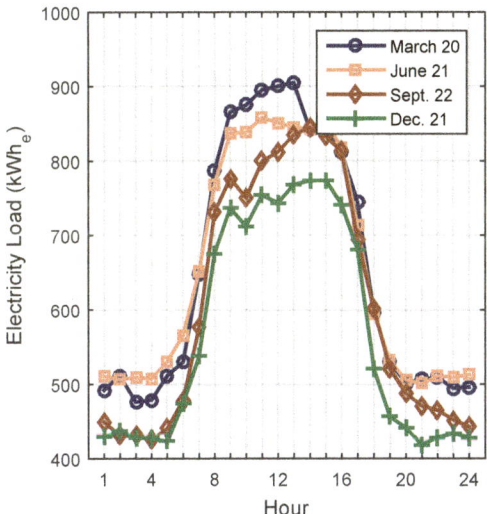

Figure 2. Average electricity load of the districts of buildings.

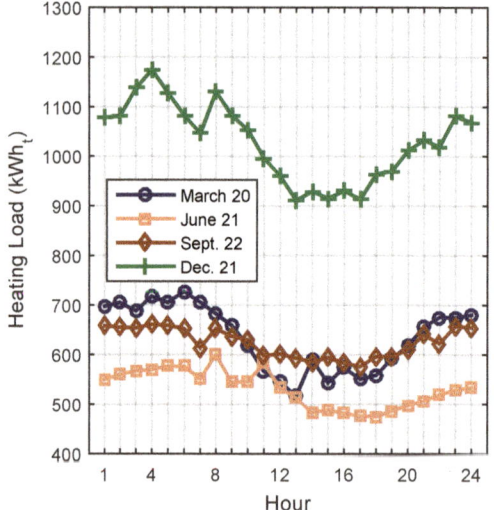

Figure 3. Average heating load of the districts of buildings.

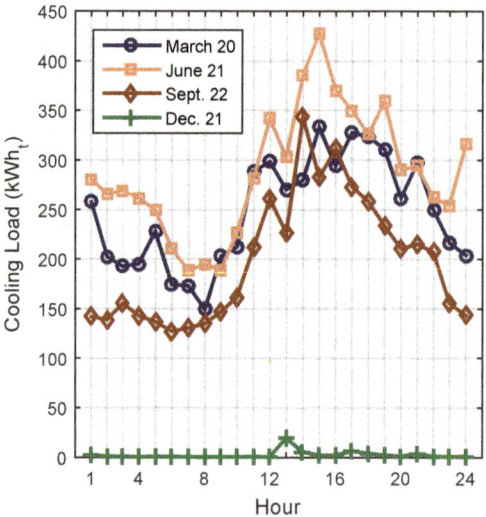

Figure 4. Average cooling load of the districts of buildings.

2.2. Proposed District Energy System

The DES utilizes the local renewable energy resources for power generation to reduce the dependence on the electrical grid. The diagram of the proposed DES is illustrated in Figure 5. In particular, solar PV and wind power are the two renewable power generation sources integrated into the DES. The National Renewable Energy Laboratory (NREL) provides hourly data over the last 15 years for wind speed [35] and solar irradiation [36] at the study location of Salt Lake City, UT, USA (40.766837, −111.846920). In addition to renewable power generation, a gas-fired microturbine CHP system is used to provide power generation during times of electrical demand when solar PV and wind power is inadequate to meet the electricity loads. A battery system is also implemented to store unconsumed renewable energy for use at a later time. Additionally, the power generated from these resources will be used to run cooling equipment to meet the cooling demand. The system will recover thermal energy from the CHP system in addition to using natural gas boilers to meet the heating demand. The technical specifications of these systems are listed in Table 2. The system sizes for the simulations presented here were chosen based on the energy demands of the buildings and the availability of renewable energy resources in the area. Different design choices will affect the on-site generation and ultimately the operating cost of the DES.

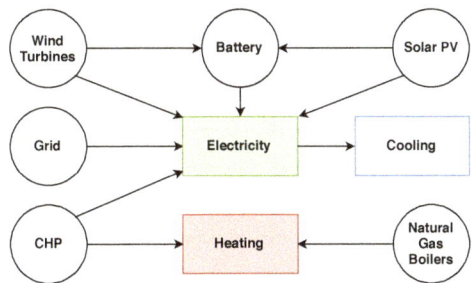

Figure 5. Diagram of the proposed district energy system management.

Table 2. Energy system specifications for the proposed DES.

Energy System	Specifications
Solar PV [37]	Capacity: 400 kW Efficiency: 15% Performance ratio: 75% Dimension: 1.6 m² for a 200 W panel Tilt angle: 40.5°
Wind Turbine [37]	Capacity: 400 kW Rotor diameter: 25 m Cut-in speed: 2.7 m/s Cut-out speed: 25 m/s Rated speed: 12 m/s
CHP [38]	Capacity: 300 kW Technology: microturbine Power to heat ratio: 0.6 Effective electrical efficiency: 55% Overall CHP efficiency: 65%
Battery [37]	Capacity: 100 kW Charging/discharging efficiency: 90% Depth of discharge: 90%
Boiler [39]	Capacity: 500 MBtu/h (146.5 kW) Fuel-to-steam efficiency: 85% Fuel: natural gas

2.3. Mathematical Model

The objective of the mathematical model is to minimize the operating cost of the DES, which consists of the operating costs of the solar PV panels, wind turbines, battery system, CHP system, natural gas boilers, and electricity purchased from the grid. This work assumes that these systems are already installed. The operating and maintenance costs of these devices can be less significant to decision makers as their variation with energy output is small. Their capital costs can be found in Appendix B. All of the variables in the mathematical model are hourly because the solar irradiation, wind speed, and energy loads are provided as hourly data. The objective function of the operating cost is detailed as follows:

$$Min\ f(X) = Min \sum_{t=1}^{T} Cost^t = Min \sum_{t=1}^{T} \left\{ \sum_{i=1}^{N_g} \left[u_i(t)E_{G,i}(t)C_{G,i}(t) \right] \right.$$

$$\left. + \sum_{j=1}^{N_s} \left[u_j(t)E_{S,j}(t)C_{S,j}(t) \right] + \sum_{k=1}^{N_h} \left[u_k(t)Q_{H,k}(t)C_{H,k}(t) \right] + u_l(t)E_{grid}(t)C_{grid}(t) \right\}. \tag{1}$$

For electricity generation (the first term in brackets on the right-hand side of Equation (1)): N_g is the number of generating units, u_i is the state of the ith unit (ON = 1, OFF = 0), $E_{G,i}$ is the electricity generation of the ith generating unit, and $C_{G,i}$ is the ith generating unit cost. The generating units consists of on-site generation from solar, wind, and CHP. For energy storage (the second term in brackets on the right-hand side of Equation (1)): N_s is the number of energy storage devices, u_j is the state of the jth unit (ON = 1, OFF = 0), $E_{S,j}$ is the energy capacity of the jth energy storage device, and $C_{S,j}$ is the jth storage device cost. For thermal heating devices, including gas-fired boilers (the third term in brackets on the right-hand side of Equation (1)): N_h is the number of heating devices, u_k is the state of the kth unit (ON = 1, OFF = 0), $Q_{H,j}$ is the thermal energy of the kth heating device, and $C_{H,k}$ is the kth heating unit cost. When purchasing electricity from the grid, u_l is the state of sending/receiving electricity from the utility (ON = 1, OFF = 0), E_{grid} is the electricity purchased (or exported) from (or to) the utility, and C_{grid} is the energy unit cost purchased from the utility.

The following equations break down the operating costs and constraints of all energy devices used in the proposed DES. The operating costs from renewable energy resources (i.e., solar and wind power) come from the operating & maintenance (O&M) costs of these systems. The fuel costs are zero since the fuels for these systems (i.e., solar irradiation and wind speed) are free to harvest. For the solar PV system, the operating cost, C_{solar}, is as follows:

$$C_{solar} = \lambda_{solar} E_{solar}, \tag{2}$$

where λ_{solar} is the solar operating cost per unit of electricity output, and E_{solar} is the solar electricity output. The solar power generation, P_{solar}, is the rate of electricity converted from solar energy (E_{solar}) per unit of time. The solar power generation is subjected to the solar power capacity constraint:

$$P_{solar,min}(t) \leq P_{solar}(t) \leq P_{solar,max}(t). \tag{3}$$

The solar electricity generation is calculated based on the specifications of the solar PV system (Table 2) and the solar irradiation available at the study location.

Similarly, the operating cost of the wind electricity generation system , C_{wind}, is as follows:

$$C_{wind} = \lambda_{wind} E_{wind}, \tag{4}$$

where λ_{wind} is the wind operating cost per unit of electricity output, and E_{wind} is the wind electricity output. The wind power generation is subjected to the wind power capacity constraint:

$$P_{wind,min}(t) \leq P_{wind}(t) \leq P_{wind,max}(t). \tag{5}$$

Wind power generation depends on the wind speed since the wind turbines impose cut-in and cut-off wind speeds as listed in Table 2.

Unlike the electricity generation from renewable energy resources, the operating cost of the CHP system comes from the fuel cost in addition to the O&M cost. The fuel cost is the cost of the natural gas used for running the microturbine in the CHP system. The operating cost of the CHP system , C_{CHP}, is detailed as follows:

$$C_{CHP} = (f_{CHP} + \gamma_{CHP})E_{CHP}, \tag{6}$$

where f_{CHP} is the CHP fuel unit cost, γ_{CHP} is O&M unit cost of the CHP system, and E_{CHP} is the electricity generation from the CHP system.

The thermal energy from the CHP system is utilized to partially meet the heating load, while the natural gas boilers are used to fulfill the remaining load. The constraint for the heating load is shown in the following equation:

$$Q_{heating} = Q_{CHP} + Q_{boiler}, \tag{7}$$

where $Q_{heating}$ is the total heating load, Q_{CHP} is the thermal energy output delivered from the CHP system, and Q_{boiler} is the thermal energy output provided by the boilers.

The operating cost for the boiler system , C_{boiler}, is as follows:

$$C_{boiler} = (f_{boiler} + \gamma_{boiler})Q_{boiler}, \tag{8}$$

where f_{boiler} is the fuel unit cost to run the boilers, γ_{boiler} is the O&M unit cost of the boilers, and Q_{boiler} is the thermal energy output of the boilers.

A battery system is also implemented as an energy storage device in the DES for events of excess electricity generation. The operating cost for the battery , C_{es}, is as follows:

$$C_{es} = \lambda_{es}E_{es}, \tag{9}$$

where λ_{es} is the unit operating cost of the battery, and E_{es} is the energy capacity of the battery. Charging occurs during events of excess power generation from the renewable energy sources, while discharging takes place if there is a lack of on-site power generation. The battery is also subjected to the charging/discharging limitations and the state of charge constraint. The depth of discharge of the battery system is 90% as it prolongs the life cycle of the battery system. If the battery system is fully charged, any further excess renewable power generation will be sold to the grid. The following equations illustrate the constraints on the battery:

$$E_{es}(t) = E_{es}(t-1) + \eta_{charge}E_{charge}\Delta t - \frac{1}{\eta_{discharge}}E_{discharge}\Delta t, \tag{10}$$

$$E_{es,min}(t) \leq E_{es}(t) \leq E_{es,max}(t). \tag{11}$$

The electrical grid is used in the event that on-site power generation is inadequate to meet the electricity load. The cost of purchasing electricity from the grid , C_{grid}, is as follows:

$$C_{grid} = \lambda_{grid}E_{grid}, \tag{12}$$

where λ_{grid} is the electricity unit cost from the grid, and E_{grid} is the purchased electricity. The unit costs of all energy systems are detailed in Table 3.

Table 3. Unit costs of energy systems [37–39].

Parameter	Symbol	Value
Solar generation operating unit cost	λ_{solar}	3.32 ¢/kWh
Wind generation operating unit cost	λ_{wind}	3.12 ¢/kWh
CHP fuel unit cost	f_{CHP}	4.54 ¢/kWh
CHP operating unit cost	γ_{CHP}	2.34 ¢/kWh
Boiler fuel unit cost	f_{boiler}	4.34 ¢/kWh
Boiler operating unit cost	γ_{boiler}	2.34 ¢/kWh
Battery operating unit cost	λ_{es}	2.67 ¢/kWh
Grid unit cost of purchasing power	λ_{grid}	7.40 ¢/kWh

The energy balance should be satisfied at all times:

$$E_{solar} + E_{wind} + E_{CHP} + E_{es} + E_{grid} = E_{load},$$ (13)

where E_{load} consists of the non-cooling electricity load and the electrical energy to run cooling equipment.

2.3.1. Uncertainty Model

Renewable energy systems in the DES are associated with uncertainties in power generation. For instance, solar PV panels and wind turbines depend on solar irradiation and wind speed, respectively. The uncertainties of these variables can be characterized by statistical probability distributions [40–42]. The Monte Carlo simulation is used for modeling and sampling the uncertainties in this study [43]. This section illustrates the uncertainty analyses of the input variables (i.e., wind speed, solar irradiation, and energy loads). Figure 6 provides the average wind speed and solar irradiation on all four example days. To determine the appropriate statistical distributions, hypothesis tests were used on the actual data. As a result, the uncertainty analysis on the wind speed is conducted based on a Weibull distribution, while the solar irradiation is analyzed based on a normal distribution.

The probability density function of wind speed based on the Weibull distribution is as follows:

$$f(x_w) = \frac{\beta_w}{\delta_w} \left(\frac{x_w}{\delta_w}\right)^{\beta_w - 1} \exp\left[-\left(\frac{x_w}{\delta_w}\right)^{\beta_w}\right],$$ (14)

where x_w is the wind speed, δ_w is the scale parameter, and β_w is the shape parameter (Table A4 in the Appendix A).

The mean and standard deviation of the wind speed (Table A5 in the Appendix A) are shown in the following equations, respectively:

$$\mu_w = \delta_w \Gamma\left(1 + \frac{1}{\beta_w}\right),$$ (15)

$$\sigma_w = \left[\delta_w^2 \Gamma\left(1 + \frac{2}{\beta_w}\right) - \delta_w^2 \left[\Gamma\left(1 + \frac{1}{\beta_w}\right)\right]^2\right]^{0.5}.$$ (16)

The cumulative distribution function of the Weibull distribution of wind speed is modeled as follows:

$$F(x_w) = 1 - \exp\left[-\left(\frac{x_w}{\delta_w}\right)^{\beta_w}\right].$$ (17)

Similarly, a normal distribution is used to model the solar irradiation. The probability density function of this normal distribution is as follows:

$$f(x_s) = \frac{1}{\sqrt{2\pi}\sigma_s} \exp\frac{-(x_s - \sigma_s)^2}{2\sigma_s^2}.$$ (18)

The mean and standard deviation of the normal distribution are μ_s and σ_s (Table A6 in the Appendix A), respectively. The cumulative distribution function of the normal distribution is as follows:

$$F(x_s) = \frac{1}{2}\left[1 + \mathrm{erf}\left(\frac{x_s - \mu_s}{\sigma_s\sqrt{2}}\right)\right].$$ (19)

Similar to the uncertainty analysis of wind speed and solar irradiation, the energy loads of the district of buildings are analyzed based on normal distributions. The probability density functions of the loads are similar to Equation (18), while their cumulative distribution functions are similar to Equation (19). Tables A1–A3 in the Appendix A represent the mean and standard deviation of electricity, heating, and cooling loads, respectively.

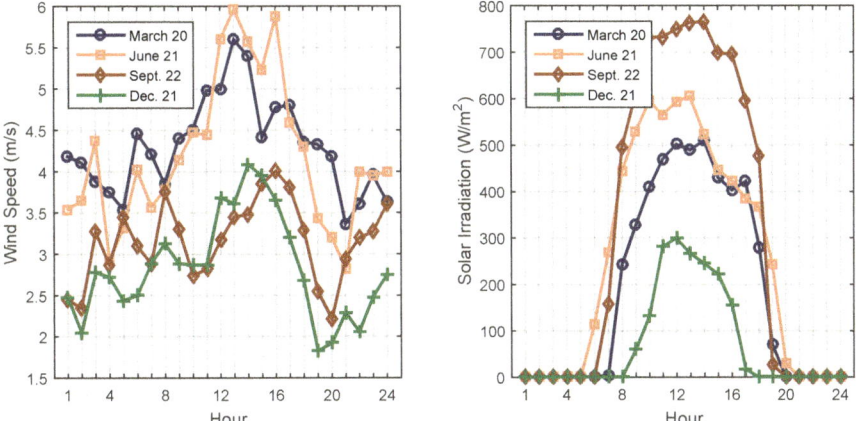

Figure 6. Average wind speed and solar irradiation at the study location (Salt Lake City).

Based on the Monte Carlo simulation, the uncertain variables can be sampled to produce deterministic inputs for the stochastic model. In this study, there are 10,000 power generation scenarios generated by the Monte Carlo simulation. Based on these, a day-ahead model of the DES is planned. Using the constraints of the electricity generation from wind turbines and solar PV panels, there are five decision variables to be considered: CHP electricity generation, battery state of charge, grid electricity purchase, CHP thermal energy, and boiler thermal energy. As a result, the total number of variables for a day-ahead model is 120 (24 h and five variables). All of these variables are linked to the objective function, which was shown in Equation (1). The decision variables are demonstrated in the following matrix:

$$\begin{bmatrix} E_{CHP}^{t1} & E_{es}^{t1} & E_{grid}^{t1} & Q_{CHP}^{t1} & Q_{boiler}^{t1} \\ E_{CHP}^{t2} & E_{es}^{t2} & E_{grid}^{t2} & Q_{CHP}^{t2} & Q_{boiler}^{t2} \\ \vdots & \vdots & \vdots & \vdots & \vdots \\ E_{CHP}^{t24} & E_{es}^{t24} & E_{grid}^{t24} & Q_{CHP}^{t24} & Q_{boiler}^{t24} \end{bmatrix}.$$

2.3.2. Stochastic Optimization Algorithm

The day-ahead model is optimized by the Particle Swarm Optimization (PSO) algorithm, which was developed by Kennedy and Eberhart in 1995 for studying bird flocking and fish schooling [44]. This optimization algorithm has been applied in many engineering applications, such as gear train design, process parameter optimization in casting, power generation scheduling, etc. [45,46]. The population-based optimization has been adopted for the stochastic optimization in this study because it offers mathematical flexibility and computational efficiency to incorporate uncertainties. Figure 7 describes the implementation of the algorithm used in this study. As illustrated in the aforementioned matrix, the PSO algorithm can search for an optimal solution (i.e., minimizing operating cost of the DES at each instant in time) based on the day-ahead model that contains a total of 120 variables. The Monte Carlo simulation method is integrated with the PSO algorithm to assess the uncertainties of renewable power generation and energy loads.

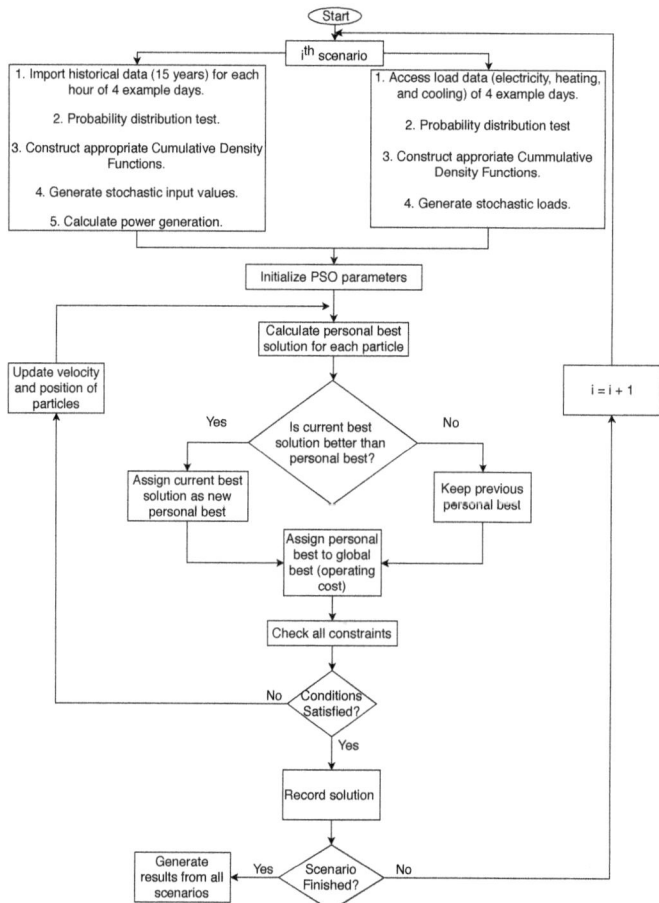

Figure 7. Flowchart of the Monte Carlo simulation and the Particle Swarm Optimization algorithm.

The PSO algorithm optimizes by allowing for communication and learning to take place among the particles in the search space. In the beginning, a group of random particles initializes the PSO algorithm and then searches for the minimum solution by updating generations of the particles. In every iteration, there are two "best" values that determine the location and velocity of each particle. The first value is the personal best solution (pBest) that each particle has achieved so far. The other "best" value is the global best (gBest) that is obtained so far by any particle in the population, which is not necessarily a global minimum in the solution space. The tolerance of the solution (i.e., the operating cost) is measured to determine the optimum population size and iteration with respect to the computational time. As a result, the population size is picked to be 50 and the number of iteration is 1000 as this gives the best trade-off between accuracy and computational time. The solution of each particle in every iteration is calculated by the objective function. The location and velocity of the particle in the search space are calculated based on the following Equations [47]:

$$x_j^{k+1} = x_j^k + v_j^{k+1}, \tag{20}$$

$$v_j^{k+1} = wv_j^k + r_1c_1\left(pBest_j^k - x_j^k\right) + r_2c_2\left(gBest_j^k - x_j^k\right), \tag{21}$$

where x is the location of the particle, v is the velocity of the particle, c_1 and c_2 are acceleration coefficients and both equal to 2.05, w is the inertia coefficient and $w = 2/[(c_1 + c_2) - 2 + [(c_1 + c_2)^2 - 4(c_1 + c_2)]^{1/2}]$, r_1 and r_2 are random numbers $\subset (0,1)$, j is the jth particle, and k is the kth iteration [45]. The three terms in Equations (21) represent inertial, cognitive, and social components, respectively. The inertial component presents the relative velocity of the particle in the search space. The cognitive component refers to the personal experience of the particle (i.e., personal best operating cost) while the social component is associated with the communication among particles (i.e., global best operating cost).

3. Results and Discussion

The simulation of the DES solves for the operating costs of the four example days, which give different operating conditions with mixed electricity, heating, and cooling demands. These four example days are associated with seasonal transitions at the study location, which offer variations in daylight hours and wind speeds. As a result, the power generation from the renewable energy resources changes throughout the year (Table 4). Furthermore, the uncertainties of renewable power generation and energy loads are shown to influence the operating costs. The mean and standard deviation of the operating cost in all hours of the four example days are shown in Table 5. The total operating cost of each day of the proposed DES is compared to the operating costs of the current energy system (relying on the electrical grid and natural gas boilers) in Table 6. These operating costs represent the average values for 10,000 power generation scenarios. The simulation time for each example day is approximately 5 h in MATLAB (2015b version by Mathworks, Natick, MA, USA) on a desktop computer with an Intel i7 processor and 16 GB of RAM.

Table 4. Purchased power and average on-site power generation on four different example days.

Hour	20 March Grid (kWh)	20 March On-Site (kWh)	21 June Grid (kWh)	21 June On-Site (kWh)	22 September Grid (kWh)	22 September On-Site (kWh)	21 December Grid (kWh)	21 December On-Site (kWh)
1	594.67	155.36	648.98	160.50	490.26	111.11	305.08	126.06
2	565.62	148.11	645.63	155.60	478.56	108.30	233.79	204.59
3	532.25	136.61	670.90	220.45	470.94	115.65	298.21	130.86
4	476.84	197.15	692.30	180.12	456.32	131.88	313.22	115.31
5	592.59	146.34	625.00	174.82	468.12	125.41	305.31	118.91
6	570.40	133.37	630.00	156.00	488.68	117.40	364.22	111.06
7	675.25	145.93	760.20	168.30	587.79	118.87	409.92	127.75
8	783.49	153.02	802.30	217.75	823.20	155.83	552.58	123.24
9	875.71	192.21	832.19	193.78	816.32	153.91	560.94	176.38
10	886.63	200.41	907.05	158.21	825.00	171.24	539.01	173.37
11	962.04	221.48	935.64	203.20	851.25	162.60	573.44	180.70
12	960.43	237.61	986.54	248.60	906.40	172.87	584.40	159.67
13	967.83	206.78	919.21	238.65	899.43	168.65	623.90	162.79
14	936.39	185.57	996.71	245.60	1028.20	180.65	620.01	159.15
15	996.27	177.40	1083.45	230.00	951.11	172.65	648.30	127.33
16	959.39	147.15	977.92	210.75	948.98	175.34	572.93	170.81
17	869.44	202.14	904.28	178.60	813.40	153.86	542.37	144.66
18	750.68	168.89	830.00	168.60	717.53	142.30	382.14	142.61
19	697.34	143.36	802.30	170.54	643.99	110.09	291.53	167.20
20	626.29	136.76	730.00	146.30	602.23	106.99	271.49	170.75
21	612.42	190.76	720.65	145.20	598.32	109.27	258.70	163.98
22	448.62	316.19	703.60	167.00	578.32	107.16	275.10	155.58
23	531.59	170.10	650.20	152.65	498.32	114.34	313.40	120.82
24	464.06	237.94	697.00	160.00	476.30	125.03	285.72	143.80
Total Daily Power Generation								
	17,336.26	4350.64	19,152.06	4451.21	16,418.98	3311.40	10,125.72	3577.41
On-site Generation Percentage								
	20.06%		18.86%		16.78%		26.11%	

Table 5. Hourly operating costs (mean and standard deviation) on four example days.

Hour	20 March μ ($)	20 March σ ($)	21 June μ ($)	21 June σ ($)	22 September μ ($)	22 September σ ($)	21 December μ ($)	21 December σ ($)
1	57.52	1.61	55.64	1.74	47.73	0.19	50.41	1.11
2	55.71	0.93	54.89	1.73	46.41	0.15	49.35	3.06
3	52.75	1.32	54.36	2.95	47.21	0.40	51.99	1.44
4	52.72	2.83	54.97	1.86	46.03	0.92	53.37	0.42
5	57.27	1.52	55.32	2.30	46.64	0.51	51.62	0.59
6	56.08	0.66	55.54	1.09	48.31	0.41	53.47	0.39
7	62.23	1.28	58.61	0.95	53.16	0.21	55.80	0.75
8	68.32	1.02	66.12	2.30	63.12	0.58	66.72	0.76
9	74.72	2.09	68.74	1.87	66.13	0.50	67.89	2.13
10	74.45	1.68	71.84	0.73	64.86	1.37	65.57	2.22
11	78.29	2.51	76.46	2.16	70.12	0.50	66.20	2.37
12	78.26	2.60	78.22	2.53	73.83	2.63	64.98	2.32
13	76.54	1.66	74.55	1.80	72.94	2.52	65.99	2.61
14	76.03	1.68	78.68	2.16	80.36	2.12	66.16	2.30
15	77.88	1.12	82.64	1.34	76.17	0.63	66.15	0.33
16	75.38	0.70	76.67	1.73	76.16	0.98	63.85	1.77
17	71.49	2.17	70.02	0.62	66.96	0.40	60.44	1.50
18	63.28	1.08	61.59	0.64	61.35	0.24	52.26	1.77
19	60.04	1.25	60.00	0.87	55.65	0.13	47.99	2.22
20	56.28	1.41	54.84	0.74	52.76	0.12	48.15	2.65
21	58.76	2.75	54.90	1.16	52.90	0.25	47.77	2.17
22	54.87	5.33	53.92	1.00	51.57	0.24	48.03	1.56
23	53.92	2.90	53.94	0.42	48.57	0.45	50.78	0.78
24	52.67	5.78	57.34	1.68	47.04	0.49	49.75	1.30

Table 6. Comparison of average operating costs.

Example Day	Proposed DES	Current System	Operating Savings	Percent Reductions
20 March	$1545.46	$1700.75	$155.29	9.13%
21 June	$1529.76	$1683.23	$153.47	9.12%
22 September	$1415.99	$1572.26	$156.27	9.94%
21 December	$1364.70	$1517.52	$152.82	10.07%

The operating costs for each hour of the four example days are illustrated in Figure 8. Due to the nature of the buildings (offices and classrooms), the majority of the energy use occurs during the day. The operating cost during occupied hours dominates the daily operating cost of the DES. The use of solar PV is beneficial for the DES since the power generation from solar PV panels can be used for fulfilling the electricity load during the day. On the other hand, the power generation from the wind turbines occurs mostly in the afternoon and at night, which can be consumed for night-time building operations and energy storage. The use of solar PV and wind power offers a balance between power generation from renewable energy resources during the day and at night. As a result, the intermittent nature of power generation from each resource can be mitigated. Furthermore, the addition of on-site generation (including wind turbines, solar PV panels, and the CHP system) reduces the dependence on the electrical grid by as much as 26%, as seen in Table 4.

For each hourly operating cost, a probability-normalized histogram is constructed to assess the influence of uncertainties. The operating costs for the three representative hours (4th hour, 12th hour, and 20th hour) in four example days are shown in Figures 9–12, respectively. The probability-normalized histograms show the potential operating costs and their probabilities in different price ranges. In other words, the distribution illustrates how probable each operating cost is for the given operating conditions.

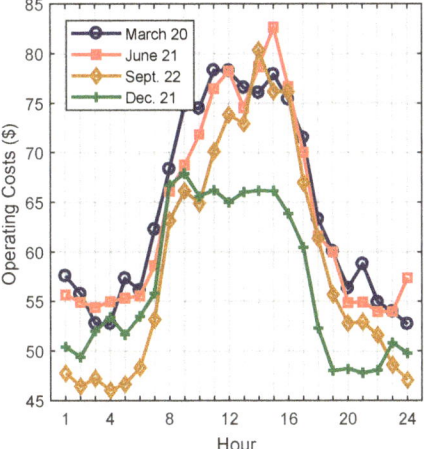

Figure 8. Operating costs of the DES of the four example days.

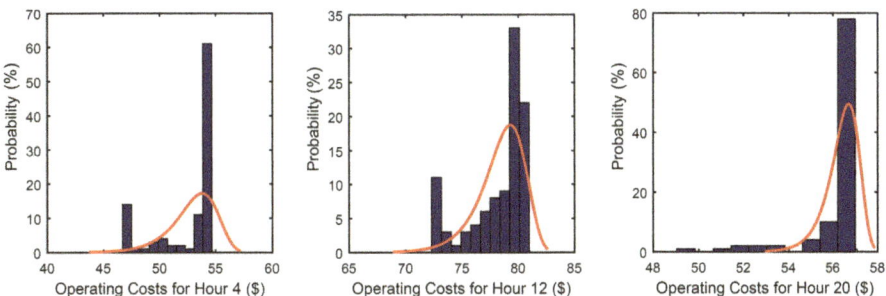

Figure 9. Probability of the operating cost on the 4th hour (**left**), 12th hour (**center**), and 20th hour (**right**) of the 20 March case study.

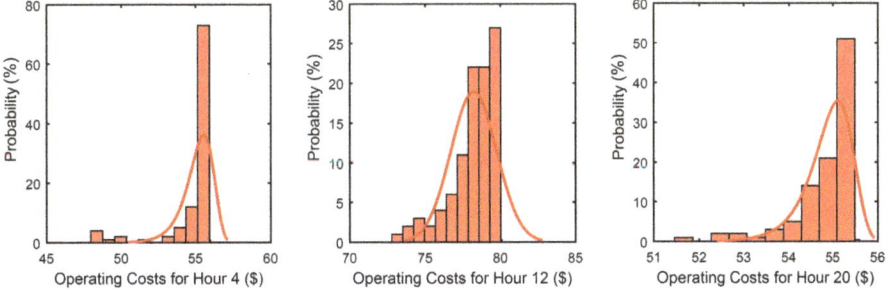

Figure 10. Probability of the operating cost on the 4th hour (**left**), 12th hour (**center**), and 20th hour (**right**) of the 21 June case study.

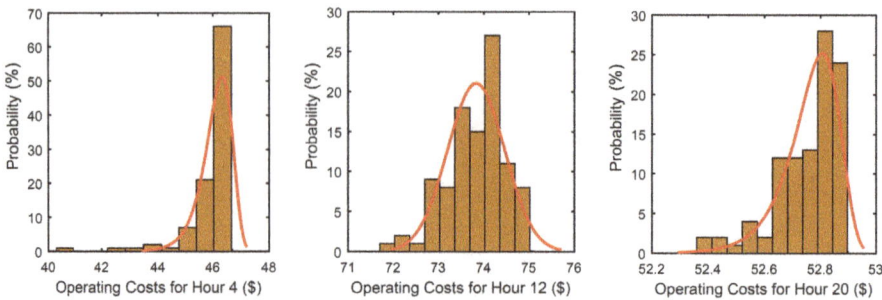

Figure 11. Probability of the operating cost on the 4th hour (**left**), 12th hour (**center**), and 20th hour (**right**) of the 22 September case study.

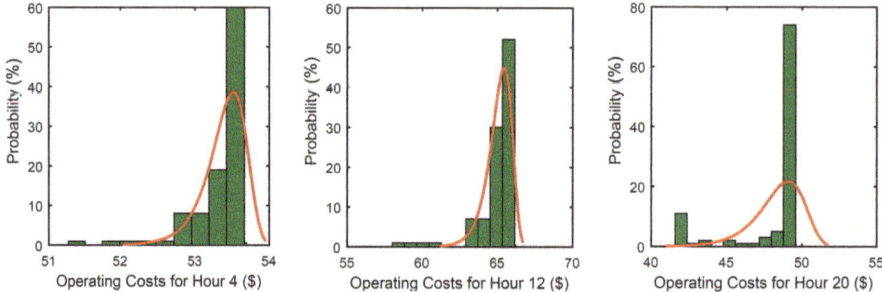

Figure 12. Probability of the operating cost on the 4th hour (**left**), 12th hour (**center**), and 20th hour (**right**) of the 21 December case study.

The uncertainties of solar PV and wind power generation drive the uncertainty in power generation overall when using renewable energy resources (Figures 13–16). It is important to note that the probability of zero power generation is also included in the plots, which can affect the general distributions. The solar irradiation values are drawn from a normal distribution while the wind speeds are from a Weibull distribution. Consequently, the potential operating costs have various distributions at different hours during the day. For instance, the operating cost follows a normal distribution during hours when the the majority of power generation comes from solar PV panels. Similarly, when wind generation dominates the makeup of electric power provided, the operating cost reflects a Weibull distribution. On the other hand, the effects from other systems are not as pronounced since energy systems such as the CHP generation, the electrical grid, and natural gas boilers are assumed to be readily available when needed. Statistical distributions of all considered power generation methods influence the type of distribution of the operating cost; however, the operating cost probability distribution tends to take the shape of the probability distribution for the source with the most uncertainty. Furthermore, the statistical distributions of energy loads are represented by normal distributions (Figures 17–20). The expected energy loads have Gaussian curves that center about their mean values. Therefore, the effect on the operating cost from the expected energy loads are determined by the means and standard deviations of the energy loads. For instance, the high variance (and, therefore, large standard deviation) of the cooling load during early afternoon hours leads to unpredictable cooling demand in those hours.

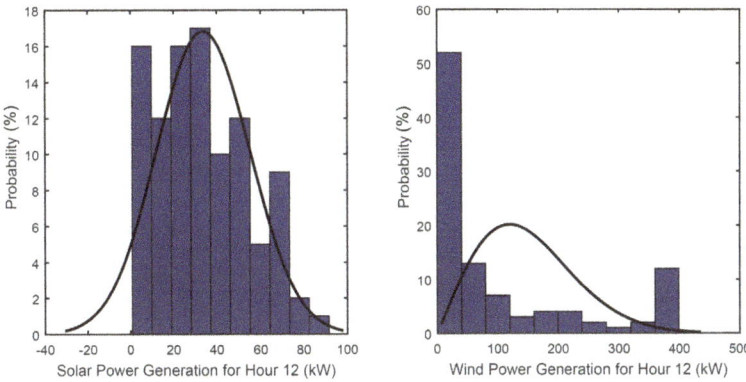

Figure 13. Probability of the power generation from solar PV (**left**) and wind (**right**) at the 12th hour on the 20 March case study.

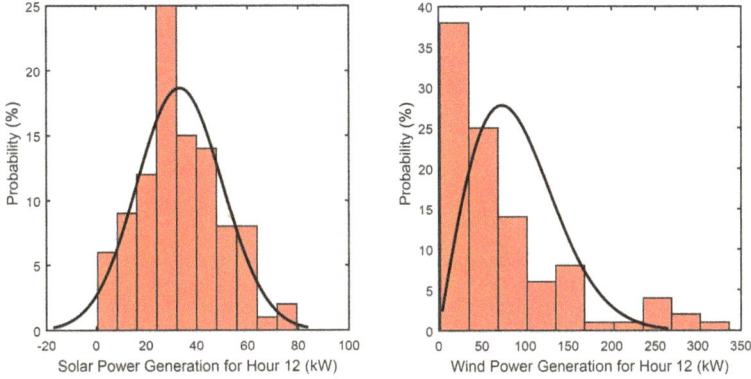

Figure 14. Probability of the power generation from solar PV (**left**) and wind (**right**) at the 12th hour on the 21 June case study.

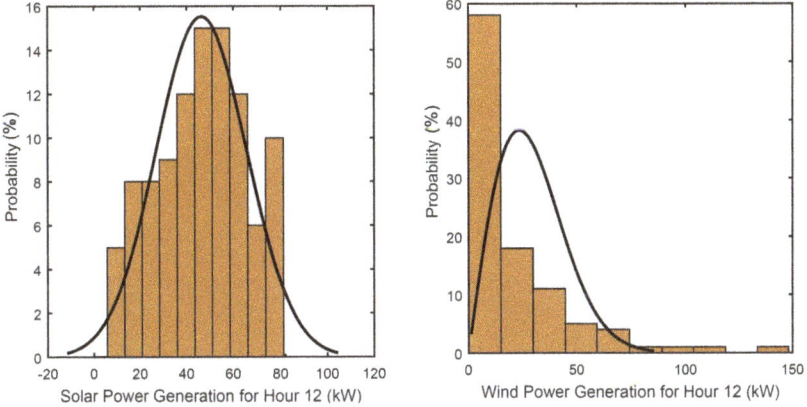

Figure 15. Probability of the power generation from solar PV (**left**) and wind (**right**) at the 12th hour on the 22 September case study.

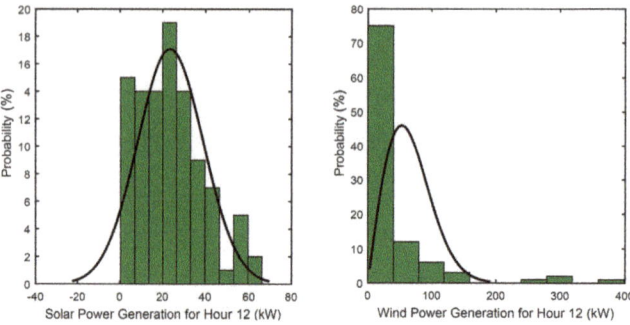

Figure 16. Probability of the power generation from solar PV (**left**) and wind (**right**) at the 12th hour on the 21 December case study.

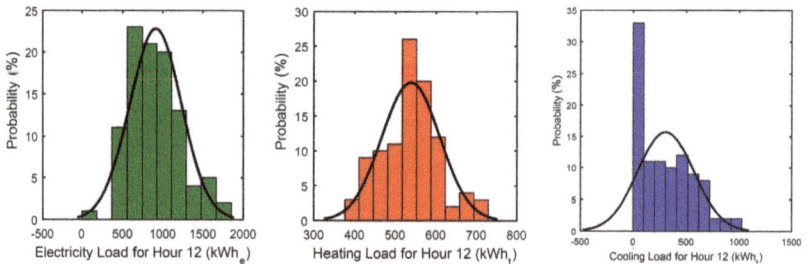

Figure 17. Probability of electricity load (**left**), heating (**center**), and cooling load (**right**) of the 12th hour on 20 March.

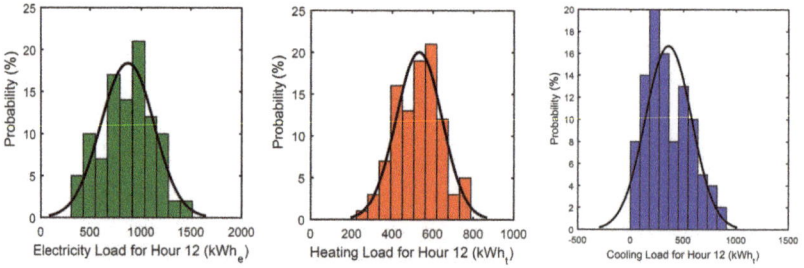

Figure 18. Probability of electricity load (**left**), heating (**center**), and cooling load (**right**) of the 12th hour on 21 June.

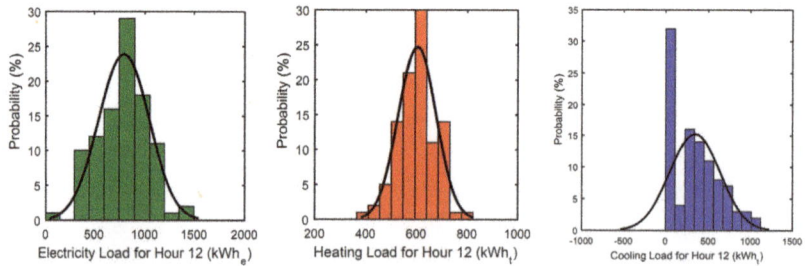

Figure 19. Probability of electricity load (**left**), heating (**center**), and cooling load (**right**) of the 12th hour on 22 September.

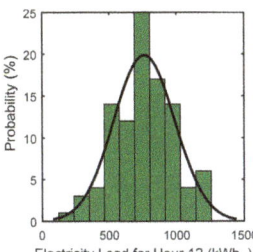 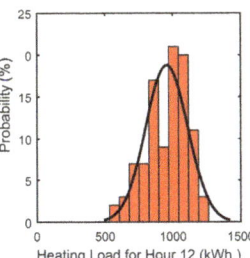 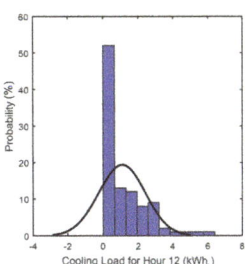

Figure 20. Probability of electricity load (**left**), heating (**center**), and cooling load (**right**) of the 12th hour on 21 December.

The addition of heating and cooling loads in the analysis of the DES has generated considerable differences compared to studies of DES without heating and cooling loads. As mentioned in the proposed DES section, all of the cooling equipment is run by electrical power. In other words, the power generation from energy systems in the DES needs to fulfill both the electricity demand and the electrical energy required by the cooling equipment. During events of high electricity and cooling demands, it is more likely for the DES to purchase power from the grid because the on-site generation is not adequate. From Table 5, the standard deviation of the operating cost often exceeds $2 during peak energy demand, which typically occurs around 12 p.m. and in the early afternoon. This observation indicates that the uncertainty of the operating cost of the DES increases during events of high energy consumption. This is due to the presence of uncertainties from electricity, heating, and cooling demands. It is also notable that the standard deviations of the operating costs on the last four hours on 20 March are considerably higher than the other hours. This is caused by the uncertainty of heating load during those hours.

The projected operating costs of the proposed DES on four example days are compared to the operating costs of the current energy system during the same periods (Table 6). Studies of DES without consideration of all three major energy loads do not provide a complete picture of the total energy consumption and operating cost. This study includes both electrical and thermal loads for more comprehensive results. The proposed DES with a mixed of power generation systems shows potential for operating cost reductions. The probabilistic operating cost savings are around $150/day (approximately 10%) compared to the actual operating cost. Overall, the strategic implementation of power generation from various sources reduces the overall operating cost of the DES.

Throughout this study, there are limitations on the applicability of this method that could be addressed by future studies on reducing the operating cost of DES with stochasticity in power generation and energy demands. Critical limitations are as follows:

- The performance of solar PV panels is assumed to be consistent throughout the lifetime of the renewable energy systems. In other words, the degradation of the energy system is not considered.
- The operating costs on the four example days are based on the specific system sizes that were detailed in Table 2. Design choices will influence the operating cost, but variations in the design of the various systems were not considered here.
- The CHP and boiler are assumed to be readily available. The uncertainties from these systems are not considered.
- The unit operating costs for the energy systems are consistent throughout the year. Monthly and seasonal changes in unit costs will influence the operating costs throughout the year.
- The start-up and shut-down costs of the energy systems are not considered.
- All of the variables are hourly. Fluctuations in power generation and energy loads on a sub-hourly basis are not accounted for here.

- The examples days do not capture extreme conditions/design days. Instead, the beginning days of four different astronomical seasons are studied, capturing a variety of conditions outside of the summer and winter design days.

4. Conclusions

DES optimization for a district of metered buildings has been investigated in this paper with consideration of uncertainties in energy loads and power generation of renewable energy resources. The optimization of this DES minimizes the operating cost, which includes the cost of meeting the electricity, heating, and cooling loads. The simulation results suggest that the operating cost of the proposed DES is less than the operating cost of the current energy system (by approximately 10%); however, the uncertainties in the system lead to unpredictability in the operating cost. Analyses based on statistical probability were demonstrated to have the capability to predict probable operating costs at a given time period.

A Monte Carlo statistical simulation has been used to incorporate the uncertainty of energy loads and power generation from renewable power generation into the DES model. The results from the case study using four example days have shown the influence of uncertain input variables. The operating cost at each hour of the four example days is heavily dependent on the sources of power generation. Even though the uncertainties of all renewable sources of power generation add to the uncertainty in the operating cost, the most dominating source of power generation at a given period determines the distribution of the operating cost. Furthermore, the uncertainty in operating cost of the electricity load is more prominent than the uncertainty in operating cost of the heating load. This is because the use of natural gas for heating is more reliable than power generation from renewable energy resources.

The low operating costs of the energy systems used in the DES contribute toward the low overall operating cost. The proposed DES incorporates solar PV and wind turbines for power generation, which can often operate at a lower cost per unit of electricity compared to purchased electricity from the grid. With on-site generation, the purchased power from the electrical grid on the four example days can be reduced by up to 26%. It is important to recognize that the percentage of on-site power generation (i.e., the reduction in purchased power) on different days of the year can vary from the on-site power generation of the four example days. Nonetheless, the buildings can be less dependent on the grid and the overall operating cost can be lowered.

In conclusion, a DES incorporating renewable energy systems offers operating cost reduction opportunities. However, the uncertainties associated with renewable energy resources can cause unreliable power generation, leading to uncertainty in operating costs. Additionally, consideration of uncertainties in the energy loads is also important. The addition of uncertainties from the electricity, heating, and cooling loads can further contribute to unpredictable operating costs. Even though the energy loads exhibit normal distributions, those with high variance can increase (or decrease) the required loads to be fulfilled. Thus, the operating cost can be highly unpredictable. A statistical analysis that incorporates the expected uncertainty is recommended for DES planning with renewable energy resources, as a deterministic calculation may give a misleading picture of the likely operating costs and potential savings.

Author Contributions: Conceptualization, T.T.D.T. and A.D.S.; Methodology, T.T.D.T. and A.D.S.; Software, T.T.D.T.; Formal Analysis, T.T.D.T.; Investigation, T.T.D.T.; Resources, T.T.D.T. and A.D.S.; Data Curation, T.T.D.T.; Writing—Original Draft Preparation, T.T.D.T. and A.D.S.; Writing—Review & Editing, T.T.D.T. and A.D.S.; Visualization, T.T.D.T.; Supervision, A.D.S.; Project Administration, T.T.D.T. and A.D.S.; Funding Acquisition, A.D.S.

Funding: This work was supported by the U.S. Department of Energy Grant No. EE0007712.

Acknowledgments: The authors would like to recognize the Facilities Management organization at the University of Utah, with particular thanks to engineer Phil Banza, for providing access to building energy data.

Conflicts of Interest: The authors declare no conflict of interest.

Abbreviations

The following abbreviations are used in this manuscript:

Abbreviations

CHP	Combined heat and power
CRF	Capital recovery factor
DES	District energy system
O&M	Operation and maintenance
PSO	Particle swarm optimization
PV	Photovoltaics

Nomenclature

β	Weibull shape parameter
γ	Unit cost of O&M ($/kWh)
σ	Standard deviation ($, m/s, W/m^2, kWh)
δ	Weibull scale parameter
λ	Unit cost of electricity generation ($/kWh)
μ	Mean ($, m/s, W/m^2, kWh)
c	Acceleration coefficient
C	Cost ($)
E	Electricity generation output (kWh$_e$)
f	Unit cost of fuel ($/kWh$_t$)
$gBest$	Global best solution
n	Project lifetime (Year)
N	Number of energy devices
$pBest$	Personal best solution
P	Power (kW)
Q	Thermal energy output (kWh$_t$)
r	Random value $\subset$ (0,1)
t	Hourly timestep (Hour)
v	Particle velocity
x	Particle position

Subscripts

ann	Annualized capital cost
boiler	Natural gas boiler
CHP	Combined heat and power
es	Energy storage
g	Electricity generation devices
grid	Electrical grid
h	Thermal heating devices
heating	Thermal heating
load	Energy load
NPC	Net present capital cost
s	Solar irradiation
solar	Solar PV power
w	Wind speed
wind	Wind power

Appendix A

Table A1. Electricity load parameters (mean and standard deviation) on four different days.

Hour	20 March		21 June		22 September		21 December	
	μ_e (kWh$_e$)	σ_e (kWh$_e$)	μ_e (kWh$_e$)	σ_e (kWh$_e$)	μ_e (kWh$_e$)	σ_e (kWh$_e$)	μ_e (kWh$_e$)	σ_e (kWh$_e$)
1	492.42	78.23	510.71	49.75	657.89	65.41	429.03	26.75
2	511.60	62.46	506.69	50.69	655.37	55.78	436.98	33.11
3	476.17	39.53	509.14	34.78	653.73	63.17	428.12	31.44
4	478.87	52.44	507.32	40.79	660.89	51.66	427.85	26.69
5	511.08	53.42	530.30	60.84	659.19	63.16	423.37	20.65
6	529.76	71.35	565.72	75.10	652.64	74.22	474.46	60.99
7	648.10	123.35	651.39	121.28	614.04	130.92	536.99	121.74
8	787.19	226.12	767.94	212.78	652.64	205.47	675.41	238.52
9	865.34	275.10	836.83	263.74	636.82	237.01	736.64	242.62
10	874.86	299.60	838.10	253.89	629.66	229.00	711.87	244.43
11	894.97	299.57	857.70	286.99	598.15	251.63	753.22	275.50
12	899.92	316.56	851.30	280.72	601.63	246.20	743.35	254.13
13	904.45	309.82	845.24	275.72	593.17	255.07	767.43	277.31
14	842.33	253.54	842.93	268.87	585.06	274.90	773.44	274.42
15	840.84	272.48	851.29	278.20	594.54	272.84	773.42	288.34
16	812.61	232.82	818.29	256.29	583.22	271.62	741.42	269.55
17	743.79	209.39	713.25	182.44	576.40	196.91	680.65	204.78
18	596.05	107.02	594.80	106.00	596.24	124.50	520.45	93.33
19	529.86	74.13	531.36	54.66	596.45	73.24	456.38	45.16
20	501.35	71.04	505.51	49.00	610.09	73.45	441.01	38.74
21	506.23	73.42	500.44	49.56	640.57	72.20	418.94	32.48
22	508.79	76.29	510.73	47.05	622.50	72.99	428.42	32.92
23	494.29	74.72	509.24	40.09	656.26	81.63	434.20	41.73
24	496.05	73.94	512.64	44.51	654.28	63.58	428.38	33.33

Table A2. Heating load parameters (mean and standard deviation) on four different days.

Hour	20 March		21 June		22 September		21 December	
	μ_h (kWh$_t$)	σ_h (kWh$_t$)	μ_h (kWh$_t$)	σ_h (kWh$_t$)	μ_h (kWh$_t$)	σ_h (kWh$_t$)	μ_h (kWh$_t$)	σ_h (kWh$_t$)
1	697.17	100.87	550.01	141.21	657.89	83.13	1078.32	142.08
2	704.61	110.71	559.89	146.39	655.37	79.77	1080.78	139.29
3	687.97	93.11	567.40	149.00	653.73	73.07	1138.20	207.98
4	716.95	115.52	570.12	149.75	660.89	77.54	1174.89	189.36
5	704.81	106.46	577.08	153.48	659.19	78.13	1127.84	173.81
6	726.84	113.43	576.53	147.03	652.64	73.65	1082.01	150.15
7	705.36	120.04	551.37	144.82	614.04	73.01	1046.14	144.54
8	682.38	126.15	600.40	169.66	652.64	78.88	1130.84	163.64
9	659.12	105.02	544.82	142.87	636.82	80.54	1082.01	162.38
10	617.38	90.15	529.00	136.83	629.66	95.09	1052.55	161.84
11	566.44	80.83	568.69	294.81	598.15	64.24	994.99	169.91
12	547.21	73.58	522.25	155.51	601.63	66.04	960.62	154.69
13	516.04	111.26	504.86	131.79	593.17	63.71	911.11	150.71
14	590.38	160.71	482.49	120.80	585.06	69.34	929.25	123.72
15	543.12	87.06	488.63	123.65	594.54	70.31	914.65	124.07
16	573.81	119.23	481.13	121.19	583.22	64.92	930.75	132.15
17	550.89	86.23	476.29	119.45	576.40	63.66	913.29	107.82
18	558.94	96.10	474.99	120.90	596.24	74.02	963.89	123.46
19	591.54	88.92	486.38	123.28	596.45	72.37	969.89	128.04
20	617.38	110.46	490.61	129.21	610.09	88.85	1010.67	129.43
21	655.57	111.91	498.24	140.00	640.57	83.58	1032.63	147.96
22	674.87	112.51	511.07	142.23	622.50	81.68	1018.04	129.34
23	674.60	102.79	528.73	139.95	656.26	87.64	1080.92	139.95
24	679.37	121.17	533.91	148.33	654.28	76.48	1068.37	139.65

Table A3. Cooling load parameters (mean and standard deviation) on four different days.

Hour	20 March		21 June		22 September		21 December	
	μ_c (kWh$_t$)	σ_c (kWh$_t$)	μ_c (kWh$_t$)	σ_c (kWh$_t$)	μ_c (kWh$_t$)	σ_c (kWh$_t$)	μ_c (kWh$_t$)	σ_c (kWh$_t$)
1	257.61	210.82	280.32	96.75	141.85	109.30	2.11	4.25
2	202.13	150.68	265.69	113.81	138.81	111.10	1.40	3.11
3	192.69	234.93	269.10	95.35	155.52	116.62	0.95	1.82
4	195.11	167.76	261.36	88.47	142.87	113.81	0.68	1.90
5	227.84	262.97	249.53	95.69	136.87	107.31	0.85	2.27
6	174.00	180.22	210.45	95.87	127.32	106.06	0.82	2.22
7	173.08	221.68	188.80	92.72	130.49	118.43	0.68	1.78
8	149.32	146.07	194.46	104.96	134.89	119.86	0.41	1.42
9	202.58	270.27	189.14	99.56	146.96	130.81	0.68	2.03
10	212.18	216.60	227.16	107.62	161.56	120.35	0.51	1.73
11	288.54	248.76	281.14	109.30	211.44	138.68	0.92	1.96
12	298.12	305.31	341.87	210.20	261.47	135.18	0.72	1.85
13	270.16	208.78	372.59	192.94	225.94	172.15	19.27	63.33
14	279.64	219.37	887.00	224.44	343.47	239.39	5.73	20.31
15	332.83	210.79	893.92	181.93	282.54	177.26	2.22	6.23
16	293.93	221.84	370.38	223.22	311.18	273.30	2.32	6.05
17	327.79	262.04	349.64	197.39	272.99	194.18	6.38	9.85
18	323.52	228.54	325.64	231.39	258.33	195.67	4.30	8.34
19	310.84	161.43	359.33	187.46	233.37	114.49	2.35	4.50
20	261.70	138.12	290.11	144.77	210.35	101.01	1.23	2.31
21	296.96	167.39	293.83	121.65	215.06	147.97	3.75	2.60
22	249.67	146.55	262.00	86.49	206.98	186.12	1.36	2.23
23	216.01	161.22	254.58	124.55	155.56	114.35	0.92	1.95
24	203.84	147.97	316.67	163.50	143.28	116.88	1.19	2.10

Table A4. Wind speed parameters (scale and shape) on four different days.

Hour	20 March		21 June		22 September		21 December	
	δ_w	β_w	δ_w	β_w	δ_w	β_w	δ_w	β_w
1	4.06	3.44	3.48	2.68	3.15	4.61	2.52	3.74
2	5.19	3.50	4.24	1.95	2.73	3.85	3.67	1.04
3	3.63	2.90	4.53	1.31	3.30	3.06	2.80	3.06
4	5.71	1.37	4.11	1.67	3.83	2.23	3.01	3.25
5	4.23	1.91	3.94	1.43	3.73	2.99	3.38	1.93
6	4.97	2.91	4.86	2.69	3.77	2.89	3.01	2.07
7	5.37	2.30	4.44	2.37	3.27	2.61	4.16	2.42
8	4.71	2.33	5.71	1.96	3.79	3.12	3.62	2.34
9	5.32	1.82	5.24	2.19	3.22	4.02	5.00	1.64
10	5.44	2.40	4.16	2.15	2.66	2.88	4.19	1.10
11	5.99	1.70	6.26	1.60	2.88	4.63	3.79	0.89
12	6.60	1.45	6.79	3.04	3.17	3.86	4.37	1.70
13	6.76	2.33	7.47	2.61	3.57	2.51	1.19	1.29
14	6.13	2.06	7.05	2.56	3.64	2.71	4.69	1.95
15	5.74	2.57	6.46	2.39	4.23	2.58	3.60	2.36
16	4.58	2.54	6.64	2.35	4.49	2.01	4.85	1.33
17	6.61	1.68	5.56	3.82	4.20	3.96	4.51	1.61
18	5.85	2.51	5.07	3.08	4.02	4.42	4.26	1.06
19	4.94	1.83	5.16	2.40	3.29	3.32	5.61	1.34
20	4.71	1.59	4.59	2.31	3.09	3.52	4.86	0.72
21	6.44	1.14	4.69	1.55	3.60	2.52	5.34	1.30
22	7.59	1.45	5.04	2.02	3.29	2.48	4.84	1.38
23	5.54	1.59	4.73	3.22	3.55	2.15	3.96	1.72
24	5.26	1.18	5.90	1.94	4.63	2.45	4.81	1.54

Table A5. Wind speed parameters (mean and standard deviation) on four different days.

Hour	20 March		21 June		22 September		21 December	
	μ_w (m/s)	σ_w (m/s)	μ_w (m/s)	σ_w (m/s)	μ_w (m/s)	σ_w (m/s)	μ_w (m/s)	σ_w (m/s)
1	4.19	1.66	3.54	2.15	2.45	1.43	2.48	1.53
2	4.11	2.00	3.65	1.10	2.35	1.18	2.05	2.02
3	3.89	1.63	4.37	2.74	3.27	1.17	2.77	1.69
4	3.75	1.84	2.92	2.35	2.87	1.84	2.73	1.83
5	3.54	1.87	3.32	2.00	3.44	1.50	2.44	1.78
6	4.46	1.59	4.02	1.88	3.10	1.64	2.50	1.59
7	4.22	1.45	3.57	1.85	2.88	1.31	2.89	2.17
8	3.85	1.52	3.78	2.31	3.75	1.18	3.13	2.16
9	4.40	2.32	4.14	2.00	3.30	1.39	2.89	2.72
10	4.51	2.87	4.47	2.30	2.74	1.67	2.87	2.06
11	4.98	3.05	4.44	2.57	2.82	1.17	2.87	2.37
12	4.99	3.87	5.60	1.79	3.17	1.64	3.68	2.61
13	5.60	2.90	5.96	2.44	3.45	1.53	3.61	2.70
14	5.39	1.93	5.57	1.73	3.48	1.30	4.09	2.29
15	4.41	1.45	5.23	1.97	3.85	1.35	3.95	2.48
16	4.77	2.28	5.88	2.65	4.01	1.35	3.65	1.91
17	4.81	1.78	4.59	1.03	3.81	0.94	3.19	1.96
18	4.36	1.62	4.30	1.30	3.29	1.14	2.68	1.89
19	4.33	2.48	3.43	1.50	2.55	0.96	1.83	2.17
20	4.19	2.51	3.20	1.80	2.21	1.06	1.92	2.02
21	3.36	2.89	2.82	1.61	2.95	1.28	2.29	2.28
22	3.61	3.18	3.99	1.31	3.19	1.19	2.06	2.04
23	3.97	2.36	3.95	1.13	3.27	1.47	2.48	1.55
24	3.63	2.54	3.99	2.41	3.60	1.55	2.75	1.94

Table A6. Solar global horizontal irradiation parameters (mean and standard deviation) on four different days.

Hour	20 March		21 June		22 September		21 December	
	μ_s (W/m^2)	σ_s (W/m^2)	μ_s (W/m^2)	σ_s (W/m^2)	μ_s (W/m^2)	σ_s (W/m^2)	μ_s (W/m^2)	σ_s (W/m^2)
1	0.00	0.00	0.00	0.00	0.00	0.00	0.00	0.00
2	0.00	0.00	0.00	0.00	0.00	0.00	0.00	0.00
3	0.00	0.00	0.00	0.00	0.00	0.00	0.00	0.00
4	0.00	0.00	0.00	0.00	0.00	0.00	0.00	0.00
5	0.00	0.00	0.33	0.82	0.00	0.00	0.00	0.00
6	0.00	0.00	114.67	112.13	0.00	0.00	0.00	0.00
7	4.87	12.76	268.80	233.49	158.40	74.93	0.00	0.00
8	242.93	226.92	442.73	283.06	494.20	201.54	0.20	0.56
9	327.73	317.12	529.40	320.47	610.40	279.02	59.40	67.32
10	409.20	334.32	594.27	331.27	731.53	260.48	132.87	164.46
11	469.47	359.90	564.60	388.64	732.73	307.62	281.80	240.79
12	501.67	400.60	593.13	338.30	749.80	325.33	298.80	270.75
13	489.47	416.97	604.87	295.30	763.87	269.62	266.47	263.38
14	509.87	390.06	523.07	383.02	764.80	236.39	245.07	261.49
15	430.53	355.43	447.47	367.61	697.60	307.82	221.47	243.97
16	403.33	363.25	423.87	366.08	696.53	238.74	155.53	175.07
17	422.20	309.26	383.60	327.20	595.60	229.72	16.60	48.49
18	280.13	213.92	365.40	269.74	476.07	125.82	0.07	0.26
19	70.07	54.09	243.80	230.65	26.27	30.89	0.00	0.00
20	0.00	0.00	28.80	51.32	0.00	0.00	0.00	0.00
21	0.00	0.00	0.33	0.90	0.00	0.00	0.00	0.00
22	0.00	0.00	0.00	0.00	0.00	0.00	0.00	0.00
23	0.00	0.00	0.00	0.00	0.00	0.00	0.00	0.00
24	0.00	0.00	0.00	0.00	0.00	0.00	0.00	0.00

Appendix B

The capital costs of different energy systems are shown in Table A7. The annualized capital costs are calculated based on the following equation:

$$C_{ann} = CRF \cdot C_{NPC},\qquad\text{(A1)}$$

where C_{ann} is the annualized capital costs, C_{NPC} is the net present capital cost, and CRF is the capital recovery factor, which can be calculated as follows:

$$CRF = \frac{i(1+i)^n}{(1+i)^n - 1},\qquad\text{(A2)}$$

where i is the discount rate and n is the project lifetime. The annualized capital costs are calculated based on a 5% discount rate and the project lifetimes of all systems are assumed to be 30 years. The annualized capital costs are also shown in Table A7.

Table A7. Capital costs of the proposed DES [37–39].

Energy System	Capital Cost
Solar PV	$997,200
Wind Turbine	$1,500,400
CHP	$1,020,000
Battery	$30,000
Boiler	$17,500
Total Capital Costs	$3,565,100
Annualized Capital Costs	$231,915

References

1. Thornton, A.; Monroy, C.R. Distributed power generation in the United States. *Renew. Sustain. Energy Rev.* **2011**, *15*, 4809–4817, doi:10.1016/j.rser.2011.07.070. [CrossRef]
2. Ehsan, A.; Yang, Q. Optimal integration and planning of renewable distributed generation in the power distribution networks: A review of analytical techniques. *Appl. Energy* **2018**, *210*, 44–59, doi:10.1016/j.apenergy.2017.10.106. [CrossRef]
3. Mokgonyana, L.; Zhang, J.; Li, H.; Hu, Y. Optimal location and capacity planning for distributed generation with independent power production and self-generation. *Appl. Energy* **2017**, *188*, 140–150, doi:10.1016/j.apenergy.2016.11.125. [CrossRef]
4. Bilgili, M.; Ozbek, A.; Sahin, B.; Kahraman, A. An overview of renewable electric power capacity and progress in new technologies in the world. *Renew. Sustain. Energy Rev.* **2015**, *49*, 323–334, doi:10.1016/j.rser.2015.04.148. [CrossRef]
5. Van der Walt, H.L.; Bansal, R.C.; Naidoo, R. PV based distributed generation power system protection: A review. *Renew. Energy Focus* **2018**, *24*, 33–40, doi:10.1016/j.ref.2017.12.002. [CrossRef]
6. Tran, T.T.; Smith, A.D. Evaluation of renewable energy technologies and their potential for technical integration and cost-effective use within the U.S. energy sector. *Renew. Sustain. Energy Rev.* **2017**, *80*, 1372–1388, doi:10.1016/j.rser.2017.05.228. [CrossRef]
7. Lorenzo, B.; Stefano, C.; Vincenzo, M.; Vittorio, R.; Luis, R.J. Hybrid renewable energy systems for renewable integration in microgrids: Influence of sizing on performance. *Energy* **2018**, *152*, e7–e133, doi:10.1016/j.energy.2018.03.165. [CrossRef]
8. Domenech, B.; Ranaboldo, M.; Ferrer-Martí, L.; Pastor, R.; Flynn, D. Local and regional microgrid models to optimise the design of isolated electrification projects. *Renew. Energy* **2018**, *119*, 795–808, doi:10.1016/j.renene.2017.10.060. [CrossRef]
9. Zachar, M.; Daoutidis, P. Understanding and predicting the impact of location and load on microgrid design. *Energy* **2015**, *90*, 1005–1023, doi:10.1016/j.energy.2015.08.010. [CrossRef]

10. Jacob, A.S.; Banerjee, R.; Ghosh, P.C. Sizing of hybrid energy storage system for a PV based microgrid through design space approach. *Appl. Energy* **2018**, *212*, 640–653, doi:10.1016/j.apenergy.2017.12.040. [CrossRef]

11. Mazzola, S.; Astolfi, M.; Macchi, E. The potential role of solid biomass for rural electrification: A techno economic analysis for a hybrid microgrid in India. *Appl. Energy* **2016**, *169*, 370–383, doi:10.1016/j.apenergy. 2016.02.051. [CrossRef]

12. Aluisio, B.; Dicorato, M.; Forte, G.; Trovato, M. An optimization procedure for Microgrid day-ahead operation in the presence of CHP facilities. *Sustain. Energy Grids Netw.* **2017**, *11*, 34–45, doi:10.1016/j.segan.2017.07.003. [CrossRef]

13. Lupangu, C.; Bansal, R.C. A review of technical issues on the development of solar photovoltaic systems. *Renew. Sustain. Energy Rev.* **2017**, *73*, 950–965, doi:10.1016/j.rser.2017.02.003. [CrossRef]

14. Pazouki, S.; Haghifam, M.R.; Moser, A. Electrical Power and Energy Systems Uncertainty modeling in optimal operation of energy hub in presence of wind, storage and demand response. *Int. J. Electr. Power Energy Syst.* **2014**, *61*, 335–345, doi:10.1016/j.ijepes.2014.03.038. [CrossRef]

15. Soroudi, A.; Amraee, T. Decision making under uncertainty in energy systems: State of the art. *Renew. Sustain. Energy Rev.* **2013**, *28*, 376–384, doi:10.1016/j.rser.2013.08.039. [CrossRef]

16. Parisio, A.; Del Vecchio, C.; Vaccaro, A. A robust optimization approach to energy hub management. *Int. J. Electr. Power Energy Syst.* **2012**, *42*, 98–104, doi:10.1016/j.ijepes.2012.03.015. [CrossRef]

17. Evins, R.; Orehounig, K.; Dorer, V.; Carmeliet, J. New formulations of the 'energy hub' model to address operational constraints. *Energy* **2014**, *73*, 387–398, doi:10.1016/j.energy.2014.06.029. [CrossRef]

18. Mavromatidis, G.; Orehounig, K.; Carmeliet, J. Design of distributed energy systems under uncertainty: A two-stage stochastic programming approach. *Appl. Energy* **2018**, doi:10.1016/j.apenergy.2018.04.019. [CrossRef]

19. Jabbari-Sabet, R.; Moghaddas-Tafreshi, S.M.; Mirhoseini, S.S. Microgrid operation and management using probabilistic reconfiguration and unit commitment. *Int. J. Electr. Power Energy Syst.* **2016**, *75*, 328–336, doi:10.1016/ j.ijepes.2015.09.012. [CrossRef]

20. Fioriti, D.; Giglioli, R.; Poli, D. Short-term operation of a hybrid minigrid under load and renewable production uncertainty. In Proceedings of the 2016 IEEE Global Humanitarian Technology Conference (GHTC), Seattle, WA, USA, 13–16 October 2016; pp. 436–443.

21. Mavromatidis, G.; Orehounig, K.; Carmeliet, J. A review of uncertainty characterisation approaches for the optimal design of distributed energy systems. *Renew. Sustain. Energy Rev.* **2018**, *88*, 258–277, doi:10.1016/ j.rser.2018.02.021. [CrossRef]

22. Olsthoorn, D.; Haghighat, F.; Mirzaei, P.A. Integration of storage and renewable energy into district heating systems: A review of modelling and optimization. *Solar Energy* **2016**, *136*, 49–64, doi:10.1016/ j.solener.2016.06.054. [CrossRef]

23. Li, Y.; Rezgui, Y.; Zhu, H. District heating and cooling optimization and enhancement—Towards integration of renewables, storage and smart grid. *Renew. Sustain. Energy Rev.* **2017**, *72*, 281–294, doi:10.1016/j.rser. 2017.01.061. [CrossRef]

24. Najibi, F.; Niknam, T. Stochastic scheduling of renewable micro-grids considering photovoltaic source uncertainties. *Energy Convers. Manag.* **2015**, *98*, 484–499, doi:10.1016/ j.enconman.2015.03.037. [CrossRef]

25. Nikmehr, N.; Najafi-Ravadanegh, S.; Khodaei, A. Probabilistic optimal scheduling of networked microgrids considering time-based demand response programs under uncertainty. *Appl. Energy* **2017**, *198*, 267–279, doi:10.1016/j.apenergy.2017.04.071. [CrossRef]

26. Yılmaz Balaman, Ş.; Selim, H. Sustainable design of renewable energy supply chains integrated with district heating systems: A fuzzy optimization approach. *J. Clean. Prod.* **2016**, *133*, 863–885, doi:10.1016/ j.jclepro.2016.06.001. [CrossRef]

27. Lu, S.; Gu, W.; Zhou, J.; Zhang, X.; Wu, C. Coordinated dispatch of Multi-Energy System with District Heating Network: Modeling and Solution Strategy. *Energy* **2018**, *152*, 358–370, doi:10.1016/j.energy.2018.03.088. [CrossRef]

28. Sameti, M.; Haghighat, F. Optimization approaches in district heating and cooling thermal network. *Energy Build.* **2017**, *140*, 121–130, doi:10.1016/j.enbuild.2017.01.062. [CrossRef]

29. Gang, W.; Augenbroe, G.; Wang, S.; Fan, C.; Xiao, F. An uncertainty-based design optimization method for district cooling systems. *Energy* **2016**, *102*, 516–527, doi:10.1016/j.energy.2016.02.107. [CrossRef]

30. Li, C.Z.; Shi, Y.M.; Liu, S.; Zheng, Z.L.; Liu, Y.C. Uncertain programming of building cooling heating and power (BCHP) system based on Monte-Carlo method. *Energy Build.* **2010**, *42*, 1369–1375, doi:10.1016/J.ENBUILD.2010.03.005. [CrossRef]

31. Mavromatidis, G.; Orehounig, K.; Carmeliet, J. Uncertainty and global sensitivity analysis for the optimal design of distributed energy systems. *Appl. Energy* **2018**, *214*, 219–238, doi:10.1016/j.apenergy.2018.01.062. [CrossRef]

32. Dantzig, G.B. Linear Programming under Uncertainty. *Manag. Sci.* **1955**, *1*, 197–206, doi:10.1287/mnsc.1.3-4.197. [CrossRef]

33. SkyFoundry. SkySpark-SkyFoundry. Available online: https://skyfoundry.com/ (accessed on 23 January 2019).

34. ASHRAE. American Society of Heating, Refrigerating and Air-Conditioning Engineers. Available online: https://www.ashrae.org/ (accessed on 23 January 2019).

35. National Renewable Energy Laboratory (NREL). Wind Data. Available online: https://www.nrel.gov/gis/data-wind.html (accessed on 23 January 2019).

36. National Renewable Energy Laboratory (NREL). *Advancing the Science of Solar Data*; NREL: Golden, CO, USA, 2018.

37. National Renewable Energy Laboratory. *Distributed Generation Renewable Energy Estimate of Costs*; Technical Report; NREL: Golden, CO, USA, 2016.

38. US Environmental Protection Agency. *Catalog of CHP Technologies*; US Environmental Protection Agency: Washington, DC, USA, 2015.

39. US Environmental Protection Agency. *Fact Sheet: CHP as a Boiler Replacement Opportunity*; US Environmental Protection Agency: Washington, DC, USA, 2013; pp. 1–6.

40. Wais, P. A review of Weibull functions in wind sector. *Renew. Sustain. Energy Rev.* **2017**, *70*, 1099–1107, doi:10.1016/j.rser.2016.12.014. [CrossRef]

41. Harris, R.I.; Cook, N.J. The parent wind speed distribution: Why Weibull? *J. Wind Eng. Ind. Aerodyn.* **2014**, *131*, 72–87, doi:10.1016/j.jweia.2014.05.005. [CrossRef]

42. Sedić, A.; Pavković, D.; Firak, M. A methodology for normal distribution-based statistical characterization of long-term insolation by means of historical data. *Sol. Energy* **2015**, *122*, 440–454, doi:10.1016/j.solener.2015.09.014. [CrossRef]

43. Tran, T.T.; Smith, A.D. Incorporating performance-based global sensitivity and uncertainty analysis into LCOE calculations for emerging renewable energy technologies. *Appl. Energy* **2018**, *216*, 157–171, doi:10.1016/j.apenergy.2018.02.024. [CrossRef]

44. Kennedy, J.; Eberhart, R.C. Particle Swarm Optimization. In Proceedings of the IEEE International Conference on Neural Networks, Perth, Australia, 27 November–1 December 1995; pp. 1942–1948.

45. Kulkarni, N.K.; Patekar, S.; Bhoskar, T.; Kulkarni, O.; Kakandikar, G.M.; Nandedkar, V.M. Particle Swarm Optimization Applications to Mechanical Engineering—A Review. *Mater. Today Proc.* **2015**, *2*, 2631–2639, doi:10.1016/j.matpr.2015.07.223. [CrossRef]

46. Mazhoud, I.; Hadj-Hamou, K.; Bigeon, J.; Joyeux, P. Particle swarm optimization for solving engineering problems: A new constraint-handling mechanism. *Eng. Appl. Artif. Intell.* **2013**, *26*, 1263–1273, doi:10.1016/j.engappai.2013.02.002. [CrossRef]

47. Marini, F.; Walczak, B. Particle swarm optimization (PSO). A tutorial. *Chemom. Intell. Lab. Syst.* **2015**, *149*, 153–165, doi:10.1016/j.chemolab.2015.08.020. [CrossRef]

Article

Effects of Producer and Transmission Reliability on the Sustainability Assessment of Power System Networks

Jose R. Vargas-Jaramillo [1], Jhon A. Montanez-Barrera [1], Michael R. von Spakovsky [2],*, Lamine Mili [3] and Sergio Cano-Andrade [1],*

[1] Department of Mechanical Engineering, Universidad de Guanajuato, Salamanca, GTO 36885, Mexico; jr.vargas.jaramillo@gmail.com (J.R.V.-J.); ja.montanezbarrera@ugto.mx (J.A.M.-B.)
[2] Department of Mechanical Engineering, Virginia Tech, Blacksburg, VA 24061, USA
[3] Bradley Department of Electrical and Computer Engineering, Northern Virginia Center, Virginia Tech, Falls Church, VA 22043, USA; lmili@vt.edu
* Correspondence: vonspako@vt.edu (M.R.v.S.); sergio.cano@ugto.mx (S.C.-A.)

Received: 4 December 2018; Accepted: 31 January 2019; Published: 10 February 2019

Abstract: Details are presented of the development and incorporation of a generation and transmission reliability approach in an upper-level sustainability assessment framework for power system planning. This application represents a quasi-stationary, multiobjective optimization problem with nonlinear constraints, load uncertainties, stochastic effects for renewable energy producers, and the propagation of uncertainties along the transmission lines. The Expected Energy Not Supplied (EENS) accounts for generation and transmission reliability and is based on a probabilistic as opposed to deterministic approach. The optimization is developed for three scenarios. The first excludes uncertainties in the load demand, while the second includes them. The third scenario accounts not only for these uncertainties, but also for the stochastic effects related to wind and photovoltaic producers. The sustainability-reliability approach is applied to the standard IEEE Reliability Test System. Results show that using a Mixture of Normals Approximation (MONA) for the EENS formulation makes the reliability analysis simpler, as well as possible within a large-scale optimization. In addition, results show that the inclusion of renewable energy producers has some positive impact on the optimal synthesis/design of power networks under sustainability considerations. Also shown is the negative impact of renewable energy producers on the reliability of the power network.

Keywords: reliability; sustainability; IEEE-RTS; uncertainties; MONA

---

1. Introduction

Power network sustainability-reliability is an important aspect of an energy generation-transmission system because the electricity demand of a group of customers needs to be assured at every instant of time with the lowest possible price and the least damage to the environment and society [1–4]. The reliability of a power network is usually dependent on the uncertainties associated with generation [5–9], transmission (and distribution) [10–12], load demand [13–15], and the presence of unexpected catastrophic events [16–18]. On the other hand, the sustainability of a power network depends on four pillars, i.e., economic, technical, environmental, and societal [19,20]. These two main characteristics, i.e., sustainability and reliability, make the planning and design/operation optimization of a power network a difficult problem to solve [21,22]. Several methodologies have been developed independently to find solutions for these problems [23–25]. However, new methodologies that capture all of the aspects of sustainability and reliability and their integration into a single framework are necessary for a more detailed understanding of the planning and operational optimization and performance analysis of power networks.

In [26], the reliability of a power network is studied when decentralized generators are added to the system. A result of this study is that because of the weather dependency of some renewable energy technologies, adding a high number of renewables may cause instabilities in the network. In [27], the reliability of the power network is also studied when renewable energy technologies are added to an existing network. The part-load performance (start and ramp rates) of the thermal plants is significantly degraded in the location where renewable energy technologies are added. In [16], a reliability and evaluation assessment of a transmission grid subject to cascading failures is presented. It focuses on the impact of extreme weather on the reliability of the network. The catastrophic events are taken into account via a stochastic model based on the annual history of weather conditions in the area under study. In [28], a methodology is proposed for quantifying the transmission reliability margin when uncertainties in the network are present due to a transfer of power. A bootstrap technique is used to generate different scenarios and to quantify the transmission reliability margin with good accuracy. In [29], a methodology is proposed for the reliability evaluation of radial distribution networks. The methodology is based on an AC multiobjective optimization of repair times, failure rates, costs, and reliability.

Bi-level models are also used for power network planning under sustainability/reliability considerations with the aim of obtaining a more detailed description of the generation/transmission infrastructure, the individual technologies, and their interactions. This type of approach results in a detailed study at two different hierarchical levels, where each level with a single objective function or multiple ones depends on a different set of independent decision variables.

In [30], a bi-level programming approach to study the vulnerability of a power network is used. The upper level analyzes the effect of outages on the network, while the lower level analyzes the effect of these outages on the system operator. In the bi-level approach of [31], the upper level optimizes the dispatch of distributed generation and the cost of market purchases, while the lower level maximizes social welfare. In [32], the upper level selects the location and contract pricing of distributed generation, while the lower level measures the reaction of the distribution company. In [33], a methodology for microgrid power and reserve capacity planning is proposed with the goal to reduce microgrid capital and operational costs and assure a reliable supply of energy to the customers. The upper level optimizes the microgrid configurations, and the lower level optimizes the reserve capacity, which directly affects the reliability of the distribution system operator. In [21], a bi-level sustainability-reliability assessment framework for power networks and their interaction with microgrids is proposed. The upper level develops the synthesis/design/operation optimization of the producers and transmission lines, while the lower level takes care of the synthesis/design/operation optimization of the producers targeted by the upper level to be part of the power network configuration.

In this paper, a probabilistic generation-transmission reliability approach is proposed to be used in the upper-level Sustainability Assessment Framework (SAF) of [21]. The reliability approach takes into account, at every node and instant of time, the propagation of uncertainties along the transmission lines, the uncertainties of the generation system (including the fluctuating effects associated with the wind and photovoltaic energy producers), and the uncertainties of the load demand. This generation-transmission reliability approach is also capable of incorporating in a straightforward manner methodologies such as that of [16] to account for the contribution of unexpected catastrophic events to the reliability of a power network.

The remainder of the paper is organized as follows. Section 2 describes the system under study, while Section 3 outlines the SAF and the generation-transmission reliability approach. Section 4 then provides a description of the different scenarios considered for the analysis followed by Section 5, which presents the results of the application of the SAF to the system under study. The paper concludes with a set of conclusions in Section 6.

2. Description of the System

The IEEE Reliability Test System (RTS) [34] is used as a study case to evaluate the methodology proposed in this work. The single-line diagram of the RTS is shown in Figure 1. The system is composed of 11 generation units and seven transmission lines, as well as one node where only generation capacity is present (Node 1), four nodes where only load demand is present (Nodes 3 to 6), and one node where generation capacity and load demand are present (Node 2). The original RTS model considers two transmission lines between Nodes 1 and 3 and two between Nodes 2 and 4. In this work, a single line with the capacity of the two lines added together is considered instead for each set of nodes.

The annual peak load for the RTS is 185 MW [35]. A Load Duration Curve (LDC) is constructed for the model based on [34,35] and is shown in Figure 2. The load in each node is a fixed percentage of the load in the RTS as given in Table 1. The installed capacity in each node is also given in Table 1. Five different types of producers are considered for the analysis, i.e., hydro, Ultra Supercritical Coal (USC), Natural Gas Combined Cycle (NGCC), Combustion Turbine (CT), and Reciprocating Internal Combustion Engine (RICE). The producer characteristics for the RTS are given in Table 2. The six nodes of the RTS are connected by seven transmission lines with a transmission voltage level of 230 kV [34]. The capacity of each of the transmission lines is given in Table 3.

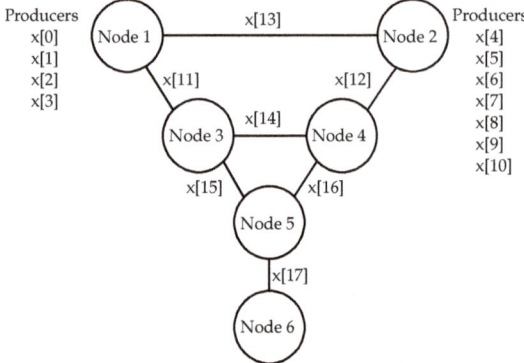

Figure 1. Schematic representation of the single-line RTS [34].

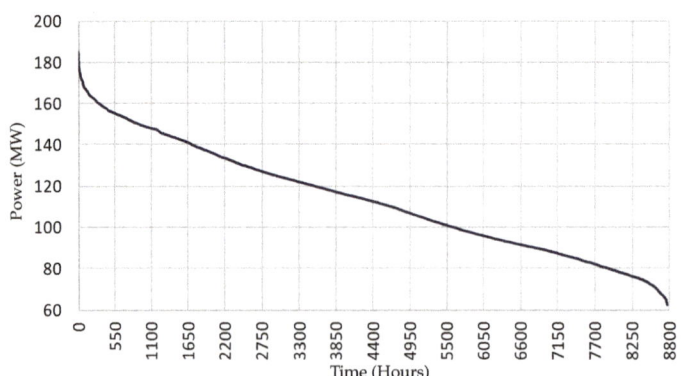

Figure 2. Load Duration Curve (LDC) for the RTS [34,35].

Table 1. Load demand and installed capacity in the nodes of the RTS [34].

Node	% of the Total Load	Installed Capacity (MW)
1	0	110
2	10.81	130
3	45.95	0
4	21.62	0
5	10.81	0
6	10.81	0

Table 2. Characteristics of the producers [34,36].

Decision Variable	Type of Producer	$P_{i,n}^{Cap}$ (MW)	$M_{i,n,\$}^{Cap}$ ($/MW-day)	$M_{i,n,\$}^{O\&M}$ ($/MW-day)	$\gamma_{i,n,\$}$ ($/MW)	$FOR_{i,n}$
x[0]	USC	40	799.35	115.34	4.60	0.030
x[1]	NGCC	40	215.01	30.14	3.50	0.030
x[2]	CT	10	242.05	47.95	3.50	0.020
x[3]	RICE	20	295.03	18.90	5.85	0.025
x[4]	Hydro	5	645.46	38.71	0.00	0.010
x[5]	Hydro	5	645.46	38.71	0.00	0.010
x[6]	Hydro	40	645.46	38.71	0.00	0.020
x[7]	Hydro	20	645.46	38.71	0.00	0.015
x[8]	Hydro	20	645.46	38.71	0.00	0.015
x[9]	Hydro	20	645.46	38.71	0.00	0.015
x[10]	Hydro	20	645.46	38.71	0.00	0.015

Table 3. Characteristics of the transmission lines [34,37].

Decision Variable	$L_{n,m}$ (km)	$P_{n,m}^{max}$ (MW)	$X_{n,m}$ (Ω)	$\gamma_{n,m,\$}$ ($/MW-day)
x[11]	300	196	380.88	112.12
x[12]	1000	196	1269.6	373.73
x[13]	200	98	253.92	74.75
x[14]	50	98	63.48	18.69
x[15]	50	98	63.48	18.69
x[16]	50	98	63.48	18.69
x[17]	50	98	63.48	18.69

3. Sustainability Assessment Approach

The upper-level SAF developed by Cano-Andrade et al. [21] is used here to optimize the design of the RTS, using a quasi-stationary, multiobjective optimization problem with nonlinear constraints. Mathematically, it is represented as follows:

Minimize:

$$\vec{C} = [C_1, C_2, \ldots, C_k]^T \tag{1}$$

with respect to non-negative $P_{i,n}^t$ and $P_{n,m}^t$ and subject to:

$$\sum_{n=1}^{N} X_{n,m,l}(P_{n,m,l}^t - P_{m,n,l}^t) = 0 \quad \text{for all } l \text{ and } t \tag{2}$$

$$P_{D_n}^t - \sum_{i=1}^{I} P_{i,n}^t - \sum_{m=1,n}^{M} [P_{m,n}^t(1 - \alpha_{m,n}X_{m,n}P_{m,n}^t) - P_{n,m}^t] \leq 0 \quad \text{for all } n \text{ and } t \tag{3}$$

$$P_{i,n}^{min} \leq P_{i,n} \leq P_{i,n}^{max} \quad \text{for all } i \text{ and } n \tag{4}$$

$$P_{n,m}^{min} \leq P_{n,m} \leq P_{n,m}^{max} \quad \text{for all } m \text{ and } n \tag{5}$$

where Equation (2) is the linearized version of Kirchoff's voltage law (KVL), which maintains the value of the phase shifter in each loop equal to zero [38]. The $P_{n,m}^t$ are the decision variables that correspond to the flow of electricity in a transmission line from node n to node m at time t; l represents a loop; and $X_{m,n}$ is the reactance of a transmission line. Equation (3) is the linearized version of Kirchhoff's current law (KCL), which assures the power balance at each node. This equation is presented as an inequality constraint in order to maintain the convexity of the solution space [38]. The $P_{i,n}^t$ are the decision variables that correspond to the generation of electricity of producer i at node n at time t; $P_{D_n}^t$ is the load demand at node n and time t; and $\alpha_{m,n} = 2.5 \times 10^{-7}$ is a constant that generates a total loss of 2% of the total generation in the network [22]. Equations (4)–(5) provide the limits for the producers and transmission lines, respectively. The maximum limit of the electricity generated by producer i at node n is given as:

$$P_{i,n}^{\max} = P_{i,n}^{\text{Cap}}(1 - \text{FOR}_{i,n}) \tag{6}$$

where $\text{FOR}_{i,n}$ and $P_{i,n}^{\text{Cap}}$ are the forced outage rate and the design capacity of a producer, respectively, given in Table 2.

Six different objective functions are used for the multiobjective optimization problem and cover the four pillars of sustainability, i.e., the total daily costs for the economic aspects, SO_2 and CO_2 daily emissions for the environmental aspects, the Disability Adjusted Life Year (DALY) for the social aspects, and the exergetic efficiency and Expected Energy Not Supplied (EENS) for the technical aspects.

The optimization problem is solved in Python 3.6.0 using the scipy.optimize library with the SLSQP algorithm [39,40].

3.1. Total Daily Costs

The variable operating and maintenance (O&M) cost, fixed O&M cost, capital cost, and cost associated with the possible construction of a new transmission line are taken into account. The total cost is, thus, written as [21]

$$C_{\$} = \sum_{t=1}^{t^{\max}} \sum_{n=1}^{N} \sum_{i=1}^{I} (M_{i,n,\$}^{\text{Cap}} + M_{i,n,\$}^{\text{O\&M}} + M_{i,n,\$}^t) + \sum_{n=1}^{N} M_{n,m,\$} \tag{7}$$

where the values of the capital cost, $M_{i,n,\$}^{\text{Cap}}$, and fixed O&M cost of production, $M_{i,n,\$}^{\text{O\&M}}$, for the different producers in the RTS are updated using [36] and are given in Table 2. The total cost associated with the possible construction of a new line from node n to node m is given as

$$M_{n,m,\$} = \gamma_{n,m,\$} L_{n,m} P_{n,m}^{\max} \tag{8}$$

where the $\gamma_{n,m,\$}$ are the effective cost of transmission coefficients particular to each transmission line and $L_{n,m}$ is the length of the transmission line. In the present work, a value of 700 $/MW-km for $\gamma_{n,m,\$}$ is considered [37]. The values of $\gamma_{n,m,\$}$, $L_{n,m}$, and $X_{n,m}$ for the different transmission lines in the RTS are given in Table 3. The capital cost, fixed O&M cost, and cost of the possible construction of a new transmission line are amortized accounting for interest, depreciation, and taxes using the annualization factor $\frac{r(1+r)^{yr}}{(1+r)^{yr}-1}$, which provides the annualized cost of a producer, where $yr = 20$ is the average useful life of a power plant, and $r = 0.05$ is the annual interest rate [21].

The variable O&M cost associated with the fuel consumption of a producer is defined as

$$M_{i,n,\$}^t = \gamma_{i,n,\$} P_{i,n}^t \tag{9}$$

where the $\gamma_{i,n,\$}$ are linear coefficients associated with the cost of fuel consumption particular to each producer and are given in Table 2. These real positive coefficients allow the model to account for the part-load behavior of the producers and to maintain the convexity of the objective function.

3.2. SO₂ Daily Emissions

SO$_2$ is considered here because of its high toxicity, and it is of particular concern when producers are close to population centers. The SO$_2$ daily emissions objective function is defined as [21]

$$C_{SO_2} = \sum_{t=1}^{t^{max}} \sum_{n=1}^{N} \sum_{i=1}^{I} E_{i,n,SO_2}^t \tag{10}$$

where the amount of SO$_2$ in kg emitted by each producer is given by

$$E_{i,n,SO_2}^t = \gamma_{i,n,SO_2} P_{i,n}^t \tag{11}$$

Here, the γ_{i,n,SO_2} are linear coefficients associated with the amount of SO$_2$ emissions particular to each producer and are given in Table 4. These real positive constants allow the model to account for the part-load behavior of the producers and to maintain the convexity of the objective function.

Table 4. Coefficients of the indicators for the different producers [21,36].

Type of Producer	γ_{i,n,SO_2} (kg/MW)	γ_{i,n,CO_2} (kg/MW)	γ_{i,n,NO_x} (kg/MW)	$\eta_{i,n}$ (%)
USC	0.154772	318.8306105	0.0928633	34
NGCC	0.001548	181.0834050	0.0116079	32
CT	0.001548	181.0834050	0.0464316	32
RICE	0.001548	181.0834050	0.1083405	31
Hydro	–	–	–	98

3.3. CO₂ Daily Emissions

CO$_2$ emissions are considered as well because of their perceived connection to the planet's greenhouse effect. The CO$_2$ daily emissions objective function is written as

$$C_{CO_2} = \sum_{t=1}^{t^{max}} \sum_{n=1}^{N} \sum_{i=1}^{I} E_{i,n,CO_2}^t \tag{12}$$

where the amount of CO$_2$ in kg emitted by each producer is defined as

$$E_{i,n,CO_2}^t = \gamma_{i,n,CO_2} P_{i,n}^t \tag{13}$$

Here, the γ_{i,n,CO_2} are linear coefficients associated with the amount of CO$_2$ emissions particular to each producer and are given in Table 4. These real positive constants allow for the model to account for the part-load behavior of the producers and to maintain the convexity of the objective function.

3.4. Disability Adjusted Loss of Life Year

The DALY quantifies the years of life lost by premature death and the years lived with a bad quality of life because of health issues associated with the emission of pollutants by the electricity producers [25,41]. The DALY objective function is expressed as

$$C_{DALY} = \sum_{t=1}^{t^{max}} \sum_{n=1}^{N} \sum_{i=1}^{I} (D_{i,n,CO_2}^t + D_{i,n,NO_x}^t) \tag{14}$$

where the contributions from the CO$_2$ and NO$_x$ emissions, respectively, to the DALY by each producer are defined as

$$D_{i,n,CO_2}^t = \gamma_{i,n,DALY}^{CO_2} \gamma_{i,n,CO_2} P_{i,n}^t \tag{15}$$

$$D_{i,n,NO_x}^t = \gamma_{i,n,DALY}^{NO_x} \gamma_{i,n,NO_x} P_{i,n}^t \tag{16}$$

Here, the γ_{i,n,NO_x} are linear coefficients associated with the amount of NO_x emissions particular to each producer and are given in Table 4. These real positive constants allow for the model to account for the part-load behavior of the producers and to maintain the convexity of the objective function. The DALY coefficients particular to each producer, $\gamma_{i,n,DALY}^{CO_2}$ and $\gamma_{i,n,DALY}^{NO_x}$, are obtained from [41] considering a hierarchical range model with a value of 0.0000570 (DALY/kg) for CO_2 and 0.0000014 (DALY/kg) for NO_x.

3.5. Exergetic Efficiency

In the present work, the only producer product considered is electricity. Thus, the exergetic efficiency in the network is written as [21]

$$C_\eta = \frac{P^{Tot}}{F^{Tot}} \tag{17}$$

where the total exergy (electricity) delivered by the producers is given as

$$P^{Tot} = \sum_{t=1}^{t^{max}} \sum_{n=1}^{N} \sum_{i=1}^{I} P_{i,n}^t \tag{18}$$

and the total chemical exergy of the fuel needed to generate P^{Tot} is written as

$$F^{Tot} = \sum_{t=1}^{t^{max}} \sum_{n=1}^{N} \sum_{i=1}^{I} F_{i,n}^t = \sum_{t=1}^{t^{max}} \sum_{n=1}^{N} \sum_{i=1}^{I} \frac{P_{i,n}^t}{\eta_{i,n}^t} \tag{19}$$

Here, $F_{i,n}^t$ is the exergy of the fuel needed by each producer, and $\eta_{i,n}^t$ is the efficiency of each producer and is given in Table 4. The fuel used by renewable producers is considered to be zero so that the exergetic efficiency for renewables is 100%.

3.6. Expected Energy Not Supplied

The reliability of the network expressed in terms of the Expected Energy Not Supplied (EENS) is written as

$$C_{EENS} = \sum_{t=1}^{t^{max}} \sum_{n=1}^{N} f_{EENS,n}^t \tag{20}$$

where $f_{EENS,n}^t$ is the expected amount of energy that is not supplied to node n at time t expressed as [23]

$$f_{EENS,n}^t = f_{LOLP,n}^t P_{D_n}^t \tag{21}$$

Here, $P_{D_n}^t$ is the electricity demand at node n and time t, and $f_{LOLP,n}^t$ is the loss-of-load probability at node n and time t, which represents the expected number of hours that the system is not able to supply the load [22]. It is given by

$$f_{LOLP,n}^t = Pr(P_{D_n}^t - \sum_{i=1}^{I} P_{i,n}^t - \sum_{m=1,n}^{M} [P_{m,n}^t(1 - \alpha_{m,n} X_{m,n} P_{m,n}^t) - P_{n,m}^t] \geq 0) \tag{22}$$

$Pr(\cdot)$ in this last expression represents the probability that $(\cdot)$ occurs and is defined by

$$Pr(\cdot) = 1 - F_n^t(X) = 1 - \int_{-\infty}^{X} f_n^t(x)dx \tag{23}$$

where $f_n^t(x)$ represents the probability density function (PDF) of the sum (convolution) of the PDF's corresponding to the generation, transmission, and load at each node, since the net power into node n is the sum of the power generated at n ($P_{gen_i} = \sum_{i=1}^{I} P_{i,n}^t$) plus the power into n ($P_{in_n} =$

$\sum_{m=1,n}^{M} P_{m,n}^t (1 - \alpha_{m,n} X_{m,n} P_{m,n}^t))$ minus the power out of n ($P_{\text{out}_n} = \sum_{m=1,n}^{M} P_{n,m}^t$) minus the load demand at n ($P_{\text{dem}_n} = P_{D_n}^t$). Therefore,

$$f_n^t(x) = f_{P_{\text{gen}_i} + P_{\text{in}_n} - P_{\text{out}_n} - P_{\text{dem}_n}}^t (x) = f_{P_{\text{gen}_i}}^t f_{P_{\text{in}_n}}^t f_{-P_{\text{out}_n}}^t f_{-P_{\text{dem}_n}}^t \tag{24}$$

Without loss of generality, normal distributions are used to represent the generation, transmission, and load demand. Thus, $F_n^t(X)$ in Equation (23) can be expressed as

$$F_n^t(X) = \frac{1}{2} \left[1 - \text{erf} \left(Z_n^t \right) \right] \tag{25}$$

where $\text{erf}(\cdot)$ is the error function of $(\cdot)$, and Z_n^t is defined as:

$$Z_n^t = -\frac{\mu_n^t}{\sqrt{2} \, \sigma_n^t} \tag{26}$$

Here μ_n^t and σ_n^t are the mean and the standard deviation, respectively, given in Equations (28) and (29) below. Thus, Equation (24) can be written as

$$f_{P_{\text{gen}_i} + P_{\text{in}_n} - P_{\text{out}_n} - P_{\text{dem}_n}}^t (x) = N_n^t \sim (\mu_n^t, \sigma_n^t) \tag{27}$$

where the mean of the normal distribution N_n^t at node n and time t is given by

$$\mu_n^t = \sum_{i=1}^{I} \mu_{P_{\text{gen}_i}}^t + \sum_{m=1,n}^{M} \mu_{P_{\text{in}_n}}^t - \sum_{m=1,n}^{M} \mu_{P_{\text{out}_n}}^t - \mu_{P_{\text{dem}_n}}^t \tag{28}$$

In this equation, $\mu_{P_{\text{gen}_i}}^t$ is the mean of the power generated by producer i at time t [22], $\mu_{P_{\text{in}_n}}^t$ is the mean of the power imported by node n at time t, $\mu_{P_{\text{out}_n}}^t$ is the mean of the power exported by node n at time t, and $\mu_{P_{\text{dem}_n}}^t$ is the mean of the load demand at time t.

The standard deviation at node n and time t is

$$\sigma_n^t = \left[\sum_{i=1}^{I} (\sigma_{P_{\text{gen}_i}}^t)^2 + \sum_{m=1,n}^{M} (\sigma_{P_{\text{in}_n}}^t)^2 + \sum_{m=1,n}^{M} (\sigma_{P_{\text{out}_n}}^t)^2 + (\sigma_{P_{\text{dem}_n}}^t)^2 \right]^{1/2} \tag{29}$$

In this expression, $\sigma_{P_{\text{gen}_i}}^t$ is the uncertainty of the power generated by producer i at time t [22], $\sigma_{P_{\text{in}_n}}^t$ is the uncertainty of the power imported by node n at time t, $\sigma_{P_{\text{out}_n}}^t$ is the uncertainty of the power exported by node n at time t, and $\sigma_{P_{\text{dem}_n}}^t$ is the uncertainty of the load demand at time t.

The EENS model as established in this work takes into account, at every node and time, the propagation of uncertainties along the transmission lines, the uncertainties of the load demand, and the uncertainties of the generation.

4. Scenarios for the Analysis

4.1. Scenario 1

For the first scenario, the quasi-stationary optimization problem is solved using a 24-h period, which represents the day of the year with the peak load demand of 185 MW. This peak load demand takes place on Thursday of Week No. 5 [34]. The peaking hour is used to fix the synthesis/design of the system, while the remaining 23 h are used to determine how well a particular synthesis/design operates relative to the objectives. The load profile for Scenario 1 is given in Figure 3.

For this scenario, the load demand at time t is assumed to be a constant so that uncertainties relative to the demand itself are not taken into account. Thus, Equation (28) reduces to

$$\mu_n^t = \sum_{i=1}^{I} \mu_{P_{\text{gen}_i}}^t + \sum_{m=1,n}^{M} \mu_{P_{\text{in}_n}}^t - \sum_{m=1,n}^{M} \mu_{P_{\text{out}_n}}^t - P_{D_n}^t \tag{30}$$

and Equation (29) to

$$\sigma_n^t = \left[\sum_{i=1}^{I} (\sigma_{P_{\text{gen}_i}}^t)^2 + \sum_{m=1,n}^{M} (\sigma_{P_{\text{in}_n}}^t)^2 + \sum_{m=1,n}^{M} (\sigma_{P_{\text{out}_n}}^t)^2 \right]^{1/2} \tag{31}$$

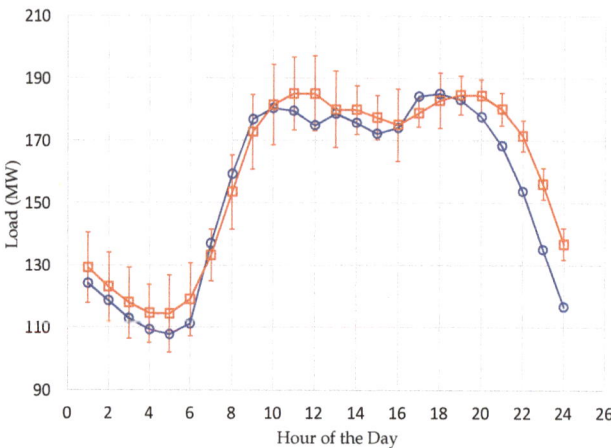

Figure 3. Load profiles for Scenario 1 (–o– blue) and for Scenario 2 and Scenario 3 (–□– red) [34,35].

4.2. Scenario 2

For Scenario 2, the first scenario is modified to take into account uncertainties in the load demand. The 365 measures corresponding to each of the 24 h, e.g., 12:00 pm, are represented by a normal distribution with the mean and standard deviation at each node. A normal distribution function fits the 365 measurements well for each hour. Using this procedure, the load demand at each hour is given by the mean of the 365 measurements, and its uncertainties are represented by its standard deviation. Since the mean of any of the 24 h of the day for this scenario is lower than that for Scenario 1, maintaining a realistic operation of the system for Scenario 2 during the year and a fair comparison with Scenario 1 requires scaling the load profile of the RTS for Scenario 2 with respect to the peaking load demand of 185 MW by a scaling factor of 1.4 [42]. The 24-h load profile, including the uncertainty for each hour, used for Scenario 2 is given in Figure 3. For this scenario, the mean and standard deviation are given by Equations (28) and (29), respectively.

4.3. Scenario 3

For Scenario 3, the second scenario is modified by introducing renewable energy technologies into the power network. At Node 1, the producer represented by the decision variable x[2] is replaced by a wind farm with a capacity of 15 MW, and the decision variable x[3] is replaced by a photovoltaic farm with a capacity of 15 MW. In this way, the total installed capacity of 30 MW at this node is not altered, and a fair comparison of this scenario with the former two scenarios is maintained. The characteristics of the wind and photovoltaic technologies are given in Table 5. Thus, for Node 1, Equation (28) is rewritten as

$$\mu_n^t = \sum_{i=1}^{I-2} \mu_{P_{\text{gen}_i}}^t + \mu_{P_{\text{gen}_{\text{wind}}}}^t + \mu_{P_{\text{gen}_{\text{pv}}}}^t + \sum_{m=1,n}^{M} \mu_{P_{\text{in}_n}}^t - \sum_{m=1,n}^{M} \mu_{P_{\text{out}_n}}^t - \mu_{P_{\text{dem}_n}}^t \tag{32}$$

and Equation (29) as

$$\sigma_n^t = \left[\sum_{i=1}^{I-2} (\sigma_{P_{gen_i}}^t)^2 + (\sigma_{P_{gen_{wind}}}^t)^2 + (\sigma_{P_{gen_{pv}}}^t)^2 + \sum_{m=1,n}^{M} (\sigma_{P_{in_n}}^t)^2 + \sum_{m=1,n}^{M} (\sigma_{P_{out_n}}^t)^2 + (\sigma_{P_{dem_n}}^t)^2 \right]^{1/2} \tag{33}$$

Table 5. Characteristics of the renewable energy technologies for Scenario 3 [36].

Decision Variable	Type of Producer	$P_{i,n}^{Cap}$ (MW)	$M_{i,n,\$}^{Cap}$ ($/MW-day)	$M_{i,n,\$}^{O\&M}$ ($/MW-day)	$M_{i,n,\$}^t$ ($/MWh)	$FOR_{i,n}$
x[2]	Wind	15	709.51	187.02	0	0.024
x[3]	Solar	15	603.63	65.90	0	0.024

The performance curves that represent the stochastic behavior of the wind and photovoltaic energy producers are approximated well using a Mixture of Normals Approximation (MONA) of three terms [43] where each term has a different weight, w_j. The weights of $w_1 = 0.31518$, $w_2 = 0.31027$, and $w_3 = 0.37456$ are used for the MONA that represents the irradiance and that which represents the wind speed, and both result in good fits of the real data. The real data of irradiance and wind velocity are obtained from [44]. The fitting of this data (irradiance and wind speed) with a MONA is also checked for the 365 measurements that represent a single hour of the 24-h used for the optimization.

For the photovoltaic farm, it is assumed that the limits on the design and operation variables are related to the value of the irradiance at each hour, which in turn is related to the capacity of the power plant, that is,

$$\mu_{P_{gen_{pv}}}^t = \frac{I^t}{I^{max}} P_{pv}^{max} \tag{34}$$

where

$$P_{pv}^{max} = I^{max} A \eta_{pv} \tag{35}$$

and

$$I^t = \sum_{j=1}^{3} w_j \mu_{irrad,j}^t \tag{36}$$

In these expressions, P_{pv}^{max} is the maximum power production of the photovoltaic farm, I^{max} is the maximum irradiance of the day used for the design, I^t is the irradiance of the day at time t, A is the total area of the photovoltaic farm, η_{pv} is the efficiency of the photovoltaic farm, and $\mu_{irrad,j}^t$ is the mean of each term of the MONA. The standard deviation is given by

$$\sigma_{P_{gen_{pv}}}^t = \frac{\sigma_{Irrad}^t}{I^{max}} P_{pv}^{max} \tag{37}$$

where

$$\sigma_{Irrad}^t = \left[\sum_{j=1}^{3} w_j (\sigma_{irrad,j}^t)^2 \right]^{1/2} \tag{38}$$

and $\sigma_{irrad,j}^t$ is the standard deviation of each term of the MONA.

The irradiance profiles for the three components of the MONA for the photovoltaic generator are given in Figure 4. The data for the irradiance is separated into three groups where each group represents a term of the MONA so that w_1 represents the first 115 days of the year, w_2 the next 113 days, and w_3 the last 137 days.

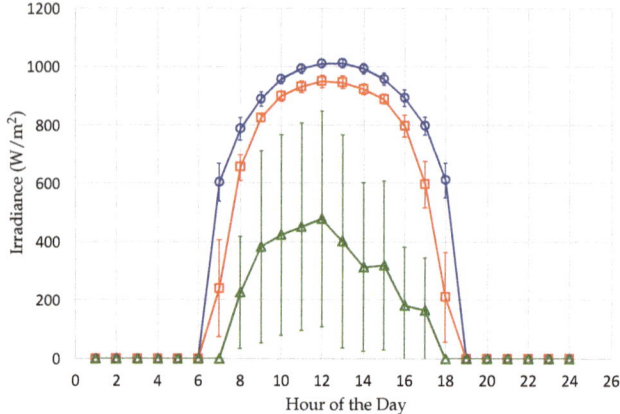

Figure 4. Irradiance profile for the three components of the MONA for the photovoltaic producer [44]. Component 1 (–o– blue), Component 2 (–□– red), and Component 3 (–△– green).

For the wind farm, it is assumed that the limits of the design and operation variables are related to the value of the wind speed at each hour, which in turn is related to the capacity of the power plant, that is,

$$\mu^t_{P_{gen_{wind}}} = \frac{(v^t)^3}{(v_{max})^3} P^{max}_{wind} \tag{39}$$

where

$$P^{max}_{wind} = \frac{1}{2}(v_{max})^3 A \, \rho \, \eta_{wind} \tag{40}$$

and

$$v^t = \sum_{j=1}^{3} w_j \mu^t_{wind,j} \tag{41}$$

Here, P^{max}_{wind} is the maximum power generated by the wind producer, v_{max} is the maximum wind speed of the day used for the design, v^t is the wind speed of the day at time t, A is the transversal area covered by the blades of the wind producer, η_{wind} is the efficiency of the wind producer, ρ is the density of the wind, and $\mu^t_{wind,j}$ is the mean of each term of the MONA. The standard deviation is given by

$$\sigma^t_{P_{gen_{wind}}} = \frac{(\sigma^t_{wind})^3}{(v_{max})^3} P^{max}_{wind} \tag{42}$$

where:

$$\sigma^t_{wind} = \left[\sum_{j=1}^{3} w_j (\sigma^t_{wind,j})^2 \right]^{1/2} \tag{43}$$

and $\sigma^t_{wind,j}$ is the standard deviation of each term of the MONA.

The wind profiles for the three components of the MONA for the wind generator are given in Figure 5. The data for the wind velocity is separated into three groups, where each group represents a term of the MONA so that w_1 represents the first 115 days of the year, w_2 the next 113 days, and w_3 the last 137 days.

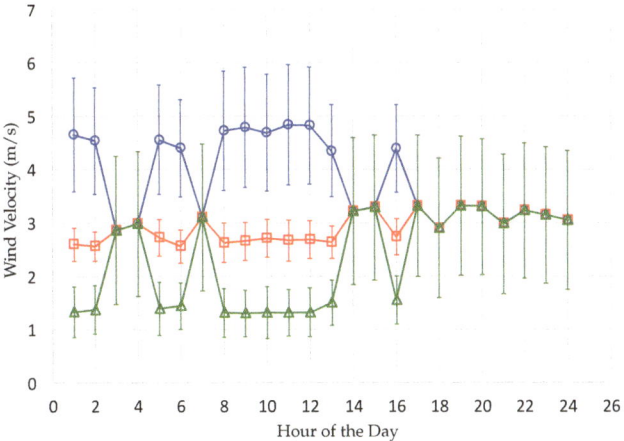

Figure 5. Velocity profile for the three components of the MONA for the wind producer [44]. Component 1 (–o– blue), Component 2 (–□– red), and Component 3 (–△– green).

5. Results and Discussion

Table 6 provides the minimum and maximum total daily costs for the three scenarios. It is seen that Scenario 3 provides the most expensive configurations even though the renewable energy producers do not contribute with variable O&M cost. However, the capital and fixed O&M costs of renewable energy producers have a significant impact on this objective function. Scenario 1 and Scenario 2 provide almost the same solution space, although a small difference between the two scenarios is observed. This difference seen in the smaller range between the minimum and maximum total daily costs for Scenario 2 can be attributed to the uncertainties in the load demand.

Table 6. Results for the total daily costs.

	Scenario 1	Scenario 2	Scenario 3
Min.	$ 141,665.9233	$ 142,041.356	$ 157,618.842
Max.	$ 199,772.624	$ 199,768.7404	$ 201,468.7844

Figure 6 shows the Pareto set of total daily costs vs. EENS for each of the three scenarios. It is observed that the reliability of the RTS increases when the cost of the configurations increases, i.e., the EENS decreases for the most expensive configurations for the three scenarios. With the exception of the least expensive configuration, Scenario 1 shows the best reliability performance followed by Scenario 2 and then Scenario 3. Thus, in Scenario 1, the configurations have a greater ability to meet the load demand than the configurations in Scenario 2 and Scenario 3. The uncertainties of the load demand in Scenario 2 affect the EENS considerably by increasing the standard deviation of the normal distribution that represents the behavior of a node, resulting in an increment of the loss-of-load probability and, as a consequence, the EENS. This effect is increased even more when the uncertainties associated with the production of electricity by the renewable energy producers, i.e., wind and photovoltaic panels, is introduced into the problem.

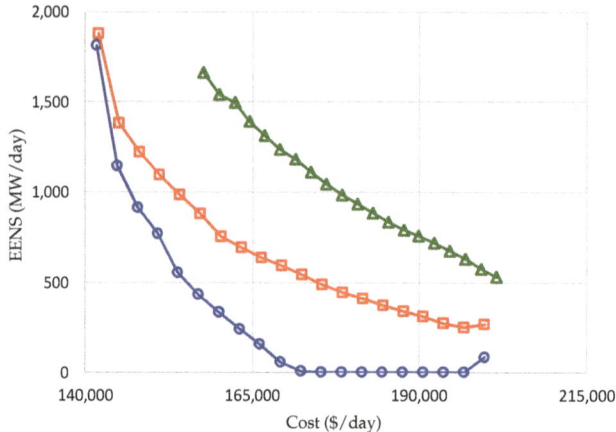

Figure 6. Cost vs. EENS Pareto set for Scenario 1 (–o– blue), Scenario 2 (–□– red), and Scenario 3 (–△– green).

Figure 7 shows the reliability of the different nodes of the RTS for the configurations in the Pareto set for the three scenarios. The value of the EENS at each node is an indicator of the ability of a node, based on the generation at the node and the power flow in and out of the node (transmission), to meet its own load demand. As is seen, the trend of the EENS at each node is similar to that of the overall system. It is observed that Node 2 is the weakest for the less expensive configurations, but is among the strongest for the most expensive configurations in the Pareto set. This behavior is observed also for Node 3. The rest of the nodes have a similar reliability in value and in trend for the different configurations in the Pareto set. For Scenario 2 and Scenario 3, the inclusion of the uncertainties in the load demand results in a faster decrease in the reliability of Node 3 than that for the rest of the nodes. The inclusion of the uncertainties associated with energy production (wind and photovoltaic) makes this effect even more significant, causing the reliability of the network for this scenario to be the lowest. It is also seen that, for Scenario 3, the reliability of Node 1, which is where the wind and photovoltaic producers are added, is not affected by the uncertainties associated with these producers. However, the neighboring nodes (2 and 3) are significantly affected by this uncertainty, showing the effect of the propagation of uncertainties through the transmission lines.

Figure 8 shows the Pareto set of total daily costs vs. SO_2 daily emissions for each of the three scenarios. It is observed that, for the three scenarios, the daily SO_2 emissions decrease for the configurations on the first part of the Pareto set and then increase as the total daily cost of the configurations increases. This is primarily due to the fact that the power production for the three scenarios is dominated by the USC and NGCC at high costs and dominated by the renewable energy producers—i.e., wind, solar, and hydro—at low costs. It is also seen that for Scenario 3, the amount of daily SO_2 emissions is higher than for the other two scenarios. This is because for this scenario, wind and photovoltaic panels produce less electricity than the technologies they replace. Thus, the USC is producing more energy for this scenario than for the other two scenarios. As also sen in this figure, the production of daily SO_2 emissions is almost the same for Scenario 1 and Scenario 2. For these scenarios, NGCC, CT, and RICE producers have a significant impact on the overall daily SO_2 emissions at low costs, and the USC dominates the daily SO_2 emissions at high costs.

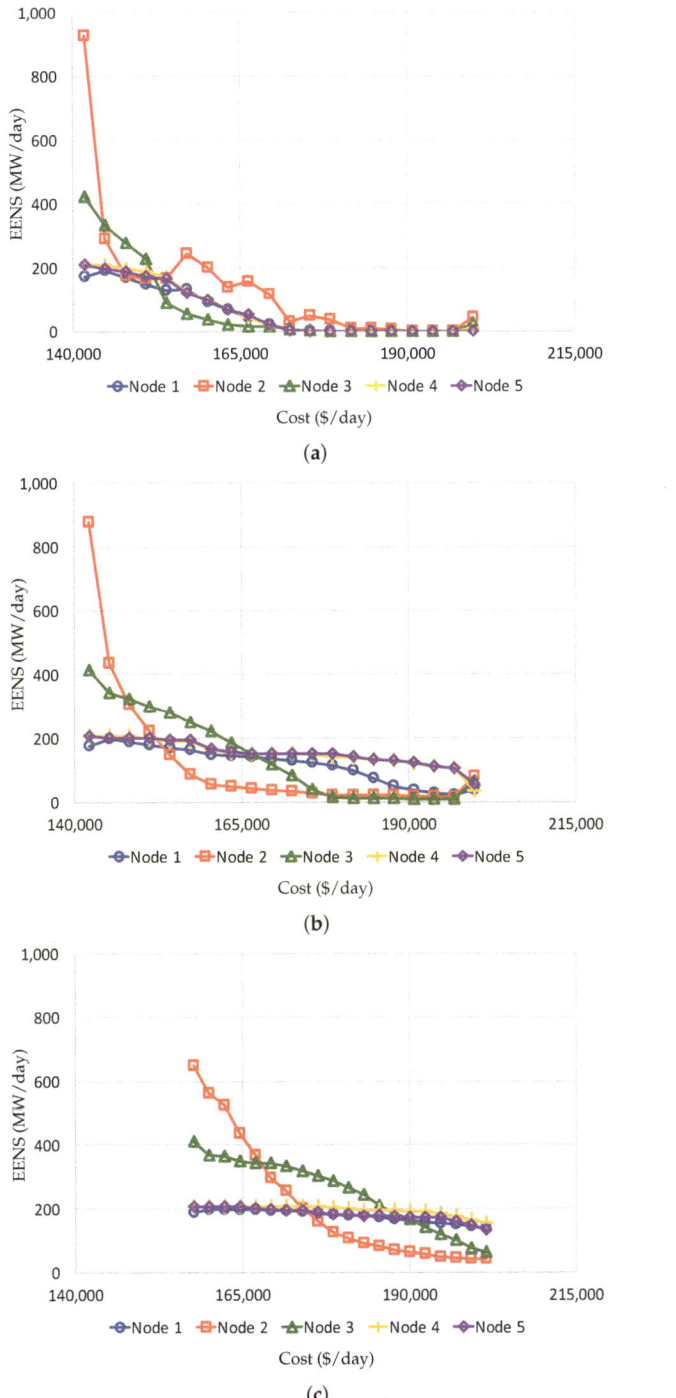

Figure 7. Cost vs. EENS Pareto set for the nodes of the RTS. (**a**) Scenario 1, (**b**) Scenario 2, and (**c**) Scenario 3.

Figure 9 shows the Pareto set of total daily costs vs. CO_2 daily emissions for each of the three scenarios. For the three scenarios, the daily CO_2 emissions decay a small amount and then start to increase as the total daily cost of the configurations increases. This is primarily because renewable energy technologies—i.e., hydro, wind, and photovoltaic producers—dominate the production at low costs and the USC and RICE dominate the production at high costs. It is also observed that for Scenario 3, the amount of daily CO_2 emissions is lower than for the other two scenarios mainly because the configurations in Scenario 3 contain a higher penetration of renewable energy producers such as hydro, wind turbines, and photovoltaic panels. As also seen, including the load uncertainties makes a difference with respect to the daily CO_2 emissions since Scenario 1 has lower values than Scenario 2.

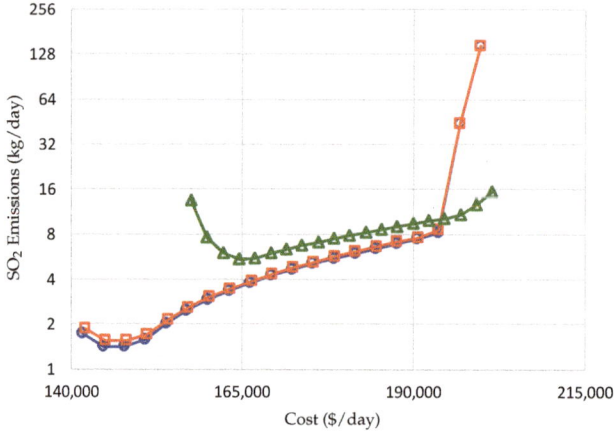

Figure 8. Cost vs. SO_2 Pareto set for Scenario 1 (–o– blue), Scenario 2 (–□– red), and Scenario 3 (–△– green).

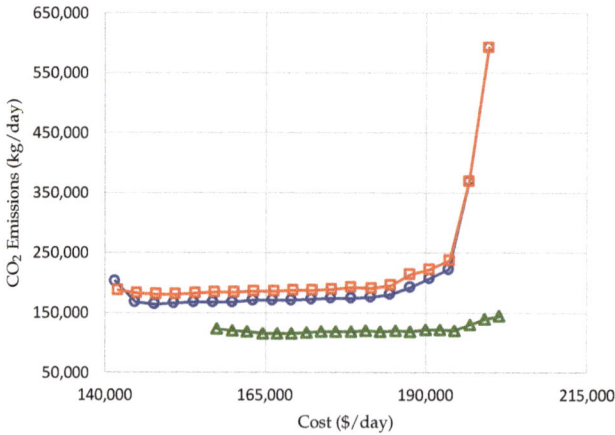

Figure 9. Cost vs. CO_2 Pareto set for Scenario 1 (–o– blue), Scenario 2 (–□– red), and Scenario 3 (–△– green).

Figure 10 shows the Pareto set of total daily costs vs. DALY for each of the three scenarios. In order to have a more representative result, the DALY is based on a one-year analysis. It is observed that, for the three scenarios, the trend of the DALY is very similar to that for CO_2 emissions, indicating that the DALY is highly dominated by this type of contaminant. Furthermore, it is seen that for Scenario 1 and Scenario 2, the DALY increases as the total daily cost of the configurations increases. This is primarily because renewable energy technologies—i.e., hydro, wind, and photovoltaic producers—dominate the production at low costs and the USC with its NO_x-DALY, and RICE and NGCC with their CO_2-DALY dominate the region of high costs. Furthermore, including the load uncertainties makes a difference with respect to the DALY since Scenario 2 has somewhat higher values than Scenario 1. It is also seen that for Scenario 3, the DALY is lower than for the other two scenarios and remains almost constant for all configurations except for the most expensive ones when the DALY increases. This is primarily because the configurations in Scenario 3 contain more renewable energy producers than do Scenario 1 and Scenario 2.

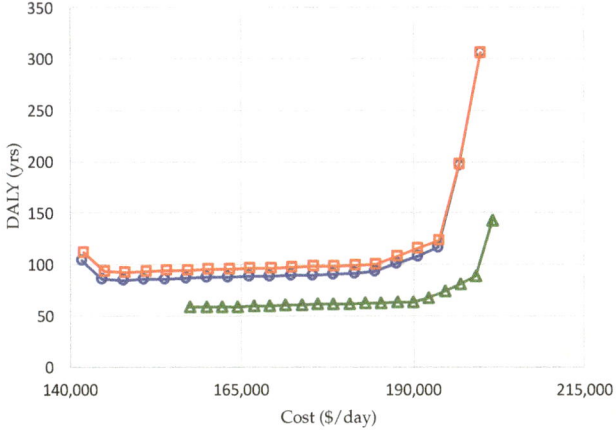

Figure 10. Cost vs. DALY Pareto set for Scenario 1 (–o– blue), Scenario 2 (–□– red), and Scenario 3 (–△– green).

Figure 11 shows the Pareto set of total daily costs vs. daily exergy production, which is directly proportional to the exergetic efficiency of the RTS, for each of the three scenarios. As seen in the figure, the daily exergy production decays a small amount in the first part of the Pareto set for the three scenarios and then starts to increase as the total daily cost of the configurations increases. This is primarily because the most efficient technologies such as NGCC and CT dominate the production at low costs and the less efficient USC and RICE dominate the production at high costs. Furthermore, as seen in the figure, including the load uncertainties makes a difference with respect to the daily exergy production since Scenario 2 has higher values than Scenario 1. It is also observed for Scenario 3 that the amount of daily exergy production is lower than for the other two scenarios mainly because the configurations in Scenario 3 contain extra renewable energy producers, which replace two fossil fuel-based producers, i.e., RICE and CT.

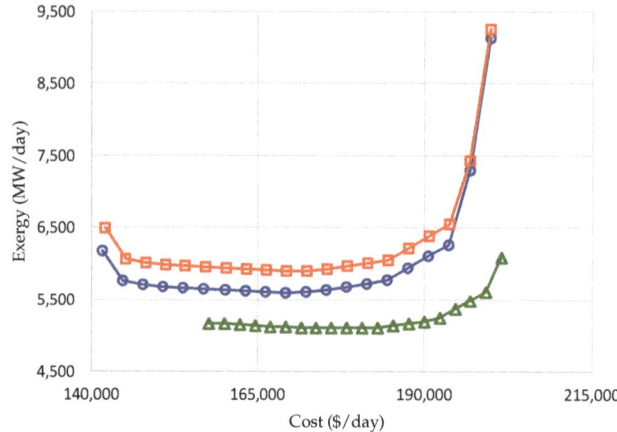

Figure 11. Cost vs. Exergy Pareto set for Scenario 1 (–○– blue), Scenario 2 (–□– red), and Scenario 3 (–△– green).

6. Conclusions

A generation-transmission reliability approach based on the mixture of normals approximation and its incorporation into a sustainability assessment framework for power network planning is proposed. The IEEE-RTS is used as a test system to evaluate the effectiveness of the methodology proposed.

The EENS objective function used here is based on a probabilistic as opposed to a deterministic approach, which can also account for the effects of catastrophic events as intrinsic to the system. The EENS provides detailed information about the producers and transmission lines during the synthesis/design optimization of power networks. Furthermore, the use of a MONA as the basis for the reliability framework makes the analysis and optimization of power networks much simpler than using a Monte Carlo approach, which in fact would render the optimization problem insoluble.

The inclusion of the cost for the possible construction of a new transmission line into the total daily cost, together with the capital cost and fixed and variable costs of producers considerably impacts the design of the power network configurations. The SO_2 daily emissions objective function favors the use of renewable energy producers and a relatively clean fossil technology such as NGCC for the most expensive configurations. However, because of the intermittent generation of wind and photovoltaic producers, the coal plant has to increase its power production for the less expensive configurations, increasing the overall production of SO_2 emissions. The CO_2 emissions objective function also gives preference to renewable energy technologies, as well as technologies such as RICE and USC. The Disability Adjusted Life Year (DALY) objective function as well prefers the use of renewable energy producers. The exergetic efficiency of the RTS objective function favors the use of producers based on natural gas such as NGCC and CT.

An increase in the penetration of renewable energy producers has some positive impact on the synthesis/design optimization of power networks under sustainability considerations. This is especially important in the reduction of CO_2 emissions (environmental pillar), in the disability adjusted life years of the population due to the emission of contaminants from local power plants (social pillar), and in the decrease of the use of fossil fuel in the network (technical pillar).

The cost (economic pillar) of using renewable energy producers is higher when compared with fossil fuel-based producers. This is not surprising since renewable energy producers are still maturing. However, it is expected that this gap will eventually be reduced.

The uncertainties associated with the generation of wind and photovoltaic producers reduce significantly the reliability of the power network. This is a disadvantage, which is only associated with

Energies **2019**, *12*, 546

these two types of renewable energy technologies due to their dependency on weather conditions. Nonetheless, it is important to consider wind and photovoltaic producers as players in a power network since they are the most mature of all the non-thermal renewable energy technologies.

Finally, the results presented here suggest that, for a reliability analysis of power networks, it is important to consider in a single methodology as many of the sources of uncertainties associated with the network as possible. These include not only those for the generation, transmission, and load demand considered here, but those attributed to other sources as well, such as, for example, those resulting from power distribution, emergency infrastructure interactions, unexpected catastrophic events, etc. Efforts to include these additional effects is left for future work.

Author Contributions: S.C.-A. and M.R.v.S. conceived of the generation-transmission reliability model, supervised the development of the project, and wrote the paper; J.R.V.-J. implemented the model as his M.S. thesis at the Universidad de Guanajuato; J.A.M.-B. obtained the final results, and L.M. contributed with the development of the generation-transmission reliability model.

Funding: This research was partially funded by the Secretariat of Public Education (SEP), Mexico, under Grant No. PRODEP/UGTO-PTC-412.

Acknowledgments: J.R. Vargas-Jaramillo and J.A. Montañez-Barrera thank the National Council of Science and Technology (CONACyT), Mexico, under Assistantships Nos. 291025 (Scholar ID 692271) and CVU 736083, respectively. S. Cano-Andrade gratefully acknowledges CONACyT, Mexico, for its financial support under the SNI program.

Conflicts of Interest: The authors declare no conflict of interest.

Abbreviations

The following nomenclature is used in this manuscript:

Symbols:

C	Objective function
P	Power flow
X	Reactance
α	power loss constant
FOR	Forced Outage Rate
M	Costs
γ	Cost/emission coefficient
L	Length
E	Emissions
D	DALY
η	Efficiency
F	Exergy of the fuel
f	Loss-of load-probability function
μ	Mean
σ	Standard deviation
σ^2	Variance
N	Normal distribution
ω	Weight of a MONA
A	Area
I	Irradiance
ρ	Density
v	Velocity

Subscripts:

k	Type of objective function
i	Generator
n	Node
m	Neighboring node
l	Loop
D	Load demand
$\$$	Costs
SO_2	Sulfur dioxide
CO_2	Carbon dioxide
DALY	Disability Adjust Loss of Life Year
EENS	Expected Energy not Supplied
LOLP	Loss of Load Probability
j	Term of an MONA
Irrad	Irradiance
pv	Photovoltaic

Superscripts:

T	Transpose
t	Time
max	Maximum
min	Minimum
Cap	Capital
N	Total number of nodes
I	Total number of producers
yr	Number of years
Tot	Total

References

1. Kjolle, G.H.; Utne, I.B.; Gjerde, O. Risk analysis of critical infrastructures emphasizing electricity supply and interdependencies. *Reliab. Eng. Syst. Saf.* **2012**, *105*, 80–89. [CrossRef]
2. Mason, K.; Duggan, J.; Howley, E. A multi-objective neural network trained with differential evolution for dynamic economic emission dispatch. International *J. Electr. Power Energy Syst.* **2018**, *100*, 201–221. [CrossRef]
3. Liu, Z.; Zeng, X.; Meng, F. An Integration Mechanism between Demand and Supply Side Management of Electricity Markets. *Energies* **2018**, *11*, 3314. [CrossRef]
4. Lin, T.Y.; Chiu, S.H. Sustainable Performance of Low-Carbon Energy Infrastructure Investment on Regional Development: Evidence from China. *Sustainability* **2018**, *1012*, 4657. [CrossRef]
5. Lin, J.; Sun, Y.; Cheng, L.; Gao, W. Assessment of the power reduction of wind farms under extreme wind condition by a high resolution simulation model. *Appl. Energy* **2012**, *96*, 21–32. [CrossRef]
6. Hong, L.; Lund, H.; Moller, B. The importance of flexible power plant operation for Jiangsu's wind integration. *Energy* **2012**, *41*, 499–507. [CrossRef]
7. Ozturk, S.; Fthenakis, V.; Faulstich, S. Assessing the Factors Impacting on the Reliability of Wind Turbines via Survival Analysis—A Case Study. *Energies* **2018**, *11*, 3034. [CrossRef]
8. Alferidi, A.; Karki, R. Development of Probabilistic Reliability Models of Photovoltaic System Topologies for System Adequacy Evaluation. *Appl. Sci.* **2017**, *7*, 176. [CrossRef]
9. Nguyen, N.; Mitra, J. Reliability of Power System with High Wind Penetration Under Frequency Stability Constraint. *IEEE Trans. Power Syst.* **2018**, *33*, 985–994. [CrossRef]
10. Akhavein, A.; Porkar, B. A composite generation and transmission reliability test system for research purposes. *Renew. Sustain. Energy Rev.* **2017**, *75*, 331–337. [CrossRef]
11. Uski, S.; Forssen, K.; Shemeikka, J. Sensitivity Assessment of Microgrid Investment Options to Guarantee Reliability of Power Supply in Rural Networks as an Alternative to Underground Cabling. *Energies* **2018**, *11*, 2831. [CrossRef]

12. Sansavini, G.; Piccinelli, R.; Golea, L.R.; Zio, E. A stochastic framework for uncertainty analysis in electric power transmission systems with wind generation. *Renew. Energy* **2014**, *64*, 71–81. [CrossRef]

13. Sanstad, A.H.; McMenamin, S.; Sukenik, A.; Barbose, G.L.; Goldman, C.A. Modeling an aggressive energy-efficiency scenario in long-range load forecasting for electric power transmission planning. *Appl. Energy* **2014**, *128*, 265–276. [CrossRef]

14. Zhou, P.; Jin, R.Y.; Fan, L.W. Reliability and economic evaluation of power system with renewables: A review. *Renew. Sustain. Energy Rev.* **2016**, *58*, 537–547. [CrossRef]

15. Dehghan, S.; Amjady, N.; Conejo, A.J. Reliability-Constrained Robust Power System Expansion Planning. *IEEE Trans. Power Syst.* **2016**, *31*, 2383–2392. [CrossRef]

16. Cadini, F.; Agliardi, G.L.; Zio, E. A modeling and simulation framework for the reliability/availability assessment of a power transmission grid subject to cascading failures under extreme weather conditions. *Appl. Energy* **2017**, *185*, 267–279. [CrossRef]

17. Wang, F.; Xu, H.; Xu, T.; Li, K.; Shafie-Khah, M.; Catalao, J.P. The values of market-based demand response on improving power system reliability under extreme circumstances. *Appl. Energy* **2017**, *193*, 220–231. [CrossRef]

18. Zhao, H.; Guo, S. External benefit evaluation of renewable energy power in China for sustainability. *Sustainability* **2015**, *7*, 4783–4805. [CrossRef]

19. World Commission on Environment and Development. *Our Common Future*; Oxford University Press: Oxford, UK, 1987; Volume 383.

20. Thrampoulidis, C.; Bose, S.; Hassibi, B. Optimal Placement of Distributed Energy Storage in Power Networks. *IEEE Trans. Autom. Control* **2016**, *61*, 416–429. [CrossRef]

21. Cano-Andrade, S.; von Spakovsky, M.R.; Fuentes, A.; Lo Prete, C.; Mili, L. Upper Level of a Sustainability Assessment Framework for Power System Planning. *ASME J. Energy Resour. Technol.* **2015**, *137*, 041601. [CrossRef]

22. Lo Prete, C.; Hobbs, B.F.; Norman, C.S.; Cano-Andrade, S.; Fuentes, A.; von Spakovsky, M.R.; Mili, L. Sustainability and Reliability Assessment of Microgrids in a Regional Electricity Market. *Energy* **2012**, *41*, 192–202. [CrossRef]

23. Billinton, R.; Allan, R.N. *Reliability Evaluation of Power Systems*, 2nd ed.; Springer: New York, NY, USA, 1996.

24. Gollwitzer, L.; Ockwell, D.; Muok, B.; Ely, A.; Ahlborg, H. Rethinking the sustainability and institutional governance of electricity access and mini-grids: Electricity as a common pool resource. *Energy Res. Soc. Sci.* **2018**, *39*, 152–161. [CrossRef]

25. Frangopuolos, C.A. Static and Dynamic Pollution and Resource Related Index. *Encicl. Life Support Syst. EOLSS* **2009**, *3*, 231.

26. Veldhuis, A.J.; Leach, M.; Yang, A. The impact of increased decentralized generation on the reliability of an existing electricity network. *Appl. Energy* **2018**, *215*, 479–502. [CrossRef]

27. Eser, P.; Singh, A.; Chokani, N.; Abhari, R.S. Effect of increased renewables generation on operation of thermal power plants. *Appl. Energy* **2016**, *164*, 723–732. [CrossRef]

28. Othman, M.M.; Musirin, I. A novel approach to determine transmission reliability margin using parametric bootstrap technique. *Electr. Power Energy Syst.* **2011**, *33*, 1666–1674. [CrossRef]

29. Canizes, B.; Soares, J.; Vale, Z.; Lobo, C. Optimal Approach for Reliability Assessment in Radial Distribution Networks. *IEEE Syst. J.* **2017**, *11*, 1846–1856. [CrossRef]

30. Arroyo, J.M. Bilevel programming applied to power system vulnerability analysis under multiple contingencies. *IET Gen. Transm. Distrib.* **2010**, *4*, 178–190. [CrossRef]

31. Haghighat, H.; Kennedy, S.W. A Bilevel Approach to Operational Decision Making of a Distribution Company in Competitive Environments. *IEEE Trans. Power Syst.* **2012**, *27*, 1797–1807. [CrossRef]

32. Rider, M.J.; Lopez-Lezama, J.M.; Contreras, J.; Padilha-Feltrin, A. Bilevel approach for optimal location and contract pricing of distributed generation in radial distribution systems using mixed-integer linear programming. *IET Gen. Transm. Distrib.* **2013**, *7*, 724–734. [CrossRef]

33. Quashie, M.; Marnay, C.; Bouffard, F.; Joos, G. Optimal planning of microgrid power and operating reserve capacity. *Appl. Energy* **2018**, *210*, 1229–1236. [CrossRef]

34. Billinton, R.; Kumar, S.; Chowdhury, N.; Chu, K.; Debnath, K.; Goel, L.; Khan, E.; Kos, P.; Nourbakhsh, G.; Oteng-Adjei, J. A Reliability Test System for Educational Purposes. *IEEE Trans. Power Syst.* **1989**, *4*, 1238–1244. [CrossRef]

35. Force, R.T. The IEEE Reliability Test System-1996. *IEEE Trans. Power Syst.* **1999**, *14*, 1010–1020.

36. U.S. Energy Information Administration (EIA). *Capital Costs Estimates for Utility Scale Electricity Generating Plants*; U.S. Energy Information Administration (EIA): Washington, DC, USA, 2016.

37. Baldick, R.; O'Neill, R.P. Estimates of Comparative Costs for Uprating Transmission Capacity. *IEEE Trans. Power Deliv.* **2009**, *24*, 961–969. [CrossRef]

38. Hobbs, B.F.; Drayton, G.; Bartholomew Fisher, E.; Lise, W. Improved Transmission Representations in Oligopolistic Market Models: Quadratic Losses, Phase Shifters, and DC Lines. *IEEE Trans. Power Syst.* **2008**, *23*, 1018–1029. [CrossRef]

39. Python Software Foundation. Available online: https://www.python.org (accesed on 23 January 2019).

40. SciPy.org. Available online: https://docs.scipy.org/doc/scipy/reference/tutorial/optimize.html (accesed on 23 January 2019).

41. Goedkoop, M.; Heijungs, R.; Huijbregts, M.; De Schryver, A.; Struijs, J.; van Zelm, R. *ReCiPe 2008: A Life Cycle Impact Assessment Method which Comprises Harmonised Category indicators at the Midpoint and the Endpoint Level*; Leiden University: Leiden, The Netherlands, 2008.

42. Kim, K. Dynamic Proton Exchange Membrane Fuel Cell System Synthesis/Design and Operation/Control Optimization under Uncertainty, Ph.D. Dissertation, Virginia Tech, Blacksburg, VA, USA, 2008.

43. Gross, G.; Garapic, N.V.; McNutt, B. The Mixture of Normals Approximation Technique for Equivalent Load Duration Curves. *IEEE Trans. Power Syst.* **1988**, *3*, 368–374. [CrossRef]

44. National Renewable Energy Laboratory (NREL). National Solar Radiation Database, 2017. Available online: https://rredc.nrel.gov/solar/old_data/nsrdb/ (accesed on 23 January 2019).

Article

Thermo-Electric Energy Storage with Solar Heat Integration: Exergy and Exergo-Economic Analysis

Daniele Fiaschi [1], Giampaolo Manfrida [1,*], Karolina Petela [2] and Lorenzo Talluri [1]

[1] Department of Industrial Engineering, Università degli Studi di Firenze, Florence, Italy; daniele.fiaschi@unifi.it (D.F.); Lorenzo.Talluri@unifi.it (L.T.)

[2] Institute of Thermal Technology, Silesian University of Technology, Gliwice, Poland; karolina.petela@polsl.pl

* Correspondence: Giampaolo.Manfrida@unifi.it; Tel.: +39-055-275-8676

Received: 17 January 2019; Accepted: 14 February 2019; Published: 17 February 2019

Abstract: A Thermo-Electric Energy Storage (TEES) system is proposed to provide peak-load support (1–2 daily hours of operation) for distributed users using small/medium-size photovoltaic systems (4 to 50 kWe). The purpose is to complement the PV with a reliable storage system that cancompensate the produc tivity/load mismatch, aiming at off-grid operation. The proposed TEES applies sensible heat storage, using insulated warm-water reservoirs at 120/160 °C, and cold storage at $-10/-20$ °C (water and ethylene glycol). The power cycle is a trans-critical CO_2 unit including recuperation; in the storage mode, a supercritical heat pump restores heat to the hot reservoir, while a cooling cycle cools the cold reservoir; both the heat pump and cooling cycle operate on photovoltaic (PV) energy, and benefit from solar heat integration at low–medium temperatures (80–120 °C). This allows the achievement of a marginal round-trip efficiency (electric-to-electric) in the range of 50% (not considering solar heat integration).The TEES system is analysed with different resource conditions and parameters settings (hot storage temperature, pressure levels for all cycles, ambient temperature, etc.), making reference to standard days of each month of the year; exergy and exergo-economic analyses are performed to identify the critical items in the complete system and the cost of stored electricity.

Keywords: energy storage; thermo-electric; supercritical CO_2; solar energy

---

1. Motivation and Introduction

The increasing market penetration of renewables is challenging the current structure of electrical grids [1]. Most renewables require a balance between production and load; production depends on highly stochastic resources (wind, wave energy) or is subject to daily cycles (solar). To solve the dispatchability issue, several countries are still obliged to use fossil fuels, with the situation aggravated by occasional use of plants and requiring operation under peak-load switch mode. Another way to solve the problem is to associate energy storage with the increasing use of renewables; today, the largest applied energy storage is still pumped hydro, which is possible only under favorable conditions of availability of sites and hydro resource.

On the other hand, the market is experiencing an increasing diffusion of off-grid or locally connected smart-grid user communities. These are at present expensive solutions, based on small-scale distributed energy systems; only for specific situations are these solutions practiced, such as on islands or in remote locations where the connected grid infrastructure is missing (and often electricity is provided by expensive diesel generators). Off-grid or small connected smart-grids today represent a challenge, where new advanced solutions trying to satisfy community energy needs (power, heat, and possibly other attractive products or services) can be investigated.

Within this context, the idea of developing medium-size energy storage (ES) systems becomes attractive. These ES should couple flexibility (providing power and possibly heat or cold), reliability, capability of load matching, and life cycle durability. These features are not easily achievable by modern batteries, which are the preferred solutions of smart-grid promoters but present several bottlenecks when applied under these stringent conditions. The consequence is that even when backing up a renewable energy system (solar, wind) with a substantial battery pack, it is in general necessary to include in the package a diesel generator for off-grid power.

The portfolio of ES technologies is quite wide, as it embraces different concepts, such as pumped-storage hydroelectricity (PHS), flywheel storage (FS), batteries, compressed air energy storage (CAES), liquid air energy storage (LAES), or other gas storages using hydrogen or CO_2, as well as super-capacitor or chemical storage [2]. All these solutions are challenged by the increasing share of electricity generation by renewables. Specifically, each technology presents advantages and drawbacks, which make the correct selection of storage systems deeply dependenton the application. Among the selection criteria are relevant aspects are the cost of the system, the total efficiency, the energy density, and the power rating. In [3], a comprehensive comparison of the various ES technologies, including both technical and economical features, is carried out.

PHS, as well as CAES, are subject to geographical constraints; therefore, they can only play a relevant role for places where the geographical sites are suitable [4]. The main issue of battery storage is the limited life cycle [5], while chemical storage certainly represents the ultimate solution for managing long-term unavailability of renewables such as wind or wave power, but is highly penalized by a limited round-trip efficiency. In this frame, Thermo-Electric Energy Storage (TEES) [4–6] represents an interesting solution in the general context of ensuring dispatchability to energy systems based on renewables. A TEES system is basically composed of two sensible heat/cold accumulators, between which temperature levels a heat engine works. The temperature levels are then re-charged by a heat pump cycle.

Multi-MW TEES cycles have been proposed, often using a trans-critical CO_2 cycle as the power cycle [5–8]. Another variant of the TEES is that using the Brayton cycle as power cycle, with air [9], Argon, or other noble gases [10,11] as working fluids.

In the literature, TEES systems are not widely studied, especially when considering the whole integration process of solar energy both as heat input (in the discharge cycle) and as electricity input (in the charging cycle, providing the work input for compressors operation). Particularly, in [12], a novel pumped thermal electricity storage (PTES) system with heat integration was proposed. The main novelty was the introduction of an auxiliary heat source, which enhances the efficiency of the system. In [13], an overview on TEES is presented, dealing also with thermo-economic evaluation. A comprehensive thermo-economic analysis of a TEES is developed in [14], where a round-trip efficiency of 64% was obtained for a 50 MW power plant with an initial investment of 34 M$. Another thermo-economic analysis is proposed in [15], where the TEES is supported by the use of solar collectors.

On the whole, relevant research work has been dedicated to TEES. Nonetheless, there is still a gap in the literature in comprehensive assessments of solar-aided TEES systems; dealing with a system using exchange of heat and power, an exergo-economic analysis is recommended as the best tool for studying parametric optimization of the system.

In the present study, a thermodynamic and exergo-economic analysis of a TEES is performed, with solar panels and photovoltaic (PV) integration in the charging and discharging cycles. Two different representative case studies of geographical sites are presented and discussed (Crotone and Pantelleria in Italy). The thermodynamic cycles were designed and optimized, from an exergy and exergo-economic perspective, to address the highest possible performance for a variable heat input depending on the availability of the solar resource.

2. TEES Description and Methods of Analysis

2.1. Description of TEES

In the present study, focusing on ESs assisting medium-size photovoltaic systems (PV), which depend on the daily availability of solar radiation, sensible heat or cold accumulation is practiced. The TEES here proposed is based on three separate systems: a power cycle, a heat pump, and a refrigeration cycle. The heat pump and the refrigeration cycle are working during the charging phase, using solar energy converted into thermal and electric energy.

As shown in Figures 1 and 2, during daylight operation the hot and cold reservoirs are charged using respectively a heat pump and a refrigeration unit. After daylight, a power cycle (Figure 3) operating between the hot, medium temperature, and cold reservoirs produces the necessary power for the community. The hot and cold reservoirs are available for flexible energy use: hot sanitary water and heating can be assisted by the hot and/or medium temperature reservoir, while in hot periods the cold reservoir can be connected to the domestic comfort cooling network.

The schematic of the heat pump cycle is shown in Figure 1. The purpose of the heat pump cycle is to restore the required sensible heat in the HWHR. To achieve the required temperature level (145 °C), the proposed heat pump works with the architecture of a supercritical CO_2 cycle. The use of a supercritical cycle, which is mostly due to the required temperature levels, allows a proper matching of the heat capacities of the stored water (pressurized at 1800 kPa) and of the heat pump working fluid (supercritical CO_2).

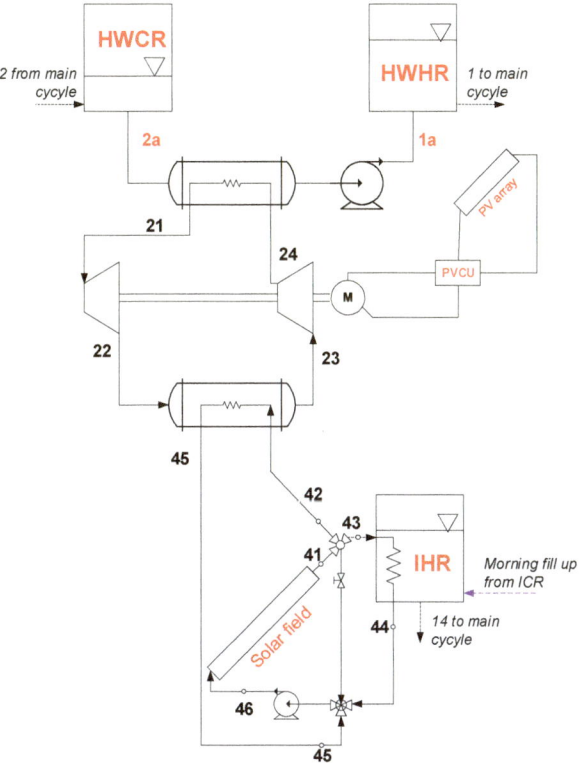

Figure 1. Solar-assisted heat pump cycle–schematic.

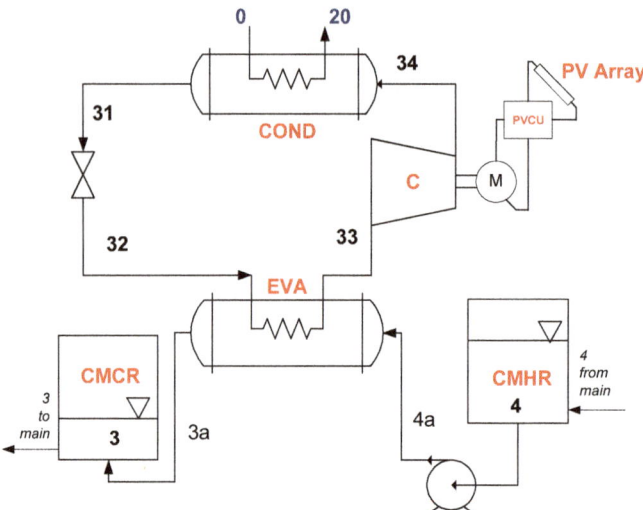

Figure 2. Refrigeration cycle—Schematic.

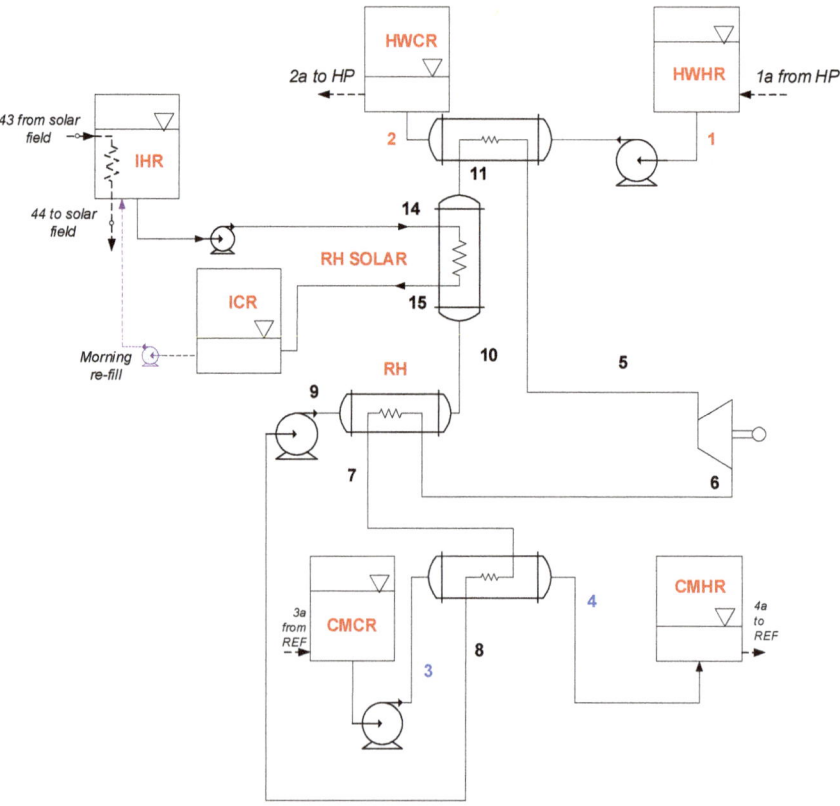

Figure 3. Trans-critical CO$_2$ power cycle—Schematic.

With respect to a traditional heat pump, an expander is proposed in place of the throttling valve (21–22), which allows an increase in COP values [16]. Moreover, the heat pump cycle is deeply integrated with solar panels and PV collectors. Specifically, the evaporator temperature is determined by the operation of the thermal collectors of the solar field, through a three-way valve, which also allows the setting of the optimal temperature of an intermediate hot reservoir (IHR).

Finally, the required compressor work (23–24, partly supported by the expander), is provided by a PV array, which thus allows a complete solar TEES integration.

The schematic of the refrigeration cycle is shown in Figure 2. The objective of the refrigeration cycle is to restore the cold energy in the CMCR, reducing the temperature of the water-ethylene glycol mixture, taken from the CMHR. The presence of a low-temperature cold storage is of paramount importance if suitable round-trip efficiency is coveted.

The condenser of the refrigeration cycle rejects heat to the environment (0–20). The ambient temperature is a very important parameter for the refrigeration cycle, as lower ambient temperatures allow higher coefficient of performance (refrigeration cycle evaporator-condenser temperature becomes closer). Different fluids (R134a, R717, R1233zd(E), R404a) were investigated as possible choices for the refrigeration cycle working fluid. The most favorable solution resulted in a subcritical R134a cycle. As for the heat pump, operation of the compressor of the refrigeration cycle is provided by surplus power available from the PV solar field.

The opportunity of using a cold storage represents a considerable advantage for the power cycle, as it can work across a higher temperature difference, therefore allowing a superior performance of the power cycle. For this purpose, the use of water mixtures with suitable anti-freeze additives (such as Ethylene Glycol or Calcium Chloride) is recommended. The cold storage is restored to the initial low temperature during the charging mode operation, using the proposed subcritical R134a refrigeration cycle.

The assumed values of the reference variables for the heat pump and the refrigeration cycles are resumed in Table 1.

Table 1. Reference design parameters for the Heat Pump and Refrigeration Cycles.

Variable	Value
Heat Pump Cycle	
$\Delta T_{CO2-HW} = T_{21} - T_{2a}$	5 °C
$P_{min, HP}$	13500 kPa
$\Delta T_{solar-CO2} = T_{42} - T_{23}$	5 °C
Refrigeration Cycle	
$\Delta T_{COLD} = T_{31} - T_0$	10 °C
$\Delta T_{EVA} = T_{3a} - T_{32}$	5 °C

The trans-critical CO_2 power cycle (Figure 3) is a common solution for TEES applications: in fact, CO_2 is particularly attractive for the temperature levels involved (high and low temperature), and the trans-critical choice allows a good matching of heat capacities for the hot resource. The basic idea is to use as far as possible the same components for the heat pump and the trans-critical CO_2 cycle through an appropriate commutation of configuration valves. This mode of operation is certainly affected by the different operational time for storage and power modes and—consequently—by the different mass flow rates; moreover, a solar-aided TEES experiences different heat input conditions throughout the year as well as throughout the day.

The proposed system uses a simple solution of sensible heat liquid reservoirs for the hot and cold storage: even though there are several limitations with this technology, use of simple insulation materials and the possibility of quickly adjusting the mass flow rates in order to match the heat capacities of both charging and discharging cycles make this solution attractive for the power size here considered. An intermediate temperature reservoir (IHR and ICR) charged by solar heat allows preheating of the working fluid, thus enhancing the efficiency of the system.

The selection of the power cycle operating temperatures, as well as the optimal conditions of the hot, intermediate and cold reservoirs (HWCR, HWHR, ICR, IHR, CMCR, CMHR), is the outcome of a parametric analysis, taking into account not only the power cycle but also the two recharging cycles (heat pump and refrigeration cycle).

Water in the additional, intermediate temperature reservoir (IHR) is warmed up to the desired temperature by solar heat during the charging phase. Exploiting heat from the IHR accumulator makes the discharging phase independent of transient meteorological conditions.

Referring to the Hot and Intermediate twin reservoirs, it is possible to place the vessels at different elevation levels to take advantage of buoyancy to design a system using no pumps for fluid displacement. However, at this stage of research, it was assumed that the circulation pumps between all twin tanks should cover a 4-m circuit head loss. The design pumping power for the assemblies was calculated accordingly, and results very limited.

The power cycle operating parameters under design conditions are summarized in Table 2.

Table 2. Power cycle design operating parameters.

Variable	Value
T_1, T_2 (HWR)	95/145 °C
P_{HWR}, P_{IHR}, P_{CMR}	1800/100/100 kPa
m_{HW}	1 kg/s
T_{14}, T_{15} (RH-SOLAR)	95/40 °C
p_5	12,000 kPa
$\Delta T_{HOT} = T_1 - T_5 = \Delta T_{SOLAR} = T_{14} - T_{11}$	5 °C
$\Delta T_{COLD} = T_8 - T_3$	10 °C
T_3, T_4 (CWR)	−20/−10 °C
ε_{RH}	0.8
η_t, η_p	0.9/0.8
Operation Time (Power Cycle)	h

A crucial issue when considering a system that relies solely on an energy source of intermittent availabilitysuch assolar radiation is its control management. In the present case, the system is preliminarily assessed assuming that it should be able to provide a constant power output for a limited time in the evening (e.g., 1 h), but control should indeed address complete daily resource and load management in a practical application. In particular, the charging phase is burdened with considerable deviations from design loads in the early morning and evening hours, as is typical for all systems relying on the solar resource. Several control paths could be tested and then potentially implemented. In the early morning and late afternoon operation, compressors could be supported by a small-capacity battery pack, or use limited grid assistance. Moreover, the compressors can be equipped with a variable-speed drive to follow the variable load conditions; and multiple parallel-arranged sets of compressors can be proposed for TEES systems of large capacity, with step-by-step load control. An automated control system based on control routines adapting the operational parameters (pressures, temperatures) to changing conditions can also be proposed. As stated before, the aim of this paper is to evaluate the possible application of solar energy-integrated TEES and to demonstrate its performance and individuate a pathway for possible improvement. In this light, control issues are not explicitly dealt at the present stage of research.

2.2. Power Cycle-Thermodynamic Model Equations

In the following, the main model equations are presented only for the power cycle, as the heat pump and refrigeration cycle are conventional units (with the main novelty of solar-thermal assistance for the heat pump, which is dealt in the following).

The operation of the power cycle is determined in terms of heat input by the pre-set conditions at the HWHR in terms of flow rate $\dot{m}_{HW}$ as shown in Equation (1):

$$\dot{Q}_{12} = \dot{m}_{HW}(h_1 - h_2) \tag{1}$$

Knowing the conditions of the hot resource and assuming the minimum possible temperature difference, it is possible to calculate the working fluid flow rate through Equation (2).

$$\dot{m}_{WF} = \frac{\dot{Q}_{12}}{(h_5 - h_{11})} \tag{2}$$

The temperature and enthalpy conditions at point 11 are defined by the heat extraction from the solar field resource Equation (3), which allows an increase of the system efficiency, as the working fluid is pre-heated before the high-temperature heat exchanger:

$$T_{11} = T_{14} - \Delta T_{solar} \tag{3}$$

The turbine power output is obtained through the application of Equation (4), assuming a turbine isentropic efficiency of 0.9 at design point:

$$\dot{W}_t = \dot{m}_{WF}(h_5 - h_6) = \dot{m}_{WF}(h_5 - h_{6s}) \cdot \eta_T \tag{4}$$

The application of energy balance at the re-heater Equation (5), assuming a re-heater efficiency of 0.8, allows setting of the re-heater heat duty:

$$\varepsilon_{RH} = \frac{\dot{Q}_{RH}}{\dot{Q}_{RH,\,MAX}} = \frac{\dot{m}_{WF} \cdot (h_6 - h_7)}{\dot{m}_{WF} \cdot (h_6 - h_{7\,min})} = \frac{\dot{m}_{WF} \cdot (h_{10} - h_9)}{\dot{m}_{WF} \cdot (h_6 - h_{7\,min})} \tag{5}$$

where $h_{7\,min}$ is evaluated at $T_{7\,min} = T_9$ and $p_7 = p_6$.

The CO_2 supercritical condenser is cooled using the cold stored in the reservoirs, thus allowing the setting of the low-pressure level of the discharging cycle Equation (6):

$$\dot{Q}_{34} = \dot{m}_{CW}(h_4 - h_3) = \dot{m}_{WF}(h_7 - h_8) \tag{6}$$

which, once T_4 and T_3 are given, can be solved for $\dot{m}_{WF}$. Finally, the calculation scheme of the cycle is closed by the calculation of the trans-critical CO_2 pump through Equation (7), assuming a pump isentropic efficiency of 0.8:

$$\dot{W}_P = \dot{m}_{WF}(h_9 - h_8) = \dot{m}_{WF}(h_{9s} - h_8)/\eta_p \tag{7}$$

Furthermore, it is possible to calculate the required volumes of the reservoirs: one hour of power cycle autonomy is assumed. Their size should satisfy the heat demand of three cycle heat exchangers. The sizing of the reservoirs volume is also fundamental for the off-design analysis.

Once all the temperatures, flow rates, and heat duties are known, the heat exchangers sizes are determined by means of the Péclet Equation (8).

$$\dot{Q}_k = (UA)_k \cdot \Delta T_{lg;k} \tag{8}$$

2.3. Solar Integration

Solar integration with the TEES uses a combination of thermal and photovoltaic conversion. Solar thermal collectors are supporting the evaporator in the heat pump system and, simultaneously, are charging the intermediate reservoir (IHR). Meanwhile, PV panels are providing electric energy to drive compressors in the heat pump and refrigeration cycles sections. At the design stage, a crucial issue is the sizing of both the solar thermal collectors and the PV fields. Their energy output is strictly dependent on the local meteorological conditions. Knowing these for a chosen location, the desired size of the two fields can be determined. The sizing was done through a one-reference-day quasi-dynamic model approach for the given location.

A commercially available flat-plate solar collector was considered for the solar-thermal field. The efficiency of solar collectors (η_{sc}) depends on incoming radiation (G_{sloped}), ambient temperature, and working fluid temperature increase; applying the typical 2nd order Bliss-equation [17]:

$$\eta_{sc} = \eta_o - (a_1 + a_2\Delta T)\frac{\Delta T}{G_{sloped}} \tag{9}$$

The coefficients η_o, a_1 and a_2 (Table 3) are commonly provided by the manufacturer of the collector. ΔT is the temperature difference between the average Heat Transfer Fluid (HTF) temperature and the ambient temperature. The HTF inlet and outlet temperatures are assumed as known at the design conditions.

Table 3. Solar-thermal collector fields operating parameters.

Variable	Description
Location	Crotone, Italy
Month for reference day	May
Slope of solar collector	45° towards South
η_0	0.719
a_1	1.45 W/(m²K)
a_2	0.0051 W/(m²K²)
A_{sc}	1.6 m²
$T_{41} = T_{42} = T_{43}$	95 °C
$\Delta T_{HTF} = T_{42} - T_{45} = T_{43} - T_{44}$	10 K
Collectors arrangement	Parallel in 10 rows

The useful heat gain from the solar field is shared between the heat pump evaporator demand and the IHR tank. In this last stage, water is warmed up to a fixed temperature to be used during the power cycle operation. The solar-thermal field arrangement was shown in Figure 1 together with the heat pump assembly.

The solar field surface A and, thus, the number of solar collectors, was found iteratively requiring that the daily solar heat yield can satisfy the heat pump and IHR tank energy needs. The applied procedure is summarized in the set of Equations (10) and (11). The procedure is formally started at 7 a.m., but depending on the month and weather conditions the effective heat accumulation can begin later. It is assumed that collectors are arranged in parallel in 10 equal rows. Thus, the task is to calculate the number of collectors in a row:

$$\int_{7:00}^{7:00+\tau_{charge}} \dot{Q}_{sc}d\tau = N_{sc}A_{sc}\int_{7:00}^{7:00+\tau_{charge}} \eta_{sc}G_{sloped}d\tau \tag{10}$$

$$\int_{7:00}^{7:00+\tau_{charge}} \dot{Q}_{sc}d\tau = \int_{7:00}^{7:00+\tau_{charge}} \dot{Q}_{evapHP}d\tau + Q_{IHR} \tag{11}$$

The amount of heat required by the heat pump evaporator is a function of HWR volume. The design conditions of the main cycle determine the need for warming and moving the total volume of water, which also depends on the charging time (τ_{charge}). On the other hand, the solar-thermal input should also assure warming the water in IHR tank to the required temperature. Considering these constraints, both the number of solar collectors and the charging time can be determined. The design assumptions for this section are collected in Table 3.

At the same time, the number of PV panels to satisfy the net compressors power must be defined. Commercially available polycrystalline modules were considered. A TRNSYS (http://www.trnsys.com/) model provided the power output distribution from one polycrystalline Schott SAPC 165 [18] PV module for the reference day of May. Knowing the electric energy needed by the compressors during the whole charging time, the number of PV panels required to produce that daily work output can be calculated.

$$\int_{7:00}^{7:00+\tau_{charge}} \left(\dot{W}_{c;ref} + \dot{W}_{c;HP} - \dot{W}_{exp;HP} \right) d\tau = n_{PV} \int_{7:00}^{7:00+\tau_{charge}} \dot{W}_{module} d\tau \tag{12}$$

2.4. Off-Design Simulation

The analysis under off-design conditions is of primary importance for solar energy conversion systems to evaluate the dynamic behaviour of the whole system and the performance over the year. Once all the components are sized, the off-design analysis investigates the capability of the charging cycles to load the reservoirs under variable meteorological conditions. The off-design analysis is built upon a time-forward simulation, which requires a time discretization. The evolutionary variable time step (τ_i) is physically determined as the time needed by the volume of HTF to flow through the solar field arranged in 10 rows (represented by the calculated length L).

$$\tau_i = \frac{L}{V_{av, i}} \tag{13}$$

The velocity $V_{av, i}$ is calculated step by step from the mass flow rate in the single collector considering an average density of the HTF and the solar collector pipe diameter, as in Equation (14). This estimate is indeed simplified, and it assumes that the HTF velocity in one single collector is maintained in the whole solar field.

$$V_{av, i} = \frac{\dot{m}_{HTF, i}}{\rho_{av, HTF} \cdot \pi \frac{d_{pipe}^2}{4}} \tag{14}$$

Knowing the hourly meteorological conditions (i.e., the solar radiation on the inclined collectors' surface), the simulation starts at 7:00 with the solar field warming-up cycle. It is assumed that the initial temperature of the HTF is 40 °C (thanks to the insulated intermediate ICR reservoir where water is stored after the last heat delivery from the power cycle). The HTF circulates in the closed-loop solar field until the outlet temperature reaches 95 °C. At this temperature level, the useful heat gain can be exploited by the heat pump evaporator and IHR tank. The solar field is simulated in a way that the 85/95 °C temperature increase is kept step by step. The temperature difference is controlled by adjusting the mass flow rate in the solar field. Since the mass flow rate is continuously changing and is different from the collector test conditions, a correction coefficient for the efficiency calculation is applied [17]. Variable mass flow rate of the HTF forces the variation of mass flow rate in the heat pump cycle, while the changing ambient temperature mainly affects the refrigeration cycle. Inlet/outlet temperatures in the heat exchangers are iteratively calculated knowing the heat exchangers geometry and estimating the heat transfer coefficients.

Once the charging period simulation is over, it is possible to apply the off-design analysis of the power cycle. Since the conditions during the off-design charging period are different from design conditions, the volumes of available fluid in the HWR and CMR tanks and the temperature of water in the IHR tank are variable. In the present simplified model (disregarding the actual daily load profile), the mass flow rate of hot water flowing from HWR was kept the same as under design conditions (1 kg/s), and the discharging time period (corresponding to the operation of the power cycle) is consequently calculated. During this time, the whole volume of water from HWR is discharged. Variable conditions at the condenser and solar pre-heater determine the parameters of the power cycle. The mass flow rate in the power cycle is also variable in time, and this influences the turbine performance: to evaluate the turbine efficiency under off-design conditions, a simplified correlation proposed by Fiaschi et al. [19] after Latimer [20] was adopted. The efficiency of the turbine is obtained calculating the off-design value of the work coefficient (ψ), through an interpolated polynomial, which was obtained from the fitting of the data provided in [20]. The work coefficient is computed through the classical non-dimensional characteristic curve of the turbine, which connects the work coefficient to the mass flow rate $\psi = f(\dot{m})$. Therefore, the off-design value of the turbine efficiency can be estimated using the correction for the input value of the ratio ψ/ψ_D, where ψ_D is the work coefficient at design point.

On the whole, the resulting off-design simulation provides an estimate of the performance of the thermo-electric storage system performance over the year; in the present case, two different Italian locations were considered: Crotone and Pantelleria. The year-round operation of the TEES is tested to evaluate how the same system (in terms of assembly, equipment sizing) would perform in 2 different locations, belonging to the same Mediterranean climate group.

2.5. Performance Indicators—Energy

Any storage system can be assessed by means of a round-trip efficiency indicating the ratio between the amount of energy delivered by the system to the amount of energy spent during the charging phase. In the system proposed, it is assumed that no non-renewable energy is consumed during operation cycles. From an overall energy balance perspective, during charging phase, the cycle uses both electric energy from PV field and heat from the solar-thermal collectors' field. The energy efficiency can be then defined as:

$$\eta_{RT} = \frac{\dot{W}_{netPC} \cdot \tau_{discharge}}{\int_{7:00}^{7:00+\tau_{charge}} \left(\dot{W}_{c;ref} + \dot{W}_{c;HP} - \dot{W}_{exp;HP} \right) d\tau + \int_{7:00}^{7:00+\tau_{charge}} \dot{Q}_{evapHP} d\tau + Q_{IHR}} \tag{15}$$

However, electricity and heat are very different forms of energy, with different economic value. This matter is correctly addressed by an exergy approach, which is described in the following. From an energy perspective, a marginal round-trip efficiency can be defined in terms of electricity only as in Equation (16):

$$\eta_{MRT} = \frac{\dot{W}_{netPC} \cdot \tau_{discharge}}{\int_{7:00}^{7:00+\tau_{charge}} \left(\dot{W}_{c;ref} + \dot{W}_{c;HP} - \dot{W}_{exp;HP} \right) d\tau} \tag{16}$$

In Equation (16), contribution from the solar-thermal field is disregarded–solar-thermal input is considered to be a secondary energy of lower quality.

2.6. Exergy and Exergo-Economic Models

The exergy analysis combines the First and Second Laws of Thermodynamics, allowing the evaluation of the efficiency of the energy system and the irreversibilities (exergy destructions) of the system components [21]. Exergy analysis has become one of the most powerful tools for the design and analysis of energy systems and powerplants [22]. Indeed, the concept of exergy can evaluate the actual thermodynamic value of energy flows.

Here, the exergy of the fluid is calculated for every point of the circuit. Exergy is generally defined as the maximum work obtainable from a system or a process through the interaction with the surrounding environment. The exergy of a j-th flow stream can be determined after [23,24] as in Equation (17):

$$\dot{Ex}_j = \dot{m}_j \left[(h_j - h_o) - T_o (s_j - s_o) \right] \tag{17}$$

Every component can be described by an exergy balance distinguishing between exergy rates connected with its fuel and product [23], according to the component exergy balance:

$$\dot{E}x_{F,k} = \dot{E}x_{P,k} + \sum \dot{E}x_{D,k} + \sum \dot{E}x_{L,k} \tag{18}$$

Equation (18) takes into account the exergy destructions and losses, which influence the irreversibility of the system operation. An exergy destruction derives from friction or irreversibility of heat transfer within a defined control volume, while an exergy loss is associated with exergy transfer (waste) to the surroundings. The directly calculated exergy efficiency of a component is defined as the ratio of the daily exergy rate of product to the daily exergy rate of fuel. The indirect definition of exergy efficiency requires the evaluation of exergy destructions and losses. The exergy efficiency can be determined by the following Equation (19):

$$\varepsilon_k = \frac{\dot{E}x_{P;k}}{\dot{E}x_{F;k}} = 1 - \frac{\dot{E}x_{D;k} + \dot{E}x_{L;k}}{\dot{E}x_{F;k}} \tag{19}$$

which can be applied both at component and system level.

In the present case, the only components producing an exergy loss in the system are the air-cooled condenser of the refrigeration cycle and the solar collectors, which represent the only point of heat transfer interaction with the environment.

The exergy analysis was performed both at design and considering the whole seasonal simulation. The round-trip efficiency calculated in terms of exergy is given by Equation (20), which includes all exergy inputs from the solar resource (solar heat as well as PV electricity):

$$\eta_{RT\,ex} = \frac{\dot{W}_{netPC} \cdot \tau_{discharge}}{\int_{7:00}^{7:00+\tau_{charge}} \left(\dot{W}_{c;ref} + \dot{W}_{c;HP} - \dot{W}_{exp;HP} \right) d\tau + \int_{7:00}^{7:00+\tau_{charge}} \dot{E}x_{F,evapHP} d\tau + \dot{E}x_{P;IHR}} \tag{20}$$

A further relevant step is to evaluate the economic profitability of the TEES; this is dealt in detail applying an exergo-economic analysis leading to evaluation of the cost of the stored electricity produced by the power cycle [23,24]. The exergo-economic approach is preferred, because exergy can be regarded in practice as the useful part of energy, and the user should pay only for this part; this is particularly true for ES devices. Consequently, rather than energy, it is useful and rational to assign a cost to exergy. This is the main characteristic of the exergo-economic analysis, which combines exergy and economic analyses by introducing costs per exergy unit [25] and following the full cost build-up along the whole process.

The approach outlined in [23,25] is applied to perform the exergo-economic analysis: for each component k, a cost balance given as in Equation (21) is formulated.

$$\begin{aligned} \dot{C}_{P,k} &= \dot{C}_{F,k} + \dot{Z}_k \\ c_{P,k}\dot{E}x_{P,k} &= c_{F,k}\dot{E}x_{F,k} + \dot{Z}_k \end{aligned} \tag{21}$$

In Equation (18) it is assumed that the cost of exergy loss is zero [23,26], as is common practice in exergo-economics. $\dot{C}_{P,k}$ and $\dot{C}_{F,k}$ represent the cost rates associated with exergy product and fuel, while $c_{P,k}$ and $c_{F,k}$ mean costs per unit of exergy of product or fuel, respectively. $\dot{Z}_k$ is the sum of cost rates associated with investment expenditures for the k-th component. Auxiliary equations needed for components balancing are written in agreement with [23,27]. Referring to a renewable resource as solar energy, it was assumed that the cost of the exergy associated with solar radiation is equal to zero (i.e., fuel for PV modules, for solar collectors' field). The cost rate connected with exergy destruction within a component can be evaluated after Equation (22):

$$\dot{C}_{D,k} = c_{F,k}\dot{E}x_{D,k} \tag{22}$$

An exergo-economic factor, relating the investment cost of component to the sum of the investment cost and the cost of exergy destruction can be calculated:

$$f_k = \frac{\dot{Z}_k}{\dot{Z}_k + \dot{C}_{D,k}} \tag{23}$$

All calculations were integrated over the day, considering the averaged reference day of each month. To estimate the daily cost of a component, the annual investment cost is first determined, as from Equation (24):

$$Z_k^{an} = \frac{ir \cdot (1 + ir)^n}{(1 + ir)^n - 1} Z_k \tag{24}$$

An interest rate ir = 8% and a 20 years lifetime were assumed. The hourly cost is a function of the annual investment cost and of the number of operation hours per year, which is different for the two locations. The daily investment cost of the component varies from month to month and was found multiplying the hourly cost by the daily operational time per day in specific month. The purchase cost of components was evaluated with the help of source data. Costs of heat exchangers, turbine, pumps, compressors were found referring to cost functions in [15,28]. Costs were updated to 2018 values, based on CEPCI indexes [29]. The solar collector cost was estimated after [30], assuming an area-dependent cost at 210 $/m^2. The PV modules purchase cost is assumed following market analyses presented in [31] as average 250 $/module. The cost functions used in the economic analysis are listed in Table 4.

Table 4. Cost functions for the equipment.

Component	Function (Units: 10^3 $, 2009)
Turbine	$1.5 \cdot \dot{W}_T^{0.6} + 10$
Compressor	$6 \cdot \dot{W}_C^{0.6} + 10$
Pump	$44 \cdot \dot{V}_{wf}^{0.75} + 20$
Heat exchanger	$0.3 \cdot A_{HE}^{0.82} + 1$
Reservoir (HWHR/HWCR, CMHR/CMCR, IHR/ICR)	$0.2 \cdot V_k^{0.785} + 2$

The yearly investment cost of the overall system also includes installation and maintenance. For the sake of simplicity, the total cost of installation, operation, and maintenance is assumed at 20% of the total purchase cost of the system [23]. The currency exchange rate applied was 0.877 €/$.

2.7. Modelling Tools

The design sizing, off-design simulation, and exergy and exergo-economic evaluation of the proposed system were performed using Engineering Equation Solver (EES) (version 9, F-chart software, Madison, WI, USA) [32] and Transient System Simulation Tool (TRNSYS) (version 17, Thermal Energy System Specialists, LLC, Madison, WI, USA) [33]. Real fluid assumption was adopted, with the numerical model solved locally through fundamental mass and energy balances, written for each specific power plant component, following general thermodynamic rules [34]. While performing the heat transfer analysis, it was possible to take advantage of the EES built-in heat transfer correlation library [34]. TRNSYS software and its Meteonorm libraries [35] were of great help to estimate the time-dependent values of incoming solar radiation and ambient temperature. The simulations were performed after processing the whole-year weather data to create average days statistically representative of specific months of the year for the 2 locations. The hourly averaged data were imported into Lookup Tables of EES and are interpolated from these tables.

3. Results

The design point simulation was performed for the assumed reference day of May in Crotone. As stated above, a reference day of the month is represented by hourly meteorological data for the location averaged over one month. Ambient temperature and solar irradiance distribution during the reference day of May in Crotone are shown in Figure 4. Off-design simulations are then repeated for reference days of summer months between April and September for both the studied locations.

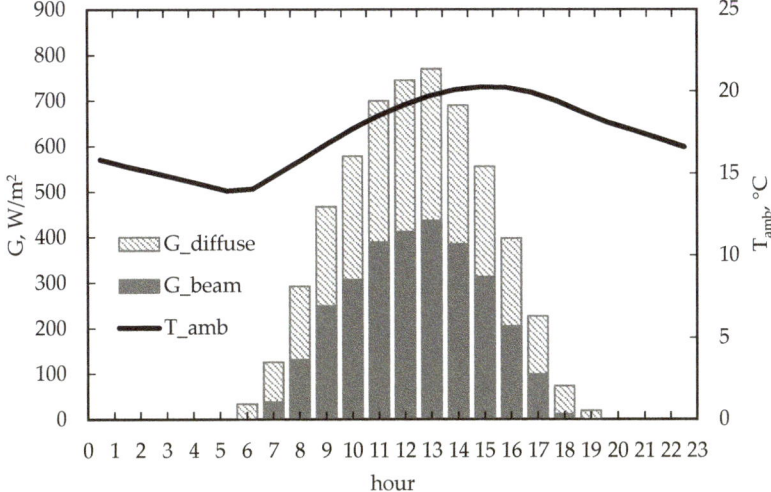

Figure 4. Distribution of meteorological data during the reference day of May in Crotone; T_{amb}—ambient temperature, $G_{diffuse}$—diffuse solar radiation, G_{beam}—direct solar radiation.

Table 5 collects the main results of the design conditions for the proposed TEES system.

Table 5. Design day analysis resulting parameters.

Variable	Value
Turbine power output $\dot{W}_T$	172.6 kW
Heat stored in HWR Q_{HWR}	212.1 kWh
Heat stored in IHR Q_{IHR}	582.8 kWh
Heat stored in CMR Q_{CMR}	650.2 kWh
Volume of HWR V_{HWR}	3.74 m^3
Volume of IHR V_{IHR}	9.175 m^3
Volume of CMR V_{CMR}	65.5 m^3
COP_{Ref}	2.96
Ref. Compressor power $\dot{W}_{C;Ref}$	14.2 kW
COP_{HP}	3.27
HP Compressor power $\dot{W}_{C;HP}$	31.38 kW
Number of solar collectors installed N_{sc}	200
Number of PV modules installed N_{pv}	224
Design charging time τ_{charge}	7 h
Design power cycle operation time τ_{PC}	1 h
Round-trip energy efficiency η_{RT}	0.14
Round-trip marginal efficiency η_{MRT}	0.51

According to the results in the table, on the reference day of May in Crotone, the designed TEES system is charged for 7 h, which allows production of a constant power of 173 kW electricity during the 1-h discharge. It was estimated that 200 solar-thermal collectors are needed to load the HWR and warm up water in IHR. Under design conditions, 40 solar collectors support the heat pump's evaporator and 160 collectors contribute heat to the IHR tank. The heat pump compressor is served from 46 PV panels, while that of the refrigeration cycle compressor relies on 178 PV panels. If a 4 m head loss between twin reservoirs is assumed, the estimated pump power between HWHR and HWCR is 0.044 kW, while the CMCR-CMHR assembly demands 0.842 kW pumping power; this confirms that the relevance of the circulation pumps in the reservoir assemblies is very small compared to the other components. Finally, the marginal round-trip efficiency results to be $\eta_{MRT} = 0.51$. If heat from solar-thermal collectors is also considered as an input, the energy efficiency is $\eta_{RT} = 0.14$.

The charging phase is dependent on variable meteorological conditions. Off-design analysis results provide insight on how the outer conditions affect system performance.

The profile of temperature in the single solar collector (thus the whole solar field) and of the water in the IHR tank during charging period is shown in Figure 5. The off-design simulation indicates that during the reference day of May in Crotone, the charging period actually lasts 7.5 h; in the same time, the water in the IHR tank is restored close to the design value of 95 °C.

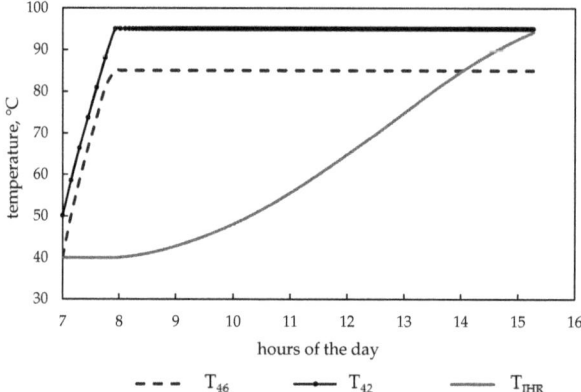

Figure 5. Off-design simulation of the charging phase (Reference day of May in Crotone)—temperature profile for the solar-thermal field and temperature increase of water in the IHR tank.

It is visible that during the first low-radiation hour of operation the HTF circulating in the solar collectors is only warming up in a closed loop, and no heat is yet transferred to the water in the IHR tank. Once the outlet temperature reaches design 95 °C, ΔT_{HTF} is kept constant as 10 K by controlling the mass flow rate and the heat is delivered to the heat pump evaporator and to IHR tank. Simultaneously, the HWHR and CMCR tanks are being charged (respectively the hot and cold reservoirs). Figure 6 presents how the volumes are increasing during the charging period.

As shown in the figure, after 7.5 h the CMCR tank (which does not rely on the solar-thermal collector output) is completely charged, and the HWHR reservoir reaches a liquid volume close to the design value of 3.6 m³. Across the year, depending on the solar radiation availability and variations of the ambient temperature, the performance of the TEES is variable: charging time is adapted, water in the IHR tank is heated up to various levels, the volume of fluids that are pumped to HWHR and CMCR reservoirs is also changing. The final volume of hot water accumulated in the HWHR (to be discharged during the power cycle operation period) determines the duration of the storage discharge time and, thus, the daily electric energy output from the power cycle. Results achieved for the reference days of the months between April and September in Crotone are summarized in Table 6. The simulations of the TEES related to the period between October and March put in evidence that the volume of charged HWHR and the temperature of water in the IHR tank were too low to allow running the power cycle with the same assumptions for the other months.

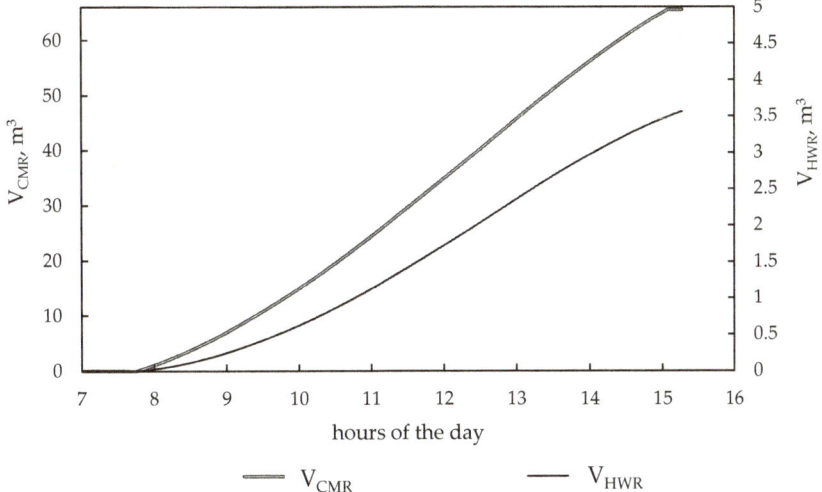

Figure 6. Off-design simulation of the charging phase (Reference day of May in Crotone)—volume increase in the storage sections: HWR (V_{HWR}) and CMR (V_{CMR}).

The highest round-trip efficiency in Crotone (0.468) is achieved for the reference day of August. On that day, the charging phase time is the shortest. Since the ambient temperature is not favorable for the refrigeration cycle, the volume of water/glycol mixture accumulated in the CMCR is relatively low (53.3 m³), but this does not affect appreciably the energy output during the discharge period.

Table 6. Off-design simulations results for system operating in Crotone.

Variable	Values for Reference Day of the Months					
	4	5	6	7	8	9
Daily charging time; h	7.5	7.5	7.5	7	6.5	7
Volume of charged HWR V_{HWR}; m³	3.4	3.6	3.7	3.7	3.6	3.7
Volume of charged CMR V_{CMR}; m³	65.5	65.5	62.8	55.1	53.3	61.4
Final temp. of water in the IHR tank; °C	91	94	100	100	97	100
Daily power cycle operation time; h	0.87	0.91	0.96	0.96	0.93	0.96
Power cycle energy output; kWh	151	157	165	165	160	166
Round-trip Marginal efficiency η_{MRT}	0.456	0.44	0.44	0.461	0.468	0.452

3.1. Exergy Analysis

The exergy efficiency of the TEES system for the reference day of May in Crotone achieved 0.603. Results of the exergy analysis are summarized in Table 7.

Based on the values of the daily exergy destruction $\dot{Ex}_{D,k}$ and of the component exergy efficiency ε_k, the solar collectors and the PV panels emerge as the most critical components. This is quite common as the system is operated at moderate temperature levels and a notable amount of exergy is wasted as the solar collector loss, or degraded from the high-temperature potential resource of extra-terrestrial solar radiation (collector exergy destruction). The balance of exergy destructions and losses is graphically summarized in Figure 7, showing the daily relative values for all components of the system. The total exergy input was assumed to be the exergy of solar radiation directed to both the PV modules and to the solar-thermal field. From a thermodynamic cycles point of view, the largest exergy destructions occur in the refrigeration cycle (compressor and throttle valve). The high exergy destruction associated with the throttling valve speaks in favor of substituting it with an expander—electrical output would partially support the compressor and thereby assist the PV field.

Other relevant contributions are represented by the turbine and by the solar pre-heater in the power cycle (respectively 1.08% and 1.07% relative exergy destruction).

Table 7. Exergy analysis results for system operating on reference day of May in Crotone.

k	Component	$\dot{Ex}_{F,k}$ (kWh/day)	$\dot{Ex}_{P,k}$ (kWh/day)	$\dot{Ex}_{D,k}$ (kWh/day)	$\dot{Ex}_{L,k}$ (kWh/day)	ε_k
1	Condenser PC	87.87	72.74	15.13	-	82.8%
2	Pump PC	27.34	21.39	5.954	-	78.2%
3	RH-int PC	3.806	1.519	2.287	-	39.9%
4	RH-mtsolar PC	78.12	48.93	29.19	-	62.6%
5	HTHE PC	49.46	45.96	3.499	-	92.9%
6	Turbine PC	186.7	157.2	29.55	-	84.2%
7	Evaporator HP	27.45	22.53	4.917	-	82.1%
8	Compressor HP	92.09	73	19.09	-	79.3%
9	Condenser HP	53.94	51.93	2.011	-	96.3%
10	Turbine HP	41.6	30.28	11.31	-	72.8%
11	Condenser RC	33.05	-	17.83	15.22	46.0%
12	Valve RC	178.8	137.7	41.11	-	77.0%
13	Evaporator RC	118.1	87.98	30.16	-	74.5%
14	Compressor RC	239.2	192.3	46.85	-	80.4%
15	Solar collectors	1433	137.3	1158	137.7	9.6%
17	IHR tank	109.8	78.12	31.69	-	71.1%
21	HWR reservoir	51.93	49.46	2.463	-	95.3%
22	CMR reservoir	87.98	87.87	0.1133	-	99.9%
23	PV panels	1304	301	1003	-	23.1%

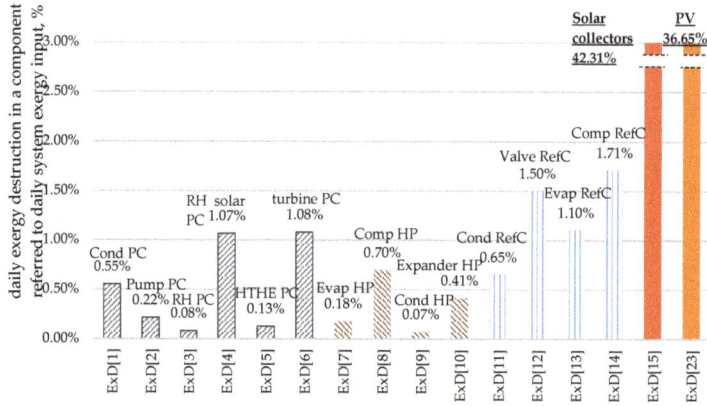

Figure 7. Relative daily exergy destruction and/or losses in all components of the TEES system (reference day of May in Crotone).

3.2. Exergo-Economic Analysis

Following the methodology given in Section 2.5, the overall specific investment cost of the whole system is 2281 €/kW. This value agrees with the results obtained by other researchers [15]. Since the power output from the cycle is strictly dependent on the daily meteorological conditions, the cost of the electric energy obtained from the TEES system is also subject to seasonal change.

Table 8 presents the results obtained from the exergo-economic analysis of the TEES system operation for the base case of the May reference day in Crotone.

In terms of economic analysis, the solar collectors and the compressors (RC and HP) are here the most expensive components. From the exergo-economic point of view, if $\dot{C}_k$ and $\dot{C}_{D,k}$ are summed, again the same components are of highest importance; however, some components with high contributions in terms of $\dot{C}_{D,k}$ emerge (notably the PC Turbine and Solar RH; and the RC Throttle Valve).

The specific cost of the exergy product c_p associated with the turbine of the power cycle represents the cost of the TEES output product, i.e., the levelized cost of electricity (LCOE). It is the fundamental result of the exergo-economic analysis, which applies Equation (21) to each component. It is calculated for each reference day of the months of operation, and it is thus subject to variation over the seasons as presented in Table 9.

Table 8. Values of selected exergo-economic variables for system operating on reference day of May in Crotone.

k	Component	PEC (€)	$\dot{Z}_k$ (€/day)	$\dot{C}_{D,k}$ (€/day)	$\dot{Z}_k+\dot{C}_{D,k}$ (€/day)	$c_{F,k}$ (€/kWh)	$c_{P,k}$ (€/kWh)	f_k (%)
1	Condenser PC	11426	0.72	9.75	10.47	0.64	0.79	7%
2	Pump PC	19523	1.22	4.64	5.87	0.78	1.05	21%
3	RH-int PC	3911	0.25	1.62	1.86	0.71	1.94	13%
4	RH-solar PC	8213	0.51	9.73	10.24	0.33	0.54	5%
5	HTHE PC	5532	0.35	1.28	1.63	0.37	0.40	21%
6	Turbine PC	40992	2.57	19.81	22.38	0.67	0.78	11%
7	Evaporator HP	2631	1.36	1.09	2.45	0.22	0.33	55%
8	Compressor HP	37590	19.38	3.81	23.19	0.20	0.27	84%
9	Condenser HP	1509	0.78	0.57	1.35	0.28	0.32	58%
10	Turbine HP	13238	6.83	3.22	10.05	0.28	0.44	68%
11	Condenser RC	4387	2.26	30.66	32.92	1.72	3.88	7%
12	Throttle Valve RC	954	0.49	13.58	14.07	0.33	0.43	3%
13	Evaporator RC	2875	1.48	13.05	14.53	0.43	0.60	10%
14	Compressor RC	54776	28.24	3.94	32.18	0.08	0.25	88%
15	Solar collectors	58934	30.39	0.00	30.39	0.00	0.22	100%
17	IHR tank	2994	1.73	7.02	8.75	0.22	0.33	20%
21	HWR reservoir	2444	1.41	0.79	2.21	0.32	0.37	64%
22	CMR reservoir	6981	4.04	0.07	4.10	0.60	0.64	98%
23	PV panels	49112	25.32	0.00	25.32	0.0	0,08	100%

Table 9. Variation of electricity yield and electricity cost over the seasons.

Location	Month for Reference Day →	4	5	6	7	8	9
Pantelleria	daily charging/discharging time, h	8/0.87	8/0.87	8/0.96	8/0.96	8/0.96	8/0.96
	daily electric energy output, kWh	150.1	149.4	165.4	165.3	165.3	165.4
	LCOE, €/kWh	0.7904	0.7945	0.7179	0.7183	0.7199	0.7179
Crotone	daily charging/discharging time, h	7.5/0.87	7.5/0.91	7.5/0.96	7/0.96	6.5/0.93	7/0.96
	daily electric energy output, kWh	150.8	157.2	165.4	165.3	160	165.5
	LCOE, €/kWh	0.8133	0.78	0.7437	0.6948	0.6672	0.694

As expected, the lowest electricity costs correspond to the highest daily electric energy output in the summer months. For all summer months in Pantelleria, the charging phase can always be 8 h per day, which allows full charging of the reservoirs. In comparison—because of different climate conditions—the off-design simulations in Crotone indicated a variation in charging time over the reference days of different months. For example, in August the charging time lasts 6.5 h. This allows production of 160 kWh over the day (a value lower than that achieved in other months), but the whole-day round-trip marginal efficiency is the highest and the exergy and exergo-economic balance determine consequently a lower cost of the stored electricity.

The main annual-averaged results for the two locations are listed in Table 10. The annual LCOE is the kWh-weighted average value of the exergo-economic cost of the turbine product c_{p6} for each month. This value takes into account the variation of electric productivity over the whole season (indicated in Table 9) and the distribution of the total plant costs over the years.

Table 10. Annual operational details for TEES systems operated in Pantelleria and Crotone.

No.	Location	Pantelleria (Italy)	Crotone (Italy)
1.	Coordinates	36.82°N 11.97°E	39.08°N 17.11°E
2.	Solar radiation on sloped surface, MWh/(m²season)	0.990	1.077
3.	Total operation time of TEES, h/year	1482	1671
4.	Annual productivity, MWh	29.307	29.413
5.	Annual average LCOE, €/kWh	0.743	0.732

4. Conclusions

A solar-integrated TEES system was presented and discussed. The system consists of three main blocks (power cycle, solar-assisted heat pump, and solar-assisted refrigeration cycle), matched through the use of properly sized reservoirs. Solar integration with the TEES uses a combination of thermal and photovoltaic conversion.

A complete energy, exergy, and exergo-economic analysis was carried out to define the system effectiveness, to assess the possibilities for design improvement and to identify the most significant contributions to the final cost of stored electricity. The proposed TEES system can deliver electric energy with a marginal 50.9% round-trip efficiency if the solar heat input to solar-thermal collectors is not considered. The exergy round-trip efficiency is 35.6%. The exergy and exergo-economic analyses suggest that the most relevant components, in terms of irreversibilities and exergo-economic costs, are always the solar energy conversion units: solar collectors field and PV panels.

Exergo-economic analyses were performed for the reference days of the months between April and August in two southern Italian locations: Crotone and Pantelleria (a small island with a present high cost of electricity). The LCOE produced by the TEES is 0.74 €/kWh for Pantelleria and 0.73 €/kWh for Crotone—values which are not at present competitive with present documented electricity costs (0.31 €/kWh [36]), but whoseresult are in line or even slightly better than documented stand-alone renewable configurations [37–39]. The result is mainly due to the high costs of the solar collectors and of the refrigeration cycle, which have a large influence on the overall exergy destructions and exergo-economic cost balance. It appears, however, that a relevant margin of improvement is possible, working both on the reduction of equipment cost and on optimized control strategies.

Author Contributions: G.M. conceived the basic idea of the TEES system and layout, D.F. proposed and suppported the introduction and dynamic analysis of solar section, L.T. gave a substantial contribution to the simulation of the power cycle components and K.P. realized the models for dynamic simulation and optimization of the overall system. K.P. and L.T. wrote the basic parts of the manuscript, G.M. and D.F. advised and revised the overall content of the paper.

Funding: This article receives no external funding.

Conflicts of Interest: The authors declare no conflict of interest.

Nomenclature

Abbreviations

CEPCI	Chemical Engineering Plant Cost Index
CMR	Cold medium reservoir (common name for CMHR and CMCR assembly)
CMHR	Cold medium hot reservoir
CMCR	Cold medium cold reservoir
COP	Coefficient of Performance (Heat Pump or Refrigeration Cycle)
ES	Energy storage
HP	Heat Pump
HWR	Hot water reservoir (common name for HWHR and HWCR assembly)
HWHR	Hot water hot reservoir
HWCR	Hot water cold reservoir
IHR	Intermediate-heat hot reservoir
ICR	Intermediate-heat cold reservoir
HTF	Heat transfer fluid
PC	Power Cycle
PVCU	PV conversion unit
RC	Refrigeration Cycle
TEES	Thermo-electric energy storage

Symbols

A	Area, m^2
$\dot{C}$	Cost rate associated with exergy transfer, €/day
η_o	Collector constant (non-dimensional)
a_1	Collector constant, $W/(m^2K)$
a_2	Collector constant, $W/(m^2K^2)$
c	Cost per unit of exergy, €/kWh
d	Diameter, m
$\dot{Ex}$	Exergy rate, kW (or kWh/day)
Ex	Exergy, kWh (or kJ)
f	Exergo-economic factor, %
G	Overall radiation, kW/m^2
h	Enthalpy, kJ/kg
ir	Interest rate, %
$\dot{m}$	Mass flow rate, kg/s
L	Length, m
LCOE	Levelized Cost of Electricity (stored), €/kWh
n	Operation period, year
N	Number
p	Pressure, kPa
$\dot{Q}$	Heat rate, kW
s	Entropy, J/(kgK)
T	Temperature, °C (or K)
UA	Heat transfer coefficient multiplied by heat exchanger area, W/K
v	Velocity, m/s
$\dot{V}$	Volumetric flow rate, m^3/s
V	Volume, m^3
$\dot{W}$	Power, kW
Z^{an}	Annual investment cost, €/year
$\dot{Z}$	Cost rate associated with capital investment and O&M costs, €/day

Greek:

Δ	Variation
ϵ	Effectiveness
h	Efficiency
r	Density, kg/m^3
τ	Time or time step (variable), h or s

Subscripts:

0	Ambient
amb	Ambient
av	Average
C	Compressor
charged	Associated with charging period
D	Destruction
evap	Evaporator
exp	Expander
F	Fuel
he	Heat exchanger
HP	Heat Pump
hw	Hot water
i	I-th time step
j	J-th flow rate
in	Inlet
inst	Plant component

L	Loss
minHP	Lower pressure part in the Heat Pump
module	PV module
MRT	Marginal Round-Trip
out	Outlet
P	Product
p	Pump
PC	Power cycle
pipe	Associated with absorber pipe
ref	Refrigeration Cycle
rel	Relative (exergy loss or destruction)
RH	Re-heater
RT	Round-Trip
SC	Solar Collector
sloped	On sloped surface
t	Turbine
tank	Tank
wf	Working fluid (CO_2 in the main power cycle)
x	Exergy

References

1. McPherson, M.; Tahseen, S. Deploying storage assets to facilitate variable renewable energy integration: The impacts of grid flexibility, renewable penetration, and market structure. *Energy* **2018**, *145*, 856–870. [CrossRef]

2. Hadjipaschalis, I.; Poulikkas, A.; Efthimiou, V. Overview of current and future energy storage technology for electric power applications. *Renew. Sustain. Energy Rev.* **2009**, *13*, 1513–1522. [CrossRef]

3. Luo, X.; Wange, J.; Dooner, M.; Clarke, J. Overview of current development in electrical energy storage technologies and the application potential in power system operation. *Appl. Energy* **2015**, *137*, 511–536. [CrossRef]

4. Benato, A.; Stoppato, A. Pumped thermal electricity storage: A technology overview. *Therm. Sci. Eng. Prog.* **2018**, *6*, 301–315. [CrossRef]

5. Ayachi, F.; Tauveron, N.; Tartière, T.; Colasson, S.; Nguyen, D. Thermo-Electric Energy Storage involving CO_2 transcritical cycles and ground heat storage. *Appl. Therm. Eng.* **2016**, *108*, 1418–1428. [CrossRef]

6. Tauveron, N.; Macchi, E.; Nguyen, D.; Tartière, T. Experimental study of supercritical CO_2 heat transfer in a Thermo-Electric Energy Storage based on Rankine and heat pump cycles. *Energy Procedia* **2017**, *129*, 939–946. [CrossRef]

7. Morandin, M.; Maréchal, F.; Mercangoz, M. Butcher, Conceptual design of a thermo-electrical energy storage system based on heat integration of thermodynamic cycles—Part A: Methodology and base case. *Energy* **2012**, *45*, 375–385. [CrossRef]

8. Morandin, M.; Maréchal, F.; Mercangoz, M. Butcher, Conceptual design of a thermo-electrical energy storage system based on heat integration of thermodynamic cycles—Part B: Alternative system configurations. *Energy* **2012**, *45*, 386–396. [CrossRef]

9. White, A.; Parks, G.; Markides, C.N. Thermodynamic analysis of pumped thermal electricity storage. *Appl. Therm. Eng.* **2013**, *53*, 291–298. [CrossRef]

10. Ruer, J. Installation and Methods for Storing and Recovering Electric Energy. WO/2008/148962, 12 December 2008.

11. McTigue, J.D.; White, A.J.; Markides, C.N. Parametric studies and optimisation of pumped thermal electricity storage. *Appl. Energy* **2015**, *137*, 800–811. [CrossRef]

12. Frate, G.F.; Antonelli, M.; Desideri, U. A novel pumped thermal electricity storage (PTES) system with thermal integration. *Appl. Therm. Eng.* **2017**, *121*, 1051–1058. [CrossRef]

13. Mercangoz, M.; Hemrle, J.; Kaufmann, L.; Z'Graggen, A.; Ohler, C. Electrothermal energy storage with transcritical CO_2 cycles. *Energy* **2012**, *45*, 407–415. [CrossRef]

14. Morandin, M.; Mercangoz, M.; Hemrle, J.; Maréchal, F.; Favrat, D. Thermoeconomic design optimization of a thermo-electric energy storage system based on transcritical CO_2 cycles. *Energy* **2013**, *58*, 517–587. [CrossRef]

15. Henchoz, S.; Buchter, F.; Favrat, D.; Morandin, M.; Mercangoz, M. Thermoeconomic analysis of a solar enhanced energy storage concept based on thermodynamic cycles. *Energy* **2012**, *45*, 358–365. [CrossRef]

16. Ferrara, G.; Ferrari, L.; Fiaschi, D.; Galoppi, G.; Karellas, S.; Secchi, R.; Tempesti, D. Energy recovery by means of a radial piston expander in CO_2 refrigeration system. *Int. J. Refrig.* **2016**, *72*, 147–155. [CrossRef]

17. Duffie, J.A.; Beckman, W.A. *Solar Engineering of Thermal Processes*, 4th ed.; John Wiley & Sons Inc.: New York, NY, USA, 2013.

18. Schott Applied Power Corporation. High Efficiency Multi-Crystal Photovoltaic Module. Available online: http://abcsolar.com/pdf/schott165.pdf/ (accessed on 13 December 2018).

19. Fiaschi, D.; Manfrida, G.; Talluri, L. Integrated model of a solar chimney equipped with axial turbines. In Proceedings of the ECOS 2015 28th International Conference on Efficiency, Cost, Optimization, Simulation and Environmental Impact of Energy Systems, Pau, France, 30 June–3 July 2015.

20. Latimer, R.J. Axial Turbine Performance Prediction. In *VKI LS Off-Design Performance of Gas Turbines*; Von Karman Institute: Sint-Genesius-Rode, Belgium, 1978.

21. Szargut, J.; Morris, D.R.; Steward, F.R. *Exergy Analysis of Thermal, Chemical, and Metallurgical Processes*; Hemisphere Publishing Corporation: Washington, DC, USA, 1988.

22. Kotas, T.J. *The Exergy Method of Thermal Plant Analysis*; Butterworths: London, UK, 1985.

23. Bejan, A.; Tsatsaronis, G.; Moran, M. *Thermal Design and Optimization*; John Wiley & Sons, Inc.: New York, NY, USA, 1996.

24. Szargut, J.; Petela, R. *Exergy*; WNT: Warsaw, Poland, 1965.

25. Tsatsaronis, G. Thermoeconomic analysis, and optimization of energy systems. *Prog. Energy Combust. Sci.* **1993**, *19*, 227–257. [CrossRef]

26. Lazzaretto, A.; Tsatsaronis, G. SPECO: A systematic and general methodology for calculating efficiencies and costs in thermal systems. *Energy* **2006**, *31*, 1257–1289. [CrossRef]

27. Morosuk, T.; Tsatsaronis, G. 3-D Exergy-based methods for improving energy-conversion systems. *Int. J. Thermodyn.* **2012**, *15*, 201–213. [CrossRef]

28. Turton, R.; Bailie, R.; Whiting, W.; Shaeiwitz, J. *Analysis, Synthesis and Design of Chemical Processes*; Prentice Hall: Englewood Cliff, NJ, USA, 2003.

29. Chemical Engineering. Economic Indicators. Available online: https://www.chemengonline.com/site/plant-cost-index/ (accessed on 13 December 2018).

30. Kalogirou, S.A. *Solar Energy Engineering, Processes and Systems*, 2nd ed.; Academic Press: New York, NY, USA, 2013.

31. Freecleansolar. 165W Module Schott SAPC-165 Poly. Available online: https://www.freecleansolar.com/165W-module-Schott-SAPC-165-poly-p/sapc-165.htm (accessed on 13 December 2018).

32. Klein, S.A.; Nellis, G.F. *Mastering EES*; f-Chart Software: Madison, WI, USA, 2012.

33. Trnsys17 Information. Available online: https://sel.me.wisc.edu/trnsys/features/ (accessed on 13 December 2018).

34. Klein, S.A.; Nellis, G.F. *Thermodynamics*; Cambridge University Press: Cambridge, UK, 2011.

35. Meteonorm Information. Available online: http://www.meteonorm.com/ (accessed on 13 December 2018).

36. Ciriminna, R.; Pagliaro, M.; Meneguzzo, F.; Pecoraino, M. Solar energy for Sicily's remote islands: On the route from fossil to renewable energy. *Int. J. Sustain. Built Environ.* **2016**, *5*, 132–140. [CrossRef]

37. Giutsos, D.M.; Blok, K.; Velzen, L.V.; Moorman, S. Cost-optimal electricity systems with increasing renewable energy penetration for islands across the globe. *Appl. Energy* **2018**, *226*, 437–449. [CrossRef]

38. Lal, S.; Raturi, A. Techno-economic analysis of a hybrid mini-grid system for Fiji Islands. *Int. J. Energy Environ. Eng.* **2013**, *3*, 10. [CrossRef]

39. Kalinci, Y.; Hepbasli, A.; Dincer, I. Techno-economic analysis of a stand-alone hybrid renewable energy system with hydrogen production and storage options. *Int. J. Hydrogen Energy* **2015**, *40*, 7652–7664. [CrossRef]

Article

A Comprehensive Methodology for the Integrated Optimal Sizing and Operation of Cogeneration Systems with Thermal Energy Storage

Luca Urbanucci *, Francesco D'Ettorre and Daniele Testi

Department of Energy, Systems, Territory, and Constructions Engineering (DESTEC), University of Pisa, Largo Lucio Lazzarino 1, 56122 Pisa, Italy; francesco.dettorre@ing.unipi.it (F.D.), daniele.testi@unipi.it (D.T.)
* Correspondence: luca.urbanucci@ing.unipi.it; Tel.: +39-050-2217134

Received: 30 December 2018; Accepted: 2 March 2019; Published: 6 March 2019

Abstract: Cogeneration systems are widely acknowledged as a viable solution to reduce energy consumption and costs, and CO_2 emissions. Nonetheless, their performance is highly dependent on their capacity and operational strategy, and optimization methods are required to fully exploit their potential. Among the available technical possibilities to maximize their performance, the integration of thermal energy storage is recognized as one of the most effective solutions. The introduction of a storage device further complicates the identification of the optimal equipment capacity and operation. This work presents a cutting-edge methodology for the optimal design and operation of cogeneration systems with thermal energy storage. A two-level algorithm is proposed to reap the benefits of the mixed integer linear programming formulation for the optimal operation problem, while overcoming its main drawbacks by means of a genetic algorithm at the design level. Part-load effects on nominal efficiency, variation of the unitary cost of the components in relation to their size, and the effect of the storage volume on its thermal losses are considered. Moreover, a novel formulation of the optimization problem is proposed to better characterize the heat losses and operation of the thermal energy storage. A rolling-horizon technique is implemented to reduce the computational time required for the optimization, without affecting the quality of the results. Furthermore, the proposed methodology is adopted to design a cogeneration system for a secondary school in San Francisco, California, which is optimized in terms of the equivalent annual cost. The results show that the optimally sized cogeneration unit directly meets around 70% of both the electric and thermal demands, while the thermal energy storage additionally covers 16% of the heat demands.

Keywords: thermal energy storage; mixed integer linear programming; rolling-horizon; combined heat and power; optimization

---

1. Introduction

Cogeneration or combined heat and power (CHP) systems are defined as energy generation units that simultaneously produce electric (or mechanical) energy and useful heat from a single fuel source [1]. The role of cogeneration in the process of reducing primary energy consumption, CO_2 emissions, and energy costs is widely acknowledged [2]. This is mainly due to the higher energy efficiency achievable by CHP systems compared to the separate generation of electricity and heat. Moreover, cogeneration systems are considered a viable solution to promote the development of microgrids with distributed renewable energy [3].

Nevertheless, the energy, environmental, and economic performances of CHP systems are highly dependent on their sizing and operational strategy. Indeed, undersizing and oversizing of cogeneration plants are frequent and may compromise the energy saving potential of those systems [4]. For this reason, several methods have been proposed in recent years to identify the integrated optimal sizing and management of CHP plants. A non-exhaustive list of selected examples is as follows. Gimelli et al. [5] presented a multi-objective optimization approach, based on a genetic algorithm, for the optimal design of modular cogeneration systems based on reciprocating gas engines. A sensitivity analysis was performed to evaluate the robustness of the optimal solution with respect to energetic, legislative, and market scenarios. Similarly, Beihong and Weiding [6] proposed a mixed integer non-linear programming (MINLP) formulation to determine the capacity of a gas turbine cogeneration plant that minimizes the annual total cost, thus proving the importance of optimization methods in the design and operation of CHP systems. On the same topic, Aviso and Tan [7] applied a different optimization tool, namely the fuzzy P-graph approach, for the optimal design of cogeneration and trigeneration systems. This approach allows the consideration of soft constraints on user demand and fuel availability. Likewise, Urbanucci and Testi [8] proposed a probabilistic methodology for the optimal sizing of CHP units under uncertainty in energy demands, based on a Monte Carlo risk analysis. Moreover, the thermoeconomic approach was also applied to cogeneration systems, as in [9], where Yokoyama et al. optimized both the exergy efficiency and the annual cost of a CHP gas turbine.

Even though the above-mentioned works are significant examples of optimization methodologies for cogeneration units, none of them consider a storage device in the energy system. Nevertheless, integrating thermal energy storages (TESs) into CHP systems can lead to higher operational flexibility and better overall performance. Indeed, as shown in [10] by Fang and Lahdelma, the decoupling of heat and electric demands by storing the exceeding CHP thermal production allows maximization of the operation of cogeneration systems, thus reducing the load share covered by back-up units. During the years, several approaches for the optimization of cogeneration systems with storage devices have been proposed. For example, Kuang et al. [11] applied a dynamic programming technique to define the most economical operation schedule of trigeneration systems. Elsido et al. [12] proposed a MINLP formulation for a district heating network, combining evolutionary algorithms with mathematical programming. The optimal design and operation for both economic and environmental objectives of a CHP system with a thermal storage tank was investigated by Fuentes-Cortés et al. [13], by exploiting commercial MINLP solvers.

The introduction of energy storage units makes the optimization of CHP systems significantly more complicated. In the absence of TES, the operation optimization problem can be considered "static", since each timestep is independent of the others. Under this condition, it can be proven that the overall optimum corresponds to the sum of the optimums of every single timestep [14]. Consequently, the optimal solution can be traced back to the solution of several optimization problems, one for each timestep, which, compared to the overall problem, are characterized by a lower dimensionality and, hence, by a lower computational cost. On the other hand, when a storage device is considered within the system, the problem becomes "dynamic", because of the dependency between the optimization variables of adjacent timesteps. This results in a higher dimensionality—thus, a higher computational cost—of the optimization problem, which calls for the adoption of decomposition strategies, like the so-called rolling-horizon approach. Rolling-horizon techniques have been proven to effectively tackle the high dimensionality of optimization problems, by decomposing the problem in shorter periods, without affecting the quality of the results [15].

The general problem of the integrated optimal sizing and operation of cogeneration systems usually results in a non-linear problem. Among the available optimization techniques, the state-of-the-art approach to solve this problem consists in linearizing the original problem to obtain a mixed integer linear programming (MILP) problem [12]. This is mainly due to the guarantee of the global optimality of the solution and to the effectiveness of the available commercial MILP solvers. Nonetheless, when dealing with the optimal design problem, with the size of each component an optimization variable, strong assumptions have to be made to retain the linearity of the model: The efficiency of the units must be kept constant, the variation of the unitary cost of the components in relation to their size must be neglected, and the temperature of the storage cannot be directly tracked [14]. To tackle those issues, a decomposition approach based on a two-level optimization algorithm is adopted in this work. At the upper level, the optimal design problem (i.e., the selection of CHP and storage capacities) is solved by means of a metaheuristic method, while at the lower level, the operational strategy is optimized by means of an MILP method. The design and operation problems must be solved simultaneously since they are intrinsically linked.

As already mentioned, the decomposition approach allows tracking of the average temperature of the storage in each timestep, and a better estimate of the thermal losses can be performed by considering also the impact of the storage geometrical characteristics. Indeed, in simpler models, the storage is seen as a simplified thermal battery, and its losses are assumed proportional to the energy content of the storage with a constant heat-loss coefficient, thus providing a poor physical representation of the system, as shown in [15,16]. In this paper, on the basis of the model proposed by Steen et al. [16], a new formulation of the optimization problem is developed, capable of considering the impact of geometrical characteristics, the tank's insulation properties, and the ambient conditions in the evaluation of TES heat losses. As well as including all the above-mentioned features, the main innovation of the proposed approach is the possibility of simulating and optimizing the operation of the storage when the temperature of the TES is below the operation threshold temperature. Furthermore, since the thermal storage makes the operation optimization problem "dynamic" and, consequently, computationally very expensive, a rolling-horizon approach is proposed.

To summarize, the main purpose and novelty of this study is to develop a comprehensive methodology for the integrated optimal sizing and operation of cogeneration systems with thermal energy storage. To the best of the authors' knowledge, such a comprehensive approach is still missing. In view of this, the present paper aims to propose a thorough methodology with the following key characteristics:

- A decomposition approach was adopted. At the upper level, the optimal design problem is solved by means of a genetic algorithm, while at the lower level, a MILP formulation is adopted to identify the optimal operation;
- A novel MILP formulation for the TES was developed to obtain a better assessment of its optimal operation;
- A rolling-horizon technique was implemented to reduce the computational time of the optimization problem.

Finally, the proposed methodology was tested in a case study. To this end, hourly electric and thermal demands of a reference building (a secondary school located in San Francisco, California) were considered, and the optimal integrated sizing and operation of a cogeneration system with thermal energy storage were identified.

The rest of the paper is structured as follows. In Section 2, the methodological framework is described: The energy system, the TES modeling, the optimization problem, and algorithm are presented in detail. Section 3 presents the case study, while an in depth-analysis of the results follows in Section 4. The last section contains concluding remarks.

2. Methodology

2.1. The Energy System

The energy system under consideration is schematically shown in Figure 1. It comprises a cogeneration unit (e.g., an internal combustion engine), a thermal energy storage, and an auxiliary boiler. Moreover, the exchange of electricity with the regional grid is allowed (both selling and buying). Thus, the thermal demand can be met partly by the CHP heat production and partly by the boiler, while the electric demand can be covered partly by the CHP electric production and partly by the electric grid. The TES can be used to store the thermal energy recovered from the cogeneration unit and, subsequently, to cover the thermal demand.

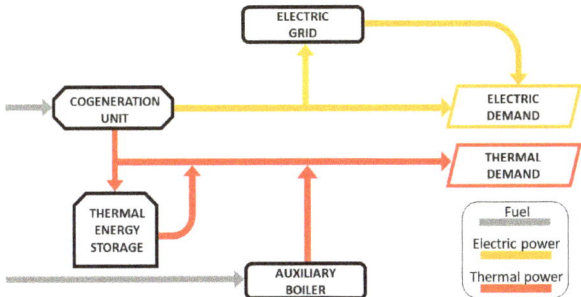

Figure 1. Schematic representation of the combined heat and power system.

An hourly timestep was adopted for the simulation of the system, as a compromise between the computational burden and temporal resolution of the scheduling problem. With this timestep length, the behavior of the machines based on the hourly energy flows can be considered, thus averaging the transient performances, which occur on shorter time scales. Indeed, among the considered generators, the internal combustion engine is the one with the longest dynamic response, with start-up times of around 2 minutes for the electric output and less than 15 minutes for the thermal output [17].

2.2. Thermal Energy Storage Modeling

A cylindrical water tank is considered as the sensible-heat TES device. At each *i-th* timestep, an equivalent TES temperature (T_{TES}) is defined on the basis of the stored energy ($H_{tot,TES}^i$), so that the following relationship holds:

$$H_{tot,TES}^i = \rho_{TES} V_{TES} c_{TES} \left(T_{TES}^i - T_{amb}^i \right) \tag{1}$$

with ρ_{TES} as the water density, V_{TES}, the storage volume, c_{TES}, the specific heat of water, and T_{TES}, the ambient temperature. In this way, the zero state of charge is the one corresponding to an equivalent TES temperature equal to the ambient temperature. The dynamic of the stored energy varies according to the following energy balance:

$$H_{tot,TES}^{i+1} - H_{tot,TES}^i = -H_{TES,loss}^i + H_{TES,charge}^i - H_{TES,discharge}^i \tag{2}$$

with $H_{TES,losses}^i$ as the storage losses, whose formulation is discussed in detail below, and $H_{TES,charge}^i$ and $H_{TES,discharge}^i$ as the energy delivered to and from the storage, respectively.

With the useful energy only a fraction of the whole TES energy content, due to the minimum temperature level that is useful to satisfy the thermal demand ($T_{u,TES}$), the energy stored is considered as the sum of two terms: The useful available energy ($H_{u,TES}^i$) and the residual energy ($H_{res,TES}^i$):

$$H_{tot,TES}^i = H_{u,TES}^i + H_{res,TES}^i \tag{3}$$

where:

$$H_{u,TES}^i = \rho_{TES} V_{TES} c_{TES} \left(T_{TES}^i - T_{u,TES} \right) \tag{4}$$

$$H_{res,TES}^i = \rho_{TES} V_{TES} c_{TES} \left(T_{u,TES} - T_{amb}^i \right) \tag{5}$$

This distinction is also useful to highlight differences between the models available in the current literature to evaluate thermal-storage losses. In simpler models, the storage is seen as a simplified thermal battery, and its losses are accounted for only on the basis of the useful storage energy content by means of a constant heat-loss coefficient [16]:

$$H^i_{TES,loss} = \theta_{TES,loss} H^i_{u,TES} \tag{6}$$

with:

$$\theta_{TES,loss} = \frac{U_{TES} A_{TES} \tau}{\rho_{TES} V_{TES} c_{TES}} \tag{7}$$

Nevertheless, Equation (6), by considering only the contribution of the useful energy ("dynamic" losses) and neglecting that related to the residual energy ("static" losses), leads to an underestimation of the losses. Indeed, it has been shown by Omu et al. [18] that considering static heat losses improves the accuracy of thermal-storage modeling, compared to a simplified thermal-battery model. To this end, different formulations that also consider the residual energy in the computation of the losses have been proposed [13]. The latter reads:

$$H^i_{TES,losses} = H^i_{dynamic,losses} + H^i_{static,losses} = \theta_{TES,loss} H^i_{u,TES} + \theta_{TES,loss} H^i_{res,TES} \tag{8}$$

The use of these models leads to an additional issue when the coupled problem of sizing and operation is considered in the MILP formulation. Indeed, as it can be seen from Equations (6) and (7), the heat losses depend on the amount of stored energy, as well as on the capacity and characteristics of the TES, such as the volume of the tank and the overall heat transfer coefficient. Consequently, by considering a constant overall heat-loss coefficient, the impact of these optimization parameters—specifically of the volume of the tank—on the storage losses is neglected, thus affecting the solution of the optimization problem. To overcome this issue, in the present work, the decomposition of the optimization problem at the design level and operation level was adopted, as described in the next section. By doing so, the effect of the volume of the tank on the heat-loss coefficient can be exactly considered. Thus, a model capable of considering both the energy and the sizing parameters in the storage heat-losses formulation is developed as follows.

From Equation (7), it can be seen that once the overall heat-transfer coefficient is identified, the value of the heat-loss coefficient can be expressed as a function of the surface of the storage tank, which in turn is a function of the storage volume. To this end, once the aspect ratio (λ) of the tank is defined, the relation between the surface and the volume of the storage tank can be formulated as follows:

$$\lambda = h_{TES}/\phi_{TES} \tag{9}$$

$$\frac{A_{TES}}{V_{TES}} = \frac{(\pi \phi_{TES} h_{TES} + \frac{\pi}{2}\phi^2_{TES})}{\pi \frac{\phi^2_{TES}}{4} h_{TES}} = \left(\frac{(4\lambda+2)}{2\lambda}\right)\left(\frac{2\pi\lambda}{V_{TES}}\right)^{1/3} \tag{10}$$

and the heat-loss coefficient can be finally expressed as a function of the volume of the storage tank:

$$\theta_{TES,loss} = \left(\frac{U_{TES}\tau}{\rho_{TES}c_{TES}}\right)\left(\frac{(4\lambda+2)}{2\lambda}\right)\left(\frac{2\pi\lambda}{V_{TES}}\right)^{1/3} \tag{11}$$

In this way, the approach of the fixed thermal efficiency of the storage, commonly found in the literature, can be overcome, and the thermal losses of the TES can be expressed as a function of the thermophysical and geometrical properties of the storage tank and medium, and the temperatures of the ambient surroundings and the storage itself.

Finally, the energy balance of the TES reads:

$$H^{i+1}_{u,TES} + H^{i+1}_{res,TES} = H^i_{res,TES} + (1 - \theta_{TES,loss})H^i_{u,TES} - \theta_{TES,loss}H^i_{res,TES} + H^i_{TES,charge} - H^i_{TES,discharge} \tag{12}$$

and the following linear constraints must be considered:

$$H^i_{u,TES} \geq -H^i_{res,TES} \tag{13}$$

$$H^i_{tot,TES} \leq H^i_{tot,TES,max} \tag{14}$$

$$0 \leq H^i_{TES,charge} \leq Q_{TES,charge,max} \cdot \tau \tag{15}$$

$$0 \leq H^i_{TES,discharge} \leq Q_{TES,discharge,max} \cdot \tau \cdot \delta^i_{TES} \tag{16}$$

$$\delta^i_{TES} \leq \left\lfloor \frac{H^{i+1}_{u,TES}}{\rho_{TES} V_{TES} c_{TES} \left(T_{u,TES} - T^i_{amb} \right)} \right\rfloor + 1 \tag{17}$$

In Equations (16) and (17), δ^i_{TES} is a binary variable, equal to 1 if the storage temperature in the i-th timestep is above its operational threshold, and, therefore, energy can be taken from the storage; otherwise, if the temperature of the storage is lower than $T_{u,TES}$, δ^i_{TES} is equal to 0 and no useful energy can be taken from the TES. The introduction of this binary variable allows a more accurate simulation and optimization of the operation of the storage. Indeed, in traditional MILP formulations, the temperature of the storage cannot drop below its operation threshold, which is unrealistic and limits the operation optimization.

Moreover, Equations (15) and (16) define the maximum amounts of energy that can be charged and discharged in a single timestep, respectively. If external plate heat exchangers are considered for both the charging and discharging the storage, those quantities coincide with the capacity of the heat exchangers themselves. Therefore, $Q_{TES,charge,max}$ and $Q_{TES,discharge,max}$ can be considered as design variables and are subject to optimization, once the investment cost of the plate heat exchangers is considered separately from that of the water storage tank.

2.3. Optimization Problem and Algorithm

The objective function to be minimized is the equivalent annual cost (K) of the system, which is calculated over the period of one year and composed of the annualized investment cost for the technologies, I, and the total annual operating cost, O:

$$K = I + O \tag{18}$$

$$I = \sum_{j=1}^{3} \alpha_j \cdot \zeta_j^{\beta_j} \cdot \Gamma_j \tag{19}$$

where ζ_j is the capacity of the j-th component to be installed (expressed in kW for the CHP, the boiler, and the plate heat exchangers, and in m³ for the TES), α_j and β_j are the correlation parameters of the equipment cost as a function of the capacity, and Γ is the capital-recovery factor [19]:

$$\Gamma_j = \frac{r(r+1)^{l_j}}{(r+1)^{l_j} - 1} \tag{20}$$

The total annual operating cost comprises the cost for purchasing electricity and natural gas and the revenue from selling electricity to the grid:

$$O = \sum_{i=1}^{8760} \left[C_F \left(F^i_{CHP} + F^i_B \right) + C_{PE} E^i_P - C_{SE} E^i_S \right] \tag{21}$$

The optimization variables can be distinguished into two main groups: Sizing variables ($P_{CHP,nom}$, V_{TES}, $Q_{B,nom}$, $Q_{TES,charge,max}$, $Q_{TES,dischharge,max}$) and operation variables (E^i_{CHP}, E^i_S, E^i_P, H^i_B, $H^i_{TES,charge}$, $H^i_{TES,discharge}$).

Demand constraints must be satisfied in each *i*-th timestep:

$$H^i_{CHP} + H^i_B + H^i_{TES,discharge} \cdot \eta_{TES,discharge} - H^i_{TES,charge} / \eta_{TES,charge} \geq H^i_d \qquad (22)$$

$$E^i_{CHP} + E^i_P - E^i_S = E^i_d \qquad (23)$$

In addition to the energy balance and constraints for the TES (Equations (12)–(17)) shown in Section 2.1), the following constraints and equations must be considered too:

$$E^i_{CHP} \leq P_{CHP,nom} \delta^i_{CHP} \qquad (24)$$

$$E^i_{CHP} \geq P_{CHP,min} \delta^i_{CHP} \qquad (25)$$

$$H^i_{CHP} = P^i_{CHP} \frac{\eta_{H,CHP}}{\eta_{E.CHP}} \qquad (26)$$

$$F^i_{CHP} = P_{CHP} / \eta_{E,CHP} \qquad (27)$$

$$H^i_B \leq Q_{B,nom} \qquad (28)$$

$$F^i_B = H^i_B / \eta_B \qquad (29)$$

$$\eta_{E,CHP} = \eta_{E,CHP,nom} (a_{E,CHP} L_{CHP} + b_{E,CHP}) \qquad (30)$$

$$\eta_{H,CHP} = \eta_{H,CHP,nom} (a_{H,CHP} L_{CHP} + b_{H,CHP}) \qquad (31)$$

where L_{CHP} is the load factor of the cogeneration unit, defined as $L_{CHP} = F_{CHP} \cdot \eta_{E,CHP,nom} / P_{CHP,nom}$, and δ^i_{CHP} is a binary variable equal to 1 when the CHP is on and equal to 0 when it is off. Equations (30) and (31) linearize the relationship between the electric and thermal efficiencies of the cogeneration unit with respect to its load factor, respectively.

Finally, the overall problem consists in the minimization of the equivalent annual cost:

$$minimize \ \{K\} \qquad (32)$$

which results in a mixed integer non-linear optimization problem, because of the non-linear variations of the unitary cost of the components in relation to their capacity, the part-load effects on CHP efficiencies, and the thermal energy storage model (the dependency of $\theta_{TES,tot,loss}$ and δ^i_{TES} on V_{TES}).

To this aim, the overall problem was decomposed into two levels and the optimization variables distinguished into two categories, as schematically summarized in Figure 2. At the upper level, the synthesis/design problem, which identifies the components that should be included in the energy system and their capacities, is addressed by a genetic algorithm (GA). At the lower level, the optimal operation problem is solved by means of an MILP technique. The two problems are nested in each other and therefore they must be solved simultaneously. For each individual solution (components size) produced by the GA, the optimal annual operation cost is identified by the MILP solver, and, thus, the total EAC is calculated. This procedure is repeated for each individual of each generation produced by the GA, until the stopping criteria is met.

As already mentioned, this decomposition allows the benefits of the MILP formulation to be reaped, while overcoming its main drawbacks. Indeed, the part-load behavior of the CHP unit is considered, the variation of the unitary cost of the components in relation to their size is not neglected, and the effect of the storage volume on its heat-loss coefficient is taken into account.

The optimization was performed by using scripts written in the MATLAB environment. The commercial solver, CPLEX [20], for the MILP optimization and the MATLAB Genetic Algorithm Solver [21] were used. The settings and parameters adopted for the optimization algorithms are shown in Table 1.

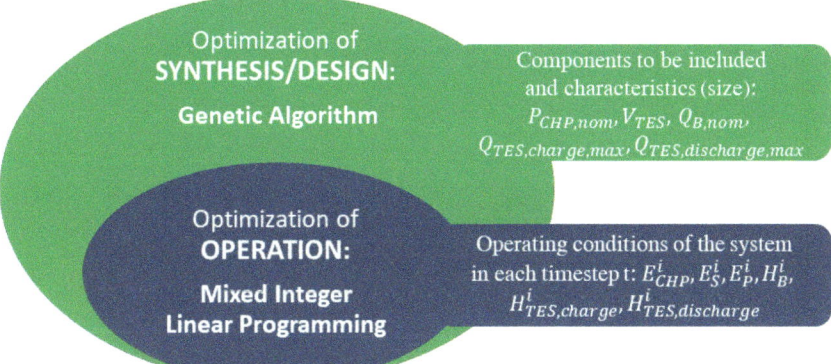

Figure 2. Outline of the bi-level optimization algorithm.

Table 1. Settings and parameters for the optimization algorithms.

MILP		Genetic Algorithm	
MaxIter:	9.2234e+18	PopulationSize:	'50'
Algorithm:	'auto'	EliteCount:	'0.05*PopulationSize'
BranchStrategy:	'maxinfeas'	CrossoverFraction:	0.8000
MaxNodes:	9.2234e+18	MigrationDirection:	'forward'
MaxTime:	1.0000e+75	MigrationInterval:	20
NodeDisplayInterval:	20	MigrationFraction:	0.2000
NodeSearchStrategy:	'bn'	Generations:	'400'
TolFun:	1.0000e-06	StallGenLimit: 50	StallTest: 'averageChange'
TolRLPFun:	1.0000e-06	TolFun:	1.0000e-06
TolXInteger:	1.0000e-05	TolCon:	1.0000e-03

2.4. Rolling-Horizon Approach for the Decomposition of the Operation Problem

As stated above, the introduction of the storage in the energy system makes the operation optimization problem "dynamic", because of the dependency between the optimization variables of adjacent timesteps. Therefore, once the sizes of the components are fixed, the problem of the minimization of the operational costs should be solved simultaneously for the whole optimization period (e.g., one year with an hourly resolution). This results in a very large number of variables and constraints, thus making the problem very challenging from the computational point of view [14]. To tackle this issue, the rolling-horizon procedure can be adopted, which consists in dividing the investigated period into smaller periods and optimizing each subproblem in sequence [22,23].

The rolling-horizon approach is schematically represented in Figure 3. The length of each sub-period of optimization is called "prediction horizon". Once the solution for a certain sub-period is found, it can be implemented for one or a few timesteps (the total duration of which is called the "control horizon"). Then, the problem is solved for the following sub-period, and so on, until the problem is solved for the whole optimization period.

Therefore, the original operation problem is divided in $N = 8760/k$ subproblems, which optimize the operation in p consecutive timesteps:

$$min\left\{\sum_{i=n\cdot c+1}^{i=n\cdot c+p}\left[C_F\left(F^i_{CHP} + F^i_B\right) + C_{PE}E^i_P - C_{SE}E^i_S\right]\right\}$$ (33)

where n goes from 0 to $(N-1)$, k is the length of the control horizon, and p is the length of the prediction horizon (both measured in the number of timesteps). All the other constraints and relationships considered for the whole operation optimization problem must still be considered and remain the same.

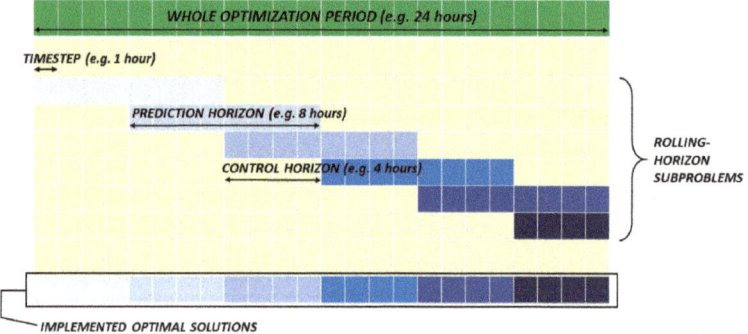

Figure 3. Rolling-horizon approach.

Two parameters must be set in the rolling-horizon decomposition of the operation problem: The length of the prediction and control horizons. As the former increases and the latter decreases, the overall solution becomes more and more accurate, until it coincides with the exact solution (i.e., the solution of the whole optimization period considered at once). On the other hand, as the prediction horizon becomes shorter and the control horizon gets longer, the computational time decreases. Therefore, a compromise must be achieved between these two conflicting objectives.

Moreover, as the size of the thermal energy storage increases, the minimum length of the rolling-horizon so that the solution is not different from the whole-period-at-once solution increases too [24]. On the contrary, when no storage is included in the system, the problem becomes "static" and each timestep can be solved separately [25].

3. Case Study Presentation

3.1. Energy Demand

The case study used for testing the methodology was chosen from the commercial reference buildings database [26] of the US Department of Energy (DOE). A secondary school located in San Francisco (California) and constructed after the year of 1980 was selected. Data about the energy load demands (whose hourly values are shown in Figure 4) were calculated by means of EnergyPlus simulation software [27] and then imported and processed in MATLAB. Hourly temperatures of the typical meteorological year of San Francisco, which are shown in Figure 5, were considered.

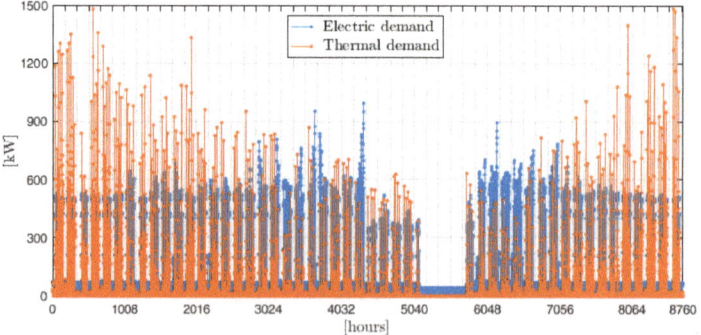

Figure 4. Energy load demand of the case study.

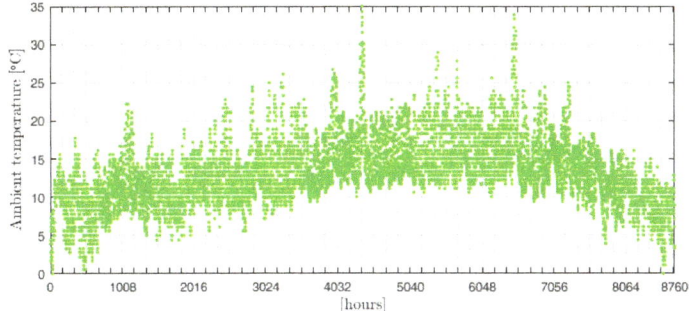

Figure 5. Hourly ambient temperature for the case study [27].

3.2. The Energy System: Technical and Economic Characterization

In this section, the values of the parameters adopted to characterize the energy system optimized in the case study are shown. Table 2 shows the technical characteristics and the thermophysical properties, while Table 3 summarizes the economic parameters. The lifetime was considered to be equal to 20 years for all the considered technologies.

Table 2. Technical characteristics and thermophysical properties.

Parameter	Value	Parameter	Value
$\eta_{E,CHP,nom}$	38.5% [8]	η_B	80% [28]
$\eta_{H,CHP,nom}$	34.4% [8]	U_{TES}	$0.5 \frac{W}{m^2 K}$ [16]
$a_{E,CHP}$	1.1260 [8]	ρ_{TES}	992 kg/m^3
$b_{E,CHP}$	−0.1260 [8]	c_{TES}	4.186 kJ/kgK
$a_{H,CHP}$	0.8253 [8]	$\eta_{TES,charge}$	96% [28]
$b_{H,CHP}$	0.1747 [8]	$\eta_{TES,discharge}$	96% [28]

Table 3. Prices of energy carriers and investment costs parameters.

Parameter	Value	Parameter	Value
C_F	40 €/MWh [8]	β_B	0.7627 [28]
C_{PE}	150 €/MWh [8]	α_{TES}	100.0 [16]
C_{SE}	50 €/MWh [8]	β_{TES}	1 [16]
α_{CHP}	15460.0 [8]	$\alpha_{TES,charge/discharge,max}$	800 [19]
β_{CHP}	0.7247 [8]	$\beta_{TES,charge/discharge,max}$	0.6 [19]
α_B	345.9 [28]	r	2% [29]

The maximum allowed temperature of the storage was set to 95 °C, while the minimum usable temperature was 60 °C. Moreover, to minimize TES heat losses, the shape of the cylindrical tank was chosen, such as to minimize its surface. Thus, the aspect ratio was set to equal 1, and the heat-loss coefficient becomes:

$$\theta_{TES,loss} = \frac{6 U_{TES} \tau}{\rho_{TES} c_{TES}} \left(\frac{\pi}{4 V_{TES}} \right)^{\frac{1}{3}} \tag{34}$$

Moreover, Table 4 shows the search space of the design variables in the optimization problem.

Table 4. Range of values of the design variables.

Parameter	Minimum Value	Maximum Value
$P_{CHP,nom}$ [kW]	200	1000
V_{TES} [m^3]	0	50
$Q_{B,nom}$ [kW]	500	1500
$Q_{TES,charge,max}$ [kW]	0	2000
$Q_{TES,discharge,max}$ [kW]	0	2000

4. Results and Discussion

In this section, the results are presented as follows. First, a preliminary parametric analysis is performed to identify the value of the rolling-horizon parameters to be adopted in the optimization. Then, the proposed methodology is applied in the case study and the results are shown and discussed.

4.1. Tuning of the Rolling-Horizon Parameters

The rolling-horizon technique adopted for the decomposition of the optimal operation problem calls for the choice of the value of both the control (k) and prediction (p) horizon, which, as stated above, should be made based on a compromise between the accuracy of the solution and the computational time. Therefore, a parametric analysis aimed at assessing the impact of k and p on the annual operating cost—and consequently on the accuracy of the optimal control solution—as well as on the computational time was carried out.

To this end, several simulations were run, by varying the values of the parameters, p and k, within the range of 3–36. Table 5 shows the size of the system components considered in this analysis. The storage capacity was chosen equal to its maximum value considered in the design stage, in order to ensure a conservative assessment of the impact of the rolling-horizon parameters on the accuracy of the solution. Indeed, it has been shown by [24] that the larger the capacity of the storage, the longer the prediction horizon should be so that the solution of the decomposed problem coincides with the solution of the original problem (i.e., the operation problem solved considering the whole-time horizon at once). In this way, when smaller storage capacities are considered in the optimal sizing problem, no deterioration of the accuracy of the solution will occur compared to that of this parametric analysis.

Table 5. Size of the components adopted in the rolling-horizon parametric analysis.

Parameter	$P_{CHP,nom}$	V_{TES}	$H_{B,nom}$	$H_{TES,charge,max}$	$H_{TES,discharge,max}$
Value	400 kW	50 m^3	1000 kW	2000 kW	2000 kW

The results are shown in Figure 6, where the annual operating cost is reported as a function of both the control and the prediction horizon values. Once the length of the control horizon was fixed, the operating cost reduces as the prediction horizon increases, and a saturation effect can be observed. When the prediction horizon increases from 24 to 36 h, with a 3-h control horizon, the cost reduction is lower than 0.1%. Indeed, a longer prediction horizon cannot be exploited because of the limited capacity of the storage. On the other hand, for a given value of p, the operating cost increases as the control horizon increases. Nonetheless, with p equal to 24 h, no relevant differences are observed when k ranges from 3 to 12 h. In fact, if the prediction horizon is long enough, longer control horizons can be adopted, thus reducing the overall number of the optimization sub-problems. Therefore, the values of $k = 12$ and $p = 24$ were chosen, since these values represent a good compromise between the solution accuracy and the computational time.

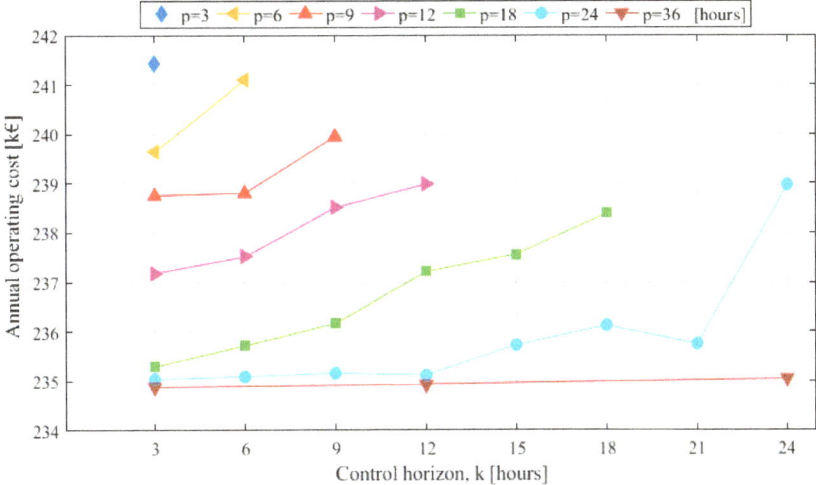

Figure 6. Impact of the prediction and control horizon lengths on the annual operating cost.

4.2. Optimization Results

Table 6 summarizes the main results from the optimization. The optimal size of each component as well as the hourly optimal operation throughout the year were identified. In Figure 7, the annual electric and thermal load shares for the optimal configuration are shown. Both the electric and thermal annual demand are directly met by the CHP for around 70%, while the TES contributes 16% to cover the heat demand.

Table 6. Results from case-study optimization.

Parameter	Value	Parameter	Value
$P_{CHP,nom}$	400 kW	$Q_{TES,discharge,max}$	425 kW
V_{TES}	12.5 m^3	EAC	317.9 k€
$Q_{B,nom}$	1000 kW	O	237.7 k€
$Q_{TES,charge,max}$	275 kW	I	80.2 k€

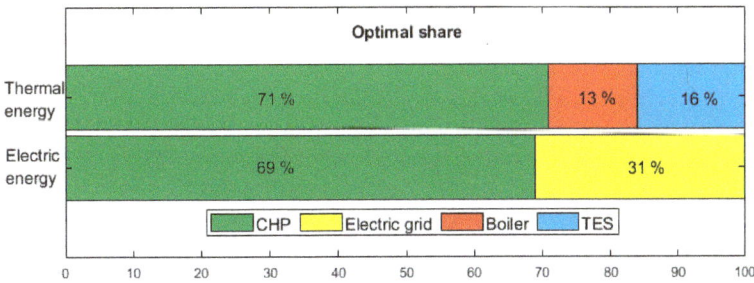

Figure 7. Optimal electric and thermal load sharing.

Figures 8–10 show examples of how the optimized energy system works and what kind of detailed outputs are available from the simulations. Figure 8 shows how the electric demand is met in a typical week. During the day, the CHP operates at full load, covering an average of 80% of the electric demand, while the remaining demand is met by purchasing electricity from the grid. During the night and the weekend, the CHP is generally off, since the electric demand is lower than its minimum capacity, and the electricity is bought from the grid. It is interesting to note that the optimal operation entails a very limited exchange with the power grid, thus avoiding stressing the stability and management of the electricity network.

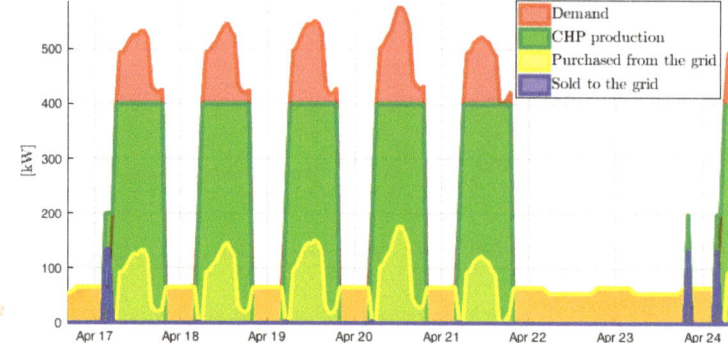

Figure 8. Optimal operation: electric demand and load sharing in a typical week.

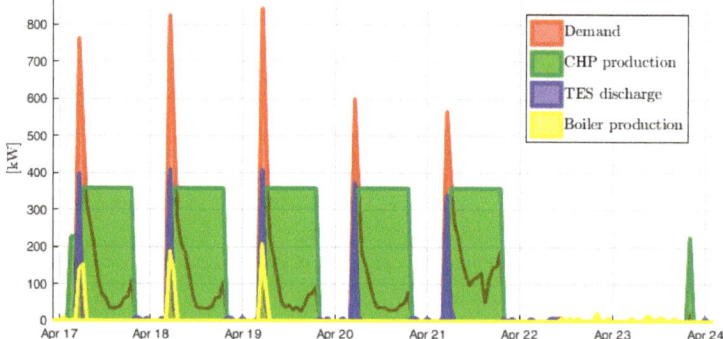

Figure 9. Optimal operation: thermal demand and load sharing in a typical week.

Figure 9 shows the same kind of results for the thermal demand. The CHP production completely covers the demand for most of the week, except when the morning peak demands occur. In those cases, the storage is discharged, and the boiler meets the remaining heat demand. On the other hand, in the evening, before the CHP is turned off, the TES is charged so that thermal energy is available in the morning. This behavior can be clearly seen in Figure 10.

As a result, the scheduling of the CHP is mainly driven by the electric demand, while the thermal output of the CHP unit is generally larger than the heat demand. Nonetheless, the thermal storage allows a fraction of the exceeding CHP thermal production to be saved, which is used to shave the heat peak demand in the early hours of the morning, when the start up of the whole heating distribution and emission system occurs. In this way, the boiler production is minimized, even though the heat capacity of the CHP is well below the thermal peak demand.

Finally, Figure 11 shows how the equivalent temperature of the storage varies throughout the year. Most of the time, its value fluctuates between 95 °C, when the storage is fully charged, and 60 °C, when the storage has just been completely discharged. Moreover, during the weekends, when there is no usage of the storage, the equivalent temperature drops below 60 °C as a result of the heat losses. This behavior is particularly emphasized during the month of August, when the school is closed and there is no thermal demand for a long period. Thus, it is more convenient to let the storage discharge due to thermal losses, rather than maintaining it at the usable temperature. As already explained, traditional approaches do not allow this behavior to be obtained, and thanks to the proposed formulation, a more comprehensive simulation of the TES has been enabled.

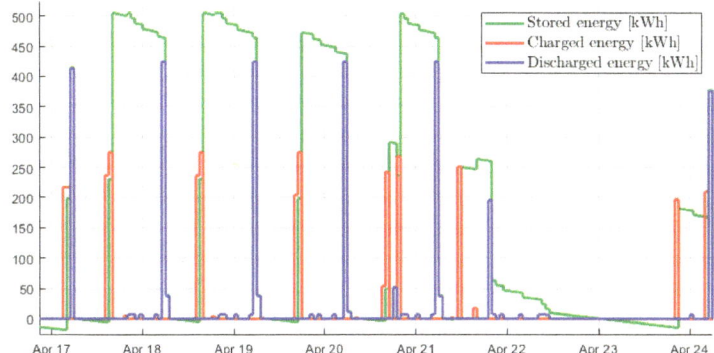

Figure 10. Optimal operation: TES stored, charged, and discharged energy in a typical week.

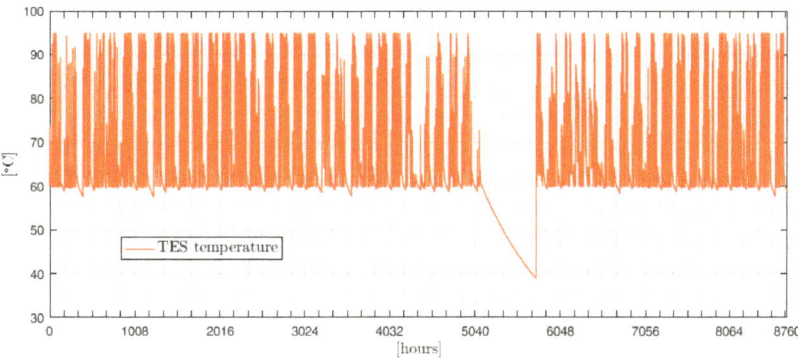

Figure 11. Hourly equivalent TES temperature during the year.

5. Conclusions

A comprehensive methodology for the optimal design of cogeneration systems with integrated thermal energy storage was proposed in this paper. The integrated optimization problem of the sizing of the equipment and operational strategy was decomposed into two levels to take into account the part-load behavior of the cogeneration unit, variation of the unitary cost of the components in relation to their size, and the effect of the storage volume on its thermal losses. The optimal operation problem was formulated so as to exploit mixed integer linear programming solvers, while the optimal sizing was tackled by means of a genetic algorithm.

A novel mixed integer linear formulation for the thermal energy storage was proposed. Both "static" and "dynamic" heat losses were considered, and the heat-loss coefficient was defined as a function of the geometrical and thermophysical properties of the storage. The introduction of a binary variable allowed a more comprehensive simulation of the operation of the storage. Furthermore, a rolling-horizon procedure was presented to reduce the computational time of the optimization problem, while preserving the quality of the results.

The methodology was tested in a case study, namely a secondary school located in San Francisco, California. The optimal size and operation of the cogeneration system with thermal energy storage were identified. Results from the optimization were presented and discussed. A parametric analysis led to the identification of the optimal values of the rolling-horizon parameters: 24 h for the prediction horizon and 12 h for the control horizon were chosen as a compromise between the accuracy of the results and the computational time. In the case study, the cogeneration unit directly covers around 70% of both the electric and thermal annual demand, while 16% of the latter is met by the storage.

Moreover, the results showed how the proposed formulation allows control strategies for the thermal energy storage to be exploited that cannot be taken into account with conventional simplified formulations. Indeed, instead of keeping the state of charge of the storage always above its minimum operational threshold, the controller allows it to drop below its threshold value. This behavior is particularly marked during extended periods of low or nil demand, as in August, when the school is closed and the temperature of the storage drops significantly below its operational threshold.

Future research may focus on: More complex multi-energy systems (modular cogeneration, heat pumps, renewable energy technologies), the effect of uncertainties on the optimal design, and multi-objective optimization with further criteria (environmental, energetic, exergetic indicators).

Author Contributions: Conceptualization, L.U., F.D. and D.T.; Data curation, L.U. and F.D.; Methodology, L.U. and F.D.; Software, L.U. and F.D.; Supervision, D.T.; Writing—original draft, L.U. and F.D.; Writing—review & editing, L.U., F.D. and D.T.

Funding: This research received no external funding.

Conflicts of Interest: The authors declare no conflict of interest.

Nomenclature

Acronyms

CHP	Combined Heat and Power
MILP	Mixed Integer Linear Programming
MINLP	Mixed Integer Non-Linear Programming
TES	Thermal Energy Storage

Parameters and Variables

A	Tank surface $[m^2]$
a	Slope of the CHP efficiency curve [-]
b	Intercept of the CHP efficiency curve [-]
C	Cost [/kWh]
c	Specific heat [kJ/kgK]
E	Electric energy [kWh]
F	Energy content of the consumed fuel [kWh]
H	Thermal energy [kWh]
h	Tank height[m]
I	Investment cost []
K	Equivalent annual cost []
k	Control horizon length $[h]$
L	CHP load factor [-]
l	Lifetime duration of the technology [years]
N	Number of rolling-horizon optimization problems
n	*n-th* rolling-horizon optimization problem
O	Total annual operating cost []
P	Electric power [kW]
p	Prediction horizon length [h]
Q	Thermal power [kW]
r	Interest rate[%]
T	Temperature [°C]
U	Heattransfer coefficient $[kW/m^2K]$
V	Volume $[m^3]$

Greek Letters

α	Coefficient of the equipment cost curve [-]
β	Exponent of the equipment cost curve [-]

Γ	Capital recovery factor [-]
δ	Binary variable
η	Storage efficiency [-]
λ	Aspect ratio [-]
θ	Heat-loss coefficient [-]
ξ	Component capacity[kW] or $\left[m^3\right]$
φ	Tank diameter [m]
ρ	Density $\left[kg/m^3\right]$
τ	Timestep length[s]

Subscripts

amb	Ambient
B	Boiler
d	Demand
E	Electric energy
F	Fuel
H	Thermal energy
max	Maximum
min	Minimum
nom	Nominal
P	Purchased
PE	Electricity purchased from the grid
res	Residual
S	Sold
SE	Electricity sold to the grid
tot	Total

Superscripts

i	*i-th* timestep

References

1. Isa, N.M.; Tan, C.W.; Yatim, A.H.M. A comprehensive review of cogeneration system in a microgrid: A perspective from architecture and operating system. *Renew. Sustain. Energy Rev.* **2018**, *81*, 2236–2263. [CrossRef]
2. Gimelli, A.; Muccillo, M. The key role of the vector optimization algorithm and robust design approach for the design of polygeneration systems. *Energies* **2018**, *11*. [CrossRef]
3. Zhang, G.; Cao, Y.; Cao, Y.; Li, D.; Wang, L. Optimal energy management for microgrids with combined heat and power (CHP) generation, energy storages, and renewable energy sources. *Energies* **2017**, *10*, 1288. [CrossRef]
4. Franco, A.; Versace, M. Optimum sizing and operational strategy of CHP plant for district heating based on the use of composite indicators. *Energy* **2017**, *124*, 258–271. [CrossRef]
5. Gimelli, A.; Muccillo, M.; Sannino, R. Optimal design of modular cogeneration plants for hospital facilities and robustness evaluation of the results. *Energy Convers. Manag.* **2017**, *134*, 20 31. [CrossRef]
6. Beihong, Z.; Weiding, L. An optimal sizing method for cogeneration plants. *Energy Build.* **2006**, *38*, 189–195. [CrossRef]
7. Aviso, K.B.; Tan, R.R. Fuzzy P-graph for optimal synthesis of cogeneration and trigeneration systems. *Energy* **2018**, *154*, 258–268. [CrossRef]
8. Urbanucci, L.; Testi, D. Optimal integrated sizing and operation of a CHP system with Monte Carlo risk analysis for long-term uncertainty in energy demands. *Energy Convers. Manag.* **2018**, *157*. [CrossRef]
9. Yokoyama, R.; Takeuchi, S.; Ito, K. Thermoeconomic analysis and optimization of a gas turbine cogeneration units by a systems approach. In Proceedings of the ASME Turbo Expo, Barcelona, Spain, 6 May 2006.
10. Fang, T.; Lahdelma, R. Optimization of combined heat and power production with heat storage based on sliding time window method. *Appl. Energy* **2016**, *162*, 723–732. [CrossRef]
11. Kuang, J.; Zhang, C.; Li, F.; Sun, B. Dynamic Optimization of Combined Cooling, Heating, and Power Systems with Energy Storage Units. *Energies* **2018**, *11*, 2288. [CrossRef]

12. Elsido, C.; Bischi, A.; Silva, P.; Martelli, E. Two-stage MINLP algorithm for the optimal synthesis and design of networks of CHP units. *Energy* **2017**, *121*, 403–426. [CrossRef]

13. Fuentes-Cortés, L.F.; Ponce-Ortega, J.M.; Nápoles-Rivera, F.; Serna-González, M.; El-Halwagi, M.M. Optimal design of integrated CHP systems for housing complexes. *Energy Convers. Manag.* **2015**, *99*, 252–263. [CrossRef]

14. Urbanucci, L. Limits and potentials of Mixed Integer Linear Programming methods for optimization of polygeneration energy systems. *Energy Procedia* **2018**, *148*, 1199–1205. [CrossRef]

15. Marquant, J.F.; Evins, R.; Carmeliet, J. Reducing computation time with a rolling horizon approach applied to a MILP formulation of multiple urban energy hub system. *Procedia Comput. Sci.* **2015**, *51*, 2137–2146. [CrossRef]

16. Steen, D.; Stadler, M.; Cardoso, G.; Groissböck, M.; DeForest, N.; Marnay, C. Modeling of thermal storage systems in MILP distributed energy resource models. *Appl. Energy* **2015**, *137*, 782–792. [CrossRef]

17. Combined Heat and Power & Trigeneration Solutions, Wartsila. Available online: https://www.wartsila.com/docs/default-source/power-plants-documents/downloads/brochures/chp-2017.pdf?sfvrsn=6c60f145_4 (accessed on 31 January 2019).

18. Omu, A.; Hsieh, S.; Orehounig, K. Mixed integer linear programming for the design of solar thermal energy systems with short-term storage. *Appl. Energy* **2016**, *180*, 313–326. [CrossRef]

19. Bejan, A.; Tsatsaronis, G.; Moran, M. *Thermal Design and Optimization*, 1st ed.; Wiley India: New Delhi, India, 1995.

20. IBM ILOG CPLEX Optimizer. Available online: http//www-01.ibm.com/software/integration/Optim (accessed on 30 December 2018).

21. MATLAB Genetic Algorithm Solver. Available online: https://it.mathworks.com/discovery/genetic-algorithm.html (accessed on 30 December 2018).

22. Ommen, T.; Markussen, W.B.W.B.; Elmegaard, B. Comparison of linear, mixed integer and non-linear programming methods in energy system dispatch modelling. *Energy* **2014**, *74*, 109–118. [CrossRef]

23. Bischi, A.; Taccari, L.; Martelli, E.; Amaldi, E.; Manzolini, G.; Silva, P.; Campanari, S.; Macchi, E. A rolling-horizon optimization algorithm for the long term operational scheduling of cogeneration systems. *Energy* **2018**. [CrossRef]

24. D'Ettorre, F.; Conti, P.; Schito, E.; Testi, D. Model predictive control of a hybrid heat pump system and impact of the prediction horizon on cost-saving potential and optimal storage capacity. *Appl. Therm. Eng.* **2019**, *148*, 524–535. [CrossRef]

25. Urbanucci, L.; Testi, D.; Bruno, J.C. An operational optimization method for a complex polygeneration plant based on real-time measurements. *Energy Convers. Manag.* **2018**, *170*. [CrossRef]

26. Office of Energy Efficiency and Renewable Energy. Commercial Reference Buildings. Available online: https://www.energy.gov/eere/buildings/commercial-reference-buildings (accessed on 30 December 2018).

27. EnergyPlus Simulation Software. Available online: https://energyplus.net (accessed on 30 December 2018).

28. Marquant, J.F.; Evins, R.; Bollinger, L.A.; Carmeliet, J. A holarchic approach for multi-scale distributed energy system optimisation. *Appl. Energy* **2017**, *208*, 935–953. [CrossRef]

29. Vögelin, P.; Koch, B.; Georges, G.; Boulouchos, K. Heuristic approach for the economic optimisation of combined heat and power (CHP) plants: Operating strategy, heat storage and power. *Energy* **2017**, *121*, 66–77. [CrossRef]

Article

Intertemporal Static and Dynamic Optimization of Synthesis, Design, and Operation of Integrated Energy Systems of Ships

George N. Sakalis, George J. Tzortzis and Christos A. Frangopoulos *

School of Naval Architecture and Marine Engineering, National Technical University of Athens, 15780 Zografou, Greece; georgesakalis@gmail.com (G.N.S.); tzortzis_georg@hotmail.com (G.J.T.)
* Correspondence: caf@naval.ntua.gr; Tel.: +30-210-772-1108

Received: 3 January 2019; Accepted: 1 March 2019; Published: 7 March 2019

Abstract: Fuel expenses constitute the largest part of the operating cost of a merchant ship. Integrated energy systems that cover all energy loads with low fuel consumption, while being economically feasible, are increasingly studied and installed. Due to the large variety of possible configurations, design specifications, and operating conditions that change with time, the application of optimization methods is imperative. Designing the system for nominal conditions only is not sufficient. Instead, intertemporal optimization needs to be performed that can be static or dynamic. In the present article, intertemporal static and dynamic optimization problems for the synthesis, design, and operation (SDO) of integrated ship energy systems are stated mathematically and the solution methods are presented, while case studies demonstrate the applicability of the methods and also reveal that the optimal solution may defer significantly from the solutions suggested with the usual practice. While in other works, the SDO optimization problems are usually solved by two- or three-level algorithms; single-level algorithms are developed and applied here, which tackle all three aspects (S, D, and O) concurrently. The methods can also be applied on land installations, e.g., power plants, cogenerations systems, etc., with proper modifications.

Keywords: energy systems; integrated ship energy systems; synthesis; design and operation optimization; intertemporal optimization; dynamic optimization

---

1. Introduction

The continuously increasing need for more efficient fuel utilization and reduction of the environmental pollution leads to the construction of integrated energy systems of increasing complexity that can produce several forms of energy, while at the same time are economically feasible. In a conventional design procedure of an energy system (power plant, propulsion plant, cogeneration system, etc.), the aim is usually to design a system that "works", i.e., a system that delivers the required energy products under certain constraints. However, the scarcity of physical and economic resources and the deterioration of the environment make it necessary to build a system that not only "works", but also it is the "best"; such a system can be designed by the application of formal optimization procedures.

During its lifetime, an energy system may operate under various conditions (load, weather state, etc.). Optimization at one set of conditions only (e.g., the "design" or "nominal" point) does not lead, in general, to the best utilization of physical and economic resources. Therefore, intertemporal optimization is needed [1], which takes into consideration the various operating conditions that the system is expected to encounter during its life time.

The optimization of an energy system can be considered at three levels—synthesis, design and operation [1,2]—which are interrelated (Figure 1). Therefore it is not correct, in general, to optimize each level in isolation from the others.

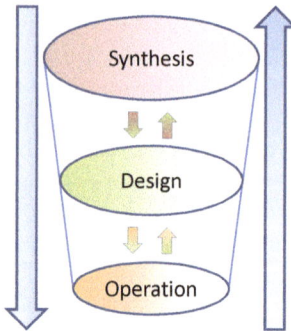

Figure 1. The three interrelated levels of optimization.

In the marine industry, the synthesis and design of energy systems is usually based on previous experience of the designer and rule-of-thumb criteria. Furthermore, the system is often designed considering full load operation only, while its off-design performance is assessed only after the system is fixed. However, the high multitude of available alternative configurations makes the study of all combinations one bvy one and the selection of the best a rather daunting task. Furthermore, it must be taken into account that the operating conditions are highly varying with time and that all modes of operation of the energy system should be considered. Thus, past experience alone is not sufficient for determining the optimum design and operation, and the development and application of mathematical optimization techniques for the synthesis, design, and operation optimization (SDOO) of marine energy systems has nowadays become a necessity with significant engineering importance.

Engineering optimization problems, involving time dependencies of the operation of the system studied, can be characterized as static or dynamic [1]. In static optimization, the values of variables that give the best value (minimum or maximum) to an objective function are requested. In dynamic optimization, the variables, the functions, and the parameters are, in general, time-dependent; therefore, the variables as functions of time that give the best value (minimum or maximum) to an objective function are requested.

It can be said that if the various modes of operation are independent of each other, i.e., the operation in a mode does not affect and is not affected by the operation in any other mode, then we have an intertemporal static optimization problem. If, however, a direct interdependency among operating modes exists (as is the case, e.g., of a system with energy storage or of a ship that needs to reach her destination at a specified time encountering various weather conditions along the way), then intertemporal dynamic optimization is needed.

In the present section, characteristic publications related to the SDOO of energy systems (including systems on ships) are presented, while a detailed literature review is beyond the limits of this article.

Several methodologies have been developed for the SDOO of energy systems [2]. Characteristic works are mentioned in the following. Pelster et al. [3] a thermoeconomic–environomic methodology that performs synthesis and design optimization via a single-level approach that utilizes a Struggle Genetic Algorithm (Str-GA) is presented and applied to a cogeneration combined cycle power plant. Mussati et al. [4] examine the synthesis and design optimization of a dual purpose desalination plant while a superconfiguration (the word 'superstructure', which is often used for systems on land, is not used here, because it has a different meaning on ships) is used in order to model all available synthesis options and a MINLP problem is stated. Another example where a superconfiguration is used and the synthesis and design aspects of the system are tackled at a single level can be found in Sun et al. [5], where a site utility system is optimized for cost minimization. Calise et al. [6] investigate the optimal synthesis and design of a hybrid solid oxide fuel cell-gas turbine power plant. Full load operation is considered and again a single-level approach for the synthesis–design levels is adopted while a genetic algorithm is used for the solution of the optimization problem.

However, a single-level approach is not always preferred for the solution of the optimization problem. In [7] the SDOO is performed on a system consisting of a set of Rankine cycles that absorb and release heat at different temperature levels; part of the absorbed heat is used for electricity production. For the solution of the problem, a bilevel hybrid algorithm is applied in which the upper level is constituted of the synthesis of the system and optimized via an evolutionary algorithm, while the lower level tackles the system design characteristics and is optimized via a traditional SQP algorithm. A common characteristic of all aforementioned studies is that only a single mode is considered for the operation of the system. Thus, optimization at the operational level is meaningless and only the synthesis and design levels are optimized. Also, since only one mode of operation is considered, the time dependency of the operation is not taken into account.

The earliest publications that address, in a concise mathematical manner, the SDOO of energy systems including time dependencies at the operational level, thus forming intertemporal SDOO problems, can be found in references [8–10]. In these studies, the optimal SDO of a cogeneration system supplying a process plant with thermal and electrical energy is investigated. The time horizon of the problem is divided into independent periods of steady state operation, thus formulating an intertemporal static SDOO problem, while a method called Intelligent Functional Approach (IFA) is used to analyze the system as a set of interrelated units [8]. The problem is solved by a three-level algorithm, which employs an iterative procedure among the three (SDO) levels of optimization until the global optimal for the objective function is found. In an application example, the internal economy of the system allows for the three-level procedure described previously to be simplified by combining the levels of synthesis and design into a single one [9]. In another example, the Thermoeconomic Functional Approach (TFA) is applied in order to divide the system into a set of interrelated units, while periods of steady state operation independent of each other are considered [10]. Again, a bi-level algorithm is preferred, in which the optimal operation is determined at the lower level while the synthesis and design are tackled simultaneously at the upper level.

Other intertemporal static SDOO studies include those in which the Local Global Optimization (LGO) and Iterative Local Global Optimization (ILGO) algorithms are implemented [11]. In LGO, the system is separated into a set of units and a nested set of optimizations is performed, with the unit level problems embedded within the problem of the overall system optimization. Based on LGO, the ILGO algorithm additionally uses shadow prices (derivatives of the optimal value of a function with respect to certain variables) to intelligently move towards the system level optimum.

Munoz and Von Spakovsky discuss the theory behind LGO and ILGO [11] and proceed with SDO optimization of a turbofan engine connected to an environmental control system for a military aircraft via the ILGO algorithm [12]. The ILGO optimization algorithm is also applied for SDOO of aircraft energy systems where a bi-level optimization approach is implemented [13], and for SDOO of a stationary total energy system (TES) for residential/commercial applications, which is based on proton exchange membrane fuel cell (PEMFC) [14]. Oyarzabal et al. [15] the optimal SDO of a PEM fuel cell cogeneration system is investigated and the LGO algorithm is utilized. Also, the trip of a military aircraft that includes many modes of operation (take-off, flight, and landing) is studied under the scope of optimizing the SDO of its energy system [16]. Transient operation of several system components is also considered and both LGO and ILGO algorithms are applied.

Not all studies involve the decomposition of the system in units via special decomposition techniques such as IFA or LGO. Olsommer et al. [17], the optimal SDO of a waste incineration system with cogeneration and a gas turbine topping cycle is under investigation for minimization of the present worth cost of the system over its entire economic lifetime. The time horizon of the system operation is divided into independent periods of steady state operation and a bilevel (synthesis/design and operation) solution procedure is applied with the utilization of an evolutionary algorithm (Str-GA).

The HEATSEP method, initially developed in order to study the heat transfer interactions in separate from the rest of the energy system [18,19], has been further developed for the SDOO of energy systems and is given the name SYNTHSEP [7,20]. It operates at two levels: The upper level, which uses an evolutionary algorithm, automatically synthesizes a basic configuration of the system consisting of elementary thermodynamic cycles and determines its intensive design parameters. The lower level, which uses a sequential quadratic programming (SQP) algorithm, determines the optimal mass flow rates of the system taking into consideration the heat transfer feasibility constraints. The method is applied for the optimization of an organic Rankine cycle (ORC) system.

Regarding the domain of marine energy systems, Dimopoulos et al. [21] the overall energy system of a cruise liner, with various technological alternatives for the synthesis, is considered and optimized for cost minimization. Time varying operational requirements are considered and an intertemporal static SDOO problem is formulated, while two levels of optimization are considered: a synthesis-design outer level and an operation inner level. The same approach is also applied for the case of a Liquefied Natural Gas (LNG) carrier [22]. In both cases a Particle Swarm Optimization (PSO) algorithm is used for the solution of the problem. In another study, the SDOO of an organic Rankine cycle system for applications on ships is performed [23]. The intertemporal static SDOO problem is tackled by a hybrid numerical scheme that combines a Genetic Algorithm (GA) and the SQP algorithm.

As mentioned in the preceding, most of the works apply a bi-level procedure for the solution of the SDOO problem, which is based on the assumption that the conditions of decomposition are strictly applicable. However if they are not, there is a danger of missing (i.e. not identifying) optimal solutions. In order to eliminate such a danger, a single-level approach has been developed and presented in Sakalis and Frangopoulos [24]: the operation optimization problem is solved for all time intervals simultaneously. At the end of this procedure, the optimal synthesis and design specifications of the system are derived. It is written in the same publication: "The single-level approach for the SDOO of systems is best suited for intertemporal optimization, as it inherently takes into account the effects that all the various operating conditions have on the synthesis of the system and the design characteristics of its components simultaneously. It also conversely takes into account the fact that the synthesis of the system and the design characteristics of the components define the possibilities for the operating options at all the instances of time during which the system is going to operate".

However, in many cases either the operating modes are not independent of each other or the whole period of operation cannot be decomposed in distinct and independent modes. In such cases, an intertemporal dynamic SDOO problem is formulated.

Very few studies on intertemporal dynamic SDOO problems can be found in the literature. Rancruel [25] and Rancruel and Von Spakovsky [26], the SDOO of an auxiliary power unit based on a solid oxide fuel cell is performed with the life cycle cost of the system as the objective function. Transient operation of certain components is considered and for the solution of the problem, the DILGO algorithm—which is the dynamic version of the ILGO algorithm—is applied. DILGO is also applied in Wang et al. [27], where the dynamic SDOO of a PEMFC energy system is performed. The same PEMFC system is examined in Kim et al. [28,29] with the additions of stochastic modeling and uncertainty analysis methodologies in order to calculate the uncertainties on the system outputs.

Arcuri et al. [30], an intertemporal dynamic SDOO problem of a small size trigeneration plant is tackled. Two levels of optimization are considered and a bi-level optimization algorithm is applied. Buoro et al. [31], the optimal SDO for advanced energy supply system for a standard and a domotic home is investigated. The annual cost minimization is set as the objective function and the whole year of operation is modeled via 12 characteristic days of operation. A superconfiguration is used and the problem is solved at a single level. Other studies that also employ a superconfiguration and formulate a single level approach to the problem can be found in Petruschke et al. [32], where intertemporal dynamic SDOO is performed in renewable energy systems via a hybrid method that exploits synergies between heuristic and optimization based approaches, Goderbauer et al. [33] where a decentralized energy supply system is optimized for an appropriate cost function via adaptive discretization algorithm and in Zhu et al. [34], where a large scale combined heat and power (CHP) system is examined. Finally, another noteworthy study can be found in Fuentes-Cortés et al. [35], where multiobjective intertemporal dynamic SDOO that encompasses economic, environmental and safety aspects, is performed for residential CHP systems.

Considering the field of marine engineering, no studies of intertemporal dynamic SDOO of energy systems of ships have been found.

In the present article, intertemporal static and intertemporal dynamic SDOO of energy systems of ships are performed. For each case, the optimization problem is stated in a suitable mathematical framework and the modeling of the energy system components is briefly presented. Also, for each problem, the solution method applied is described in brief and a numerical example is presented, which demonstrates the applicability of the method and also reveals that the optimal solution may defer significantly from the solutions suggested in the usual practice. It is important to highlight that the problems are formulated and solved in an appropriate manner so that the SDO aspects of optimization are treated simultaneously via a single level approach.

It is noted that the general mathematical statement as well as a collection of several solution approaches for the intertemporal static and dynamic SDOO problems can be found in Frangopoulos [1].

2. Intertemporal Static SDOO of an Energy System of Ship with Gas Turbines as Main Engines

2.1. Studies on Gas Turbines as Ship Propulsion Engines

Due to the relatively low thermal efficiency of gas turbines in comparison with Diesel engines, which are most usually installed on ships, heat recovery may be of utmost importance, in order for a system to be an economically viable alternative to Diesel engines. Furthermore, the high flow rate and temperature of the exhaust gases make gas turbines ideal for combined cycle systems.

In the present work, a novel approach of the SDO optimization problem, initially appearing in [24] for the case of integrated ship energy systems with Diesel main engines is extended to the case of gas turbine systems, as it is considered that the utilization of gas turbines on ships is an important subject attracting a continuous research interest.

A thorough review of the possibility of using gas turbine-based combined cycles on merchant ships has been reported in the series of works [36–38]. The possibility of using such systems in place of Diesel engines is investigated as a means of reducing pollutants emissions and their environmental and health impacts, while at the same time complying with more and more strict emission regulations. It is indicated that gas turbine combined cycles can very well satisfy these regulations. Furthermore, the benefits of lower volume and weight of these systems on commercial vessels is assessed as an extra motive for their utilization.

Altosole et al. [39] a case study is conducted for the possibility of the application of a gas turbine-based combined cycle power plant instead of two-stroke Diesel engines on a large containership, after optimization for three different bottoming steam cycle designs. In addition to the benefits related to the overall weight and volume decrease of the machinery, a significant decrease in fuel consumption in comparison with the fuel consumption decrease achieved by bottoming cycles based on Diesel engines is reported.

A comparison between the thermodynamic performance of systems using gas turbines or low-speed Diesel engines with steam bottoming cycles is presented in Dzida [40]. Performance data of commercially available engines of both types are used and it is concluded that both types of the overall systems can achieve comparable efficiencies with the employment of the steam bottoming cycle.

The majority of modern combined cycle applications employ variable geometry gas turbines for better partial load performance. The effects of variable geometry inlet guide vanes and the fuel feeding regulation on the thermal efficiency and the overall performance of the prime movers during partial load operation are studied in Hanglid [41] for the cases of single-shaft and two-shaft marine gas turbines, while the efficiency of the overall system is studied in Hanglid [42]. The results suggest that, even though the efficiency of the gas turbine itself tends to generally deteriorate, especially at low loads, the use of variable geometry gas turbines is evidently beneficial for the thermal efficiency of an appropriately designed combined cycle. Another possibility of variable geometry gas turbines studied specifically for use in marine applications appears in Wang et al. [43], where the off-design performance of a marine gas turbine with compressor variable stator vanes is studied, and appropriate control strategies are proposed.

Other possibilities of integrating gas turbines with other technologies for marine applications have also been reported. Besides the utilization of water/steam in bottoming cycles, alternative waste heat cycles and configurations, possibly more suitable for ship applications, have been proposed, as for example in Sharma et l. [44] and Hou et al. [45], where supercritical CO_2 waste heat recovery cycles are proposed. Both studies suggest significant power enhancement and a very important increase of the thermal efficiency of the overall power plant. The improved partial load performance of such cycles is also highlighted [45].

Wang et al. [46], a system based on the waste heat recovery using both a standard steam bottoming cycle and an organic Rankine cycle operating in a cascaded way for the construction of a cogeneration system is studied in various operating conditions, and the improvements in comparison with a sole water/steam cycle are quantified.

Other works suggest the integration of gas turbines with fuel cells in energy systems of ships [47]. In such systems, the waste heat of the exhaust gas is used to preheat the fuel used in the fuel cell to the required temperature of operation. Tse et al. [48], a system combining fuel cell and gas turbine modules is extended with the use of absorption heat pumps for the production of cooling power, constructing a trigeneration or CCHP system studied for marine applications.

Apart from the cases where gas turbines are used in conjunction with steam bottoming cycles or other waste heat recovery configurations, studies have also appeared in which gas turbine configurations are used solely for the production of mechanical power in ship energy systems. Armellini et al. [49,50], a comparison is made between the alternatives of using (a) gas turbines as main engines, (b) Diesel engines with no pollution abatement, and (c) Diesel engines complemented with pollutant emission control devices (SCR, scrubber). These three different systems are simulated and optimized for the case of a cruise ship with the aim of maximizing the overall energy efficiency in several operating conditions, while the pollutants emissions are afterwards quantified. The results show that the employment of gas turbines leads to important environmental benefits, comparable with the alternative of using emission control devices in a Diesel engine-based system, while at the same time the complexity of the engine room is avoided.

Doulgeris et al. [51], gas turbine-based systems are assessed as an alternative for installation on a RoPax fast ferry ship. Simple cycle and intercooled–recuperated configurations are studied. In the method presented, several technical, economic, and environmental parameters concerning the operation of the system during the whole life cycle of the ship are taken into account. The study reports the benefits of using intercooled–recuperated gas turbines in comparison with simple cycle configurations. De Leon et al. [52], the development of a computer simulation framework is described which is used for assessing the differences of the thermodynamic efficiency and other performance characteristics of intercooled−recuperated, intercooled−reheated, and intercooled–reheated–recuperated configurations.

The need for enhanced performance characteristics throughout the whole operating power range of gas turbines used in marine applications, has led to the study of several advanced gas turbine thermodynamic cycles. The off-design operation and performance of an intercooled two-stage compression configuration is studied and optimized for certain operating states in Ji et al. [53].

An important factor considering the possibility of employing gas turbines in the energy system of ships is the potential of using natural gas as a fuel. In the study presented in El-Gohary and

Seddiek [54] and further extended in El-Gohary and Ammar [55], a comparison is made regarding the utilization of this type of fuel instead of diesel oil, demonstrating that the thermodynamic performance of the gas turbine operating on natural gas is very close to the case in which diesel oil is used, and that the natural gas can be thought of as a very appealing replacement for diesel oil, taking also into account the other advantages related to the economic and environmental benefits.

Natural gas is ideal for gas turbines and, as a consequence, gas turbines are very good candidates for LNG carriers, where they operate on the boil off gas. A technoeconomic study is presented in El-Gohary [56], where the potential economic benefits of using a gas turbine-based power plant burning LNG instead of reciprocating engines operating on HFO are demonstrated. The possibilities of using combined cycles for power plants of LNG carriers are also examined in Fernández et al. [57], where alternative configurations are proposed as potential solutions for the overall energy system.

In this section, the SDOO of an integrated energy system of ship comprising gas turbines and the possibility of combined cycle is performed. In contrast with the works presented in Dimopoulos et al. [21,22], where the solution is obtained with a two-level approach (level A for synthesis and design and level B for operation), as described in the preceding, a unified approach for the solution of the complete SDOO problem is applied. The general method and the pertaining mathematical formulation are presented in detail in Sakalis and Frangopoulos [24], where a generic type of main engines is considered.

The SDOO problem is initially formulated and solved considering three different types of gas turbine configurations and two types of fuels. Afterwards, the effects that the fuel price and the capital cost have on the optimal solutions for the best performing gas turbine configuration are studied.

It has to be noted that for the present application, the system is considered to be operating in static conditions. This means that the energy profile of the ship is assumed to be adequately approximated by considering a predetermined number of operating modes, which also are characterized by predetermined magnitudes of the loads to be covered and their respective duration during a typical year of ship operation.

2.2. Description of the System and Formulation of The Optimization Problem

The system is used for covering the demands for propulsion ($\dot{W}_p$), electrical ($\dot{W}_e$) and thermal power ($\dot{Q}_{hl}$), during different operating modes of a ship. The number of the operating modes is considered predetermined and equal to N_T. The superconfiguration of the system considered is presented in Figure 2.

The gas turbines (GT) are coupled to the propellers by means of a speed reducing gearbox. The exhaust gases of the gas turbines are fed into heat recovery steam generators which produce superheated steam at two pressure levels and saturated steam for potentially covering thermal loads. The superheated steam drives steam turbines; their power outputs can be fed to the propeller and/or to electric generators. The proper allocation of the steam turbine power between the propulsion and the electrical loads is among the results of optimization.

Provision is taken for the possibility that the employment of a steam bottoming cycle may not be an optimal solution. For this reason, the potential inclusion of Diesel generator sets (DG) and fuel fed auxiliary boilers for covering the electrical and thermal loads is considered, which will cover electric and thermal loads also in port, where the main engines and, consequently, the bottoming cycle, do not operate. In any case, the proper allocation of the energy loads among the bottoming cycle components and the independently operating components (that is, the Diesel generator sets and the auxiliary boilers) is to be determined by the optimization procedure. An exhaust gas boiler (EGB) may also be included in the system for covering thermal loads when the exhaust gas flows are not exploited in heat recovery steam generators (HRSGs).

The steam is produced in the existing HRSGs at common pressure levels and it is delivered to collectors, one for each pressure level (only one collector is depicted in Figure 2 for simplicity).

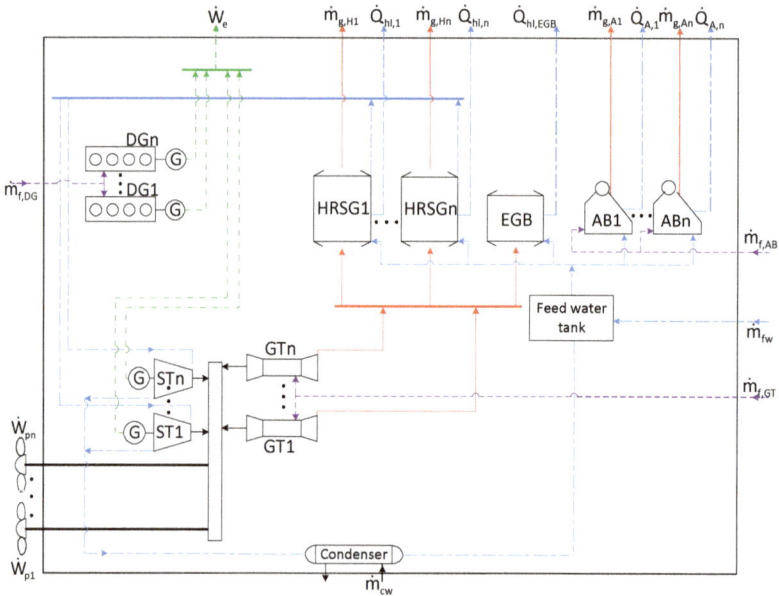

Figure 2. Superconfiguration of the energy system.

In Figure 2, the dots among components imply that the final number of each type of component present will be decided by the optimization procedure.

Overall, the number of each type of components present in the system and the physical and functional interconnections between them, as also their design characteristics and operating point at each instant of time will collectively be determined by the solution of the optimization problem.

The minimization of the present worth cost (*PWC*) of building and operating the energy system for a predetermined number of years is selected as the optimization objective:

$$
\begin{aligned}
\min PWC = &\sum_{k=GT,DG,HRSG,ST,AB,EGB} \left[\sum_{x=1}^{n_k} C_{c,k,x} \right] \\
&+ PWF(N_Y, f, i) \sum_{y=1}^{N_T} \left\{ \sum_{k=GT,DG,AB} \left[\sum_{x=1}^{x=n_k} \dot{m}_{f,kxy} t_{kxy} c_{f,k} \right] \right\} + \\
&+ PWF(N_Y, f, i) \sum_{y=1}^{N_T} \left\{ \sum_{k=GT,DG,ST} \left[\sum_{x=1}^{n_k} c_{om,kxy} \dot{W}_{kxy} t_{kxy} \right] + \sum_{k=HRSG,AB,EGB} \left[\sum_{x=1}^{n_k} c_{om,kxy} \dot{Q}_{kxy} t_{kxy} \right] \right\}
\end{aligned}
\tag{1}
$$

In Equation (1), the 1st line includes the capital costs of the components, the 2nd line consists of the fuel costs, and the 3rd line consists of the operating and maintenance costs.

The simulation procedure of the system as a whole is carried out with the purpose of calculating the value of the objective function, which expresses the complete SDOO problem, in a single computational step. In this simulation procedure, proper variables (that are to be used as independent variables of the optimization problem) determine the number of operating components during each operating mode and their proper functional interconnections among them in order for the loads to be covered. This modeling procedure is presented in Sakalis and Frangopoulos [24] and is briefly repeated in the present section for convenience, while the details of the modeling procedure of certain individual components are presented in Section 2.3 and in the aforementioned publication.

For the problem formulation, the operating profile of the energy system is represented (with an acceptable degree of approximation) by a number N_T of modes, during which steady state operation is assumed. Each mode y ($y = 1, 2, \ldots, N_T$) has a predetermined duration t_y. During each operating mode, the energy demands for propulsion ($\dot{W}_{p,y}$), electricity ($\dot{W}_{e,y}$), and heat ($\dot{Q}_{hl,y}$) are also predetermined and have constant values.

The power balance equations are valid at each instance of time:

$$\sum_{x=1}^{x=n_{GT,y}} \dot{W}_{GT,x,y} + \sum_{v=1}^{v=n_{ST,y}} \dot{W}_{STp,v,y} = \dot{W}_{p,y}, \; y = 1, \ldots, N_T \tag{2}$$

$$\sum_{x=1}^{x=n_{DG,y}} \dot{W}_{DG,x,y} + \sum_{v=1}^{v=n_{ST,y}} \dot{W}_{STe,v,y} = \dot{W}_{e,y}, \; y = 1, \ldots, N_T \tag{3}$$

$$\sum_{z=1}^{z=n_{HRSG,y}} \dot{Q}_{hl,z,y} + \sum_{u=1}^{u=n_{AB,y}} \dot{Q}_{AB,u,y} + \dot{Q}_{EGB,y} = \dot{Q}_{hl,y}, \; y = 1, \ldots, N_T \tag{4}$$

In Equations (2)–(4), the n_i ($i = GT, ST, DC, HRSG, AB$) symbols represent the number of operating components according to their type during mode y, $\dot{W}_{GT,x,y}$ is the power delivered by gas turbine x to the propeller ($x = 1, \ldots, n_{GT,y}$), $\dot{W}_{STp,v,y}$ is the propulsion power part of steam turbine v ($v = 1, \ldots, n_{ST,y}$), $\dot{W}_{DG,x,y}$ is the output of Diesel generator set x ($x = 1, \ldots, n_{DG,y}$), $\dot{W}_{STe,v,y}$ is the electrical power part delivered by steam turbine generator v, $\dot{Q}_{hl,z,y}$ is the thermal power covered by the HRSG z to thermal loads ($z = 1, \ldots, n_{HRSG,y}$), and $\dot{Q}_{AB,u,y}$ is the output of auxiliary boiler u ($u = 1, \ldots, n_{AB,y}$). Due to the fact that the propulsion power may be partially covered by steam turbines, the total power delivered by the gas turbines can be lower than the total power required, and thus it holds that

$$\sum_{x=1}^{n_{GT,y}} \dot{W}_{GT,x,y} = \lambda_{GT,y}\dot{W}_{p,y}, \quad \lambda_{GT,y} \leq 1, y = 1, \ldots, N_T \tag{5}$$

The power output of each of the operating $n_{GT,y}$ engines during mode y is calculated as follows

$$\dot{W}_{GT,1,y} = \widetilde{W}_{GT,1,y}\lambda_{GT,y}\dot{W}_{p,y}, \quad \widetilde{W}_{GT,1,y} \leq 1 \tag{6}$$

$$\dot{W}_{GT,x,y} = \widetilde{W}_{GT,x,y}\left(\lambda_{GT,y}\dot{W}_{p,y} - \sum_{i=1}^{x-1} \dot{W}_{GT,i,y}\right), \quad \widetilde{W}_{GT,x,y} \leq 1, \quad 1 < x < n_{GT,y} \tag{7}$$

$$\dot{W}_{GT,n_{GT,y},y} = \lambda_{GT,y}\dot{W}_{p,y} - \sum_{i=1}^{n_{GT,y}-1} \dot{W}_{GT,i,y}, \quad x = n_{GT,y} \tag{8}$$

The maximum of the values of $n_{GT,y}$ among all operating modes y will also determine the final number of main engines that will be present in the system:

$$n_{GT} = \max\left(n_{GT,1}, n_{GT,2}, \ldots, n_{GT,y} \ldots, n_{GT,N_T}\right) \tag{9}$$

The nominal power of each main engine is temporarily set as

$$\dot{W}_{GTx,N,temp} = \max\left(\dot{W}_{GT,x,1}, \dot{W}_{GT,x,2}, \ldots, \dot{W}_{GT,x,y}, \ldots, \dot{W}_{GT,x,N_T}\right) \tag{10}$$

The nominal power of the main engines on ships is usually slightly oversized (sea margin), in order for the upcoming hull and propeller fouling effects to be counteracted appropriately, as also for the conditions that the ship may operate in adverse weather conditions. In a gas turbine combined cycle, the steam turbine power production is expected to be, in general, much higher than in the case of combined cycle based on Diesel engines, due to the favorable exhaust gas characteristics. By an appropriate design of the bottoming cycle, it is thus possible that the steam turbine may have quite a significant contribution to the propulsion load, affecting, in this way, the appropriate (optimal) operational and nominal characteristics of the gas turbines. The need for sea margin is considered in the determination of the nominal power output of the system in the following way.

Among the operational modes, one will present the highest propulsion load, which is symbolized with $\dot{W}_{p,max}$. The sea margin excess power requirement is herein expressed with Equation (11), which relates the sum of the nominal power rating of the operating gas turbines and the sum of steam turbine propulsion powers symbolized with $\dot{W}_{ST,p,ml}$.

$$\sum_{i=1}^{n_{GT,ml}} \dot{W}_{GTi,N,temp} + \dot{W}_{ST,p,ml} \geq \dot{W}_{p,max}/\mu_s \tag{11}$$

where the index *ml* implies the aforementioned operating mode in which $\dot{W}_{p,max}$ appears and μ_s is the sea margin factor, usually taken equal to 0.85.

For the sum of the steam turbine propulsion powers in mode *ml*, the following equation must hold.

$$\dot{W}_{ST,p,ml} = (1 - \lambda_{GT,ml})\dot{W}_{p,max} \tag{12}$$

where $\lambda_{GT,ml}$ is the fraction of propulsion power $\dot{W}_{p,max}$ delivered by the gas turbines and is an independent variable of the optimization problem. If the characteristics of the steam produced cannot result in an $\dot{W}_{ST,p,ml}$ sufficient for covering $\dot{W}_{p,max}$, then the candidate solution is discarded by the optimization procedure as nonfeasible.

Equations (11) and (12) lead to inequality (13), which expresses the requirement for the sum of nominal power ratings of the gas turbines:

$$\sum_{i=1}^{n_{GT}} \dot{W}_{GTi,N,temp} \geq \dot{W}_{p,max}(\lambda_{GT,ml} + 1/\mu_s - 1) \tag{13}$$

If inequality (13) does not hold, the values of $\dot{W}_{GTi,N,temp}$ are proportionally increased until (13) holds as an equality, and the temporary values $\dot{W}_{GTi,N,temp,sm}$ are obtained.

The nominal power rating for each gas turbine *x* is finally determined by the equation

$$\dot{W}_{GTx,N} = W_{N,x,mult}\dot{W}_{GTx,N,temp,sm}, \quad W_{N,x,mult} \geq 1 \tag{14}$$

The fuel consumption $\dot{m}_{fGTx,y}$ and exhaust gas properties (mass flow rate $\dot{m}_{gGTx,y}$ and temperature $T_{gGTx,y}$) of each of the main engines can afterwards be calculated (as the nominal power rating and partial load brake powers are already determined) for each mode by applying the computational simulation procedures of gas turbines described in Section 2.3.

In each operating mode *y*, the exhaust gas inlet in the HRSG *z* is determined according to

$$\dot{m}_{gz,y} = \sum_{x=1}^{n_{GT,y}} \zeta_{x,y}\dot{m}_{gGTx,y}, \quad \zeta_{x,y} = 1 \quad if \quad g_{x,y} = z \tag{15}$$
$$\zeta_{x,y} = 0 \quad if \quad g_{x,y} \neq z$$

where each $g_{x,y}$ variable refers to gas turbine *x* and denotes the number of the HRSG towards which its exhaust gas is driven.

The nominal mass flow rate $\dot{m}_{gz}$ and temperature T_{gz}, for which the HRSG *z* is designed, are calculated as

$$\dot{m}_{gz} = m_{gz,mult}\frac{\sum_{y=1}^{N_T} \dot{m}_{gz,y}t_y}{\sum_{y=1}^{N_T} t_y} \tag{16}$$

$$T_{gz} = T_{gz,mult} \frac{\sum\limits_{y=1}^{N_T} T_{gz,y} t_y}{\sum\limits_{y=1}^{N_T} t_y} \tag{17}$$

where $m_{g,mult}$ and $T_{g,mult}$ are intended to be used as independent optimization variables.

The bottoming cycle operates at two pressure levels—P_{HP} and P_{LP}—common among the HRSGs. In nominal conditions of operation for HRSG z, during which the exhaust gas characteristics are determined by Equations (16) and (17), the steam produced in the two pressure levels for feeding the turbines will have mass flow rates $\dot{m}_{HPz}$, $\dot{m}_{LPz}$, and temperatures T_{HPz} and T_{LPz}.

The HRSGs are of double-pressure and the nominal values of P_{HP}, P_{LP}, $\dot{m}_{HPz}$, $\dot{m}_{LPz}$, T_{HPz}, and T_{LPz} for HRSG z are to be used as inputs for simulating the integrated energy system. With values for these variables set, the design procedure described in Section 2.3.2 can be initiated for each HRSG. The design procedure is used for the determination of the heat exchange areas throughout the HRSG and with these areas determined, the off-design operation properties of steam (that is, mass flow rates $\dot{m}_{HPz,y}$ and $\dot{m}_{LPz,y}$ and temperatures $T_{HPz,y}$ and $T_{LPz,y}$) can also be calculated.

The total mass flow rates $\dot{m}_{HP,COL,y}$ and $\dot{m}_{LP,COL,y}$, in each steam collector, before feeding the turbines, are readily calculated with mass balances, and the respective temperatures $T_{HP,COL,y}$ and $T_{LP,COL,y}$ after stream mixing is determined by energy balances.

For the determination of the steam mass flow rate delivered to each turbine, the following equations hold (for the high-pressure level).

$$\dot{m}_{HP,1,y} = \tilde{m}_{HP,1,y} \dot{m}_{HP,COL,y}, \quad \tilde{m}_{HP,1,y} \leq 1, \tag{18}$$

$$\dot{m}_{HP,v,y} = \tilde{m}_{HP,v,y} \left(\dot{m}_{HP,COL,y} - \sum_{i=1}^{v-1} \dot{m}_{HP,i,y} \right), \quad \tilde{m}_{HP,v,y} \leq 1, \quad 1 < v < n_{ST,y} \tag{19}$$

$$\dot{m}_{HP,n_{ST,y},y} = \dot{m}_{HP,COL,y} - \sum_{i=1}^{n_{ST,y}-1} \dot{m}_{HP,i,y}, \quad v = n_{ST,y} \tag{20}$$

where $n_{ST,y}$ represents the number of steam turbines that operate during mode y.

The mas flow rate $\dot{m}_{HPv}$ and temperature T_{HPv} for the design of steam turbine v are calculated as

$$\dot{m}_{HPv} = \dot{m}_{HPv,mult} \frac{\sum\limits_{y=1}^{N_T} \dot{m}_{HPv,y} t_y}{\sum\limits_{y=1}^{N_T} t_y} \tag{21}$$

$$T_{HPv} = T_{HPv,mult} \frac{\sum\limits_{y=1}^{N_T} T_{HPv,y} t_y}{\sum\limits_{y=1}^{N_T} t_y} \tag{22}$$

Similar equations are used for the low pressure level. During design point operation, the pressure levels at the steam turbine inlets are the same with the ones of HRSGs.

With the values of intensive and extensive thermodynamic properties of the steam feeding the steam turbines at the regarded as design point operation, the steam turbine design procedure described in detail in Sakalis and Frangopoulos [24], is applied, the design power production is calculated, and off-design power assessment can also be carried out.

The power $\dot{W}_{STv,y}$ produced by steam turbine v is allocated between propulsion and electrical loads during each mode y. The following equation must hold for the propulsion parts.

$$\dot{W}_{STp,y} = \sum_{v=1}^{v=n_{ST,y}} \dot{W}_{STp,v,y} = (1 - \lambda_{GT,y})\dot{W}_{p,y} \tag{23}$$

If during mode y the number of operating steam turbines $n_{ST,y}$ is higher than one, their total propulsion power is allocated among them in proportion to the total power output of each one:

$$\frac{\dot{W}_{STp,v,y}}{\sum_{v=1}^{v=n_{ST,y}} \dot{W}_{STp,v,y}} = \frac{\dot{W}_{STv,y}}{\sum_{v=1}^{v=n_{ST,y}} \dot{W}_{STv,y}} \tag{24}$$

The total electrical power produced by the steam turbine generators, is calculated as the power that remains (if any) after the covering of the propulsion load:

$$\dot{W}_{STe,y} = \sum_{v=1}^{v=n_{ST,y}} \dot{W}_{STe,v,y} = \eta_G \left(\sum_{v=1}^{v=n_{ST,y}} \dot{W}_{STv,y} - \sum_{v=1}^{v=n_{ST,y}} \dot{W}_{STp,v,y} \right) = \eta_G \left(\sum_{v=1}^{v=n_{ST,y}} \dot{W}_{STn,y} - (1 - \lambda_{GT,y})\dot{W}_{p,y} \right) \tag{25}$$

The total power delivered by the Diesel generator sets $\dot{W}_{DG,TOT,y}$ is readily calculated with the following equation, in case that total electrical power delivered by the steam turbines is not sufficient to cover the loads:

$$\dot{W}_{DG,TOT,y} = \dot{W}_{e,y} - \dot{W}_{STe,y} \tag{26}$$

In Equation (26), $\dot{W}_{e,y}$ is the total electric load during mode y.

During mode y, $n_{DG,y}$ Diesel generator sets will be operating. If $n_{DG,y} > 1$, the power delivered by each Diesel generator set x, $\dot{W}_{DG,x,y}$, will be calculated with the following equations.

$$\dot{W}_{DG,1,y} = \tilde{W}_{DG,1,y}\dot{W}_{DG,TOT,y}, \quad \tilde{W}_{DG,1,y} \leq 1, \tag{27}$$

$$\dot{W}_{DG,x,y} = \tilde{W}_{DG,x,y}\left(\dot{W}_{DG,TOT,y} - \sum_{i=1}^{x-1} \dot{W}_{DG,i,y} \right), \quad \tilde{W}_{DG,x,y} \leq 1, \quad 1 < x < n_{DG,y} \tag{28}$$

$$\dot{W}_{DG,n_{DG,y},y} = \dot{W}_{DG,TOT,y} - \sum_{i=1}^{n_{DG,y}-1} \dot{W}_{DG,i,y}, \quad x = n_{DG,y} \tag{29}$$

Variables $\dot{W}_{DGN,x,mult}$ are used for the determination of the nominal power rating of the Diesel generator x, similarly to the case of main engines.

The thermal load during mode y is allocated between the HRSGs and the auxiliary boiler as follows

$$\dot{Q}_{hl,HRSG,TOT,y} = \lambda_{Q,y}\dot{Q}_{hl,y} \, 0 \leq \lambda_{Q,y} \leq 1 \tag{30}$$

$$\dot{Q}_{AB,TOT,y} = (1 - \lambda_{Q,y})\dot{Q}_{hl,y} \, 0 \leq \lambda_{Q,y} \leq 1 \tag{31}$$

The $\dot{Q}_{hl,HRSG,TOT,y}$ is allocated among the HRSGs accordingly:

$$\dot{Q}_{hl,1,y} = \tilde{Q}_{hl,1,y}\dot{Q}_{hl,HRSG,TOT,y}, \quad \tilde{Q}_{hl,1,y} \leq 1, v \tag{32}$$

$$\dot{Q}_{hl,z,y} = \tilde{Q}_{hl,z,y}\left(\dot{Q}_{hl,HRSG,TOT,y} - \sum_{i=1}^{z-1} \dot{Q}_{hl,i,y} \right), \quad \tilde{Q}_{hl,z,y} \leq 1, \quad 1 < z < n_{HRSG,y} \tag{33}$$

$$\dot{Q}_{hl,n_{HRSG,y},y} = \dot{Q}_{hl,HRSG,TOT,y} - \sum_{i=1}^{n_{HRSG,y}-1} \dot{Q}_{hl,i,y}, \quad z = n_{HRSG,y} \tag{34}$$

The EGB, which may be included, is employed only in cases in which exhaust gas is available from any engine because it is not exploited for the production of superheated steam (due to technical reasons or because this could be dictated by an optimal solution), so that its heat content can be used for covering thermal loads only.

Equations (2)–(34) are a closed form set of equations which is used for the determination of the number of operating components as well as heir functional interconnections that should exist in order for the energy system to fulfill its purpose. This number and the functional interconnections may be different among different operating modes. The final synthesis of the system is thus dependent on the "temporary" syntheses during each mode. Furthermore, the component design characteristics are determined in a procedure that takes into account the different values of the loads to be covered during all of the operating modes.

Generally, the inputs to the model of the overall system are intended to be used as independent variables of the optimization problem, which are collectively reported in Table 1. The optimization problem formulated is of the mixed integer nonlinear programing type and is solved with the use of genetic algorithms (the number of variables for voyage operating modes is $N_T - 1$, while number N_T is reserved for harbor operating mode).

Table 1. Independent variables of the optimization problem.

$n_{GT,y},\quad y = 1, 2, \ldots, N_T - 1$	$n_{DG,y},\quad y = 1, 2, \ldots, N_T$
$g_{x,y},\quad x = 1, 2, \ldots n_{HRSG,max},\quad y = 1, 2, \ldots, N_T - 1$	$\lambda_{GT,y},\quad \lambda_{Q,y},\quad y = 1, 2, \ldots, N_T - 1$
$\widetilde{W}_{N,x,mult},\quad x = 1, 2, \ldots, n_{GT,max}$	$\widetilde{W}_{DGN,x,mult},\quad x = 1, 2, \ldots, n_{DG,max}$
$\widetilde{W}_{GT,x,y},\quad x = 1, 2, \ldots, n_{GT,max} - 1,\quad y = 1, 2, \ldots, N_T - 1$	$\widetilde{W}_{DG,x,y},\quad x = 1, 2, \ldots, n_{DG,max} - 1,\quad y = 1, 2, \ldots, N_T$
$m_{gz,mult},\quad T_{gz,mult},\quad z = 1, 2, \ldots, n_{HRSG,max}$	$P_{HP},\quad P_{LP}$
$\dot{m}_{k,z},\quad T_{k,z},\quad z = 1, 2, \ldots, n_{HRSG,max},\quad k = HP, LP$	
$\widetilde{Q}_{hl,z,y},\quad z = 1, 2, \ldots, n_{HRSG,max} - 1,\quad y = 1, 2, \ldots, N_T - 1$	
$n_{ST,y},\quad y = 1, 2, \ldots, N_T - 1$	
$\widetilde{m}_{STk,v,y},\quad v = 1, 2, \ldots, n_{ST,max} - 1,\quad y = 1, 2, \ldots, N_T - 1,\quad k = HP, LP$	
$m_{kv,mult},\quad T_{kv,mult},\quad v = 1, 2, \ldots, n_{ST,max},\quad k = HP, LP$	
$n_{AB,y},\quad y = 1, 2, \ldots, N_T$	

More details concerning the nature of the independent variables and the mathematical form of the objective function, as well as the tuning parameters and the application of the genetic algorithm can be found in Sakalis and Frangopoulos [24].

2.3. Modeling of Individual Components

In the present section, the simulation models used for the gas turbine configurations and the HRSGs operating in conjunction with this type of main engines are presented. Modeling of other components, as well as the individual heat exchangers appearing in the HRSGs, is presented in detail in Sakalis and Frangopoulos [24].

2.3.1. Modeling of Gas Turbines

Three different gas turbine types depicted in Figure 3 are considered as main engines.

All three have a separate power turbine coupled to the propeller. Types with a separate power turbine are favorable for mechanical ship propulsion, as the rotational speed of the propeller is low and highly variant, and particularly so in case of a fixed pitch propeller.

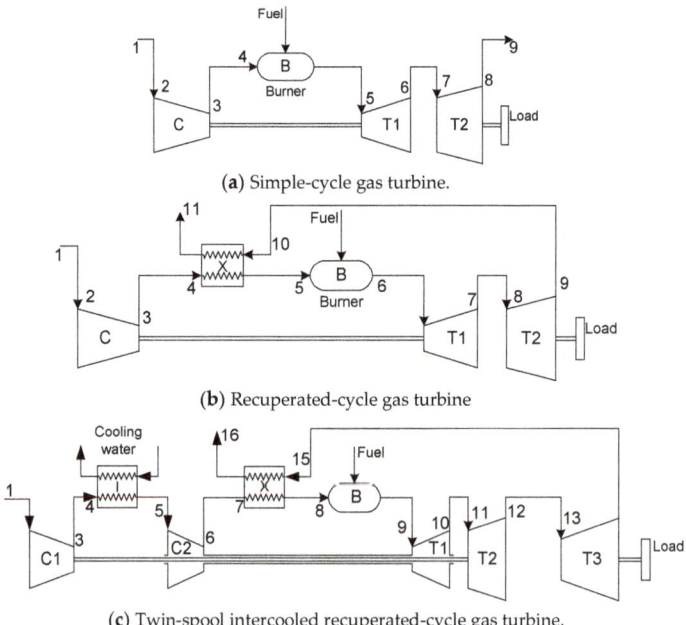

(a) Simple-cycle gas turbine.

(b) Recuperated-cycle gas turbine

(c) Twin-spool intercooled recuperated-cycle gas turbine.

Figure 3. The three gas turbine types considered as main engines.

Modeling and optimization is performed with two alternative fuels for each configuration: Marine Diesel Oil (MDO), with a lower heating value LHV_{MDO} = 42,500 kJ/kg, and natural gas (NG), with LHV_{NG} = 47,100 kJ/kg and composition as presented in Table 2.

Table 2. Natural gas composition.

Component	Composition % Volume
CH_4	88.5
C_2H_6	4.7
C_3H_8	1.6
C_4H_{10}	0.2
N_2	5.0

For the simulation of the three gas turbine types, a dedicated software has been developed by the NTUA Laboratory of Thermal Turbomachines [58], which calculates all the intensive and extensive thermodynamic properties of the working medium throughout the system, according to design point specifications and the operation point (off-design operation). Real gas properties are used throughout the configurations; isentropic efficiencies are calculated according to the gas properties and pressure ratios with the incorporation of loss models, the combustion process is based on specialized simulation procedures according to chemical kinetics and, regarding the off-design operation, appropriate maps are generated for the compressors and turbines.

For the integration of gas turbines in the simulation of the overall superconfiguration of the energy system, this simulation program is used for the calculation of the specific fuel consumption *SFC*, the mass flow rate $\dot{m}_g$ and the temperature T_g of the exhaust gases as functions of the nominal power output and the load factor. The following general mathematical form is thus obtained:

$$\Phi_i = \Phi_i\left(\dot{W}_{GT,N}, f_L\right), \quad \Phi = SFC, \dot{m}_g, T_g, \quad i = A, B, C \tag{35}$$

The capital cost of the gas turbines is estimated as described in Appendix A, based on the cost model presented in Frangopoulos [59].

2.3.2. Modeling of Heat Recovery Steam Generators

Each HRSG consists of a water preheater, low pressure economizer, evaporator and superheater, and high-pressure economizer, evaporator, and superheater (Figure 4), which are multipass heat exchangers. The HRSG feeds the steam turbine (points 14 and 25 in Figure 4 with mass flow rates $\dot{m}_{LP}$ and $\dot{m}_{HP}$, respectively), while a fraction of the saturated low pressure steam is used for thermal loads (point 17 with mass flow rate $\dot{m}_{hl}$). A deaerator is also integrated with the HRSG, and a heating stream (point 31, $\dot{m}_{da}$) originating from the low pressure drum is used, if necessary, for heating the feed water to the appropriate conditions for deaeration.

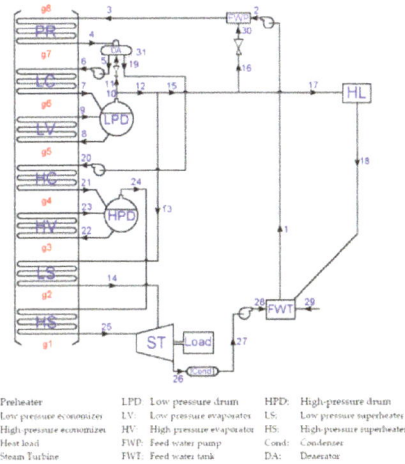

PR: Preheater	LPD: Low pressure drum	HPD: High-pressure drum	
LC: Low pressure economizer	LV: Low pressure evaporator	LS: Low pressure superheater	
HC: High pressure economizer	HV: High pressure evaporator	HS: High-pressure superheater	
HL: Heat load	FWP: Feed water pump	Cond: Condenser	
ST: Steam Turbine	FWT: Feed water tank	DA: Deaerator	

Figure 4. Bottoming cycle and internal structure of the HRSG.

Each HRSG is designed according to the procedure described in Sakalis and Frangopoulos [24]. The required inputs include the nominal exhaust gas properties (mass flow rate $\dot{m}_{gz}$ and temperature T_{gz}, which correspond to the point g1 in Figure 4), steam pressure levels (P_{HP} and P_{LP}), and steam mass flow rates and temperatures ($\dot{m}_{HPz}$ and $\dot{m}_{LPz}$ and T_{HPz}, and T_{LPz}). These quantities are calculated before the HRSG design algorithm is applied, during the modeling of the system as a whole, or they are used as independent variables of the SDOO problem [24].

In the design algorithm, mass and energy balances are initially performed throughout the HRSG, which give the thermodynamic state of the fluids at the various points. Several checks are performed for ensuring feasibility of the design with the inputs given, which can be thought of as constraints of the SDOO of the system. Examples of constraints are the minimum temperature difference between fluids in each heat exchanger, the minimum exhaust gas temperature at the exit of the HRSG (specified at 130 °C for MDO and 100 °C for natural gas), the minimum temperature of water at the inlet of the HRSG (specified at 105 °C for MDO and 75 °C for natural gas), and the requirement that no steaming will be induced in the economizers and the preheater by an excessive heat transfer.

After the initial mass and energy balance calculations, each heat exchanger is designed with the P−NTU or the ε − NTU method according to its type, with the procedure described in Sakalis and Frangopoulos [24]. Among the results are the structural characteristics and the heat exchange surface area of each heat exchanger, which are also required for the simulation of off-design operation.

In operating mode y, HRSG z will be fed with exhaust gas of mass flow rate $\dot{m}_{gz,y}$ and temperature $T_{gz,y}$ that will be, in general, different from the nominal ones ($\dot{m}_{gz}$ and T_{gz} referred above). The off-design operation is simulated with a computational algorithm developed for this purpose. Heat balances and heat transfer calculations with the P−NTU or the $\varepsilon-$ NTU method for each heat exchanger are again performed, resulting in a system of nonlinear equations, in which the values of heat transfer areas are now fixed. By solving this system of equations, the feasibility of off-design operation with the particular values of $\dot{m}_{gz,y}$ and $T_{gz,y}$ is investigated for each operating mode y. If the operation is feasible, the final outcomes of the off-design simulation algorithm include the mass flow rates $\dot{m}_{HPz,y}$, $\dot{m}_{LPz,y}$, and temperatures $T_{HPz,y}$ and $T_{LPz,y}$ of the steam (points 25 and 14, respectively), during each operating mode y of the system.

2.3.3. Other Components

The steam turbines included in the system are designed according to the procedure described in Sakalis and Frangopoulos [24]. The main required inputs for the design, from the point of view of the integrated system, are the mass flow rates and the thermodynamic properties of the steam streams feeding the turbines. After the design has taken place, the off-design performance and power production can be calculated with dedicated simulation algorithms also presented in the aforementioned publication.

The Diesel generator sets are simulated with regression models developed from data available from manufactures. Essentially, for the purposes of the present work, the quantity that has to be calculated is the fuel consumption; the related regression models are functions of the design power and the load factor (in the same sense as in Equation (35)).

The EGB consists of an economizer, an evaporator and a steam drum only, and the procedures for its design and operation are similar to those of the HRSGs. For the auxiliary boilers, it is considered that they operate with constant thermal efficiency and their design power output is equal to the maximum operating that is presented to each of them.

2.4. Application Examples

The SDOO problem for the system of Figure 2 with minimization of the present worth cost as objective is first solved for each one of the six combinations of gas turbine type (Figure 3) and fuel (MDO, NG). Then a parametric study with respect to fuel price and capital cost is presented.

2.4.1. Data and Assumptions

The annual energy profile of the ship is assumed to be satisfactorily represented with three voyage modes and one harbor mode, with energy needs and duration as given in Table 3. The values of pertinent economic parameters, including nominal prices of fuels and operation and maintenance unit costs (excluding fuel), are given in Table 4 (O&M costs are estimated with adaptation of data presented initially in Dimopoulos and Frangopoulos [22]). In Table 4, N_Y is the number of years of operation of the system, f is the inflation rate, and i is the market interest rate.

Table 3. Annual energy profile of the ship.

Mode y	$\dot{W}_{p,y}$ (kW)	$\dot{W}_{e,y}$ (kW)	$\dot{W}_{hl,y}$ (kW)	t_y (Hours)
1	26,000	1500	400	2690
2	22,000	1500	300	1575
3	14,000	700	200	1620
4	0	1200	150	1000

Table 4. Values of economic parameters.

Parameter	Value	Parameter	Value
$c_{f,MDO}$	400\$/ton	$c_{om,GT}$	0.006\$/kWh
$c_{f,NG}$	150\$/ton	$c_{om,DG}$	0.007\$/kWh
N_Y	20	$c_{om,HRSG}$, $c_{om,AB}$	0.005\$/kWh
f	3%	$c_{om,ST}$	0.004\$/kWh
i	8%	$c_{om,GT}$	0.006\$/kWh

2.4.2. Optimization Results for the Nominal Values of Parameters

The solution of the SDOO problem results in the same optimal synthesis of the system for all six combinations of gas turbine type and fuels: it consists of one unit of each type, as given in Table 5. It is noted that, as an optimization constraint, the maximum number for each type of units was set equal to two. It is noted that one Diesel generator set is included in the optimal configuration, as also one auxiliary boiler, because they are needed for port operation (mode 4), while during the three voyage modes, the electrical and thermal loads are covered by the steam bottoming cycle.

Table 5. Optimal synthesis of the system.

$n_{GT} = 1$	$n_{HRSG} = 1$	$n_{ST} = 1$
	$n_{DG} = 1$	$n_{AB} = 1$

The optimal design characteristics of the components and their capital cost for the six cases are presented in Table 6.

Table 6. Optimal design characteristics and capital cost of components for the six combinations of gas turbine type and fuel.

Type/Fuel	(a)/MDO	(b)/MDO	(c)/MDO	(a)/NG	(b)/NG	(c)/NG
$\dot{W}_{GT}$ (kW)	23828	26941	28656	24662	27373	29379
$\dot{Q}_{HRSG}$ (kW)	23663	19708	13765	20623	15799	9605
$\dot{m}_{g,HRSG}$ (kg/s)	47.36	77.57	62.27	54.42	67.48	54.47
$\dot{m}_{HP,HRSG}$ (kg/s)	5.561	4.532	3.091	5.565	3.904	2.663
$\dot{m}_{LP,HRSG}$ (kg/s)	1.631	2.136	1.604	0.608	1.263	0.442
P_{HP} (bar)	63.58	17.65	19.23	63.67	21.66	23.29
P_{LP} (bar)	6.12	4.71	4.27	9.00	7.71	7.25
$T_{g,in,HRSG}$ (°C)	584.90	377.12	348.63	554.08	420.34	387.15
$T_{g,out,HRSG}$ (°C)	153.57	157.77	157.8	226.91	218.22	234.92
$T_{HP,HRSG}$ (°C)	552.46	347.26	316.43	522.89	387.34	354.53
$T_{LP,HRSG}$ (°C)	182.85	170.55	174.51	199.51	187.29	188.97
$\dot{W}_{ST}$ (kW)	8624	5009	3376	7589	4925	2706
$\dot{m}_{HP,ST}$ (kg/s)	6.431	4.696	3.139	6.276	4.505	2.944
$\dot{m}_{LP,ST}$ (kg/s)	1.608	2.111	1.590	0.549	1.320	0.458
$T_{HP,ST}$ (°C)	543.53	327.49	307.88	513.43	386.07	324.17
$T_{LP,ST}$ (°C)	179.87	162.31	171.1	187.96	207.95	193.76
$\dot{W}_{DG}$ (kW)	1206	1210	1203	1207	1209	1207
$C_{c,GT}$ (\$)	10,718,229	13,107,957	15,083,626	10,878,489	13,197,730	15,245,075
$C_{c,HRSG}$ (\$)	2,684,082	2,344,735	1,905,809	2,332,648	1,621,354	1,110,526
$C_{c,ST}$ (\$)	1,498,472	1,202,482	993,061	1,425,071	1,151,211	887,139
$C_{c,DG}$ (\$)	874,689	875,808	873,849	874,958	875,529	874,969

Figure 5 depicts the simulation results for the specific fuel consumption and exhaust gas mass flow rate and temperature as functions of the load factor, for the three gas turbine types operating on MDO presented in Table 6. The curves for operation with natural gas have similar forms. Type (a) (simple gas turbine) has the largest specific fuel consumption, which also exhibits a larger increase as the load factor decreases.

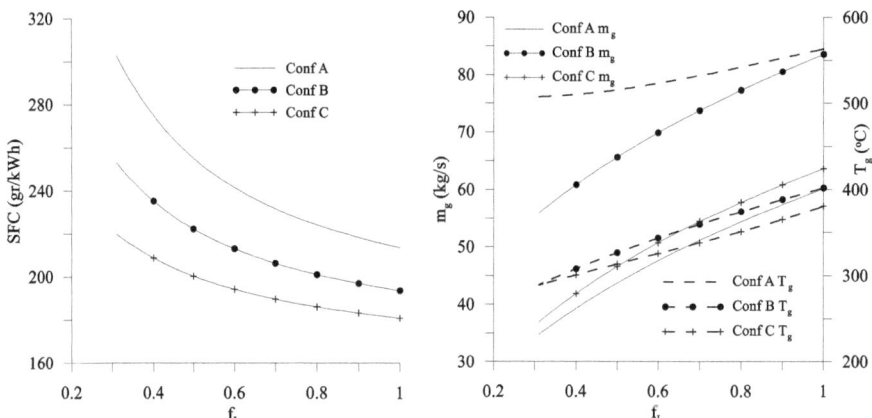

Figure 5. Variation of specific fuel consumption (SFC) and exhaust gas characteristics in partial load operation (fuel MDO).

The results in Table 6 show that, in the optimal design, the power capacity of the steam bottoming cycle (thermal power of HRSG and mechanical power of the steam turbine) decreases as the complexity of the gas turbine type increases from (a) to (c) (Figure 3). This tendency can be attributed to the fact that the thermal efficiency of the gas turbine unit increases as its thermodynamic cycle becomes more advanced (Figure 5), with consequence the production of exhaust gases with decreasing energy content, i.e., decreasing capacity for additional power production.

The operational technical and economic characteristics of the six combinations are presented in Tables 7–12, where SFC_{GT} and SFC_{CC} refer to the specific fuel consumption of the gas turbine unit and the combined cycle, respectively. In the general case where more than one main engines or steam turbines operate in any mode y, the SFC_{CC} can be defined as follows

$$SFC_{CC} = \frac{\sum\limits_{x=1}^{n_{GT,y}} \dot{W}_{GT,x} \times SFC_{GT,x}}{\sum\limits_{x=1}^{n_{GT,y}} \dot{W}_{GTx} + \sum\limits_{v=1}^{n_{ST,y}} \dot{W}_{ST,v}} \tag{36}$$

As seen from the design and operational characteristics of the systems, gas turbine type (a) gives the largest potential of power production with a bottoming cycle for both fuels. Its higher specific fuel consumption is counterbalanced by the exploitation of the thermal energy content of the exhaust gases, which results in a higher contribution of the steam turbine power to the propulsion load and to the lowest annual fuel cost among the three types. Furthermore, the combined cycle specific fuel consumption is the lowest when type (a) is used.

Table 7. Operational technical and economic characteristics for the system with gas turbine type (a) and fuel MDO.

Mode 1	Mode 2	Mode 3
	$\dot{W}_{GT}$ (kW)	
19,240.01	16,256.68	9491.33
	SFC_{GT} (gr/kWh)	
223.808	233.3587	275.5277
	$\dot{Q}_{HRSG}$ (kW)	
25,774.74	22,915.2	16,841.91
	$\dot{m}_{g,HRSG}$ (kg/s)	
54.59	50.52	39.2
	$T_{g,in,HRSG}/T_{g,out,HRSG}$ (°C)	
543.07/135.48	530.50/138.90	510.07/139.16
	$\dot{m}_{HP,HRSG}$ (kg/s)	
6.329	5.635	4.186
	$\dot{m}_{LP,HRSG}$ (kg/s)	
1.632	1.515	1.129
	$\dot{m}_{hl}$ (kg/s)	
0.1545	0.1159	0.0772
	$\dot{W}_{ST}$ (kW)	
8265.38	7274.15	5226.8
	SFC_{CC} (gr/kWh)	
156.5536	161.2199	177.6805
	GT annual fuel cost ($)	
463,3329.06	2,389,991.03	1,694,600.04
	GT annual O&M cost ($)	
310,533.75	153,625.61	92,255.71
	HRSG O&M cost ($)	
350,171.93	182,279.96	137,797.45
	Steam turbine O&M cost $	
88,935.51	46,041.35	33,999.27

Table 8. Operational technical and economic characteristics for the system with gas turbine type (b) and fuel MDO.

Mode 1	Mode 2	Mode 3
	$\dot{W}_{GT}$ (kW)	
22,324.94	18,887.21	11,776.75
	SFC_{GT} (gr/kWh)	
199.9417	206.3503	229.9452
	$\dot{Q}_{HRSG}$ (kW)	
21,188.97	18,445.27	12,476.87
	$\dot{m}_{g,HRSG}$ (kg/s)	
78.21	73.76	62.69
	$T_{g,in,HRSG}/T_{g,out,HRSG}$ (°C)	
378.40/144.49	359.51/143.62	314.96/143.13
	$\dot{m}_{HP,HRSG}$ (kg/s)	
5.05	4.381	2.972
	$\dot{m}_{LP,HRSG}$ (kg/s)	
2.12	1.928	1.379
	$\dot{m}_{hl}$ (kg/s)	
0.1552	0.1164	0.0776
	$\dot{W}_{ST}$ (kW)	
5175.14	4643.9	2938.45
	SFC_{CC} (gr/kWh)	
162.3154	165.6268	184.0279
	GT annual fuel cost ($)	
4,802,925.31	2,455,350.95	1,754,787.9
	GT annual O&M cost ($)	
360,324.5	178,484.18	114,469.96
	HRSG annual O&M cost ($)	
287,870.41	146,723.72	102,083.46
	Steam turbine annual O&M cost ($)	
55,684.52	29,256.6	19,365.14

Table 9. Operational technical and economic characteristics for the system with gas turbine type (c) and fuel MDO.

Mode 1	Mode 2	Mode 3
	$\dot{W}_{GT}$ (kW)	
24,058.85	20,517.82	12,884.59
	SFC_{GT} (gr/kWh)	
185.0074	189.1652	204.3804
	$\dot{Q}_{HRSG}$ (kW)	
14,274.47	12,308.24	8190.35
	$\dot{m}_{g,HRSG}$ (kg/s)	
58.99	54.96	44.28
	$T_{g,in,HRSG}/T_{g,out,HRSG}$ (°C)	
356.47/147.56	340.05/146.72	307.06/147.37
	$\dot{m}_{HP,HRSG}$ (kg/s)	
3.304	2.81	1.84
	$\dot{m}_{LP,HRSG}$ (kg/s)	
1.557	1.427	1.017
	$\dot{m}_{hl}$ (kg/s)	
0.1554	0.1166	0.0777
	$\dot{W}_{ST}$ (kW)	
3444.36	2997.19	1833.82
	SFC_{CC} (gr/kWh)	
161.8380	165.0545	178.9159
	GT annual fuel cost ($)	
4,789,345.29	2,445,191.84	1,706,415.29
	GT annual O&M cost ($)	
388,309.76	193,893.47	125,238.22
	HRSG annual O&M cost ($)	
193,930.97	97,906.42	67,011.93
	Steam turbine O&M cost ($)	
37,061.28	18,882.29	12,207.16

Table 10. Operational technical and economic characteristics for the system with gas turbine type (a) and fuel NG.

Mode 1	Mode 2	Mode 3
	$\dot{W}_{GT}$ (kW)	
20,074.11	17,020.17	10,089.86
	SFC_{GT} (gr/kWh)	
206.1992	213.9852	248.7073
	$\dot{Q}_{HRSG}$ (kW)	
22,434.93	19,618.54	14,238.04
	$\dot{m}_{g,HRSG}$ (kg/s)	
55.03	51.31	40.34
	$T_{g,in,HRSG}/T_{g,out,HRSG}$ (°C)	
541.77/189.81	525.31/195.20	500.82/196.08
	$\dot{m}_{HP,HRSG}$ (kg/s)	
6.186	5.472	4.037
	$\dot{m}_{LP,HRSG}$ (kg/s)	
0.592	0.521	0.374
	$\dot{m}_{hl}$ (kg/s)	
0.1535	0.1151	0.0768
	$\dot{W}_{ST}$ (kW)	
7427.69	6482.47	4615.36
	SFC_{CC} (gr/kWh)	
150.5089	154.9641	170.6484
	GT annual fuel cost ($)	
1,670,193.45	860,437.60	609,789.37
	GT annual O&M cost ($)	
323,996.13	160,840.64	98,073.42
	HRSG annual O&M cost ($)	
304,797.79	156,056.61	116,493.10
	Steam turbine O&M cost ($)	
79,921.97	40,839.54	29,907.51

Table 11. Operational technical and economic characteristics for the system with gas turbine type (b) and fuel NG.

Mode 1	Mode 2	Mode 3
	$\dot{W}_{GT}$ (kW)	
22,785.05	19,424.37	12,170.46
	SFC_{GT} (gr/kWh)	
185.0029	190.456	211.1788
	$\dot{Q}_{HRSG}$ (kW)	
17,376.5	15,159.49	9817.69
	$\dot{m}_{g,HRSG}$ (kg/s)	
77.38	73.12	62.1
	$T_{g,in,HRSG}/T_{g,out,HRSG}$ (°C)	
382.16/188.30	363.97/184.99	319.25/182.76
	$\dot{m}_{HP,HRSG}$ (kg/s)	
4.449	3.921	2.56
	$\dot{m}_{LP,HRSG}$ (kg/s)	
1.349	1.196	0.828
	$\dot{m}_{hl}$ (kg/s)	
0.1539	0.1154	0.0769
	$\dot{W}_{ST}$ (kW)	
4716.93	4092.76	2532.37
	SFC_{CC} (gr/kWh)	
154.2726	157.3103	174.8060
	GT annual fuel cost ($)	
1,700,873.65	874,003.99	624,544.51
	GT annual O&M cost ($)	
367,750.71	183,560.28	118,296.86
	HRSG annual O&M cost ($)	
236,074.62	120,586.84	80,326.59
	Steam turbine annual O&M cost ($)	
50,754.12	25,784.36	16,409.77

Table 12. Operational technical and economic characteristics for the system with gas turbine type (c) and fuel NG.

Mode 1	Mode 2	Mode 3
	$\dot{W}_{GT}$ (kW)	
24,790.93	21,155.23	132,36.94
	SFC_{GT} (gr/kWh)	
171.9147	175.8823	190.2175
	$\dot{Q}_{HRSG}$ (kW)	
10,311.13	8965.8	5865.57
	$\dot{m}_{g,HRSG}$ (kg/s)	
59.51	55.89	45.2
	$T_{g,in,HRSG}/T_{g,out,HRSG}$ (°C)	
360.39/210.81	342.95/204.46	308.26/196.24
	$\dot{m}_{HP,HRSG}$ (kg/s)	
2.934	2.584	1.706
	$\dot{m}_{LP,HRSG}$ (kg/s)	
0.459	0.409	0.286
	$\dot{m}_{hl}$ (kg/s)	
0.154	0.1155	0.077
	$\dot{W}_{ST}$ (kW)	
2709.48	2348.92	1471.93
	SFC_{CC} (gr/kWh)	
153.9768	158.3053	171.1823
	GT annual fuel cost ($)	
1,719,687.12	879,046.09	611,849.37
	GT annual O&M cost ($)	
400,125.68	199,916.93	128,663.12
	HRSG annual O&M cost ($)	
140,085.56	71,318.87	47,991.06
	Steam turbine annual O&M cost ($)	
29,154.02	14,798.17	9538.08

The PWC (objective function) for each one of the six combinations is presented in Table 13, and is lower when type (a) is used for both fuels. As seen, the simplest of the gas turbine types is the best choice in terms of PWC, even if the specific fuel consumption of the gas turbine itself is higher. Furthermore, the simplicity of construction of this type of gas turbine in comparison to the two other types studied probably makes it more appealing for application in integrated ship energy systems. For these reasons, the effects that important parameters have on the optimal solution are investigated for energy systems in which gas turbines of type (a) are used.

Table 13. Optimal PWC for the six combinations of gas turbines and fuels.

Type of Gas Turbine	PWC	
	MDO	NG
A	141,171,375	71,545,007
B	145,300,524	72,237,706
C	143,784,701	71,711,376

2.4.3. Effect of Fuel Price on Optimal Solutions

For the system with gas turbine type (a), which has the best economic performance, the variations of the optimal solution with varying fuel price have been investigated. For this purpose, the SDOO problem has been solved for price of MDO in the range of 300 to 700 $/ton and natural gas price in the range of 100 to 300 $/ton, while the rest of parameters remain at their nominal values. The synthesis of the system remains unaltered and is the same as reported in Table 5, i.e., the inclusion of steam bottoming cycle is economically feasible in all cases. The design characteristics of the system components are given in Tables 14 and 15 for the various prices of MDO and natural gas, respectively.

Table 14. Optimal design characteristics and capital cost of the system components for various MDO prices.

Fuel Price ($/ton)	300	400	500	600	700
$\dot{W}_{GT}$ (kW)	23,976	23,828	23,803	23,739	23,731
$\dot{Q}_{HRSG}$ (kW)	22,688	23,663	24,293	25,893	26,579
$\dot{m}_{g,HRSG}$ (kg/s)	47.53	47.36	48.22	51.07	51.82
$\dot{m}_{HP,HRSG}$ (kg/s)	5.461	5.561	5.708	6.078	6.224
$\dot{m}_{LP,HRSG}$ (kg/s)	1.351	1.631	1.696	1.799	1.838
P_{HP} (bar)	65.34	63.58	66.25	66.62	65.62
P_{LP} (bar)	8.24	6.12	5.73	5.67	5.68
$T_{g,in,HRSG}$ (°C)	592.00	584.90	585.18	585.72	590.07
$T_{g,out,HRSG}$ (°C)	179.90	153.57	150.20	148.03	147.24
$T_{HP,HRSG}$ (°C)	560.26	552.46	550.63	555.46	561.56
$T_{LP,HRSG}$ (°C)	193.51	182.85	180.53	179.56	176.17
$\dot{W}_{ST}$ (kW)	8677	8624	8613	8731	8636
$\dot{m}_{HP,ST}$ (kg/s)	6.380	6.431	6.412	6.396	6.384
$\dot{m}_{LP,ST}$ (kg/s)	1.320	1.608	1.658	1.670	1.730
$T_{HP,ST}$ (°C)	567.72	543.53	542.84	553.76	541.26
$T_{LP,ST}$ (°C)	188.78	179.87	178.17	182.69	176.23
$\dot{W}_{DG}$ (kW)	1205	1206	1204	1210	1205
$C_{c,GT}$ ($)	10,746,899	10,718,229	10,713,354	10,700,901	10,699,234
$C_{c,HRSG}$ ($)	2,495,601	2,684,082	2,777,585	2,982,490	3,043,967
$C_{c,ST}$ ($)	1,485,766	1,498,472	1,500,705	1,506,214	1,507,158
$C_{c,DG}$ ($)	874,410	874,689	874,130	875,808	874,410

Table 15. Optimal design characteristics and capital cost of the system components for various prices of natural gas.

Fuel Price ($/ton)	100	150	200	250	300
$\dot{W}_{GT}$ (kW)	25,167	24,662	24,380	24,170	23,831
$\dot{Q}_{HRSG}$ (kW)	19,119	20,623	22,119	23,320	23,551
$\dot{m}_{g,HRSG}$ (kg/s)	57.50	54.42	50.92	50.17	47.30
$\dot{m}_{HP,HRSG}$ (kg/s)	5.396	5.565	5.464	5.582	5.498
$\dot{m}_{LP,HRSG}$ (kg/s)	0.325	0.608	1.196	1.503	1.634
P_{HP} (bar)	62.85	63.67	65.69	65.60	67.69
P_{LP} (bar)	7.54	9.00	9.45	7.15	6.44
$T_{g,in,HRSG}$ (°C)	538.97	554.08	575.65	575.85	591.51
$T_{g,out,HRSG}$ (°C)	251.91	226.91	200.61	174.60	161.73
$T_{HP,HRSG}$ (°C)	507.53	522.89	544.63	544.11	559.62
$T_{LP,HRSG}$ (°C)	191.47	199.51	199.82	189.16	185.16
$\dot{W}_{ST}$ (kW)	6983	7589	7943	8174	8491
$\dot{m}_{HP,ST}$ (kg/s)	6.254	6.276	6.152	6.231	6.329
$\dot{m}_{LP,ST}$ (kg/s)	0.290	0.549	1.096	1.386	1.531
$T_{HP,ST}$ (°C)	476.26	513.43	529.52	530.73	544.12
$T_{LP,ST}$ (°C)	180.24	187.96	185.65	167.34	162.92
$\dot{W}_{DG}$ (kW)	1205	1207	1208	1209	1207
$C_{c,GT}$ ($)	10,973,925	10,878,489	10,824,556	10,784,367	10,718,674
$C_{c,HRSG}$ ($)	2,149,778	2,332,648	2,454,784	2,643,177	2,663,628
$C_{c,ST}$ ($)	1,378,655	1,425,071	1,452,170	1,469,284	1,469,208
$C_{c,DG}$ ($)	874,410	874,958	875,249	875,529	874,969

With increasing fuel price, the power capacity of the steam bottoming cycle generally also increases. For a better visualization of the operating performance of the bottoming cycle, the variation of $\dot{Q}_{HRSG}$ and $\dot{W}_{ST}$ and of the fraction $\dot{W}_{ST}/\dot{W}_{GT}$ is diagrammatically presented in Figure 6a,b for three voyage modes as functions of the fuel price. It is generally noticed that as the fuel price increases, the steam bottoming cycle recovers more thermal energy from the exhaust gas and produces more mechanical power during all operating modes, with this trend being more evident in the case of natural gas.

With both fuels, and in the whole range of fuel prices examined, the fraction $\dot{W}_{ST}/\dot{W}_{GT}$ has a significantly large value, indicating the importance of the steam bottoming cycle in energy systems where gas turbines are used as main engines. One more important attribute observed in Figure 6a,b is that the value of the fraction $\dot{W}_{ST}/\dot{W}_{GT}$ increases significantly in operating mode 3; mode 2 is higher than in mode 1, as in mode 1 the gas turbine operates closer to the nominal power rating and has higher thermal efficiency. This means that the design of the bottoming cycle is carried out in a way that the need for increasing the thermal efficiency of the overall energy system in modes where the main engine does not operate quite efficiently is taken into account in the optimization procedure.

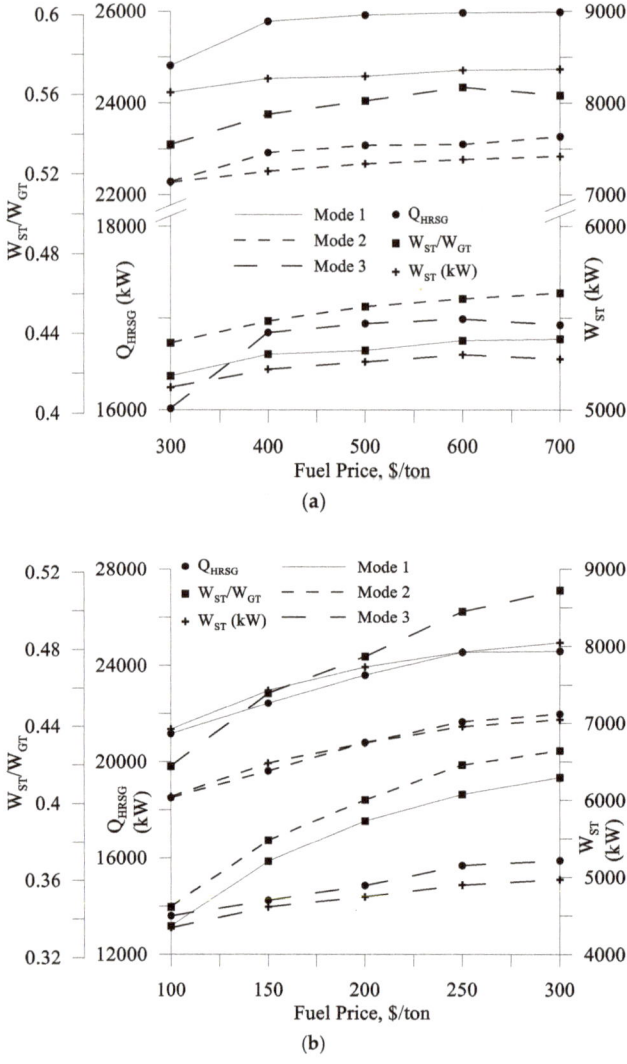

Figure 6. (a) Effect of MDO price on the optimal values of $\dot{Q}_{HRSG}$, $\dot{W}_{ST}$ and (b) the effect of natural gas price on the optimal values of $\dot{Q}_{HRSG}$, $\dot{W}_{ST}$, and $\dot{W}_{ST}/\dot{W}_{GT}$.

The optimal PWC of the investment for varying fuel price is presented in Table 16. The variation of the PWC with the fuel price is nearly linear for both fuels, indicating the major contribution that the cost of fuel has on the objective function.

Table 16. Optimal PWC for varying fuel price.

MDO		Natural Gas	
Fuel Price ($/ton)	PWC ($)	Fuel Price ($/ton)	PWC ($)
300	114,302,675	100	58,889,296
400	141,171,375	150	71,545,007
500	167,995,232	200	84,112,723
600	194,661,060	250	96,522,638
700	221,544,251	300	110,365,050

2.4.4. Effect of Capital Cost on Optimal Solutions

For the system with gas turbine type (a), the effect of the capital cost on the optimal solution has also been investigated. For this purpose, the components capital costs were multiplied with a capital cost factor, which was given the values 0.5 and 2, and the optimization problems were solved for both fuels. The optimal synthesis of the system again remains unaltered. The variation of the design characteristics of the components is reported in Table 17.

Table 17. Optimal design characteristics and capital cost of components for various capital costs.

Fuel	MDO			Natural Gas		
Capital cost factor	0.5	1	2	0.5	1	2
$\dot{W}_{GT}$ (kW)	23,811	23,828	24,014	24,420	24,662	24,793
$\dot{Q}_{HRSG}$ (kW)	24,659	23,663	23,489	22,124	20,623	20,427
$\dot{m}_{g,HRSG}$ (kg/s)	47.05	47.36	50.84	49.36	54.42	54.84
$\dot{m}_{HP,HRSG}$ (kg/s)	5.720	5.561	5.437	5.652	5.565	5.560
$\dot{m}_{LP,HRSG}$ (kg/s)	1.716	1.631	1.762	0.890	0.608	0.506
P_{HP} (bar)	65.52	63.58	57.47	64.09	63.67	52.50
P_{LP} (bar)	6.72	6.12	5.06	9.12	9.00	7.44
$T_{g,in,HRSG}$ (°C)	602.23	584.90	571.27	586.45	554.08	554.76
$T_{g,out,HRSG}$ (°C)	149.76	153.57	172.40	199.45	226.91	233.17
$T_{HP,HRSG}$ (°C)	571.52	552.46	538.83	557.20	522.89	522.39
$T_{LP,HRSG}$ (°C)	180.02	182.85	177.36	201.83	199.51	193.40
$\dot{W}_{ST}$ (kW)	9085	8624	8323	8471	7589	7491
$\dot{m}_{HP,ST}$ (kg/s)	6.264	6.431	6.386	6.281	6.276	6.296
$\dot{m}_{LP,ST}$ (kg/s)	1.701	1.608	1.625	0.879	0.549	0.494
$T_{HP,ST}$ (°C)	607.69	543.53	575.96	590.38	513.43	535.96
$T_{LP,ST}$ (°C)	191.89	179.87	175.92	204.76	187.96	200.54
$\dot{W}_{DG}$ (kW)	1206	1206	1209	1204	1207	1209
$C_{c,GT}$ ($)	5,357,419	10,718,229	21,508,254	5,416,119	10,878,489	21,806,551
$C_{c,HRSG}$ ($)	1,398,610	2,684,082	4,927,116	1,251,530	2,332,648	4,315,563
$C_{c,ST}$ ($)	767,846	1,498,472	2,947,318	750,324	1,425,071	2,832,784
$C_{c,DG}$ ($)	437,345	874,689	1,751,058	437,065	874,958	1,751,058

Figure 7a,b depicts the variation of $\dot{Q}_{HRSG}$ and $\dot{W}_{ST}$ and of the fraction $\dot{W}_{ST}/\dot{W}_{GT}$ with the capital cost in the three sailing modes. The variation of $\dot{Q}_{HRSG}$ and $\dot{W}_{ST}$ with the capital cost is not significant in the case of MDO, with a slight reduction of the mechanical power production being observed as the capital cost increases. More noticeable is the effect of capital cost in case of natural gas, with a significant increase of the contribution of the bottoming cycle when the capital costs are decreased to half the nominal values.

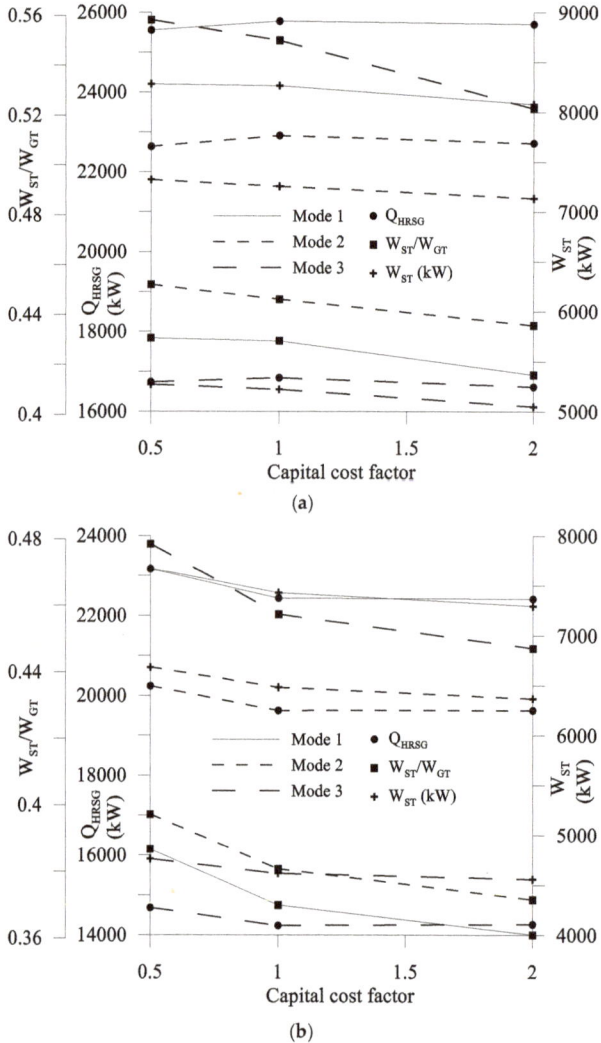

Figure 7. (**a**) Effect of capital cost on the optimal values of $\dot{Q}_{HRSG}$ and $\dot{W}_{ST}$ and on the fraction $\dot{W}_{ST}/\dot{W}_{GT}$ with MDO. (**b**) Effect of capital cost on the optimal values of $\dot{Q}_{HRSG}$ and $\dot{W}_{ST}$ and on the fraction $\dot{W}_{ST}/\dot{W}_{GT}$ for natural gas.

The capital cost of the bottoming cycle components is given in Table 18, along with the O&M PWC of the bottoming cycle. The increased power production in the case of natural gas for capital cost factor 0.5 is reflected in the increased related O&M cost.

Table 18. Optimal PWC and bottoming cycle (BC) costs for varying capital cost factors.

Fuel	Capital Cost Factor	Total PWC ($)	Capital Cost of BC ($)	PWC of O&M of BC ($)
	0.5	132,975,803	2,148,514	10,310,416
MDO	1	141,171,375	4,182,555	10,280,546
	2	157,641,585	7,892,692	10,269,102
	0.5	63,561,140	1,974,960	9,204,124
NG	1	71,545,007	3,757,720	8,918,232
	2	87,119,816	7,141,844	8,915,494

2.5. General Comments on the Results of Section 2

The method for the synthesis, design, and operation optimization of integrated energy systems of ships, presented in a preceding paper, has been applied here properly supplemented with additional steps for the SDOO of a system with gas turbines in three different types as main engines operating either on MDO or natural gas. It is found that, with minimization of the present worth cost of the system as objective and for the values of parameters considered in this study, a steam bottoming cycle is always feasible, while the simple gas turbine configuration results in the lowest value of the present worth cost. The effect of varying fuel prices and capital costs on the optimal synthesis, design and operation of the system is further investigated.

The increase of fuel price results in a steam bottoming cycle of, generally, increased power capacity. Also, the fact that the MDO price is, in general, higher than the price of natural gas (per unit of fuel energy) results in a bottoming cycle of higher capacity. The capital costs also affect the optimization results with the effect being stronger in case of natural gas.

The modeling of the system and the procedure for solution of the synthesis, design and operation optimization problem allow for taking into consideration the effects of all the operating modes simultaneously.

3. Intertemporal Dynamic SDOO of an Energy System of Ship based on Gas Turbines, 2-X Diesel Engines, and 4-X Diesel Engines as Main Engines

3.1. Description of the System

In this problem, the optimal SDO of an integrated energy system of a ship that will serve all energy demands is requested, taking into consideration weather conditions changing with space and time.

The problem is specified in an appropriate way that simultaneously the time horizon of a whole year of operation is considered. Specifically, the ship performs a characteristic round trip (between ports A and B) which includes the necessary amount of time (and service of energy needs) that is required while staying at both ports. The duration of each trip (in all round trips) for each season is variable and under optimization. In that way, the number of round trips per season and consequently the total number of round trips per year is not fixed, but it is also optimized. It is noted that the number of round trips for each season can be a decimal number so as to model the (possible) passage from one season to the next in the same round trip. However, the problem is set in an appropriate manner so as the total number of annual round trips is an integer.

The propulsion power demand is not prespecified, because it is a function of speed and weather conditions. The ship speed at any instant of time is an optimization variable. Also, the wind speed and direction encountered by the ship during each trip, for each season, are given as inputs. The wave height and direction are then calculated, since they are correlated with the wind speed with the help of the Beaufort scale. Once these parameters are determined, the resistance and propulsion power calculations are performed.

The electrical and thermal loads are also parameters of the problem and are given as inputs. They are defined to vary with time but in a different manner for each of the four seasons.

In Figure 8, a superconfiguration of the ship energy system is presented. Three types of gas turbines, 4-stroke Diesel engines and 2-stroke Diesel engines are available as technology alternatives for the synthesis of the propulsion plant. The number and type of propulsion engines that will be installed is determined by the optimization and they will drive a single propeller. Furthermore, single-pressure HRSGs and steam turbine(s) may be installed. Saturated steam extraction from the drum of the HRSGs will be used to, completely or partly, serve the ship thermal demands, while the superheated steam produced by the HRSGs will be led to the steam turbine(s). The number of HRSGs and STs that will be installed is again determined by the optimization. The produced steam turbine(s) power will be distributed between the propeller and a generator, which will supply electric power for the electric loads. Finally, a number of (decided by the optimization) Diesel generator sets and an auxiliary boiler will be included in the system, in order to supply electric and thermal power during voyages, if the STGs and HRSGs cannot completely satisfy the demands. Also, the thermal and electric demands in ports will be completely served by the auxiliary boiler and Diesel generator set(s).

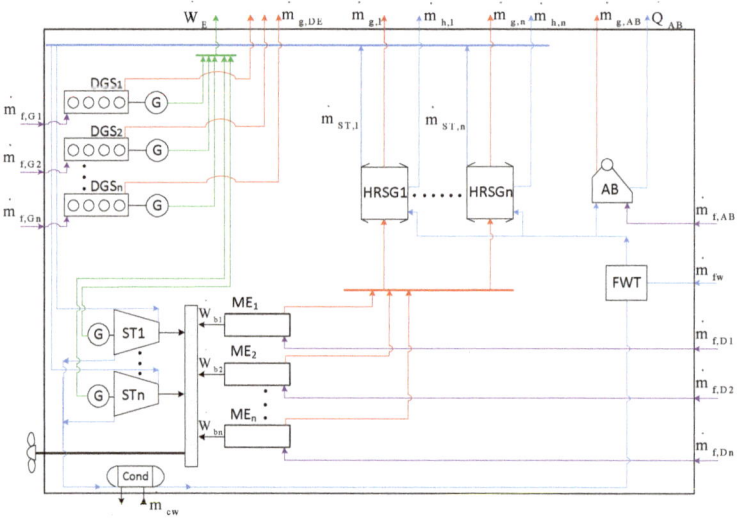

Figure 8. Superconfiguration of the generic energy system. (AB: auxiliary boiler; Cond: condenser; DGS: Diesel generator set; FWT: feed water tank, G: generator; HRSG: heat recovery steam generator; ME: main engine, ST: steam turbine).

For each trip, a suitable freight rate is defined and, thus, the corresponding revenue can be calculated. The economic criterion that serves as the objective function of the optimization problem is the Net Present Value (NPV) after 20 years of operation and the goal is its maximization. The problem is solved for the nominal parameter values; then, a parametric study for the fuel price and the freight rate is performed.

3.2. Mathematical Statement of the Optimization Problem

The dynamic optimization problem can be mathematically stated using a Differential Algebraic Equation (DAE) formulation. The objective function, maximization of the NPV, is stated mathematically as

$$\max_{\vec{x},\ \vec{t}_f} NPV = PWR - PWC \tag{37}$$

where

PWR: present worth of revenue

$$PWR = f_r C_{load}(2d_{AB})TEU \cdot N_{trips,a} \cdot PWF(N_n, i) \tag{38}$$

f_r: freight rate (in €/nm/TEU); C_{load}: safety loading factor of containership; d_{AB}: distance between ports A and B; TEU: containership cargo capacity; $N_{trips,a}$: total (annual) number of round trips; N_t: nominal technical life of the system; i: market interest rate; PWF: Present Worth Factor.

$$PWF(N_t, i) = \frac{(1+i)^{N_t} - 1}{i \cdot (1+i)^{N_t}} \tag{39}$$

PWC present worth cost:

$$PWC = PWC_c + PWC_f + PWC_{om} \tag{40}$$

PWC_c capital present worth cost,
PWC_f fuel present worth cost,
PWC_{om} operation and maintenance present worth cost,
$\vec{t}_f$ and $\vec{x}$ vectors containing the control (optimization) variables of the problem. Vector $\vec{t}_f$ consists of the single trip durations for each season:

$$\vec{t}_f = (t_{trip,AB,s}, t_{trip,BA,s}) \text{ for } s = 1,2,3,4 \tag{41}$$

Based on this definition, the total annual number of round trips can be given as the sum of the round trips of each season:

$$N_{trips,a} = \sum_{s=1}^{4} N_{trips,s} = \sum_{s=1}^{4} \left(\frac{\tau_s}{t_{trip,AB,s} + t_{trip,BA,s} + t_{port,A,s} + t_{port,B,s}} \right) \tag{42}$$

where
τ_s maximum permissible annual hours of operation for season s,
$t_{trip,AB,s}$ duration of trip from port A to B for season s,
$t_{trip,BA,s}$ duration of trip from port B to A for season s.
$t_{port,A,s}$ time spend in port A for season s (in a round trip),
$t_{port,B,s}$ time spend in port B for season s (in a round trip).
Vector $\vec{x}$ consists of the vectors of synthesis, design, and operation optimization variables ($\vec{v}, \vec{w}$ and $\vec{z}$, respectively):

$$\vec{x} = (\vec{v}, \vec{w}, \vec{z}) \tag{43}$$

with

$$\vec{v} = (z_{D,2-X}, z_{D,4-X}, z_{GT1}, z_{GT2}, z_{GT3}, z_B, z_{ST}, z_{DG}, y_{AB}) \tag{44}$$

$$\vec{w} = \left(\dot{W}_{bn,i,j}, \dot{m}_{gn,k}, T_{gn,k}, \dot{m}_{sn,k}, \dot{m}_{STn,l}, \dot{W}_{DGn,m}, \dot{Q}_{ABn} \right) \tag{45}$$

$$\vec{z} = \left(\dot{W}_{b,i,j}, \lambda_{h,k}, \lambda_{e,l}, \dot{W}_{DG,m} \right) \tag{46}$$

where

$$j = \begin{cases} 0, \dots, z_{D,2-X} & \text{for } i = D, 2-X \\ 0, \dots, z_{D,4-X} & \text{for } i = D, 4-X \\ 0, \dots, z_{GT1} & \text{for } i = GT1 \\ 0, \dots, z_{GT2} & \text{for } i = GT2 \\ 0, \dots, z_{GT3} & \text{for } i = GT3 \end{cases} \tag{47}$$

$$k = 0, \ldots, z_B \tag{48}$$

$$l = 0, \ldots, z_{ST} \tag{49}$$

$$m = 0, \ldots, z_{DG} \tag{50}$$

and

$z_{D,2-X}$ number of two stroke Diesel engines (integer variable),

$z_{D,4-X}$ number of four stroke Diesel engines (integer variable),

z_{GT1} number of type (a) gas turbines (integer variable),

z_{GT2} number of type (b) gas turbines (integer variable),

z_{GT3} number of type (c) gas turbines (integer variable),

z_B number of heat recovery steam generators (integer variable),

z_{ST} number steam turbines (integer variable),

z_{DG} number of Diesel generator sets (integer variable),

y_{AB} variable determining the existence of the auxiliary boiler (binary variable),

$W_{bn,i,j}$ nominal brake power output of jth engine of type i (invariant, i.e., time-independent optimization variable),

$\dot{m}_{gn,k}$ nominal exhaust gas mass flow rate of kth HRSG (invariant),

$T_{gn,k}$ nominal exhaust gas temperature of kth HRSG (invariant),

$\dot{m}_{sn,k}$ nominal steam mass flow rate of kth HRSG (invariant),

$\dot{m}_{ST_n,l}$ nominal steam mass flow rate of lth ST (invariant),

$\dot{W}_{DG_n,m}$ nominal power output of mth generator set (invariant),

$\dot{Q}_{AB_n}$ nominal thermal power output of auxiliary boiler (invariant),

$W_{b,i,j}$ brake power output of jth engine of type i,

$\lambda_{h,k}$ fraction of kth HRSG steam mass flow rate for serving thermal loads:

$$\dot{m}_{s,h,k} = \lambda_{h,k} \cdot \dot{m}_{s,k} \tag{51}$$

$\dot{m}_{s,h,k}$ steam mass flow rate drawn from kth HRSG drum for serving thermal loads,

$\dot{m}_{s,k}$ steam mass flow rate of kth HRSG unit,

$\lambda_{e,l}$ fraction of lth steam turbine power output delivered to generator:

$$\dot{W}_{STG,l} = \lambda_{e,l} \cdot \dot{W}_{ST,l} \tag{52}$$

$\dot{W}_{STG,l}$ lth steam turbine generator power for serving electric loads,

$\dot{W}_{ST,l}$ lth steam turbine power output,

$\dot{W}_{DG,m}$ mth Diesel generator set power output.

Indexes j, k, l, and m run through all the values from 0 up to an upper value. At the beginning of the optimization, the upper values of indexes j, k, l, and m are not fixed, since they are in fact defined by the values of their respective integer synthesis variables. However, they are bound from above with the same upper bounds of these respective integer synthesis variables, which must be well determined and fixed at the start of the optimization. Specifically, as can be seen from Equation (47), the upper value of index j depends on the index i which determines the type of propulsion equipment and from the respective value of the, under optimization, integer variable that determines how many components of type i will be installed. Thus, the integer values of the synthesis control variables dictate the number of the design and operation variables for the components. Variable y_{AB} that determines the existence of the auxiliary boiler is binary. In both cases of integer and binary variables, value 0 denotes that the unit is not installed.

The main differential variables for the specific problem are as follows. The distance travelled by the ship:

$$\frac{d}{dt}D_{traveled} = V \tag{53}$$

The fuel consumption of the propulsion engines and Diesel generator sets, given by the product of the Specific Fuel Consumption (SFC) with the produced brake power as

$$\frac{d}{dt}m_f = b_f \dot{W}_b \tag{54}$$

and the fuel consumption of the auxiliary boiler

$$\frac{d}{dt}m_{f,AB} = \frac{\dot{Q}_{AB}}{\eta_{AB} \cdot H_u} \tag{55}$$

where the efficiency of the boiler, η_{AB}, is considered a constant parameter.

Another family of differential variables is derived from the energy output of each component, which is generally given as

$$\frac{d}{dt}E = \dot{Y}, \quad \dot{Y} = \dot{W}, \dot{Q} \tag{56}$$

The operational costs of the system, such as the fuel and operation and maintenance costs for each component, are calculated based on the energy output and the fuel consumption of each component. The capital costs for each component are calculated using the values of the design variables. For the gas turbines, the cost function given in Appendix A has been used. For the remaining components, the capital costs are calculated as described in Tzortzis and Frangopoulos [60].

Since the propulsion plant characteristics, i.e., type, number, and nominal power of the engines, are not known in advance but they are derived by optimization, the Specific Fuel Consumption (SFC) should not be given only as a function of the load factor. Thus, based on appropriate models and manufacturer data, SFC surfaces are constructed, where the SFC for each engine of type i as well as the exhaust gas properties (mass flow rate and temperature) are given as functions of the nominal power and the load factor. The same procedure is also applied in the modeling of Diesel generator sets SFC and exhaust gas properties. The models used for the propulsion engines and the Diesel generator sets are presented in Section 3.3.

The sum of the brake power of the main engines and the steam turbine(s) must be equal to the brake power demand, as stated by the equality constraint of Equation (57). Also, the electric and thermal power produced by the integrated system must be equal to the electric and thermal demands of the ship, as stated in Equations (58) and (59), respectively.

$$\sum_{i,j} \dot{W}_{b,i,j} + \sum_l \dot{W}_{ST,p,l} = \sum_{i,j} \dot{W}_{b,i,j} + \sum_l (1 - \lambda_{e,l}) \cdot \dot{W}_{ST,l} = \dot{W}_b \tag{57}$$

$$\sum_l \dot{W}_{STG,l} + \sum_m \dot{W}_{DG,m} = \sum_l \lambda_{e,l} \cdot \dot{W}_{ST,l} + \sum_m \dot{W}_{DG,m} = \dot{W}_e \tag{58}$$

$$\sum_k \dot{Q}_{B,k} + \dot{Q}_{AB} = \dot{Q} \tag{59}$$

where $\dot{W}_{ST,p,l}$: propulsion power from lth ST, $\dot{W}_b$: required brake power from the engines, $\dot{W}_e$: electric load, $\dot{Q}_{B,k}$: heat drawn from kth HRSG drum for serving thermal loads, $\dot{Q}$: thermal load.

It is noted that the brake power demand is calculated as a function of the ship resistance, R_{tot}, propulsive efficiency, η_{prop}, and ship speed, V, as

$$\dot{W}_b = \frac{V \cdot R_{tot}(V, WS, \mathbf{p})}{\eta_{prop}(V, WS, \mathbf{p})} \tag{60}$$

where, WS: weather state and; $\mathbf{p}$: constant parameters describing the vessel.

Also, there are a significant number of equalities and inequalities related to the simulation of each component, but their full presentation is beyond the limits of this text. Noteworthy inequality constraints include the bounds imposed on the speed of the ship and the load factor, f_L, of all components (main engines, steam turbines, Diesel generator sets, etc.) that ensure the compliance with the operational limits specified by the manufacturer:

$$V_{min} \leq V \leq V_{max} \tag{61}$$

$$f_{L_{min}} \leq f_L \leq f_{L_{max}} \tag{62}$$

Of course, all control variables are accompanied by upper and lower bounds. However, the upper and lower bounds may not be necessary for all state variables.

Finally, additional constraints can easily be imposed by emission regulations if, for example, the ship travels within emission controlled areas (ECAs). However, such a scenario is not studied in this work.

3.3. Modeling of Main Components

For the simulation of the individual components (at both design and off-design operating points) that constitute the integrated marine energy system presented in Figure 8, specific models are used that can be divided in two categories. Those that have been developed using a first principles approach combined with literature data, such as the models for the single pressure HRSG, the ST and the resistance–propulsion model, and those that are based on regression analysis of data, such as the models for the Diesel engines (two and four stroke), the auxiliary boiler and the DG sets. All details considering the mathematical formulation of these models, the specific values of all model parameters and the relative references can be found in Tzortzis and Frangopoulos [60].

For each of these components, apart from performance models, cost models for the capital cost of equipment and for the operation and maintenance costs are also developed. Again, the corresponding equations, values of parameters and references can be found in Tzortzis and Frangopoulos [60].

Considering the GTs, the same modeling procedure, in terms of performance and in terms of (capital and operation and maintenance) costs that was described in Section 2, is applied.

3.4. Treatment of Synthesis Variables, Solution Procedure and Related Software

According to the mathematical statement presented in Section 3.2, the problem is stated (and consequently treated by the optimization procedure) in a single level as a Mixed Integer Dynamic Optimization Problem (MIDO). The distinction between the three levels of synthesis, design, and operation is only conceptual; however, it is reflected in the general mathematical formulation in terms of the type of variables used to describe each level.

For the level of operation continuous real variables are used that change at each instant of time, for the level of design "static" or invariant real variables are used, and for the level of synthesis integer and binary static variables are used.

The values of the integer synthesis variables have a tremendous effect on the whole problem, since the specific value of each variable affects the underlying design and operation levels in terms of the number of (design and operation) variables that should be present in the problem as well as in terms of the underlying system of equations. Essentially, this means that each time one integer variable changes value, the optimization problem must be reformulated either by adding the necessary extra variables and their related equations or by subtracting them, depending on the increase or the decrease of the value of the integer variable.

Of course, this adversity could be treated by using a conventional "if … then … else" custom algorithmic formulation for each integer variable, where for each value of the variable the underlying system (variables and equations) would be reformulated. However, this would not be a true single-level treatment of the problem, and it would be impossible to apply any gradient based dynamic optimization method for the solution of the problem. Furthermore, the complexity of the required code would be highly increased.

In order to tackle with this specific difficulty, a transcription technique of integer variables to binary variables based on the idea of the superconfiguration (Figure 8) has been applied. The idea is to simultaneously consider all possible technology alternatives of the system and for each alternative to consider the maximum number of units, given by the upper bound of the respective integer variable. Then, each unit can be represented by a binary variable that determines the existence or not of the said unit. In this way, each integer variable that is present in the formulation of Section 3.2 is translated into a series of binary variables and thus all integer variables are eliminated from the system. In other words, each value of each integer variable now corresponds to a binary variable.

Furthermore, since now only binary variables are used, it can be arranged so that the value 1 corresponds to the existence of the specific component while the value 0 will correspond to the exclusion of the specific component from the system. This feature can be used to our advantage, since now, instead of using an "if … then … else" strategy, a more compact formulation can be applied. The problem can be stated with the maximum possible number of design and operation variables with all their accompanying equations (model equations, constraints, costs, etc.) multiplied by the respected binary variable. The idea is that, if the optimizer dictates the installation of a component (thus it will set the relative binary variable equal to 1) the accompanying system of equations will not be affected. The cost calculations, pertinent to the component, will participate in the objective function calculations and the relative gradients will not be zero. However, if the relative binary variable is set to zero, although all relative to the component variables and equations will still be present in the system, they will not affect the optimization.

In order to solve the MIDO problem posed in this study, a direct sequential method (Figure 9) is selected and implemented. The basic procedure is presented in Figure 9 and is summarized as follows.

1. Insertion of the initial value of the duration of each control interval and the initial values of the control variables over each interval.
2. Integration of the dynamic system model over the entire time horizon and determination of the variation (with time) of all state variables in the system.
3. Calculation of the values of the objective function and constraints as well as the values of their partial derivatives (sensitivities) with respect to all quantities specified.
4. Revision of the choices made on step 1 by a suitable NLP optimizer and repetition of the procedure until convergence criteria are met.

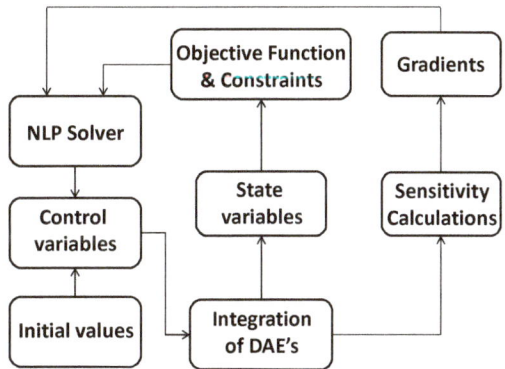

Figure 9. Sketch of the sequential approach used.

The application of the sequential method and the dynamic simulation of the models used to calculate ship resistance, propulsion, performance of main engines, HRSGs, steam turbines, and Diesel generator set units and their interconnections, as well as the effects of the dynamically varying weather and loads, was implemented in the commercial gPROMS® software. For the control variables a piecewise constant parameterization scheme, over equally spaced time intervals, is selected. In the gPROMS® environment, the overall direct sequential method is implemented via the solver CVP_SS. The user imports the control variable parameterization and CVP_SS links everything to the NLPSQP solver which handles the NLP optimization problem. The DAE problem is tackled by the DASOLV solver, which also performs the computation of sensitivities. The BDNLSOL solver performs the initialization and re-initialization activities when DASOLV is used for simulation. Finally, the OAERAP solver handles the mixed integer part of the problem (i.e., the binary variables). Details on all solvers are available in the gPROMS®user guides [61], which can be downloaded from the PSE website.

Often in the literature, as well as in the gPROMS® documentation, the sequential method presented above is referred to as a "single shooting" method. The term is derived from step 2 of the algorithm presented above, which involves a single integration of the dynamic model (DAE system of equations) over the entire time horizon. Further details about the single shooting methods can be found in the literature [62].

3.5. Data and Assumptions

A containership with carrying capacity of 9572 TEU and DWT of 111,529 MT is considered that for each season performs a characteristic round trip between ports A and B−$d_{AB} = 3000$ nm. All vessel characteristics, such as ship dimensions and coefficients, which are used in order to provide an accurate calculation of ship resistance and required propulsion power, are given in Table 19.

Table 19. Vessel dimensions propulsion power and related coefficients for containership.

Parameter	Symbol	Value
Ship Dimensions		
Overall length (m)	L_{OA}	336
Length between perpendiculars (m)	L_{pp}	321
Length at the waterline (m)	L_{WL}	317
Breadth (m)	B	45
Draught (m)	T	15
Forward molded draught (m)	T_F	14.7
Aft molded draught (m)	T_A	15.2
Draught at midship (m)	T_M	15
Wetted volume (m^2)	∇	146,491
Wetted Surface	S	19,029
Ship Hull Coefficients		
Block coefficient	C_B	0.6506
Prismatic coefficient	C_P	0.6605
Waterplane area coefficient	C_{WP}	0.8560
Midship section coefficient	C_M	0.9850
Longitudinal position at the centre of buoyancy (m)	lcb	152.7
Height at the centre of transverse area (m)	h_B	22
Propulsion Power Coefficients		
Bearing efficiency	η_b	0.98
Stern-tube efficiency	η_{st}	0.97
Gearing efficiency	η_g	0.99
Rotative efficiency	η_r	0.98
Open water efficiency	η_o	0.99
Service speed (kn)	V_S	24
Brake power at service speed (kW)		69,439

The operational profile of the ship is approximated with four modes of operation (loading and off-loading at ports, loaded trip from port A to port B and loaded trip for port B to port A). In this study, maneuvering periods are not considered since their duration is much shorter compared to the duration of the whole round trip. Thus, their effect on the objective function can be neglected. The time schedule of the ship for each round trip and all seasons is given in Table 20.

Table 20. Time schedule of the ship.

Mode	Description	Duration
1	Off-loading and loading at port A (all seasons)	2 days
2	Loaded trip from port A to port B (all seasons)	Variable
3	Off-loading and loading at port B (all seasons)	2 days
4	Loaded trip from port B to port A (all seasons)	Variable
	Total round trip	Variable

The electric and thermal loads used as inputs are given in Tables 21 and 22. They are defined as functions of time for an 8-day time horizon, differing for each season; also, they are considered constant at ports.

Table 21. Thermal power demands for each season.

Time (Days)	Thermal Power (kW)			
	Summer	Fall	Winter	Spring
From port A to port B				
1	850	860	1000	990
2	880	900	1050	980
3	860	950	1080	1010
4	900	970	1100	1020
5	840	930	1060	950
6	850	960	1100	970
7	845	959	1100	980
8	851	963	1090	970
From port B to port A				
1	860	930	1100	1040
2	870	950	1080	970
3	870	980	1060	960
4	890	970	990	950
5	860	890	1040	980
6	870	910	1070	960
7	880	920	1080	970
8	870	910	1050	960
Ports				
A	950	950	950	950
B	950	950	950	950

Table 22. Electric power demands for each season.

Time (Days)	Electric Power (kW)			
	Summer	Fall	Winter	Spring
From port A to port B				
1	1495	1508	1482	1625
2	1599	1573	1586	1599
3	1560	1547	1482	1547
4	1514	1560	1508	1495
5	1508	1495	1586	1534
6	1495	1495	1495	1521
7	1490	1495	1485	1525
8	1495	1495	1490	1520
From port B to port A				
1	1625	1547	1534	1521
2	1560	1625	1521	1625
3	1534	1664	1625	1573
4	1521	1508	1651	1625
5	1560	1501	1495	1599
6	1521	1547	1521	1586
7	1520	1539	1520	1580
8	1525	1550	1521	1580
Ports				
A	1500	1500	1500	1500
B	1500	1500	1500	1500

The wind speed is a function of time and space for each season and can be found in Tables A2–A5 (Appendix B). The wind direction is given in Table A6, also as a function of time and space, but is assumed to remain constant over all seasons. All data have been gathered from internet sites that are accessible to anyone, which are specialized in accurate real-time, as well as historical weather data, for any region (sea or land) [63].

Values of certain cost parameters that are used for the PWC and NPV calculations are given in Table 23. For the gas turbines, MDO is considered as fuel with a Lower Heating Value (LHV) of 42,700 kJ/kg, while for the 2-X Diesel engines, the 4-X Diesel engines, the Diesel generator sets, and the auxiliary boiler, HFO is considered as a fuel with a lower heating value (LHV) of 39,550 kJ/kg.

Table 23. Economic parameters.

Parameter	Symbol	Value
MDO fuel cost nominal value	$c_{f,MDO}$	0.450 €/kg
HFO fuel cost nominal value	$c_{f,HFO}$	0.300 €/kg
Technical life of the system	N_t	20 years
Maximum permissible hours of operation—Summer	τ_1	2160
Maximum permissible hours of operation—Fall	τ_2	2160
Maximum permissible hours of operation—Winter	τ_3	1450
Maximum permissible hours of operation—Spring	τ_4	2160
Market interest rate	i	10%
Freight rate (nominal value)	f_r	0.0326€/TEU nm
Loading factor of containership	C_{load}	0.85

In Table 24, a list of lower and upper bounds of certain SDO variables is presented. Details regarding several numerical solution parameters are given in Table 25. It is noted that each control variable is essentially decomposed to a number of static variables; in fact, as many as the number of intervals used for the time horizon discretization, which leads to a significant increase in the scale of the problem and, consequently, the computing time.

Table 24. Bounds on synthesis, design, and operation variables.

Variable	Lower Value	Upper Value
Number of gas turbines of (each type)	0	2
Number of 2-X Diesel engines	0	3
Number of 4-X Diesel engines	0	4
Number of total propulsion engines	1	6
Number of DG sets	0	4
Number of HRSGs	0	2
Number of STs	0	2
Gas turbine nominal power output (kW) (any type)	3000	90,000
2-X Diesel engines nominal power output (kW)	3500	90,000
4-X Diesel engines nominal power output (kW)	3500	21,000
Generator set nominal power output (kW)	300	5000
Load factors (all equipment)	0.15	1
Ship speed (kn)	14	25.4
Single trip duration (days)	5	8

Table 25. Numerical solution parameters.

Parameter	Value
Single trip distance	3000 nm
Round trip distance	6000 nm
Single trip duration from port A to B (all seasons)	variable
Single trip duration from port B to A (all seasons)	variable
Total round trip duration	variable
Length of time intervals on trips	1 day
Length of time intervals in ports	2 days
Number of time intervals used	66
Optimization convergence tolerance	10^{-4}

3.6. Numerical Results for the Nominal Values of Fuel Price and Freight Rate

The optimal synthesis and design of the system are presented in Tables 26 and 27. Optimal round trip durations and number of round trips for each season are given in Table 28. Optimal cost values for each component of the system and revenues are given in Table 29. Optimal values of certain control variables per time interval and season are presented in Tables 30–33.

Table 26. Optimal synthesis of the system.

Type of Propulsion Engines	2–Stroke Diesel
Number of Diesel engines (prime movers)	1
Number of HRSGs	1
Number of steam turbines	1
Number of DG sets	1
Number of auxiliary boilers	1

Table 27. Optimal design specifications of the system components.

Variable	Optimal Value
Main engine nominal brake power (kW)	22,411
DG set 1 nominal electric power (kW)	1500
DG set 2 nominal electric power (kW)	–
Heat recovery steam generator	
Thermal power (kW)	4115
Exhaust gas mass flow rate (kg/s)	41.8
Nominal inlet exhaust gas temperature (°C)	252.5
Auxiliary boiler nominal thermal power (kW)	950
Steam-turbine	
Nominal power (kW)	836
Nominal steam mass flow rate (kg/s)	1.24

Table 28. Trip durations and round trips per season.

Season	Summer	Fall	Winter	Spring
Trip from port A to B duration (days)	7.61	7.62	7.68	7.59
Trip from port B to A duration (days)	7.81	7.87	8	7.79
Round trip duration (days)	19.42	19.49	19.68	19.38
Number of round trips	4.6	4.6	3.1	4.7
Total round trips per year		17		

Table 29. Cost items (in €).

Season	Summer	Fall	Winter	Spring
Present worth of revenue	61,917,011	61,917,011	41,726,681	63,263,033
Present worth cost of fuel	16,481,054	16,587,940	11,901,074	16,789,753
Present worth cost of O&M	1,927,287	1,942,087	1,383,655	1,969,489
Capital cost		10,665,380		
Total PWC		79,647,719		
Total present worth of revenue		228,823,736		
Net Present Value (objective)		149,176,017		

Table 30. Optimal ship speed versus time (in kn).

Summer		Fall		Winter		Spring	
Day	V	Day	V	Day	V	Day	V
From port A to port B							
1	16.59	1	16.55	1	16.51	1	16.58
2	16.47	2	16.46	2	16.40	2	16.51
3	16.54	3	16.52	3	16.44	3	16.55
4	16.57	4	16.55	4	16.51	4	16.55
5	16.18	5	16.17	5	16.00	5	16.33
6	16.35	6	16.33	6	16.08	6	16.44
7	16.33	7	16.34	7	16.01	7	16.35
7.61	16.36	7.62	16.32	7.68	16.18	7.59	16.50
From port B to port A							
1	15.80	1	15.68	1	15.49	1	15.98
2	14.82	2	14.40	2	13.58	2	15.09
3	16.55	3	16.55	3	16.68	3	16.43
4	16.37	4	16.30	4	16.34	4	16.31
5	16.06	5	15.97	5	15.78	5	16.18
6	16.11	6	16.06	6	15.71	6	16.13
7	16.05	7	15.95	7	15.69	7	16.10
7.81	16.17	7.87	16.11	8	15.73	7.79	16.18

Table 31. Propulsion power from Diesel engine(s) and ST versus time (in kW).

	Summer			Fall			Winter			Spring	
Day	$\dot{W}_b$	$\dot{W}_{STP}$	Day	$\dot{W}_b$	$\dot{W}_{STP}$	Day	$\dot{W}_b$	$\dot{W}_{STP}$	Day	$\dot{W}_b$	$\dot{W}_{STP}$
					From port A to port B						
1	19,000	0	1	18,991	0	1	18,924	0	1	18,935	0
2	19,108	0	2	19,080	0	2	19,026	0	2	18,995	0
3	19,043	0	3	19,020	0	3	18,989	0	3	18,955	0
4	19,015	0	4	18,991	0	4	18,924	0	4	18,960	0
5	19,398	0	5	19,362	0	5	19,418	0	5	19,171	0
6	19,232	0	6	19,206	0	6	19,345	0	6	19,069	0
7	19,240	0	7	19,225	0	7	19,314	0	7	19,100	0
7.61	19,290	0	7.62	19,208	0	7.68	19,423	0	7.59	19,050	0
					From port B to port A						
1	20,188	53	1	20,356	0	1	21,889	358	1	20,158	0
2	20,310	776	2	21,047	128	2	22,131	422	2	20,762	45
3	19,536	0	3	19,549	0	3	21,145	122	3	19,731	0
4	19,717	0	4	19,802	0	4	21,415	174	4	19,847	0
5	20,011	0	5	20,113	0	5	21,683	373	5	19,973	0
6	19,962	0	6	20,024	0	6	21,752	352	6	20,020	0
7	19,980	0	7	20,100	0	7	21,761	345	7	20,010	0
7.81	19,960	0	7.87	20,050	0	8	21,751	353	7.79	20,030	0

Table 32. Contribution of the HRSG and auxiliary boiler to thermal loads versus time for all seasons (in kW).

	Summer			Fall			Winter			Spring	
Day	$\dot{Q}_h$	$\dot{Q}_{AB}$	Day	$\dot{Q}_h$	$\dot{Q}_{AB}$	Day	$\dot{Q}_h$	$\dot{Q}_{AB}$	Day	$\dot{Q}_h$	$\dot{Q}_{AB}$
					From port A to port B						
1	850	0	1	860	0	1	1000	0	1	990	0
2	880	0	2	900	0	2	1050	0	2	980	0
3	860	0	3	950	0	3	1080	0	3	1010	0
4	900	0	4	970	0	4	1100	0	4	1020	0
5	840	0	5	930	0	5	1060	0	5	950	0
6	850	0	6	960	0	6	1100	0	6	970	0
7	845	0	7	959	0	7	1100	0	7	980	0
7.61	851	0	7.62	963	0	7.68	1090	0	7.59	970	0
					From port B to port A						
1	860	0	1	930	0	1	1100	0	1	1040	0
2	870	0	2	950	0	2	1080	0	2	970	0
3	870	0	3	980	0	3	1060	0	3	960	0
4	890	0	4	970	0	4	990	0	4	950	0
5	860	0	5	890	0	5	1040	0	5	980	0
6	870	0	6	910	0	6	1070	0	6	960	0
7	880	0	7	920	0	7	1080	0	7	970	0
7.81	870	0	7.87	910	0	8	1050	0	7.79	960	0
					Ports						
A	0	950	A	0	950	A	0	950	A	0	950
B	0	950	B	0	950	B	0	950	B	0	950

Table 33. Electric power of STG and DG versus time for all seasons (in kW).

	Summer			Fall			Winter			Spring	
Day	$\dot{W}_{STG}$	$\dot{W}_{G1}$	**Day**	$\dot{W}_{STG}$	$\dot{W}_{G1}$	**Day**	$\dot{W}_{STG}$	$\dot{W}_{G1}$	**Day**	$\dot{W}_{STG}$	$\dot{W}_{G1}$
					From port A to port B						
1	786	709	1	784	724	1	745	737	1	748	877
2	781	818	2	775	798	2	734	852	2	752	847
3	785	775	3	760	787	3	725	757	3	743	804
4	774	740	4	755	805	4	719	789	4	741	754
5	798	710	5	774	721	5	741	845	5	764	770
6	792	703	6	762	733	6	728	767	6	756	765
7	789	701	7	762	733	7	721	763	7	757	769
7.61	792	703	7.62	762	733	7.68	725	765	7.59	755	765
					From port B to port A						
1	758	867	1	796	751	1	425	1109	1	763	758
2	35	1525	2	678	947	2	371	1150	2	749	876
3	794	740	3	765	899	3	656	969	3	774	799
4	792	729	4	773	735	4	628	1023	4	779	846
5	807	753	5	801	700	5	422	1073	5	774	825
6	803	718	6	794	753	6	436	1085	6	781	805
7	801	719	7	790	749	7	434	1086	7	780	804
7.81	803	722	7.87	793	757	8	436	1085	7.79	780	802
					Ports						
A	0	1500	A	0	1500	A	0	1500	A	0	1500
B	0	1500	B	0	1500	B	0	1500	B	0	1500

The optimal NPV after 20 years of operation is 149,176,017€. The optimal number of total round trips per year is 17. The optimization was completed in 11,060 s, performing 41 major NLP iterations.

For the nominal values of fuel price and freight rate, one 2-stroke Diesel engine with a single HRSG, a single ST, and one Diesel generator set are installed.

Thermal loads are always fully covered by the bottoming cycle and the ST power output is given to serve the electric loads with the exception of the return trip in winter, when brake power demand is the highest and 35%—on average—of the ST power output is directed to the propeller.

3.7. Parametric Study for Fuel Cost and Freight Rate

For the sensitivity analysis, variation of the fuel price and the freight rate is considered. For the fuel price, four more values were considered: 200, 400, 500, and 600 €/ton, in addition to the nominal value of 300 €/ton. For the freight rate, apart from the nominal, the double price is also considered. Sensitivity analysis results for the optimal synthesis and design characteristics of the system are presented in Tables 34–37. The variation of the optimal NPV is given in Tables 38 and 39. Tables 40–43 summarize the effect of fuel price and freight rate on the optimal trip durations and number of round trips per season for the whole year.

Table 34. Effect of fuel price on the optimal synthesis of the system for nominal freight rate.

Fuel Price (€/ton):	200	300	400	500	600
DEs	1	1	1	1	1
HRSGs	–	1	1	1	1
STs	–	1	1	1	1
DG sets	1	1	1	1	1
AB	1	1	1	1	1

Table 35. Effect of fuel price on the optimal synthesis of the system for double freight rate.

Fuel Price (€/ton):	200	300	400	500	600
DE	1	1	1	1	1
HRSG	–	1	1	1	1
ST	–	1	1	1	1
DG sets	1	2	2	2	2
AB	1	1	1	1	1

Table 36. Effect of fuel price on the optimal design specifications of the system for nominal freight rate price (the numbers give the nominal power in kW).

Fuel Price (€/ton):	200	300	400	500	600
DE	31,051	22,411	21,797	22,665	23,414
HRSG	–	4115	3588	3611	3629
ST	–	836	744	750	762
DG 1	1664	1500	1500	1500	1500
DG 2	–	–	–	–	–
AB	1100	950	950	950	950

Table 37. Effect of fuel price on the optimal design specifications of the system for double freight rate (the numbers give the nominal power in kW).

Fuel Price (€/ton)	200	300	400	500	600
DE	64,117	48,114	38,359	32,019	27,224
HRSG	–	8456	6713	5364	4562
ST	–	1882	1477	1150	934
DG 1	1664	490	326	610	697
DG 2	–	1020	1180	900	812
AB	1100	950	950	950	950

Table 38. Effect of fuel price on the optimal NPV for nominal freight rate (values in €).

Item	Fuel Price (€/ton)				
	200	300	400	500	600
Capital Cost	10,954,420	10,665,380	10,404,230	10,603,600	10,769,740
Fuel PWC (total)	56,663,379	61,759,821	73,607,171	91,767,488	109,912,009
OPM PWC (total)	8,073,774	7,222,518	6,432,542	6,440,651	6,450,140
Total PWC	75,691,573	79,647,719	90,443,943	108,811,739	127,131,889
Total PWR	242,283,956	228,823,736	215,363,516	215,363,516	215,363,516
NPV (objective)	166,592,383	149,176,017	124,919,573	106,551,776	88,231,627

Table 39. Effect of fuel price on the optimal NPV for double freight rate (values in €).

Item	Fuel Price (€/ton)				
	200	300	400	500	600
Capital Cost	16,525,690	17,161,440	14,994,690	13,501,900	12,315,590
Fuel PWC (total)	102,168,004	110,804,019	119,978,572	125,697,852	128,678,938
OPM PWC (total)	14,573,599	12,949,546	10,548,733	8,856,380	7,566,637
Total PWC	133,267,293	140,915,005	145,521,995	148,056,132	148,561,165
Total PWR	565,329,230	538,408,790	511,488,351	484,567,911	457,647,472
NPV (objective)	432,061,936	397,493,785	365,966,355	336,511,779	309,086,306

Table 40. Effect of fuel price on the optimal trip durations for nominal freight rate (numbers in days).

Trip	Fuel Price (€/ton)				
	200	300	400	500	600
Summer 1	6.78	7.61	8	8	8
Summer 2	6.93	7.81	8	8	8
Fall 1	6.79	7.62	8	8	8
Fall 2	6.97	7.87	8	8	8
Winter 1	6.82	7.68	8	8	8
Winter 2	7.14	8	8	8	8
Spring 1	6.77	7.59	8	8	8
Spring 2	6.99	7.79	8	8	8

Table 41. Effect of fuel price on the optimal trip durations for double freight rate (numbers in days).

Trip	Fuel Price (€/ton)				
	200	300	400	500	600
Summer 1	5.57	6.08	6.54	6.97	7.38
Summer 2	5.64	6.18	6.66	7.12	7.56
Fall 1	5.57	6.09	6.56	6.97	7.39
Fall 2	5.65	6.20	6.69	7.15	7.61
Winter 1	5.58	6.11	6.57	7.01	7.44
Winter 2	5.72	6.30	6.83	7.34	7.83
Spring 1	5.56	6.08	6.53	6.95	7.36
Spring 2	5.63	6.16	6.65	7.10	7.54

Table 42. Effect of fuel price on the optimal number of round trips for nominal freight rate.

Season	Fuel Price (€/ton)				
	200	300	400	500	600
Summer	4.9	4.6	4.3	4.3	4.3
Fall	4.8	4.6	4.3	4.3	4.3
Winter	3.4	3.1	3	3	3
Spring	4.9	4.7	4.4	4.4	4.4
Total per year	18	17	16	16	16

Table 43. Effect of fuel price on the optimal number of round trips for double freight rate.

Season	Fuel Price (€/ton)				
	200	300	400	500	600
Summer	5.7	5.3	5.1	4.8	4.7
Fall	5.6	5.5	5.2	4.9	4.5
Winter	4	3.7	3.5	3.3	3.1
Spring	5.7	5.5	5.2	5.0	4.7
Total per year	21	20	19	18	17

For all fuel price and freight rate values a single two-stroke Diesel engine was installed. For both freight rates and fuel prices 300 €/ton and above, a bottoming cycle was installed with a single HRSG and ST, while for fuel price 200 €/ton and all freight rates no bottoming cycle was installed. For double freight rate and a fuel price equal to or higher than 300 €/ton, two Diesel generator sets were installed, while in all other cases a single Diesel generator set was selected. Thermal loads are always fully covered by the bottoming cycle, when installed; alternatively an auxiliary boiler of higher nominal power output is installed.

Trip durations generally seem to increase as fuel price rises (need for cost effective system) and for nominal freight rate and fuel price 400 €/ton and above they reach their upper limit. Also, it is interesting the fact that for double freight rate and fuel price at 200 €/ton, all trip durations fall under the 6-day period.

For nominal freight rate and fuel price values 300 €/ton and above, the Diesel engine nominal power was low (21–24 MW) since speeds decreased, thus reducing the available thermal energy of the exhaust gases. As a result a ST of ~750–850 kW was designed in all cases. Due to the small power output of the ST, the installation of one Diesel generator set is optimal since no more than 50% of the electric loads can be covered during trips by the STG even if all the ST power output was directed to the electric loads.

For double freight rate, ship speeds are higher and thus the nominal power output is higher (28–64 MW). This means that the bottoming cycle system is of higher nominal power output too and can serve a larger percentage of electric loads during trips (when compared to the nominal freight rate cases) leaving only a small remainder that is covered by a DG set of low power output. Thus, two Diesel generator sets are installed: one of relatively low power output (350–700 kW), that covers the remainder electric loads that the STG cannot cover during trips and one of higher power output (800–1200 kW) that operates at ports, in parallel with the first one. Also, the second DG set operates during winter (trip from port A to port B) since then the ST power is mainly diverted to the propeller in order to accommodate the high brake power demands. If a single DG set was to be installed, it would have to be of adequate power output in order to serve the electric demand at ports (1500 kW) and thus it would operate in very low load factors in order to cover the low remaining electric demands during trips. This would be very inefficient in terms of SFC and thus not optimal.

Considering the number of total round trips per year, it is noted that the maximum number of round trips per year is observed in the smallest fuel price for both freight rates. For nominal freight rate, the number of round trips per year remains the same (at its lower limit) for fuel price 400 €/ton and above, while for double freight rate, the number of round trips per year decreases as the fuel price increases. It is evident that, as freight rates increase, the need for more trips (and more revenue) becomes more important than cost effectiveness.

The required computational time for all optimizations varied between 11,500 and 12,700 s.

3.8. General Comments on the Results of Section 3

In this study, a general mathematical framework appropriate for the statement and solution of synthesis, design and operation intertemporal dynamic optimization problems on marine energy systems is introduced. The mathematical formulation of the complete three-level optimization problem is presented and a solution procedure that treats all three levels in a single-step is proposed. The method is properly applied for an energy system of a containership with gas turbines, four-stroke and two-stroke Diesel engines allowed as propulsion alternatives. Also, the impact of fuel price and freight rate on the optimal solution is investigated.

In each case, the optimal solution for the SDO of the energy system is achieved in reasonable computational times. It is noted that, the optimal system, for all fuel prices and freight rates, consists of a single two-stroke Diesel engine. In all cases, with the exception of those with the lowest considered fuel price, a steam bottoming cycle is always installed, while, the number, design characteristics, and operational strategy of the DG set(s) vary with the variation of the fuel price and freight rate.

Also, the increase of freight rate or decrease of fuel price leads to the increase of total annual round trips of the ship and vice versa. Moreover, in all cases, the optimization results reveal the optimum ship speed profile that minimizes the fuel consumption, which would be impossible to be identified by experience alone.

In the present work, the goal was to focus on the presentation of the methodology. Thus, certain simplifying assumptions were made in order to avoid needless complexities. For example, the weather conditions throughout the ship route(s) have been considered known (deterministic), while in reality they are stochastic; component degradation of machinery during the operation of the ship and hull and propeller fouling have also been ignored. In case more accurate results are needed, these simplifications should be relaxed.

4. A General Remark

We consider it important to copy here a remark written in Reference [24]: "The energy system of a ship has to comply with rules and regulations of classification societies and national agencies. For example, the number and nominal power of the generator sets have to be determined so that sufficient redundancy exists and, under emergency conditions, at least the critical loads are covered. In some of the solutions presented here, the energy system contains only one Diesel generator set. It goes without saying that any result of the optimization procedure will form the basis, which will lead to the final selection by taking into consideration pertinent rules and regulations".

5. Closure

General comments coming out of the particular examples have been written at the end of Sections 2 and 3. It is worth noting that the installation of a bottoming cycle for better exploitation of fuel energy is economically feasible in all cases, except if the price of fuel is below a certain threshold revealed by the optimization which, however, is unrealistically low.

As demonstrated in Sections 2 and 3, optimization can reveal solutions (design and operation) for energy systems that cannot be identified by experience alone and they are often different from those appearing in common practice. Thus, the extra effort required by an engineer in order to obtain the knowledge and apply optimization is rewarding in many respects.

Mathematical optimization of energy systems has been the subject of systematic research over the last thirty years and the journal articles have an impressive increase in quantity and improvement in quality with time.

The problem of triple optimization—synthesis, design, and operation—is still challenging, and is more so if dynamic conditions are considered. A contribution to the field has been attempted with the present article and, even though the examples are related to marine energy systems, the methods described can also be applied for optimization of energy systems on land.

As indicated in reference [1], there are many subjects still open for further investigation.

Author Contributions: G.N.S. and G.J.T. performed the simulations and optimizations and had the primary role in writing Sections 2 and 3, respectively, while C.A.F. supervised the works. All three authors cooperated in writing the current text.

Funding: This research received no external funding.

Conflicts of Interest: The authors declare no conflicts of interest.

Nomenclature

C_c	Capital cost ($)
c_f	Fuel cost ($/ton)
c_{om}	Operation and maintenance (O&M) cost ($/kWh)
CCHP	Combined cooling, heat and power
CHP	Combined heat and power
f	Inflation ratio
f_L	Load factor
$g_{x,y}$	Number of the HRSG receiving exhaust gas of gas turbine x during operating mode y
HFO	Heavy fuel oil
i	Market interest rate
LHV	Fuel lower heating value
LNG	Liquefied natural gas
$\dot{m}_{da}$	Heating steam mass flow rate for deaerator
$\dot{m}_{kv}$	Nominal steam mass flow rate of steam turbine v, $k = HP, LP$ (kg/s)

$\dot{m}_{kz}$	Nominal steam mass flow rate of HRSG z, $k = HP, LP$ (kg/s)
$\dot{m}_{k,COL}$	Collector steam mass flow rate, $k = HP, LP$ (kg/s)
$\dot{m}_f$	Fuel consumption (kg/s)
$\dot{m}_{gz}$	Nominal exhaust gas mass flow rate of HRSG z (kg/s)
$m_{gz,mult}$	Variable for determination of $\dot{m}_{gz}$
$\dot{m}_{pr}$	Heating steam for preheating HRSG feed water
$\widetilde{m}_{STkv,y}$	Variable for determination of $\dot{m}_{k,v}$ $k = HP, LP$
n_k	Number of components of type k in the system
N_T	Number of operation modes
N_Y	Number of years of the investment
O&M	Operation and maintenance
P_{HP}	High-pressure level of steam production (bar)
P_{LP}	Low pressure level of steam production (bar)
PWC	Present worth cost
PWF	Present worth factor
$\dot{Q}_{AB,u}$	Thermal power of the auxiliary boiler u (kW)
$\dot{Q}_{hl}$	Thermal load (kW)
$\dot{Q}_{hl,z}$	Thermal load covered by HRSG z (kW)
SCR	Selective catalytic reduction
SDO	Synthesis, design and operation
SFC	Specific fuel consumption (gr/kWh)
t_y	Duration of operating mode y (h)
T_{gz}	Nominal exhaust gas temperature of HRSG z (°C)
$T_{gz,mult}$	Variable for determination of T_{gz}
$\dot{W}_{DGx,N}$	Nominal power rating of Diesel generator set x (kW)
$\dot{W}_{DG,TOT,y}$	Total power delivered by the Diesel generator sets in mode y
$\widetilde{W}_{DG}$	Variable for determination of Diesel generator set power
$\dot{W}_{GT,N}$	Nominal power rating of gas turbine (kW)
$\widetilde{W}_{GT}$	Variable for determination of main engine brake power
$\dot{W}_e$	Electrical load (kW)
$W_{N,x,mult}$	Variable for determination of $\dot{W}_{MEx,N}$
$\dot{W}_p$	Propulsion load (kW)
$\dot{W}_{ST}$	Steam turbine power (kW)
$\dot{W}_{STe}$	Part of $\dot{W}_{ST}$ used for electrical loads (kW)
$\dot{W}_{STp}$	Part of $\dot{W}_{ST}$ used for propulsion (kW)

Subscripts

AB	Auxiliary boiler
DG	Diesel generator Set
GT	Gas turbine
hl	Heat load serving mass flow rate
HRSG	Heat recovery steam generator
k	General index for component of type k, or pressure levels HP/LP
max	Maximum propulsion load
ME	Main Engine
sm	Sea margin
ST	Steam turbine
x	Main engine x index
y	Operating mode y
z	HRSG z index

	Hellenic symbols
η_G	Electrical efficiency of generator
λ_{GT}	Fraction of $\dot{W}_p$ covered by the main engine
λ_Q	Fraction of $\dot{Q}_{hl}$ covered by the HRSGs
μ_s	Sea margin coefficient

Appendix A

The capital cost of the gas turbines is estimated with the related cost model presented in Frangopoulos [59], with the unit cost parameters modified for taking into account the current level of technology and to approximate capital cost data available from various sources. A common regression formula has been developed for the gas turbine configurations examined in the present work and has the general mathematical expression of Equation (A1). The values of the coefficients for each gas turbine types are presented in Table A1.

$$C_{c,GT}(\$) = \left(\dot{W}_{GT,N}(kW)\right)^a \exp\left(b - c\dot{W}_{GT,N}(kW)\right) \tag{A1}$$

Table A1. Coefficients for Equation (A1).

Coefficient	GT Type (a)	GT Type (b)	GT Type (c)
a	0.451124718450259	0.451124718937363	0.45112471762403
b	11.6601660998132	11.8085861004949	11.9225303717006
c	$8.15305188415185 \times 10^{-7}$	$8.15305206333862 \times 10^{-7}$	$8.15305153736658 \times 10^{-7}$

The capital costs of other components of the system are estimated as explained in Sakalis and Frangopoulos [24].

Appendix B

Wind speed and direction data for the numerical example of Section 3.

Table A2. Wind speed (in kn) as a function of time and space in Summer.

Time (Days)	Distance from Port A (nm)					
	513	1026	1539	2052	2565	3078
1	4.43	4.91	4.32	5.45	4.91	6.05
2	13.61	15.87	16.41	13.71	13.12	16.36
3	17.87	16.25	17.44	18.03	16.79	15.71
4	9.18	11.39	9.40	9.94	10.48	17.87
5	24.95	19.22	16.79	24.19	26.30	24.35
6	21.65	23.00	24.62	26.84	26.35	22.89
7	22.73	21.06	27.21	24.62	26.89	23.98
8	23.65	23.97	22.08	22.30	22.89	25.11

Table A3. Wind speed (in kn) as a function of time and space in Fall.

Time (Days)	Distance from Port A (nm)					
	513	1026	1539	2052	2565	3078
1	4.75	5.35	4.75	5.94	5.35	6.53
2	14.85	17.22	17.82	14.85	14.25	17.82
3	19.60	17.82	19.01	19.60	18.41	17.22
4	10.10	12.47	10.10	10.69	11.28	19.60
5	27.32	20.79	18.41	26.13	28.51	27.73
6	23.76	24.95	26.73	29.10	28.51	26.95
7	24.95	23.16	29.70	26.73	29.10	23.76
8	25.54	26.13	23.76	24.35	24.95	24.95

Table A4. Wind speed (in kn) as a function of time and space in Winter.

Time (Days)	Distance from Port A (nm)					
	513	1026	1539	2052	2565	3078
1	5.23	5.88	5.23	6.53	5.88	7.19
2	16.33	18.95	19.60	16.33	15.68	19.60
3	21.56	19.60	20.91	21.56	20.25	18.95
4	11.11	13.72	11.11	11.76	12.41	21.56
5	30.05	22.87	20.25	28.75	31.36	29.40
6	26.13	27.44	29.40	32.01	31.36	28.44
7	27.44	25.48	32.67	29.40	32.01	28.13
8	28.09	28.75	26.13	26.79	28.44	29.44

Table A5. Wind speed (in kn) as a function of time and space in Spring.

Time (Days)	Distance from Port A (nm)					
	513	1026	1539	2052	2565	3078
1	3.89	4.37	3.89	4.86	4.37	5.35
2	12.15	14.09	14.58	12.15	11.66	14.58
3	16.04	14.58	15.55	16.04	15.06	14.09
4	8.26	10.21	8.26	8.75	9.23	16.04
5	22.35	17.01	15.06	21.38	23.33	21.87
6	19.44	20.41	21.87	23.81	23.33	20.41
7	20.41	18.95	24.30	21.87	23.81	19.94
8	20.90	21.38	19.44	19.92	22.41	21.41

Table A6. Wind direction in degrees ($°$) with respect to north, counting counterclockwise, as a function of time and space for all seasons.

Time (Days)	Distance from Port A (nm)					
	513	1026	1539	2052	2565	3078
1	318	320	310	345	300	260
2	315	330	330	330	300	250
3	315	325	334	300	285	260
4	320	325	330	250	265	255
5	321	328	345	260	244	230
6	317	320	305	255	230	225
7	323	333	300	245	250	228
8	320	330	328	260	242	230

References

1. Frangopoulos, C.A. Recent developments and trends in optimization of energy systems. *Energy* **2018**, *164*, 1011–1020. [CrossRef]
2. Frangopoulos, C.A.; Von Spakovsky, M.R.; Sciubba, E.A. Brief Review of Methods for the Design and Synthesis Optimization of Energy Systems. *Int. J. Appl. Thermodyn.* **2002**, *5*, 151–160.
3. Pelster, S.; Favrat, D.; Von Spakovsky, M.R. The Thermoeconomic and Environomic Modeling and Optimization of the Synthesis, Design, and Operation of Combined Cycles with Advanced Options. *J. Eng. Gas Turbines Power* **2001**, *123*, 717–726. [CrossRef]
4. Mussati, S.F.; Aguirre, P.A.; Scenna, N.J. A Rigorous, Mixed-integer, Nonlineal Programming Model (MINLP) for Synthesis and Optimal Operation of Cogeneration Seawater Desalination Plants. *Desalination* **2004**, *166*, 339–345. [CrossRef]
5. Sun, L.; Gai, L.; Smith, R. Site Utility System Optimization with Operation Adjustment under Uncertainty. *Appl. Energy* **2017**, *186*, 450–456.
6. Calise, F.; Dentice d' Accadia, M.; Vanoli, L.; Von Spakovsky, M.R. Full Load Synthesis/Design Optimization of a Hybrid SOFC–GT Power Plant. *Energy* **2007**, *32*, 446–458. [CrossRef]

7.   Toffolo, A. A Synthesis/Design Optimization Algorithm for Rankine Cycle Based Energy Systems. *Energy* **2014**, *66*, 115–127. [CrossRef]

8.   Frangopoulos, C.A. Intelligent Functional Approach: A Method for Analysis and Optimal Synthesis–Design–Operation of Complex Systems. *Int. J. Energy Environ. Econ.* **1991**, *1*, 267–274.

9.   Frangopoulos, C.A. Optimization of Synthesis–Design–Operation of a Cogeneration System by the Intelligent Functional Approach. *Int. J. Energy Environ. Econ.* **1991**, *1*, 275–287.

10.   Frangopoulos, C.A. Optimal Synthesis and Operation of Thermal Systems by the Thermoeconomic Functional Approach. *J. Eng. Gas Turbines Power* **1992**, *114*, 707–714. [CrossRef]

11.   Munoz, J.R.; Von Spakovsky, M.R. A Decomposition Approach for the Large Scale Synthesis/Design Optimization of Highly Coupled, Highly Dynamic Energy Systems. *Int. J. Appl. Thermodyn.* **2001**, *4*, 19–33.

12.   Munoz, J.R.; Von Spakovsky, M.R. The Application of Decomposition to the Large Scale Synthesis/Design Optimization of Aircraft Energy Systems. *Int. J. Appl. Thermodyn.* **2001**, *4*, 61–76.

13.   Rancruel, D.F.; Von Spakovsky, M.R. Decomposition with Thermoeconomic Isolation Applied to the Optimal Synthesis/Design of an Advanced Tactical Aircraft System. *Int. J. Thermodyn.* **2003**, *6*, 93–105. [CrossRef]

14.   Georgopoulos, N.G.; Von Spakovsky, M.R.; Muñoz, J.R. A Decomposition Strategy Based on Thermoeconomic Isolation Applied to the Optimal Synthesis/Design and Operation of a Fuel Cell Based Total Energy System. In Proceedings of the IMECE2002 ASME, International Mechanical Engineering Congress & Exposition, New Orleans, LA, USA, 17–22 November 2002.

15.   Oyarzabal, B.; Von Spakovsky, M.R.; Ellis, M.W. Optimal Synthesis/Design of a PEM Fuel Cell Cogeneration System for Multi-unit Residential Applications–Application of a Decomposition Strategy. *J. Energy Resour. Technol.* **2004**, *126*, 30–39. [CrossRef]

16.   Munoz, J.R.; Von Spakovsky, M.R. Decomposition in Energy System Synthesis/Design Optimization for Stationary and Aerospace Applications. *J. Aircr.* **2003**, *40*, 35–42.

17.   Olsommer, B.; Favrat, D.; Von Spakovsky, M.R. An Approach for the Time-dependent Thermoeconomic Modeling and Optimization of Energy System Synthesis, Design and Operation Part I: Methodology and Results. *Int. J. Appl. Thermodyn.* **1999**, *2*, 97–114.

18.   Lazzaretto, A.; Toffolo, A. A method to separate the problem of heat transfer interactions in the synthesis of thermal systems. *Energy* **2008**, *33*, 163–170. [CrossRef]

19.   Toffolo, A.; Lazzaretto, A.; Morandin, M. The HEATSEP method for the synthesis of thermal systems: An application to the S-Graz cycle. *Energy* **2010**, *35*, 976–981. [CrossRef]

20.   Lazzaretto, A.; Manente, G.; Toffolo, A. SYNTHSEP: A general methodology for the synthesis of energy system configurations beyond superstructures. *Energy* **2018**, *147*, 924–949. [CrossRef]

21.   Dimopoulos, G.G.; Kougioufas, A.V.; Frangopoulos, C.A. Synthesis, Design and Operation Optimization of a Marine Energy System. *Energy* **2008**, *33*, 180–188. [CrossRef]

22.   Dimopoulos, G.G.; Frangopoulos, C.A. Synthesis, Design and Operation Optimization of the Marine Energy System for a Liquefied Natural Gas Carrier. *Int. J. Thermodyn.* **2008**, *11*, 203–211.

23.   Kalikatzarakis, M.; Frangopoulos, C.A. Thermo-economic Optimization of Synthesis, Design and Operation of a Marine Organic Rankine Cycle System. *Proc Inst. Mech. Eng. Part M J. Eng. Marit. Environ.* **2017**, *231*, 137–152. [CrossRef]

24.   Sakalis, G.N.; Frangopoulos, C.A. Intertemporal optimization of synthesis, design and operation of integrated energy systems of ships: General method and application on a system with Diesel main engines. *Appl. Energy* **2018**, *226*, 991–1008. [CrossRef]

25.   Rancruel, D.F. Dynamic Synthesis/Design and Operational/Control Optimization Approach Applied to a Solid Oxide Fuel Cell Based Auxiliary Power Unit Under Transient Conditions. Ph.D. Dissertation, Virginia Polytechnic Institute and State University, Blacksburg, VA, USA, 2005.

26.   Rancruel, D.F.; Von Spakovsky, M.R. Development and Application of a Dynamic Decomposition Strategy for the Optimal Synthesis/Design and Operational/Control of a SOFC Based APU under Transient Conditions. In Proceedings of the International Mechanical Engineering Congress and Exposition—IMECE, Orlando, FL, USA, 5–11 November 2005; ASME: New York, NY, USA, 2005. Paper No. 82986.

27.   Wang, M.; Kim, K.; Von Spakovsky, M.R.; Nelson, D. Multi- versus Single-level of Dynamic Synthesis/Design and Operation/Control Optimizations of a PEMFC System. In Proceeding of the IMECE2008—ASME International Mechanical Engineering Congress and Exposition, Boston, MA, USA, 31 October–6 November 2008.

28. Kim, K.; Von Spakovsky, M.R.; Wang, M.; Nelson, D.J. A Hybrid Multi-level Optimization Approach for the Dynamic Synthesis/Design and Operation/Control under Uncertainty of a Fuel Cell System. *Energy* **2011**, *36*, 3933–3943. [CrossRef]

29. Kim, K.; Wang, M.; Von Spakovsky, M.R.; Nelson, D.J. Dynamic Synthesis/Design and Operation/Control Optimization under Uncertainty of a PEMFC System. In Proceedings of the IMECE 2008 ASME International Mechanical Engineering Congress and Exposition, Boston, MA, USA, 31 October–6 November 2008; ASME: New York, NY, USA, 2008. paper no. 68070. pp. 679–689.

30. Arcuri, P.; Beraldi, P.; Florio, F.; Fragiacomo, P. Optimal Design of a Small Size Trigeneration Plant in Civil Users: A MINLP (Mixed Integer Non Linear Programming) Model. *Energy* **2015**, *80*, 628–641. [CrossRef]

31. Buoro, D.; Casisi, M.; Pinamonti, P.; Reini, M. Optimal Synthesis and Operation of Advanced Energy Supply Systems for Standard and Domotic Home. *Energy Convers. Manag.* **2012**, *60*, 96–105. [CrossRef]

32. Petruschke, P.; Gasparovic, G.; Voll, P.; Krajacic, G.; Duic, N.; Bardow, A. A Hybrid Approach for the Efficient Synthesis of Renewable Energy Systems. *Appl. Energy* **2014**, *135*, 625–633. [CrossRef]

33. Goderbauer, S.; Bahl, B.; Voll, P.; Lübbeckeb, M.; Bardow, A.; Koster, A.M.C.A. An Adaptive Discretization MINLP Algorithm for Optimal Synthesis of Decentralized Energy Supply Systems. *Comput. Chem. Eng.* **2016**, *95*, 38–48. [CrossRef]

34. Zhu, Q.; Luo, X.; Zhang, B.; Chen, Y. Mathematical Modeling and Optimization of a Large–scale Combined Cooling, Heat and Power System that Incorporates Unit Changeover and Time–of–use Electricity Price. *Energy Convers. Manag.* **2017**, *133*, 385–398.

35. Fuentes–Cortés, L.F.; Ponce–Ortega, J.M.; Nápoles–Rivera, F.; Serna–González, M.; El–Halwagi, M. Optimal Design of Integrated CHP Systems for Housing Complexes. *Energy Convers. Manag.* **2015**, *99*, 252–263. [CrossRef]

36. Hanglid, F. A review on the use of gas and steam turbine combined cycles as prime movers for large ships. Part I: Background and design. *Energy Convers. Manag.* **2008**, *49*, 3458–3467.

37. Hanglid, F. A review on the use of gas and steam turbine combined cycles as prime movers for large ships. Part II: Previous work and implications. *Energy Convers. Manag.* **2008**, *49*, 3468–3475.

38. Hanglid, F. A review on the use of gas and steam turbine combined cycles as prime movers for large ships. Part III: Fuels and emissions. *Energy Convers. Manag.* **2008**, *49*, 3476–3482.

39. Altosole, M.; Benvenuto, G.; Campora, U.; Laviola, M.; Trucco, A. Waste Heat Recovery from Marine Gas Turbines and Diesel Engines. *Energies* **2017**, *10*, 718. [CrossRef]

40. Dzida, M. Comparing combined gas turbine/steam turbine and marine low speed piston engine/steam turbine systems in naval applications. *Pol. Marit. Res.* **2011**, *18*, 43–48. [CrossRef]

41. Hanglid, F. Variable geometry gas turbines for improving the part–load performance of marine combined cycles—Gas turbine performance. *Energy* **2010**, *35*, 562–570.

42. Hanglid, F. Variable geometry gas turbines for improving the part–load performance of marine combined cycles—Combined cycle performance. *Appl. Therm. Eng.* **2011**, *31*, 467–476.

43. Wang, Z.; Li, J.; Fan, K.; Li, S. The Off–Design Performance Simulation of Marine Gas Turbine Based on Optimum Scheduling of Variable Stator Vanes. *Math. Probl. Eng.* **2017**, *2671251*. [CrossRef]

44. Sharma, O.P.; Kaushik, S.C.; Manjunath, K. Thermodynamic analysis and optimization of a supercritical CO_2 regenerative recompression Brayton cycle coupled with a marine gas turbine for shipboard waste heat recovery. *Therm. Sci. Eng. Prog.* **2017**, *3*, 62–74. [CrossRef]

45. Hou, S.; Wu, Y.; Zhou, Y.; Yu, L. Performance analysis of the combined supercritical CO_2 recompression and regenerative cycle used in waste heat recovery of marine gas turbine. *Energy Convers. Manag.* **2017**, *151*, 73–85. [CrossRef]

46. Wang, Z.; Li, Y.G.; Li, S. Performance simulation of 3–stage gas turbine CHP plant for marine applications. In Proceedings of the ASME Turbo Expo 2016: Turbomachinery Technical Conference and Exposition GT2016, Seoul, Korea, 13–17 June 2016. GT2016–56312.

47. Welaya, Y.M.A.; Mosleh, M.; Ammar, N.R. Thermodynamic analysis of a combined gas turbine power plant with a solid oxide fuel cell for marine applications. *Int. J. Nav. Archit. Ocean Eng.* **2013**, *5*, 529–545. [CrossRef]

48. Tse, L.K.C.; Wilkins, S.; McGlashana, N.; Urbanb, B.; Martinez–Botasa, R. Solid oxide fuel cell/gas turbine trigeneration system for marine applications. *J. Power Sources* **2011**, *196*, 3149–3162. [CrossRef]

49. Armellini, A.; Daniotti, S.; Pinamonti, P. Gas Turbines for power generation on board of cruise ships: A possible solution to meet the new IMO regulations? *Energy Procedia* **2015**, *81*, 540–547. [CrossRef]

50. Armellini, A.; Daniotti, S.; Pinamonti, P.; Reini, M. Evaluation of gas turbines as alternative energy production systems for a large cruise ship to meet new maritime regulations. *Appl. Energy* **2018**, *211*, 306–317. [CrossRef]

51. Doulgeris, G.; Korakianitis, T.; Pilidis, P.; Tsoudis, E. Techno–economic and environmental risk analysis for advanced marine propulsion systems. *Appl. Energy* **2012**, *99*, 1–12. [CrossRef]

52. De Leon, L.S.; Zachos, P.K.; Pachidis, V. A comparative assessment of dry gas turbine cycles for marine applications. In Proceedings of the ASME Turbo Expo 2013: Turbine Technical Conference and Exposition GT2013, San Antonio, TX, USA, 3–7 June 2013. GT2013–95321.

53. Ji, N.; Li, S.; Wang, Z.; Zhao, N. Off–Design Behavior Analysis and Operating Curve Design of Marine Intercooled Gas Turbine. *Math. Probl. Eng.* **2017**, 8325040. [CrossRef]

54. El-Gohary, M.M.; Seddiek, I.S. Utilization of alternative marine fuels for gas turbine power plant onboard ships. *Int. J. Nav. Arch. Ocean Eng.* **2013**, *5*, 21–32. [CrossRef]

55. El-Gohary, M.M.; Ammar, N.R. Thermodynamic analysis of alternative marine fuels for marine gas turbine power plants. *J. Mar. Sci. Appl.* **2016**, *15*, 95–103. [CrossRef]

56. El-Gohary, M.M. The future of natural gas as a fuel in marine gas turbine for LNG carriers. *Proc. Inst. Mech. Eng. Part M J. Eng. Marit. Environ.* **2012**, *226*, 371–377. [CrossRef]

57. Fernández, I.A.; Gómez, M.R.; Gómez, J.R.; Insuab, Á.B. Review of propulsion systems on LNG carriers. *Renew. Sustain. Energy Rev.* **2017**, *67*, 1395–1411. [CrossRef]

58. Software MarineGTs, Laboratory of Thermal Turbomachines, National Technical University of Athens, Greece. Information. Available online: https://www.ltt.ntua.gr/index.php/en/softwaremn/marine-gts (accessed on 2 January 2019).

59. Frangopoulos, C.A. Application of the thermoeconomic functional approach to the CGAM problem. *Energy* **1994**, *19*, 323–342. [CrossRef]

60. Tzortzis, G.J.; Frangopoulos, C.A. Dynamic optimization of synthesis, design and operation of marine energy systems. *J. Eng. Marit. Environ.* **2018**. [CrossRef]

61. *gPROMS. User Guide*, version 4.1.1; Process Systems Enterprise Ltd.: London, UK, 2016.

62. Bard, Y. *Nonlinear Parameter Estimation*; Academic Press: New York, NY, USA, 1974.

63. Internet Sites with Weather Data. Available online: https://earth.nullschool.net and http://www.meteoearth.com and http://www.accuweather.com and http://enterprisesolutions.accuweather.com (accessed on 15 July 2017).

Article

Smart Energy Systems: Guidelines for Modelling and Optimizing a Fleet of Units of Different Configurations

Sergio Rech

Veil Energy Srl–San Pietro in Gu, 35010 Padova, Italy; s.rech@veil-energy.eu; Tel.: +39-049-827-7478

Received: 7 February 2019; Accepted: 4 April 2019; Published: 6 April 2019

Abstract: The need to reduce fossil fuels consumption and polluting emissions pushes towards the search of systems that combine traditional and renewable energy conversion units efficiently. The design and management of such systems are not easy tasks because of the high level of integration between energy conversion units of different types and the need of storage units to match the availability of renewables with users' requirements properly. This paper summarizes the basic theoretical and practical concepts that are required to simulate and optimize the design and operation of fleet of energy units of different configurations. In particular, the paper presents variables and equations that are required to simulate the dynamic behavior of the system, the operational constraints that allow each unit to operate correctly, and a suitable objective function based on economic profit. A general Combined Heat-and-Power (CHP) fleet of units is taken as an example to show how to build the dynamic model and formulate the optimization problem. The goal is to provide a "recipe" to choose the number, type, and interconnection of energy conversion and storage units that are able to exploit the available sources to fulfill the users' demands in an optimal, and therefore "smart", way.

Keywords: smart energy systems; Mixed-Integer NonLinear/Linear Programming (MINLP/MILP); dynamic modelling; design and operation optimization; fleet of energy conversion and storage units

---

1. Introduction

The design of new energy systems and the management of the existing ones are tasks of increasing complexity due to the urgent need to replace fossil fuels with renewable energy sources, with stricter constraints deriving from evolving energy markets rules and the change over time in users' demands. In fact, units fed by fossil fuels are generally able to generate the desired energy products (electricity, thermal energy, cooling, and fuels) more or less promptly depending on the unit type and size (e.g., a set of small-to-medium size internal combustion engines are more suitable to respond promptly to variable energy requirements than a big size steam power station). In contrast, the energy generated by units converting non-dispatchable renewables such as solar and wind strongly depends on the meteorological conditions and site features. Thus, the choice of type, number and size of these units becomes a more challenging task, with the possible inclusion of storage units that may become necessary to fulfill the users' requirements, or convenient when the system is connected to the energy grids, because of the variable price at which the surplus and deficit of energy can be traded.

In this framework, the search of the best configuration of a fleet of energy units (which may serve, for instance, an industrial district, a municipality, or a whole country) cannot disregard the definition of its best operation in order to meet, at any time, both the users' requirements and the operating constraints of the energy conversion and storage units. The experience of the designer is crucial in this process but usually not sufficient to define the optimum number, type, size, and loads of the units under variable users' requirements, primary energy sources availability and market costs/prices. So, the application of adequate modeling and optimization approaches are needed to design optimum solutions that, when built, can actually fulfill the users' demands in the best possible way.

Recently, two very relevant reviews on development and tools for modelling [1] and optimization [2] of energy systems have been published. The former compares a large amount of modelling tools proposed in the literature to analyze electric and CHP (Combined Heat-and-Power) fleets of units having large share of non-dispatchable renewables (up to 100%), ranging from small-scale fleet of power units and a temporal resolution of seconds (or subseconds) to the worldwide energy system and temporal resolution of decades. In medium-to-large scale fleet of energy units (i.e., energy systems including one or more tens of energy units) most of the models use a time discretization of one hour to simulate the dynamic behavior of the system. The latter provides a very clear statement and mathematical formulation of the static and dynamic optimization of energy systems of different dimension, complexity, and detail (see also [3]). Three optimization levels are considered: synthesis (choice of the energy units and interconnections that appear in the system), design (definition of technical characteristic of the energy units and properties of the substances entering and exiting each unit at nominal load) and operation (definition of the operation of each chosen and designed energy unit under specified external conditions). In general, these three levels are to be inter-related to obtain a complete optimization of the system. A brief but comprehensive review on solution methods is also presented focusing on the synthesis optimization and providing examples of suitable objective functions.

In the optimization of fleet of energy units (energy systems), the Mixed Integer NonLinear Programming (MINLP) is widely acknowledged as one of the best approaches in terms of both simplicity and accuracy. In particular, integer or binary variables are used to include or not an energy unit in the optimum configuration or to model its on/off status. On the other hand, state of the art MINLP solution algorithms may require a very high computational effort to solve optimization problems including a very high number of real and integer decision variables associated with the design and operation of systems including several energy units and interconnections. Thus, in most of the works in the literature, the optimization problem is reduced to a Mixed Integer Linear Programming (MILP) one by considering linear characteristic maps of all energy conversion units and linearizing all the other nonlinear constraints. This approach was applied for the first time in the late 1970s in [4] to determine the unit commitment (optimum operation) of a group of power units. Multi-product systems including few CHP units have been analyzed about ten years later considering first the optimum operation only (see, e.g., [5,6]), and then also the optimum design (see, e.g., [7]). From then on, the MILP approach has been applied to more complex systems including also thermal storage units (see, e.g., [8,9]) or other types of storage units (such as hydroelectric and electric storages [10]). When thermal storage units are considered, it is generally assumed that the thermal energy is stored at constant temperature.

The MINLP (or MILP) design optimization of an energy system is closely linked to the concept of superstructure (proposed for the first time in [11]) by which the space of possible configurations of the system is explicitly defined a priori. Each solution, including the optimum one, is extracted from the superstructure by excluding parts of it (see, e.g., [12,13]). Superstructure-free methods have been also proposed (see, e.g., [14]) which start from a first-attempt system configuration and add, remove or modify parts of it to define new design alternatives.

Various techniques have been introduced to limit the increasing computational effort required by the MINLP (or MILP) design and operation optimization of energy systems deriving from the increasing number of units, longer optimization periods or more complex energy markets rules and incentive mechanisms. Among these decomposition methods [15] and rolling-horizon methods [16,17] are noteworthy.

An alternative to the MINLP approach is proposed in [18] to optimize the capacity of a variable temperature thermal storage unit according to the optimum operation of the CHP system in which it is included. A two-step optimization method has been developed to reduce the problem complexity resulting from the variable temperature of the storage unit.

Despite the high interest of the literature in this topic, all works concern specific problems and a general approach to "guide" the designer in formulating dynamic optimization problems of this type of systems is still missing. An interesting contribution in this direction is [19] which, however, only considers the short-term operation of Combined Cooling, Heat-and-Power (CCHP) energy systems.

This work provides guidelines to formulate the problem of the dynamic optimization of the design and operation of an energy system consisting in a fleet of energy conversion and storage units having different and complex configurations. The goal is to provide the reader with all the necessary information for this mathematical formulation starting from scratch, i.e., the number and form of the equations, the type and numbers of variables, the "shape" of the objective function/functions, the choice of the decision variables, and the required input data.

The problem is formulated using a dynamic modeling approach based on Mixed-Integer NonLinear Programming (MINLP), which is able to describe the behavior of the system at any load with the minimum loss of information. The maximum profit is considered as example of objective function. The set of design decision variables includes binary variables, which are used to decide about the inclusion/exclusion of an energy conversion unit in the optimum configuration, and real variables to choose the optimum capacity of the storage units. The set of operation decision variables includes binary variables to decide about the on/off status of the energy conversion units and real variables to define their load. The model and the optimization problem are structured to be *general*, *simple* and to require a *low computational effort*:

- The *generality* of the problem is given by considering a general energy system configuration (Section 2) that includes both energy conversion units and storage units, which may have multiple and different inputs (renewable and fossil primary energy sources, electricity, thermal energy and cooling) and outputs (i.e., electricity, thermal energy, cooling and synthetic- or derived-fuels). In this general configuration, each unit is seen as a black box. This type of schematic allows units of very different type to be modeled and analyzed using the same type of equations but, on the other hand, does not permit to improve the "internal" configuration of the unit (which is out of the scope of this work).

- To keep the problem *simple*, the number of variables and equations of the model (Section 3.1) is kept as small as possible while maintaining a good accuracy in the simulation of the dynamic behavior of the energy system units. To this end, only variables associated with power streams and energy quantities are considered. Thus, mass balances and equations of state are not included in the model so that also intensive and extensive variables such as mass flow rates, pressures and temperatures do not appear explicitly in the model (Section 3.1.1). However, the operation of all system units is kept within the operating boundaries (feasible operation) by considering the values of some of these parameters in the equations describing the behavior of the units (characteristic maps, Section 3.1.2). Moreover, a criterion to define the type and number of equations is proposed to build the model by simply "assembling" the same types of equations for units having different types and numbers of input and output streams.

- *Low computational effort* in optimizing the operation of the energy system is obtained by reducing the MINLP problem to a linear (MILP) one in which, when possible, linear equations are used to describe the behavior of the system units (Section 3.2). In all other cases, linearization techniques are applied, which, however, require the inclusion of auxiliary variables. In the search for the optimal system design, a two-step decomposition technique is proposed to further reduce the computational effort.

The formulation of the optimization problem is applied to a general energy system including all the most common units for generation of power, heat, and Combined Heat-and-Power (CHP), and electric and thermal storage units (Section 4). This example can be used as a basic database from which equations and variables can be "extracted" to formulate optimization problems for more specific system configurations. Finally, a couple of numerical applications that were solved using the presented general formulation and equations are shown to demonstrate the potential of the suggested guidelines in the optimization of this type of system (Section 5).

2. General Energy System Made up of a Fleet of Energy Conversion and Storage Units

The energy system considered here includes a fleet of energy conversion and storage units that serves different energy users (Figure 1). To simplify the modelling of the system, the units are considered as black boxes, so their operation is described considering only the input and output flows. Dotted lines identify the control volumes of total system, units and environment. The units are connected with each other and with the external environment by arrows in the points in which mass and energy transfers take place. The solid arrows represent desired mass and/or energy inputs (*Fuels*) or desired outputs (*Products*) while dotted arrows identify undesired outputs (mass and energy *Losses* or emissions). Variables φ are associated with generic flow variables (mass flow rate or power) entering/exiting the control volume of a unit whereas variables Φ are associated with quantities contained in the control volume of a unit (mass or energy). The symbol t within brackets means that the value of the variable is a function of time. The numerical subscripts (1, 2, etc.) identify the number of the unit and the subscripts *in*, *out*, and *L* refer to the flows of fuel, product, and loss/emissions, respectively. When a unit receives or releases more than one input or output, an additional numeric subscript is combined with *in* or *out*. Number 1 refers to the main product stream of the unit (e.g., electrical power in cogeneration systems) and the other numbers to secondary or recovery products (e.g., thermal power in cogeneration systems). No specific order is assigned to the fuel streams (i.e., they are considered to be of equal importance). The external environment is considered as a unit both on the primary energy source side (*ES*) and users' demand side (*UD*). On *ES*-side there are only product streams (φ_{ES,out_q}), on the *UD*-side only fuel streams (φ_{UD,in_p}).

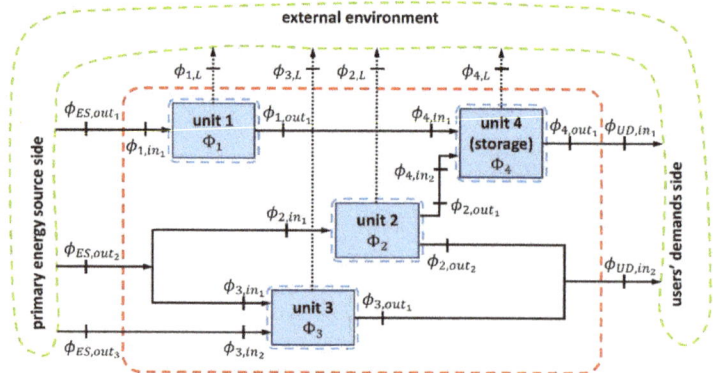

Figure 1. General configuration of a fleet of energy units.

3. Methodology

In general, the energy system behavior can be described by a mathematical model. This Section first presents the dynamic off-design model of a fleet of units operating at variable load and then the general formulation for the optimization of the design and operation of such systems, taking as reference the system in Figure 1. The model equations are subdivided into categories and rearranged to keep the construction of the model as simple as possible. In particular, a Mixed-Integer NonLinear Programming (MINLP) approach is applied using binary variables to identify the on/off status of the energy conversion units. Simplifying assumptions are also introduced to reduce the computational effort in the optimization of the design and operation of the system while maintaining feasibility of the solution.

3.1. Dynamic Off-Design Model

In general, the off-design model of an energy system includes mass and energy balances and other equations required to solve them, which relate energy to mass (and power to mass flow rate) through intensive variables (equations of state). Other relationships define the possible ranges of variation of the model variables according to technological or other constraints. Among these last relationships there are the so-called characteristic maps of the units which link the fuels of a unit (φ_{i,in_p} in Figure 1) to its products (φ_{i,out_q} in Figure 1).

To keep the number of equations and variables small, only power streams and energy amounts are considered in the model using the variables φ and Φ in Figure 1, respectively. This is an acceptable simplification, since the evaluation or maximization of the economic profit is the final goal of the analysis (Equation (20) in Section 3.2). In fact, this profit depends on costs (investment and operation) and revenues, which can both be evaluated starting from the variables associated with power (sizes of the energy conversion units or their fuel consumption, sold outputs, and emissions) and with amounts of energy (capacity of thermal storage units) only.

Both φ and Φ depend, in general, on intensive and extensive quantities which can be included in a single array $\mathbf{x}$ (i.e., $\varphi(t) = \varphi(\mathbf{x}(t))$ and $\Phi(t) = \Phi(\mathbf{x}(t))$). For instance, in thermal units, these quantities are pressures, temperatures, mass flow rates, and fluids properties. These quantities do not have the same influence on the behavior of a unit, so only the quantities in $\mathbf{x}$ having the strongest impact can be considered and included in an array $\tilde{\mathbf{x}}$ of reduced length. As discussed below (see Equations (9) to (16)), the dependence of φ and Φ on the quantities included in the array $\tilde{\mathbf{x}}$ does not appear in the model explicitly, although it is indirectly considered with the aid of specific parameters k_i, which permit to modify the relationships between $\varphi_{i,in}$ and $\varphi_{i,out}$ (characteristic maps) in the possible off-design operation according with the unit characteristics. This approach allows the use of equations having the same form in the description of the behavior of units with very different characteristics, belonging to both categories of storage or energy conversion units.

The general criterion used in the following to keep the model simple consists in:

- Using only the energy balance equation to describe the dynamic behavior of a storage unit. To this end, this equation is rearranged to include a variable describing the functional characteristics of the storage unit;
- Including the minimum number of characteristic maps to describe the behavior of the energy conversion units. The energy balance equations of these units are considered only when the calculation of the loss/emissions streams ($\varphi_{i,L}$ in Figure 1) is required.

3.1.1. Energy Balance Equations

The energy balance equations of the dynamic model of the general energy system in Figure 1 are subdivided here into the following three categories:

1. Interconnections between units
2. Interconnections between units and the external environment
3. Energy conservation within the units.

Equations in category 1 (Equations (i) and (ii) in Table 1) simply state that the output power stream exiting a unit is equal to the same stream entering the unit downstream. These equations are directly derived from the configuration of the energy system.

Table 1. Equations of the dynamic model of the fleet of energy units in Figure 1.

Equation Category	Units	Equation	Reference
Balance equation at the interconnection between units	1 – 4	$\varphi_{1,out_1}(t) = \varphi_{4,in_1}(t)$	(i)
	2 – 4	$\varphi_{2,out_1}(t) = \varphi_{4,in_2}(t)$	(ii)
	ES – 1	$\varphi_{ES,out_1}(t) = \varphi_{1,in_1}(t)$	(iii)
Balance equation at the interconnection between units and environment	ES – 2,3	$\varphi_{ES,out_2}(t) = \varphi_{2,in_1}(t) + \varphi_{3,in_1}(t)$	(iv)
	ES – 3	$\varphi_{ES,out_3}(t) = \varphi_{3,in_2}(t)$	(v)
	4 – UD	$\varphi_{4,out_1}(t) = \varphi_{UD,in_1}(t)$	(vi)
	2,3 – UD	$\varphi_{2,out_2}(t) + \varphi_{3,out_1}(t) = \varphi_{UD,in_2}(t)$	(vii)
Balance equation expressing the energy conservation within the units	1	$\varphi_{1,in_1}(t) - \varphi_{1,out_1}(t) - \varphi_{1,L}(t) = 0$	(viii)
	2	$\varphi_{2,in_1}(t) - \varphi_{2,out_1}(t) - \varphi_{2,out_2}(t) - \varphi_{2,L}(t) = 0$	(ix)
	3	$\varphi_{3,in_1}(t) + \varphi_{3,in_2}(t) - \varphi_{3,out_1}(t) - \varphi_{3,L}(t) = 0$	(x)
	4 (storage)	$\varphi_{4,in_1}(t) + \varphi_{4,in_2}(t) - \varphi_{4,out_1}(t) - \varphi_{4,L}(t) = \frac{d\Phi_4(t)}{dt}$	(xi)
Characteristic maps of the units	1 (in–out)	$\varphi_{1,in_1}(t) = k_{1,0}{\cdot}\delta_1(t) + k_{1,1}{\cdot}\varphi_{1,out_1}(t)$	(xii)
	2 (in–out)	$\varphi_{2,in_1}(t) = k_{2,0}{\cdot}\delta_2(t) + k_{2,1}{\cdot}\varphi_{2,out_1}(t) + k_{2,2}{\cdot}\varphi_{2,out_2}(t)$	(xiii)
	2 (out–out)	$\varphi_{2,out_2}(t) = k_{2,3}{\cdot}\delta_2(t) + k_{2,4}{\cdot}\varphi_{2,out_1}(t)$	(xiv)
	3 (in–out)	$\varphi_{3,in_1}(t) + k_{3,2}{\cdot}\varphi_{3,in_2}(t) = k_{3,0}{\cdot}\delta_3(t) + k_{3,1}{\cdot}\varphi_{3,out_1}(t)$	(xv)
	3 (in–in)	$\varphi_{3,in_2}(t) = k_{3,3}{\cdot}\delta_3(t) + k_{3,4}{\cdot}\varphi_{3,in_1}(t)$	(xvi)
	4 (storage)	$\varphi_{4,L}(t) = k_{4,0} + k_{4,1}{\cdot}\Phi_4(t)$	(xvii)

Equations in category 2 (Equations (iii) to (vii) in Table 1) refer to the streams linking the system with the external environment, both on the input side (primary energy sources–ES) and output side (useful products–UD and emissions/losses–L). For simplicity, when different units require streams of the same primary energy source (e.g., natural gas from the gas distribution network), a single stream exiting the ES-side is considered (i.e., the streams entering the units result from a splitter, Equation (iv) in Table 1). When the total availability of primary sources is limited (by the fixed size of the existing distribution networks, e.g., natural gas), variable (e.g., biomass, water), or fixed over time (e.g., sun, wind), the following constraints are added to the model

$$\text{limited} \qquad\qquad \text{variable} \qquad\qquad \text{fixed}$$
$$\varphi_{ES,out_q}(t) \le \varphi_{ES,out_q}^{MAX} \qquad \varphi_{ES,out_q}(t) \le \varphi_{ES,out_q}^{MAX}(t) \qquad \varphi_{ES,out_q}(t) = \varphi_{ES,out_q}^{MAX}(t) \tag{1}$$

Similarly, when different units generate the same type of final product (e.g., thermal power sent to a district heating network), a single stream entering the UD-side is considered (i.e., the streams exiting the units are mixed, Equation (vii) in Table 1).

The balance equations belonging to category 3 (Equations (viii) to (xi) in Table 1) assume the general differential form

$$\frac{d\Phi_j(t)}{dt} = \sum_p \varphi_{j,in_p}(t) - \sum_q \varphi_{j,out_q}(t) - \varphi_{j,L}(t). \tag{2}$$

This differential form is necessary to describe correctly the behavior of a storage unit (Equation (xi) of unit 4 in Table 1). The amount of energy contained in the storage unit at any time τ is then calculated as

$$\Phi_j(\tau) = \Phi_j(0) + \underbrace{\int_{t=0}^{\tau} \sum_p \varphi_{j,in_p}(t){\cdot}dt}_{\Phi_{j,in}^{\tau}} - \underbrace{\int_{t=0}^{\tau} \sum_q \varphi_{j,out_q}(t){\cdot}dt}_{\Phi_{j,out}^{\tau}} - \underbrace{\int_{t=0}^{\tau} \varphi_{j,L}(t){\cdot}dt}_{\Phi_{j,L}^{\tau}}, \tag{3}$$

where $\Phi_j(0)$ is the initial value of the energy contained in the storage unit, and $\Phi_{j,in}^\tau$, $\Phi_{j,out}^\tau$ and $\Phi_{j,L}^\tau$ are the total amount of energy sent to, taken from and lost by the storage unit in the time period 0 to τ, respectively. In general, the loss stream ($\varphi_{j,L}(t)$) and the total energy lost ($\Phi_{j,L} = \int \varphi_{j,L}(t){\cdot}dt$), depend on both the characteristics of the storage unit and the intensive and extensive quantities in the array x_j. To simplify the model, Equation (3) is expressed as a function of the round-trip efficiency ($\eta_{j,RT}$ in Equation (4)), which indirectly takes into account the total energy lost ($\varphi_{j,L}(t)$). This parameter is in fact defined as the ratio of energy entering ($\Phi_{j,in}^{RT}$) to energy exiting ($\Phi_{j,out}^{RT}$) the storage unit in an average charging/discharging process $\left(\eta_{j,RT} = \dfrac{\Phi_{j,out}^{RT}}{\Phi_{j,in}^{RT}} = \dfrac{\int_{RT} \Sigma \varphi_{j,out}{\cdot}dt}{\int_{RT} \Sigma \varphi_{j,in}{\cdot}dt}\right)$. Thus, $\eta_{j,RT}$ contains information on the amount of energy lost during the process being this amount the difference between energy entering and exiting the storage unit ($\Phi_{j,L}^{RT} = \int_{RT} \varphi_{j,L}{\cdot}dt = \Phi_{j,in}^{RT} - \Phi_{j,out}^{RT}$). For simplicity it is assumed that the effect of $\eta_{j,RT}$ is equally subdivided between the charging and discharging phases of the storage unit.

$$\Phi_j(\tau) = \Phi_j(0) + \int_{t=0}^{\tau} \sum_p \sqrt{\eta_{j,RT}}{\cdot}\varphi_{j,in_p}(t){\cdot}dt - \int_{t=0}^{\tau} \sum_q \frac{1}{\sqrt{\eta_{j,RT}}}{\cdot}\varphi_{out,j_q}(t){\cdot}dt \tag{4}$$

The following two inequalities are required to take into account the complete filling and emptying of the storage unit

$$\Phi_j(t) \leq \Phi_j^{MAX} \text{ or } \Phi_j(t) \leq v_j{\cdot}\Phi_j^{MAX} \tag{5}$$

$$\Phi_j(t) \geq \Phi_j^{MIN} \text{ or } \Phi_j(t) \geq 0 \tag{6}$$

where Φ_j^{MAX} is the capacity of the storage unit, Φ_j^{MIN} is the minimum amount of energy that can be contained in the storage unit (which can be equal to zero). An oversizing coefficient ($v_j > 1$) can be considered to guarantee the correct operation of the storage unit (e.g., to maintain the thermal stratification within thermocline thermal energy storage tanks, to avoid overcharge in batteries, etc.).

The energy balances of the energy conversion units should be included in the model only if the loss/emissions streams ($\varphi_{i,L}$ in Figure 1) are to be known (e.g., when these streams are associated with an emission cost, see Equation (20) in Section 3.2). In this case, a steady-state form is adopted (Equation (i) to (iii) in Table 1) because the inertia of energy conversion units is usually negligible with respect to that of the storage units

$$\varphi_{i,L}(t) = \sum_p \varphi_{i,in_p}(t) - \sum_q \varphi_{i,out_q}(t). \tag{7}$$

3.1.2. Characteristic Maps

The characteristic maps describe the behavior of each energy conversion unit (Equation (xiv) to (xvii) in Table 1). Considering a unit having one fuel stream and one product stream (e.g., unit 1 in Figure 1) one characteristic map is sufficient to model the unit behavior, which can be expressed as input–output relationship

$$\varphi_{i,in}(t) = f(\varphi_{i,out}(t)) \tag{8}$$

This relationship (Figure 2a) is typically nonlinear and depends on the array $\tilde{x}_i$ of intensive and extensive quantities mentioned above. For constant values of $\tilde{x}_i$ the relationship can be linearized (Figure 2b,c) for most of the existing thermal systems in the usual range of possible loads with a loss of accuracy well compensated by a strong simplification of the model. For instance, the comparison between results of the linear model and data from the manufacturer of a 427 kWe CHP internal combustion engine showed [10] errors between −0.43% (at nominal power) and 0.76% (at minimum load, i.e., 50% of the nominal power) in the prediction of the fuel consumption, and between −0.27% (at 73% of nominal power) and 0.52% (at minimum load) in the prediction of thermal power output. Higher maximum errors (up to −7.58%) were found [20] for coal fired steam units, gas turbines, and combined cycle units. However, the average errors in the entire range of possible loads were found to be much smaller (−0.01 to 0.27%). Linear models introduce acceptable errors also in simulating the behavior of hydroelectric units (e.g., errors in between −2.07% and 2.77% were observed at various water mass flow rates and available heads in [10]). Conversely, solar and wind units do not behave linearly, but their nonlinear models can be solved independently of the optimization problem as shown in Section 3.2, and therefore they do not need to be linearized.

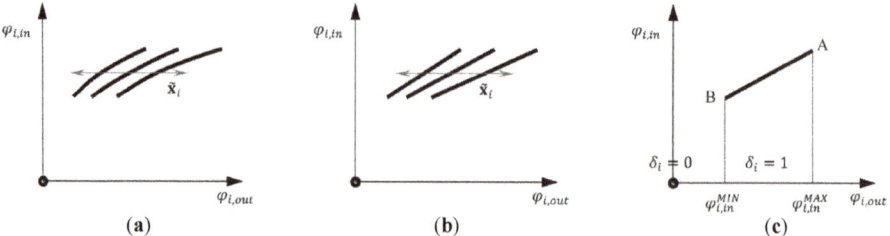

Figure 2. Characteristic map of an energy conversion unit having one fuel ($\varphi_{i,in}$) and one product ($\varphi_{i,out}$): (**a**) nonlinear input–output relationship, (**b**) linearized input–output relationship and (**c**) use of a binary variable δ_i to identify the on/off status of the unit.

The linear form of Equation (8) is

$$\varphi_{i,in}(t) = k_{i,0}\delta_i(t) + k_{i,1}\cdot\varphi_{i,out}(t) \tag{9}$$

$$\varphi_{i,out}(t) \leq k_{i,3}\cdot\varphi_{i,out}^{MAX}\cdot\delta_i(t) \tag{10}$$

$$\varphi_{i,out}(t) \geq k_{i,4}\cdot\varphi_{i,out}^{MIN}\cdot\delta_i(t) \tag{11}$$

where $k_{i,0}$ to $k_{i,4}$ are (usually positive) parameters depending on the type and features of the energy conversion unit, and $\varphi_{i,out}^{MAX}$ and $\varphi_{i,out}^{MIN}$ are the maximum and minimum unit loads at nominal value of $\tilde{x}_i$, respectively. The maximum load of a unit at nominal conditions ($\varphi_{i,out}^{MAX}$) is also meant here as the unit size. The value of the four parameters $k_{i,0}$ to $k_{i,4}$ depends on $\tilde{x}_i$ only (i.e., $k_{i,0} = k_{i,0}(\tilde{x}_i)$, etc.). This dependence is not shown in Equations (9) to (11) for clarity. The two inequalities in Equations (10) and (11) are necessary to fix the extreme points of the characteristic map (A and B in Figure 2c, i.e., the operating boundaries of the energy conversion unit), which in general depend on $\tilde{x}_i$ as shown in Figure 2b. The binary variable δ_i is used to identify the on/off status of the unit:

- When δ_i is equal to zero, Equations (10) and (11) give $\varphi_{i,out} = 0$ and so Equation (9) gives $\varphi_{i,in} = 0$ (the unit is off)
- Conversely, when δ_i is equal to one, Equations (10) and (11) let $\varphi_{i,out}$ vary within the range of possible loads and the fuel consumption is calculated by Equation (9) (the unit is on).

Note that the unit efficiency (η_i) resulting from Equation (9) varies according to both the unit load ($\varphi_{i,out}$) and the array $\tilde{x}_i$. In particular, for fixed $\tilde{x}_i$ and $\delta_i(t) = 1$ (the definition of efficiency is meaningful only when the unit is operating):

$$\eta_i(t) = \frac{\varphi_{i,out}(t)}{\varphi_{i,in}(t)} = \frac{1}{k_{i,0}}\cdot\left(1 - \frac{k_{i,1}}{k_{i,1}\cdot\varphi_{i,out}(t) + k_{i,1}}\right) \tag{12}$$

In many types of energy conversion units (e.g., internal combustion engines), the unit behavior is only marginally affected by all quantities in x_i, and so the dependence on $\tilde{x}_i$ in the unit characteristic map can be neglected to further simplify the model. In these cases, in fact, the parameters $k_{i,0}$, $k_{i,1}$, $\varphi_{i,out}^{MAX}$ and $\varphi_{i,out}^{MIN}$ in Equations (5) to (7) are constants (see Equations (37) and (39) in Section 4.2.2).

In energy conversion units generating more than one product stream, the additional products can be generated

i) by recovering waste streams (e.g., in a CHP gas turbine the thermal power output is recovered from the exhaust gases), or

ii) by consuming a part of the streams used to generate the main product (e.g., in the extraction-condensing CHP steam turbines, at constant fuel a higher steam extraction for thermal use results in a lower power output).

In case i), the fuel consumption is not affected by the additional products and it can still be calculated using Equation (9). Conversely, in case ii), other terms must be added in Equation (9) to take into account the increase in fuel consumption due to the additional products. Considering a unit which generates two products (e.g., unit 2 in Figure 1) the characteristic map becomes

$$\varphi_{i,in}(t) = k_{i,0}\cdot\delta_i(t) + k_{i,1}\cdot\varphi_{i,out_1}(t) + k_{i,2}\cdot\varphi_{i,out_2}(t) \tag{13}$$

Equation (13) can be easily generalized for units generating a number of products equal to n_{out} being $k_{i,0}$ usually independent from the product streams (φ_{i,out_q})

$$\varphi_{i,in}(t) = k_{i,0}\cdot\delta_i(t) + \sum_{q=1}^{n_{out}} k_{i,q}\cdot\varphi_{i,out_q}(t) \tag{14}$$

In both cases i) and ii) above an additional characteristic map is to be included in the model of the system for each additional product q (Figure 3), which generally links the additional product (φ_{i,out_q} for $q = 2,\ldots,n_{out}$) to the main product (φ_{i,out_1}). Again, in most cases, these output–output maps can be well approximated by linear relationships (for constant values of $\tilde{x}_i$).

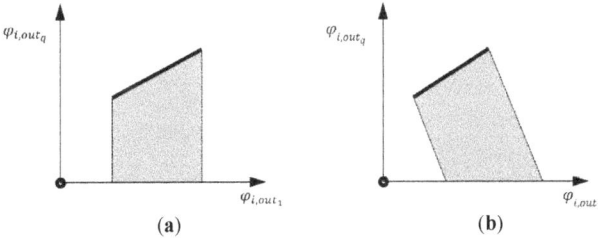

(a) (b)

Figure 3. Additional characteristic maps of an energy conversion unit having more products: (**a**) the additional product (φ_{i,out_q}) is generated by recovering waste streams, (**b**) the additional product (φ_{i,out_q}) is generated consuming a part of the streams used to generate the main product (φ_{i,out_1}).

In case i) (additional product from waste streams recovery, Figure 3a), the output–output map is described by the following equation

$$\varphi_{i,out_q}(t) = k_{i,3}\cdot\delta_i(t) + k_{i,4}\cdot\varphi_{i,out_1}(t), \tag{15}$$

where δ_i are the same binary variables describing the on/off status in Equations (9) to (11), and $k_{i,3}$ and $k_{i,4}$ are parameters which depend on the type and features of the energy conversion unit, and may or not depend on $\tilde{x}_i$. If the recovery system of the unit can be bypassed (e.g., a part of the flue gas exiting a CHP gas turbine can bypass the heat recovery heat exchanger), the "=" sign in Equation (15) is substituted with the "≤" sign. In this case, the maximum value of φ_{i,out_q} is defined by Equation (15) and the minimum value by the following additional constraint (the gray area in Figure 3a represents all the feasible operating points of the energy conversion unit)

$$\varphi_{i,out_q}(t) \geq \varphi_{i,out_q}^{MIN} \text{ or } \varphi_{i,out_q}(t) \geq 0 \tag{16}$$

In case ii) (additional product from useful streams, Figure 3b) the feasible area is described by more than two relationships [9], examples are provided in Section 4.2.2 (Equations (44) and (47)).

Similarly, the modeling of energy conversion units having more than one fuel stream (less frequent) requires an additional characteristic map for each additional fuel stream. Again, in general the additional maps link the additional fuel stream (φ_{i,in_p} for $p > 1$) to the main one (φ_{i,in_1}) (input–input characteristic map, see, e.g., Equation (38) in Section 4.2.2).

In conclusion, the total number "n_{eq}" of characteristic maps which are required to describe the behavior of an energy conversion unit having "n_{in}" inputs and "n_{out}" output is in general equal to the sum of the number of inputs and output of the unit minus one ($n_{eq} = n_{in} + n_{out} - 1$).

3.2. Optimization Problem

The optimization of the design of a fleet of energy units is meant here as the search of the number, type, size, and interconnection of energy conversion and storage units which maximize the economic profit during a reference period of time T (entire system lifespan, a reference year, representative days of the different seasons, etc.). To obtain an optimum system design able to satisfy the users' demands and all other constraints associated with the operational characteristics of the system units and the variability of primary resources availability, the design optimization is performed in combination with the optimization of the operation of each system unit.

A binary variable β_i (constant in the total period of time T of the analysis) is used to choose whether to include or not in the optimum configuration an energy conversion unit i belonging to a predefined set of available units of fixed type and size. This new binary variable β_i is associated with the binary variable $\delta_i(t)$ (time-varying) describing the on/off status of the unit (see Section 3.1.2) as follows

$$\delta_i(t) \leq \beta_i. \tag{17}$$

When β_i is equal to zero, δ_i is always equal to zero, i.e., the unit is not included because it does not contribute to the power generation during the period T. Conversely, when β_i is equal to one, Equation (17) allows the unit to be turned on ($\delta_i = 1$) or off ($\delta_i = 0$), i.e., the unit is included.

The size (maximum capacity, φ_{i,out_1}^{MAX}) of the energy conversion units is not included in the set of the design decision variables because most of the energy conversion units (internal combustion engines, steam power station, boilers, wind turbines, etc.) are available only in specific sizes according to manufacturers' catalogues. Moreover, the use of the actual performance of the energy conversion units available in the market guarantees more reliable results. In fact, the efficiency and other operating characteristics (e.g., maximum load ramps, time required for a start-up, etc.) of an energy conversion unit depend on its size, but accurate relationships that link these performances to the unit size are usually not available.

Different considerations apply when solar units (photovoltaic or solar thermal units) are considered, which are built by assembling components of very small size (photovoltaic modules or solar thermal panels). In this case, the performance of the total unit is almost independent of the number of the components that make it up, and so from the size of the total unit, but depends on the components type only (e.g., PV cells of different materials). So, when solar units are considered, their size can be left free to vary and no binary variables β_i are used, as shown in Section 4.1.

Similarly, no binary variables are added for the design choices related to the storage units: the capacity of the storage unit Φ_j^{MAX} is not fixed but is a direct outcome of the optimization procedure according to the optimum operation strategy of the total system. An optimum capacity equal to zero corresponds to the exclusion of the storage unit from the optimum configuration.

In general, the optimization of the design and operation of the energy system in Figure 1 corresponds to

$$\text{Find } \mathbf{y}_D^* \text{ and } \mathbf{y}_O^*(t) \in \mathcal{R}^n \text{ or } \mathfrak{I}^m \text{ which maximizes } Z(T) = Z(\mathbf{y}_D, \mathbf{y}_O(t))_T$$
$$\text{subject to } \begin{array}{l} \mathbf{g}(\mathbf{y}_D, \mathbf{y}_O(t)) = 0 \\ \mathbf{l}(\mathbf{y}_D, \mathbf{y}_O(t)) \geq 0 \end{array} \tag{18}$$

where $\mathbf{y}_D$ and $\mathbf{y}_O(t)$ are the arrays of the decision variables associated with the energy system design (D) and operation (O), respectively, $Z(\mathbf{y}_D, \mathbf{y}_O(t))_T$ is the objective function, and $\mathbf{g}(\mathbf{y}_D, \mathbf{y}_O(t))$ and $\mathbf{l}(\mathbf{y}_D, \mathbf{y}_O(t))$ are all the equality and inequality constraints of the system model (Section 3.1). Among the inequalities $\mathbf{l}(\mathbf{y}_D, \mathbf{y}_O(t))$, constraints on the maximum load ramp rate and minimum uptime and downtime of energy conversion units are also included to avoid solutions that are not feasible or that could, if implemented, lead to unit malfunctions (caused, e.g., by excessive thermal stress or intermittent operation of the unit). The constraint on the maximum load ramp rate is expressed as in Equation (19), whereas the constraints on minimum uptime and downtime are shown in Equations (55) to (58) in Section 4.3.

$$\frac{d\varphi_{i,out_1}(t)}{dt} \leq \Delta\varphi_{i,out_1}^{MAX}, \tag{19}$$

where $\Delta\varphi_{i,out}^{MAX}$ is the maximum rate at which the unit can increase or decrease its main output. In some types of units, the value of $\Delta\varphi_{i,out}^{MAX}$ depends on the operating constraints (i.e., normal operation, start-up, and shutdown) as shown in Equations (53) and (54) in Section 4.2.2.

The design decision variables ($\mathbf{y}_D$, array of constant values in the period T) are the binary variables β_i and the storage capacities Φ_j^{MAX}. The operation decision variables ($\mathbf{y}_O$, array of time profiles) are the real variables defining the load of each energy conversion ($\varphi_{i,out}(t)$) and storage ($\varphi_{j,out}(t)$) unit, and the binary variables $\delta_i(t)$ describing the on/off status of the energy conversion units. When the optimization procedure is limited only to the operation of the system, the design variables β_i and Φ_j^{MAX} are fixed parameters, whereas $\varphi_{i,out}(t)$, $\varphi_{j,out}(t)$ and $\delta_i(t)$ and are free to vary.

If non-dispatchable primary energy sources (sun, wind, and run-of-river hydropower) are considered, no choices can be made on the operation of the units that convert these sources (e.g., photovoltaic plants, wind turbines or farms, run-of-river hydroelectric plants). Accordingly, the operation variables of these units are no more included in the decision variables set but provided as inputs to the optimization problem. To do this, the models of the units are solved independently of the optimization problem, i.e., their relationships does not appear among the constraints $\mathbf{g}(\mathbf{y}_D, \mathbf{y}_O(t))$ and $\mathbf{l}(\mathbf{y}_D, \mathbf{y}_O(t))$. The design binary variables β_i remain among the decision variables to let the optimization procedure choose about the inclusion (or not) of the units in the final configuration of the total system.

The optimization of the design and operation of a fleet of energy units generally aims at maximizing the economic profit because it is the main driving force behind investments and business decisions. Thus, the economic profit is considered here as example of objective function. Different objectives can be considered without substantially modifying the relationships of the model ($\mathbf{g}(\mathbf{y}_D, \mathbf{y}_O(t))$ and $\mathbf{l}(\mathbf{y}_D, \mathbf{y}_O(t))$) and the choice of the decision variables ($\mathbf{y}_D, \mathbf{y}_O(t)$ in Equation (18)). In fact, the equations and variables presented in Section 3.1 are generally sufficient to evaluate the performance and economic parameters that are broadly used to judge the operation of an energy system (e.g., annual average energy or exergy efficiency, total emissions in the year of operation, etc.). Similarly, multiple objective functions can be considered to find the best trade-offs between different objectives, such as the minimization of both costs and emissions. This multi-objective approach, which is out of the scope of the work, allows to take both objectives into consideration while avoiding solutions with low cost, due to underestimated emission cost, but high environmental impact.

The economic profit to be maximized is calculated as the difference between revenues and expenditures (Equation (20)).

$$
\begin{aligned}
Z\left(\beta_i, \Phi_j^{MAX}, \varphi_{i,\,out}(t), \varphi_{j,\,out}(t), \delta_i(t)\right)_T = & \int_0^T \Sigma_p\, \varphi_{UD,in_p}(t) \cdot p_{UD,in_p}(t) \cdot dt + \int_0^T \Sigma_{i,q}\, \varphi_{i,out_q}(t) \cdot s_{i,out_q}(t) \cdot dt \\
& + \Sigma_i\, \beta_i \cdot \varphi_{i,in_1}^{MAX} \cdot s_i \cdot T^{[y]} + \Sigma_j\, \Phi_j^{MAX} \cdot s_j \cdot T^{[y]} \\
& - \int_0^T \Sigma_q\, \varphi_{ES,out_q}(t) \cdot c_{ES_q}(t) \cdot dt - \int_0^T \Sigma_r\, \varphi_{r,L}(t) \cdot c_{r,L}(t) \cdot dt \\
& - \Sigma_i\, \beta_i \cdot \left(\varphi_{i,in_1}^{MAX} \cdot c_{i,O\&M} \cdot T^{[y]} + C_{i,O\&M}\left(\varphi_{i,in_1}^* \; (\forall t \in [0,T])\right)\right) \\
& - \Sigma_j \left(\Phi_j^{MAX} \cdot c_{j,O\&M} \cdot T^{[y]} + C_{i,O\&M}\left(\varphi_{j,in_1}^* \; (\forall t \in [0,T])\right)\right) \\
& - \Sigma_i\, N_{i,SU} \cdot C_{i,SU} - \Sigma_i\, \beta_i \cdot \varphi_{i,in_1}^{MAX} \cdot a_i \cdot T^{[y]} - \Sigma_j\, \Phi_j^{MAX} \cdot a_j \cdot T^{[y]}
\end{aligned}
\tag{20}
$$

Revenues are received from:

- Selling the streams $\varphi_{UD,in_p}(t)$ to the users at the unit prices $p_{UD,in_p}(t)$. The unit prices can be variable or constant over the period T depending on the sale contracts and any feed-in tariffs established by law.
- Incentives to support generation and investments. The former consist in providing premium feed-in tariffs $s_{i,out_q}(t)$ (e.g., the green certificates), which typically decline over time to track and encourage technological changes, to specific products $\varphi_{i,out_q}(t)$ (e.g., electric power from renewable sources, thermal power from CHP units). The latter are direct subsidies or tax credits which are calculated as the product of the size of the energy conversion (φ_{i,in_1}^{MAX}) or storage unit (Φ_j^{MAX}) that receives the incentive and a grant per unit of installed capacity (s_i and s_j). In Equation (20) the duration of the optimization period expressed in years ($T^{[y]}$) is also included because the unit grants s_i and s_j are supposed to be provided on year basis. The investment incentive of each energy conversion unit is multiplied by the corresponding binary decision variable β_i because no incentive is received if the unit is not included in the optimum configuration ($\beta_i = 0$).

Expenditures derive from:

- Consumption of the primary energy sources streams ($\varphi_{ES,out_q}(t)$) at unit costs $c_{ES_q}(t)$ and charges for emission of the streams $\varphi_{r,L}(t)$ at unit costs $c_{r,L}(t)$. Both unit costs can be variable or constant over the period T depending on purchase contracts and emission trading markets (e.g., CO_2 emission allowances market [21]).
- Operation and Maintenance (O&M) costs of the units. These costs depend on both the size of the unit and its operating profile (i.e., number of hours of operation, load factor, load variation, etc.). Accordingly, they are estimated in Equation (20) using annual costs per unit of installed capacity $c_{O\&M_i}$ and $c_{O\&M_j}$ (independent of the unit operation) and total costs $C_{i,O\&M}$ and $C_{j,O\&M}$, which depend on the optimum operating profile of the unit in the total period T (φ_{i,in_1}^* ($\forall t \in [0,T]$) and φ_{j,in_1}^* ($\forall t \in [0,T]$) in Equation (20)). The latter are known only after the optimization run is completed, so guess values of $C_{i,O\&M}$ and $C_{j,O\&M}$ are to be chosen and the procedure iterated using updated values of these costs until convergence. For this reason, in stationary applications in which the load scheduling of the units does not generally show frequent and sudden variations, it is acceptable to incorporate all O&M costs in constant annual costs per unit of installed capacity ($\bar{c}_{r,O\&M}$ in Equation (60), Section 4.4), so as to avoid iterative optimization runs. As for the incentives, the O&M costs of the energy conversion unit is multiplied by the associated binary variable β_i.
- Start-up costs (considered only for the energy conversion units), which are calculated by multiplying the total number of start-ups in the period T ($N_{i,SU}$) by the cost of each start-up ($C_{i,SU}$). $N_{i,SU}$ (integer quantity) can be easily obtained from the binary variables δ_i as shown in Section 4.3 (Equation (59)).

- Amortization of purchase and installation costs of the units calculated as size of the units multiplied by the annual amortization costs per unit of capacity (a_i and a_j). Again, the binary variables β_i are used to include only the amortization costs of the units belonging to the optimal configuration.

Unit costs and prices in Equation (20) are in €/kWh (or homogeneous quantities), and feed-in tariffs and O&M costs per unit of capacity are in (€/kW·y) (or homogeneous quantities) for the energy conversion units and in (€/kWh·y) (or homogeneous quantities) for the storage units.

The input data to the optimization procedure are:

- The time profile of the intensive and extensive quantities in the array $\tilde{x}_i(t)$ (see Section 3.1).
- The time profiles of the user's demands ($\varphi_{UD,in_q}(t)$), availability of primary energy sources ($\varphi^{MAX}_{ES,out_p}(t)$ in Equation (1)) and the generation from non-dispatchable primary energy sources.
- All variable and constant prices, costs, and feed-if tariffs ($p_{in_p}(t)$, $c_{ES_q}(t)$, $c_{r,L}(t)$, $c_{i,O\&M}$, $c_{j,O\&M}$, $C_{i,O\&M}$, $C_{j,O\&M}$, $C_{i,SU}$, a_i, a_j, $s_{i,out_q}(t)$, s_i, s_j in Equation (20)).
- The maximum and minimum load (φ^{MAX}_{i,in_1} and φ^{MIN}_{i,in_1}) and the other parameters used to model the behavior of the energy conversion units belonging to the predefined set of available units (i.e., $k_{i,q}$ in Equations (9), (10), (11), (14) and (15), and $\Delta\varphi^{MAX}_{i,out_1}$ in Equation (19)).
- The round-trip efficiencies of the storage units ($\eta_{j,RT}$ in Equation (4)).

The approach can be easily adapted to accept values of all input data having stochastic variability, just feeding it with all possible scenarios of these variables (see, e.g., [22]). This possibility is out of the scope of this work, which aims to supply in a simple way the most basic guidelines.

The optimization problem qualifies as a dynamic MINLP problem because of the inclusion of binary decision variables (β_i and $\delta_i(t)$) and constraints connecting each time interval dt to the previous one (energy balance of storage units in Equation (4), maximum load ramp rate in Equation (19) and minimum uptime and downtime in Equations (55) to (58) in Section 4.3). To reduce the computational effort required to solve this problem all nonlinear relationships including one or more decision variables are linearized, as discussed in the Section 3.1.2.

On the other hand, the search for the optimum design of a fleet of energy units may involve a very large number of binary and real decision variables even if the number of units included in the optimal configuration is not excessive. This is because the predefined set of candidate energy conversion units to be included in the system should contain a sufficient number of units of different type and size to explore as completely as possible the space of possible solutions. For instance, if this set includes:

- ten power units and ten heat units (for both of them the decision variables are β_i, $\delta_i(t)$, and $\varphi_{i,out}(t)$),
- ten CHP units (where the decision variables are β_i, $\delta_i(t)$, φ_{i,out_1} and $\varphi_{i,out_2}(t)$),
- both electric and thermal storage units (where the decision variables are Φ^{MAX}_j and $\varphi_{j,out}(t)$),

the optimization problem of the design and operation of the fleet of energy units in the whole year of operation divided into hours will include more than $630\,k$ decision variables ($\underbrace{2 \times 10 \times (1 + 2 \times 8760)}_{\text{power and heat units}} +$

$\underbrace{10 \times (1 + 3 \times 8760)}_{\text{CHP units}} + \underbrace{2 \times (1 + 8760)}_{\text{storage units}})$.

So, further simplifications may be required to keep the computational effort that is necessary to solve the MILP optimization problem within acceptable limits. One possible simplification consists in subdividing the optimization problem in two steps (Figure 4). In the first step, the optimum design of the system is obtained by solving the MILP optimization problem in Equation (18) in a reduced period of time $\tilde{T}$ which is representative of the total period T (e.g., twelve typical days which are representative of each month, or three days representative of summer, winter, and midseason). In the second step, the resulting values of the design variables $\tilde{\beta}_i^*$ and Φ_j^{*MAX} are used to optimize only the operation of the best design obtained in the first step in the total period T. In so doing, the number of decision variables is significantly reduced compared to the nonsimplified optimization both in the first step (a lower number of time instants is considered) and in the second one (all design variables are fixed, and the variables and equations associated with the units excluded from the optimum configuration obtained in the first step are no more considered).

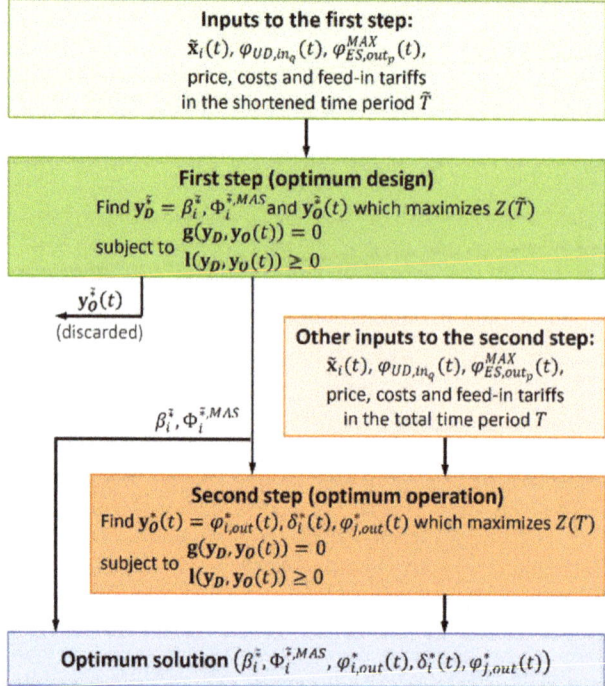

Figure 4. Two-step decomposed optimization procedure.

The values of the input data in the reduced period $\widetilde{T}$ are to be carefully selected since they have a strong influence on the system design. Different sets of values can be considered and the different designs resulting from the first step (Figure 4) compared on the basis of the optimum operation obtained in the second step to find the one which behaves better in the total period T.

4. General Application: Design and Operation Optimization of a CHP Fleet of Energy Units

The guidelines proposed in Section 3 are here used to set the optimum design and operation of a fleet of units serving electrical and thermal energy users (Figure 5).

4.1. General CHP Fleet of Energy Units

Fossil fuels (coal and natural gas) and both non-dispatchable (sun and wind) and dispatchable (storage hydropower and biomass—wood and bio-oil) renewables are considered as available primary energy sources.

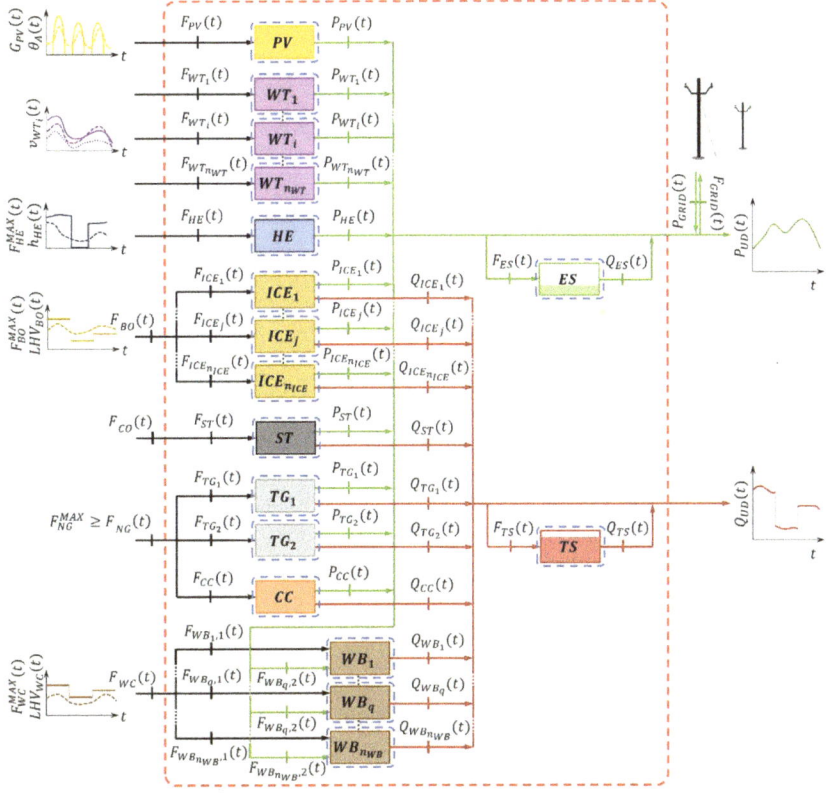

Figure 5. Fleet of units serving electrical and thermal energy users.

The related distribution and supply networks are assumed as existing as well as the electricity grid to which the system is connected (electricity can be either sold or bought from the grid). The time trends of users' demands and energy sources availability are also drawn in Figure 5 to highlight the variability and general non-contemporaneity of these quantities. The aim is twofold: i) showing how a MILP approach easily adapts to model and optimize energy systems that include units of very different types and sizes, and ii) providing the equations to describe the off-design behavior of the main types of energy conversion and storage units.

The types and number of energy conversion units that are candidates for the optimal configuration are:

- one photovoltaic power station (*PV*) of size P_{PV}^{MAX},
- a set of n_{WT} wind turbines (WT_i) having the same fixed sizes (P_{WT}^{MAX}),
- one storage hydroelectric plant (*HE*) of fixed size (P_{HE}^{MAX}), this unit is considered as existing and the associated purchase and installation costs are already amortized,
- a set of n_{ICE} bio-oil fueled CHP internal combustion engines (*ICE*) of various sizes ($P_{ICE_j}^{MAX}$),
- one coal fired CHP steam unit (*ST*) of fixed size (P_{ST}^{MAX}),
- two ($n_{GT} = 2$) natural gas fueled CHP gas turbines (*GT*) of different sizes ($P_{GT_1}^{MAX}$ and $P_{GT_2}^{MAX}$),
- one natural gas fired CHP combined cycle units (*CC*) of fixed size (P_{CC}^{MAX}),
- a set of n_{WB} woodchip boilers (*WB_i*) having the same fixed sizes (Q_{WB}^{MAX}).

Note that n_{WT}, n_{ICE}, n_{GT}, and n_{WB} are not the number of units included in the optimal configuration, but the number of units included in the set of candidate ones. These numbers are required for the mathematical formulation and implementation of the optimization problem because some arrays and summations use them as the maximum value of the index/set.

A thermal (*TS*) and an electric (*ES*) storage unit are also considered, the capacity of which are free to vary in the optimization procedure. The size of the PV power station (P_{PV}^{MAX}) is also left free to vary being this type of unit designed by "assembling" modules of very small size having almost constant performance and unit cost for very different sizes of the total plant.

4.2. Off-Design Model

To simplify the calculations, the total period of analysis T is generally subdivided into one-hour-long time intervals ($\Delta t = 1$ h) identified in the following using the set $t = 1, 2, \ldots, T$. Accordingly, each variable of the model assumes a discrete value in the entire hour of each time interval t.

For clarity, the stream variables in the Figure 1 associated with fuels and desired products ($\varphi_{i,in}$ and $\varphi_{i,out}$, respectively) are renamed as F (power inputs to the units), P and Q (electric and thermal power outputs), whereas the losses/emissions stream ($\varphi_{i,L}$) are not explicitly considered in the model. Accordingly, the electric (thermal) power entering the electric (thermal) storage unit is identified by the variable F_{ES} (F_{TS}). Note that the electric consumption of woodchip boilers is also taken into account using the variable $F_{WB_q,2}$ ($F_{WB_q,1}$ refers to the consumption of woodchips).

The power inputs associated with the different primary energy sources are calculated as follows:

- Solar energy (PV)

$$F_{PV}(t) = G_{PV}(t) \cdot A_{PV}, \tag{21}$$

where A_{PV} is the total active aperture area of solar energy conversion unit (PV), and $G_{PV}(t)$ is the global solar irradiance on the PV modules plane.

- Wind energy

$$F_{WT_i}(t) = \frac{1}{2} \cdot v_{WT_i}^3(t) \cdot \rho_{air} \cdot A_{WT_i}, \tag{22}$$

where $v_{WT_i}(t)$ is the velocity of the free airstream, ρ_{air} is the ambient air density (constant) and A_{WT_i} is the swept area of the wind turbine (e.g., $\pi \cdot r_{WT_i}^2$ for a horizontal axis turbine).

- Hydropower

$$F_{HE}(t) = \dot{V}_{HE}(t) \cdot \rho_{water} \cdot h_{HE}(t) \cdot g = \dot{m}_{HE}(t) \cdot h_{HE}(t) \cdot g, \tag{23}$$

where $\dot{V}_{HE_i}(t)$ and $\dot{m}_{HE_i}(t)$ are the volumetric and mass flow rates of water entering the hydroelectric unit, respectively, ρ_{water} is the water density (constant) and $h_{HE_i}(t)$ is the available water head and g is the standard gravity.

- Solid (woodchip and coal), liquid (bio-oil), or gaseous (natural gas) fossil or renewable fuels

$$F_r(t) = \dot{m}_{r,f}(t) \cdot LHV_{r,f}(t), \tag{24}$$

where $\dot{m}_{r,f}(t)$ is the fuel mass flow rate and $LHV_{r,f}(t)$ is its lower heating value (which may vary with time depending on the fuel type).

The interconnections between units and between units and the external environment are fixed. All thermal streams $Q_i(t)$ generated by the energy conversion units (boilers and CHP) are collected and sent to the users or stored (totally or partially) in the thermal storage unit to be used at a later time. The corresponding energy balance is

$$Q_{UD}(t) = \sum_{q=1}^{n_{WB}} Q_{WB_q}(t) + \sum_{j=1}^{n_{ICE}} Q_{ICE_j}(t) + Q_{ST}(t) + \sum_{p=1}^{2} Q_{GT_p}(t) + Q_{CC}(t) + Q_{TS}(t) - F_{TS}(t). \tag{25}$$

Similarly, all electric streams $P_i(t)$ generated by power or CHP units (after deduction of the electric consumption of the woodchip boilers) can be stored, sent to the users or to the grid.

The electrical power can also be taken from the grid to satisfy the users' requirement or charge the electric storage unit:

$$
\begin{aligned}
P_{UD}(t) &= P_{PV}(t) + \sum_{i=1}^{n_{WT}} P_{WT_i}(t) + P_{HE}(t) + \sum_{j=1}^{n_{ICE}} P_{ICE_j}(t) + P_{ST}(t) + \sum_{p=1}^{2} P_{GT_p}(t) + P_{CC}(t) \\
&\quad - \sum_{q=1}^{n_{WB}} F_{WB_q,2}(t) + P_{ES}(t) - F_{ES}(t) + P_{GRID}(t) - F_{GRID}(t),
\end{aligned}
\tag{26}
$$

where $P_{GRID}(t)$ and $F_{GRID}(t)$ are the electric power taken and sent to the grid, respectively.

The total consumption of woodchips (WC), bio-oil (BO), coal (CO) and natural gas (NG) are obtained from the energy balance on primary energy sources side

$$
F_{WC}(t) = \sum_{q=1}^{n_{WB}} F_{WB_q}(t), \; F_{BO}(t) = \sum_{j=1}^{n_{ICE}} F_{ICE_j}(t), \; F_{CO}(t) = F_{ST}(t), \; F_{NG}(t) = \sum_{p=1}^{2} F_{GT_p}(t) + F_{CC}(t). \tag{27}
$$

The total availability of hydropower, bio-oil and woodchips (variable with time) and the maximum mass flow rate of natural gas (limited by the existing network) are expressed in terms of power

$$
F_{HE}(t) \leq F_{HE}^{MAX}(t), \; F_{WC}(t) \leq F_{WC}^{MAX}(t), \; F_{BO}(t) \leq F_{BO}^{MAX}(t), \; F_{NG}(t) \leq F_{NG}^{MAX}, \tag{28}
$$

where the values of the maximum power ($F_{HE}^{MAX}(t)$, $F_{WC}^{MAX}(t)$, $F_{BO}^{MAX}(t)$ and F_{NG}^{MAX}) can be easily calculated by Equations (23) and (24) from the associated maximum mass or volumetric flow rate.

4.2.1. Dynamic Model of the Storage Units

The dynamic behavior of the storage units is described by their energy balances only (see Section 3.1). The electric energy contained in the electric storage unit is calculated from its energy balance (see Equation (4)). An oversizing coefficient ($v_{ES,1}$) is considered to prevent overcharging and the minimum amount of electric energy that can be contained in the electric storage is calculated as a fraction of the (unknown) storage maximum capacity $\left(v_{ES,2} = \dfrac{E_{ES}^{MIN}}{E_{ES}^{MAX}} \right)$.

$$
\begin{aligned}
E_{ES}(t) &= E_{ES}(t-1) + \left(\sqrt{\eta_{ES,RT}} \cdot F_{ES}(t) - \frac{1}{\sqrt{\eta_{ES,RT}}} \cdot P_{ES}(t) \right) \cdot \Delta t \\
E_{ES}(t) &\leq v_{ES,1} \cdot E_{ES}^{MAX} \\
E_{ES}(t) &\geq v_{ES,2} \cdot E_{ES}^{MAX}
\end{aligned}
\tag{29}
$$

$$
E_{ES}(0) = E_{ES}(T) = E_{ES}^{MIN}. \tag{30}
$$

The thermal storage unit consists of one or more thermocline tanks in which hot and cool water is stored in the upper and lower part, respectively. For simplicity, the thermal energy contained in this unit is expressed in terms of total volume of hot water stored at time t assuming constant values of hot ($\theta_{TS,hot}$) and cold ($\theta_{TS,cold}$) water temperatures, density (ρ_{TS}), and specific heat ($c_{p,TS}$). The two coefficients $v_{TS,1}$ and $v_{TS,2}$ in Equation (31) are introduced to maintain the thermal stratification within the thermocline tanks.

$$
\begin{aligned}
V_{TS}(t) &= V_{TS}(t-1) + \frac{1}{\rho_{TS} \cdot c_{p,TS} \cdot (\theta_{TS,hot} - \theta_{TS,cold})} \cdot \left(\sqrt{\eta_{TS,RT}} \cdot F_{TS}(t) - \frac{1}{\sqrt{\eta_{TS,RT}}} \cdot Q_{TS}(t) \right) \cdot \Delta t \\
V_{TS}(t) &\leq v_{TS,1} \cdot V_{TS}^{MAX} \\
V_{TS}(t) &\geq v_{TS,2} \cdot V_{TS}^{MAX}
\end{aligned}
\tag{31}
$$

$$
V_{TS}(0) = V_{TS}(T) = V_{TS}^{MIN}. \tag{32}
$$

The energy contained in the storage units at the beginning and end of the total period T is set equal to the minimum value by means of the additional constraint in Equations (30) and (32). This guarantees that the electric and thermal energy generated by the energy conversion units in the total period T is equal to the sum of the energy consumed by the users plus the losses of the storage units (and plus the net energy exchanged with the grid in case of electricity), i.e., there is no "free" energy taken from the storage units.

4.2.2. Model of the Energy Conversion Units

For simplicity, the energy balances of the energy conversion units are not considered. The CO_2 emission allowances costs are computed in the total profit by using an emission factor per unit of fuel (α_{i,CO_2} in Equation (60) in Section 4.4).

As explained in Section 3.2, the behavior of the energy conversion units fueled by non-dispatchable primary energy sources (PV and WT_i) are simulated independently from the optimization procedure starting from historical or forecasted data of the ambient conditions (i.e., solar irradiance $G_{PV}(t)$, ambient temperature $\theta_A(t)$ and wind speed $v_{WT_i}(t)$). The approach used in the optimization process in order to consider the different sizes of the photovoltaic power station and the inclusion (or not) of the wind turbines is presented in the following. The specific models of these units are instead referred to the literature being beyond the scope of this paper.

The power output of the photovoltaic power station is expressed as

$$P_{PV}(t) = \pi_{PV}(t) \cdot P_{PV}^{MAX}, \tag{33}$$

where $\pi_{PV}(t)$ is the power output per unit of nominal power (P_{PV}^{MAX}) which can be calculated from the type of photovoltaic modules and auxiliary devices (energy inversion and conditioning system) included in the power station:

$$\pi_{PV}(t) = \frac{\eta_{PV}(t)}{\eta_{PV}^{MAX}} \cdot G_{PV}(t) = \frac{\eta_{PV,m}(t)}{\eta_{PV,m}^{MAX}} \cdot \frac{\eta_{aux}(t)}{\eta_{aux}^{MAX}} \cdot G_{PV}(t) \tag{34}$$

In Equation (34) $\eta_{PV,m}^{MAX}$, η_{aux}^{MAX}, $\eta_{PV,m}(t)$ and $\eta_{aux}(t)$ are the design (constant) and off-design (time-varying) efficiencies of the PV modules and auxiliary devices for power conditioning, respectively, which can be calculated independently from P_{PV}^{MAX} from nominal and historical or forecasted data of solar irradiance ($G_{PV}(t)$) and ambient temperature ($\theta_A(t)$), as suggested in [10,23]. The resulting time profile of $\pi_{PV}(t)$ is an input data of the optimization procedure which calculates $P_{PV}(t)$ by Equation (33) for the candidate value of the decision variable P_{PV}^{MAX}.

The time profile of the power output that each wind turbine of fixed size ($P_{WT_i}^{MAX}$) could produce if included in the optimum system configuration ($\check{P}_{WT_i}(t)$) is directly obtained from historical or forecasted data of the wind speed ($v_{WT_i}(t)$) using mathematical models (see, e.g., [24]) or experimental characteristic maps provided by the manufacturers (which are generally sufficiently accurate). The "actual" power generation of each turbine is then calculated in the optimization procedure by multiplying $\check{P}_{WT_i}(t)$ by the binary decision variable β_{WT_i} which defines the inclusion or not of the turbine in the system:

$$P_{WT_i}(t) = \beta_{WT_i} \cdot \check{P}_{WT_i}(t). \tag{35}$$

Note that Equations (33) and (35) are linear because $\pi_{PV}(t)$ and $\check{P}_{WT_i}(t)$ do not depend on any decision variable.

The behavior of all the energy conversion units fueled by dispatchable energy sources is modelled by their characteristic maps only. As discussed in Section 3.1.2, one characteristic map is sufficient to model the unit having one input and one desired output (HE), whereas two characteristic maps are required to model the CHP units (ICE_i, TG_i, ST and CC which have one input and two desired outputs) and the woodchip boilers (WB_i, which have two inputs and one desired output). For simplicity, the minimum load of each unit ($\varphi^{MIN}_{i,out}$ in Equation (11)) is expressed as a fraction of the maximum load (e.g., 30% for the woodchip boilers and 50% for the other units) and the thermal power is assumed to be generated at the constant temperature $\theta_{TS,hot}$.

A multi-jet Pelton turbine is installed in the existing hydroelectric power plant. The behavior of this machine is affected by the available water head $h_{HE}(t)$ (i.e., the water pressure within the nozzle), as shown in Figure 6a. This behavior is generally modelled with good accuracy by varying the value of the parameter $k_{HE,0}$ in the equation describing the characteristic according with $h_{HE}(t)$:

$$F_{HE}(t) = k_{HE,0}(h_{HE}(t)) \cdot \delta_{HE}(t) + k_{HE,1} \cdot P_{HE}(t)$$
$$P_{HE}(t) \geq k_{HE,2} \cdot k_{HE,3}(h_{HE}(t)) \cdot P^{MAX}_{HE} \cdot \delta_{HE}(t) \tag{36}$$
$$P_{HE}(t) \leq k_{HE,3}(h_{HE}(t)) \cdot P^{MAX}_{HE} \cdot \delta_{HE}(t).$$

Note that also the maximum and minimum power output of the unit varies with $h_{HE}(t)$ by means of the parameter $k_{HE,3}$ ($k_{HE,2} = 1$ at nominal h_{HE} and $k_{HE,2} < 1$ for lower h_{HE}).

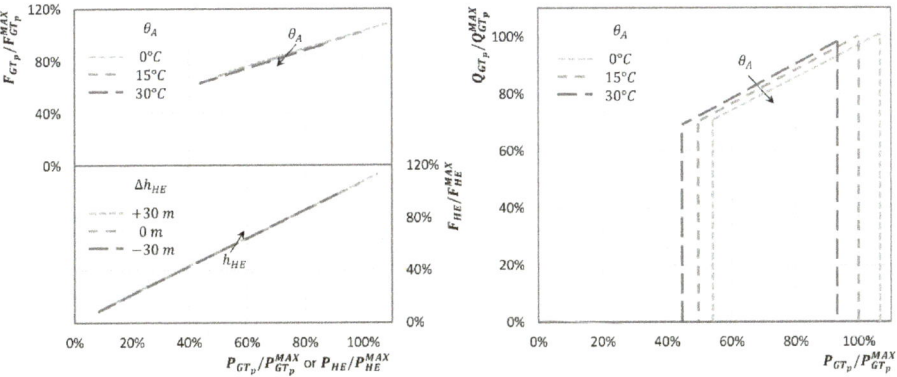

Figure 6. Characteristic maps of a CHP gas turbine (GT) for different ambient temperatures (θ_A): (a) $F - P$ and (b) $Q - P$ map. Figure 6a also represents the map of the hydroelectric unit (HE) for different available water head (Δh_{HE} is the difference between the off-design and design values of h_{HE}).

No modifications of the two characteristic maps (input–output relationship in Equation (37) and input–input relationship in Equation (38)) are considered for the woodchip boilers (WB_q):

$$F_{WB_q,1}(t) = k_{WB_q,0} \cdot \delta_{WB_q}(t) + k_{WB_q,1} \cdot Q_{WB_q}(t)$$
$$Q_{WB_q}(t) \geq k_{WB_q,2} \cdot Q^{MAX}_{WB} \cdot \delta_{WB_q}(t) \tag{37}$$
$$Q_{WB_q}(t) \leq Q^{MAX}_{WB} \cdot \delta_{WB_q}(t),$$

$$F_{WB_q,2} = k_{WB_q,3} \cdot \delta_{WB_q}(t) + k_{WB_q,4} \cdot F_{WB_q,1}(t). \tag{38}$$

In the input–input characteristic map (Equation (38)) the electricity consumption ($F_{WB_q,2}(t)$) is directly linked to the woodchips consumption ($F_{WB_q,1}(t)$), being $F_{WB_q,2}(t)$ mainly due to combustion air blowing.

In the CHP internal combustion engines (ICE_j), the thermal power is generated by recovering the waste heat (from exhaust gasses, charging air, lubricating oil, and jacket water) in a heat recovery system. An auxiliary cooling system is always included in this type of units to dissipate heat in case of failure of the heat recovery system or absence of thermal demand. However, here it is considered that the heat can only be recovered (the heat recovery system cannot be bypassed) to exclude solutions entailing heat dissipations. Accordingly, the input–output characteristic map (Equation (39)) does not include $Q_{ICE_j}(t)$ and the "=" sign appears in the output–output characteristic map (Equation (40)).

$$F_{ICE_j}(t) = k_{ICE_j,0} \cdot \delta_{ICE_j}(t) + k_{ICE_j,1} \cdot P_{ICE_j}(t)$$
$$P_{ICE_j}(t) \geq k_{ICE_j,2} \cdot P_{ICE_j}^{MAX} \cdot \delta_{ICE_j}(t) \tag{39}$$
$$P_{ICE_j}(t) \leq P_{ICE_j}^{MAX} \cdot \delta_{ICE_j}(t),$$

$$Q_{ICE_j}(t) = k_{ICE_j,3} \cdot \delta_{ICE_j}(t) + k_{ICE_j,4} \cdot P_{ICE_j}(t). \tag{40}$$

The bypass of the heat recovery systems is instead considered ("$\leq$" sign in Equation (42)) in the two CHP gas turbines. The operation of these units is usually strongly affected by the ambient temperature ($\theta_A(t)$, Figure 6). This phenomenon is described by varying the values of the parameters $k_{GT_p,0}$, $k_{GT_p,3}$ and $k_{GT_p,4}$ in the characteristic maps in Equations (41) and (42) according with $\theta_A(t)$.

$$F_{GTp}(t) = k_{GT_p,0}(\theta_A(t)) \cdot \delta_{GT_p}(t) + k_{GT_p,1} \cdot P_{GT_p}(t)$$
$$P_{GT_p}(t) \geq k_{GT_p,2} \cdot k_{GT_p,3}(\theta_A(t)) \cdot P_{GT_p}^{MAX} \cdot \delta_{GT_p}(t) \tag{41}$$
$$P_{GT_p}(t) \leq k_{GT_p,3}(\theta_A(t)) \cdot P_{GT_p}^{MAX} \cdot \delta_{GT_p}(t),$$

$$Q_{GT_p}(t) \leq k_{GT_p,4}(\theta_A(t)) \cdot \delta_{GT_p}(t) + k_{GT_p,5} \cdot P_{GT_p}(t) \quad Q_{GT_p}(t) \geq 0. \tag{42}$$

The coal-fired CHP steam unit (*ST*) includes an extraction-condensing steam turbine. Accordingly, the input–output characteristic map (Equation (43)) of the unit calculates the fuel consumption (F_{ST}) starting from both desired output streams (P_{ST} and Q_{ST}, see Section 3.1.2). The turbine has two extraction points at different pressures, i.e., at different temperatures (θ_{ST}^{MAX} and θ_{ST}^{MIN}), so the thermal power $Q_{ST}(t)$ can be generated at any temperature θ_{ST} in the range θ_{ST}^{MIN} to θ_{ST}^{MAX} by mixing properly the mass flow rates of the two extracted streams. Figure 7a shows the input–output characteristic map at minimum (zero) and maximum (Q_{ST}^{MAX}) thermal power output for fixed values of θ_{ST}: for each value of θ_{ST} the characteristic map can move from the rightmost black thick line ($Q_{ST} = 0$) to the line corresponding to the maximum thermal power output (for example, when $\theta = 130\,°C$ the map can move up to the leftmost grey dotted line in Figure 7a). The dependence on θ_{ST} can be generally taken into account only with the parameter k which multiplies the thermal power output (i.e., $k_{ST,2} = k_{ST,2}(\theta_{ST})$ in Equation (43)). The maximum (and minimum) power output is also affected by Q_{ST} and θ_{ST} because of the limit on the maximum mass flow rate of steam that can be produced in the steam generator.

$$F_{ST}(t) = k_{ST,0} \cdot \delta_{ST}(t) + k_{ST,1} \cdot P_{ST}(t) + k_{ST,2}(\theta_{ST}(t)) \cdot Q_{ST}(t)$$
$$P_{ST}(t) \geq k_{ST,3} \cdot k_{ST,4}(Q_{ST}(t), \theta_{ST}(t)) \cdot P_{ST}^{MAX} \cdot \delta_{ST}(t) \tag{43}$$
$$P_{ST}(t) \leq k_{ST,4}(Q_{ST}(t), \theta_{ST}(t)) \cdot P_{ST}^{MAX} \cdot \delta_{ST}(t).$$

The output–output characteristic map is a feasibility area (Figure 7b) bounded by the following inequalities, which vary according with generation temperature θ_{ST} [25]:

line 1 $\left(\text{min } \dot{m} \text{ in the low-pressure turbine}\right)$: $Q_{ST}(t) \leq k_{ST,5}(\theta_{ST}(t)) \cdot \delta_{ST}(t) + k_{ST,6} \cdot P_{ST}(t)$

ine 2 (min load of the steam generator) : $Q_{ST}(t) \geq k_{ST,7} \cdot \delta_{ST}(t) + k_{ST,8}(\theta_{ST}(t)) \cdot P_{ST}(t)$

line 3 (max load of the steam generator) : $Q_{ST}(t) \leq k_{ST,9} \cdot \delta_{ST}(t) + k_{ST,10}(\theta_{ST}(t)) \cdot P_{ST}(t)$ (44)

line 4 $\left(\text{max } \dot{m} \text{ in the higher temperature steam extraction}\right)$: $Q_{ST}(t) \leq k_{ST,11}(\theta_{ST}(t)) \cdot P_{ST}(t)$

line 5 $\left(\text{min } \dot{m} \text{ in the higher temperature steam extraction}\right)$: $Q_{ST}(t) \geq 0$

Here, the generation temperature is constant ($\theta_{ST}(t) = \theta_{TS,hot}$), so the input–output characteristic map moves only according with Q, while the output–output map is fixed.

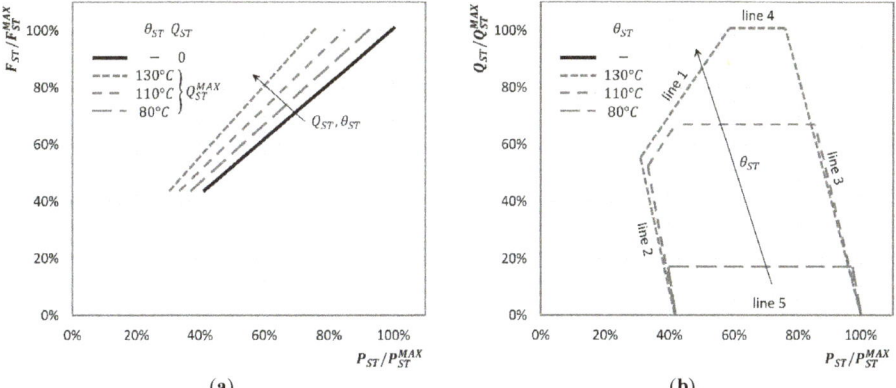

Figure 7. Characteristic maps of the CHP steam unit (*ST*) for different generation temperatures (θ_{ST}):
(a) $F - P$ map when $Q_{ST} = 0$ and $Q_{ST} = Q_{ST}^{MAX}$, and (b) $Q - P$ map.

The CHP combined cycle (*CC*) includes two identical couples gas turbine/HRSH (Heat Recovery Steam Generator) operating in conjunction with a single steam turbine. To extend the range of possible loads, the *CC* can operate with one turbine only (one GT mode in Figure 8) or both turbines (two GT mode in Figure 8) on. The behavior of the *CC* in both modes is described by two different characteristic maps (Figure 8). Two binary variables ($\delta_{CC1GT}(t)$ and $\delta_{CC2GT}(t)$ in Equations (45) to (47)) and the following constraint are used to identify the two alternative operating modes [25,26]:

$$\delta_{CC1GT}(t) + \delta_{CC2GT}(t) \leq 1 \tag{45}$$

As for the other units, when either $\delta_{CC1GT}(t)$ or $\delta_{CC2GT}(t)$ are equal to one the combined cycle is operating in the one or two GT mode, respectively, when $\delta_{CC1GT}(t) = \delta_{CC2GT}(t) = 0$ the CHP combined cycle is off.

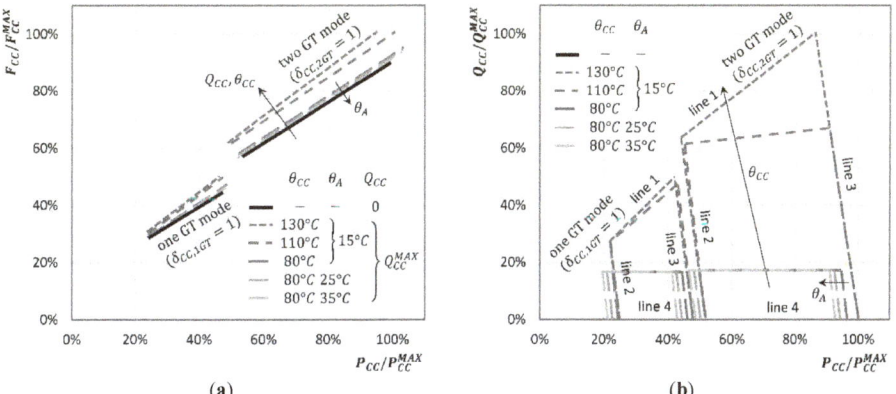

Figure 8. Characteristic maps of the CHP combined cycle unit (*CC*) for different generation (θ_{CC}) and ambient (θ_A) temperatures: (a) $F - P$ for $Q_{CC} = 0$ and $Q_{CC} = Q_{ST}^{MAX}$ (b) $Q - P$ map. For simplicity, the dependence on θ_A is shown only for $\theta_{CC} = 80°$.

The steam section includes an extraction-condensing steam turbine having two extraction points, so the thermal energy can be generated at different temperatures ($\theta_{CC}^{MIN} \leq \theta_{CC}(t) \leq \theta_{CC}^{MAX}$). Accordingly, in both operating modes the input–input characteristic map of the *CC* in Equation (47) is modified according with $\theta_{CC}(t)$ (the *CC* unit behaves in a similar way to the *ST* unit). Moreover, the ambient temperature $\theta_A(t)$ affects the behavior of the gas turbines, and therefore that of the total unit, as shown in Figure 8.

The characteristic maps of the *CC* unit in one and two GT modes are described by the following relationships

$$
\begin{aligned}
F_{CC1GT}(t) &= k_{CC1GT,0}(\theta_A(t)) \cdot \delta_{CC1GT}(t) + k_{CC1GT,1} \cdot P_{CC1GT}(t) + k_{CC1GT,2}(\theta_{CC}(t)) \cdot Q_{CC1GT}(t) \\
P_{CC1GT}(t) &\geq k_{CC1GT,3} \cdot k_{CC1GT,4}(Q_{CC1GT}(t), \theta_A(t), \theta_{CC}(t)) \cdot P_{CC1GT}^{MAX} \cdot \delta_{CC1GT}(t) \\
P_{CC1GT}(t) &\leq k_{CC1GT,4}(Q_{CC1GT}(t), \theta_A(t), \theta_{CC}(t)) \cdot P_{CC1GT}^{MAX} \cdot \delta_{CC1GT}(t), \\
F_{CC2GT}(t) &= k_{CC2GT,0}(\theta_A(t)) \cdot \delta_{CC2GT}(t) + k_{CC2GT,1} \cdot P_{CC2GT}(t) + k_{CC2GT,2}(\theta_{CC}(t)) \cdot Q_{CC2GT}(t) \\
P_{CC2GT}(t) &\geq k_{CC2GT,3} \cdot k_{CC2GT,4}(Q_{CC2GT}(t), \theta_A(t), \theta_{CC}(t)) \cdot P_{CC2GT}^{MAX} \cdot \delta_{CC2GT}(t) \\
P_{CC2GT}(t) &\leq k_{CC2GT,4}(Q_{CC2GT}(t), \theta_A(t), \theta_{CC}(t)) \cdot P_{CC2GT}^{MAX} \cdot \delta_{CC2GT}(t);
\end{aligned}
\tag{46}
$$

$$
\begin{aligned}
&\text{line 1 (1GT)}: Q_{CC1GT}(t) \leq k_{CC1GT,5}(\theta_A(t), \theta_{CC}(t)) \cdot \delta_{CC1GT}(t) + k_{CC1GT,6}(\theta_{CC}(t)) \cdot P_{CC1GT}(t) \\
&\text{line 2 (1GT)}: Q_{CC1GT}(t) \geq k_{CC1GT,7}(\theta_A(t)) \cdot \delta_{CC1GT}(t) + k_{CC1GT,8} \cdot P_{CC1GT}(t) \\
&\text{line 3 (1GT)}: Q_{CC1GT}(t) \leq k_{CC1GT,9}(\theta_A(t), \theta_{CC}(t)) \cdot \delta_{CC1GT}(t) + k_{CC1GT,10} \cdot P_{CC1GT}(t) \\
&\text{line 4 (1GT)}: Q_{CC1GT}(t) \geq 0, \\
&\text{line 1 (2GT)}: Q_{CC2GT}(t) \leq k_{CC2GT,5}(\theta_A(t), \theta_{CC}(t)) \cdot \delta_{CC2GT}(t) + k_{CC2GT,6}(\theta_{CC}(t)) \cdot P_{CC2GT}(t) \\
&\text{line 2 (2GT)}: Q_{CC2GT}(t) \geq k_{CC2GT,7}(\theta_A(t)) \cdot \delta_{CC2GT}(t) + k_{CC2GT,8} \cdot P_{CC2GT}(t) \\
&\text{line 3 (1GT)}: Q_{CC2GT}(t) \leq k_{CC2GT,9}(\theta_A(t), \theta_{CC}(t)) \cdot \delta_{CC2GT}(t) + k_{CC2GT,10} \cdot P_{CC2GT}(t) \\
&\text{line 4 (2GT)}: Q_{CC2GT}(t) \geq 0.
\end{aligned}
\tag{47}
$$

The effective fuel input, and electric and thermal power output of the *CC* unit are:

$$
F_{CC}(t) = F_{CC1GT}(t) + F_{CC2GT}(t), \ P_{CC}(t) = P_{CC1GT}(t) + P_{CC2GT}(t), \ Q(t) = Q_{CC1GT}(t) + Q_{CC2GT}(t).
\tag{48}
$$

4.3. Additional Dynamic Constraints

Constraints on maximum load ramp rate, and minimum uptime and downtime of energy conversion units involving combustion processes (of biomass or fossil fuels) are added to avoid failures or breakdowns caused by excessive thermal stress. These constraints allow for identifying the start-ups of the units with the help of the binary variables δ_i, which define the on/off (see Equation (9) in Section 3.1.2). The total number of start-ups is also calculated to evaluate the associated costs.

A start-up of the CHP internal combustion engines (ICE_r) and woodchip boilers (WB_r) at time t is identified by the additional binary variable $\sigma_r(t)$ and the following relationship:

$$
\delta_r(t-1) - \delta_r(t) + \sigma_r(t) \geq 0
\tag{49}
$$
$$
\text{for } r = ICE_1, \ldots, ICE_{n_{ICE}}, WB_1, \ldots, WB_{n_{WB}}.
$$

In fact, Equation (49) forces $\sigma_r(t)$ to turn its value to one when $\delta_r(t-1) = 0$ and $\delta_r(t) = 1$, i.e., when the unit is turned on. In all other cases $\sigma_r(t) = 0$.

The constraints on the operation of steam and gas turbines generally depend on the temperature of casing and rotor. Three different types of start-up are considered for the CHP steam unit (ST) according to the time elapsed from the latest shutdown of this unit, which in turn determines the casing and rotor temperature [26–28]: a hot start-up is defined when the time period between shut-down and start-up is shorter than $T_{ST,HS}$ (3 to 5 hours depending on the steam unit type and size), a warm start-up when it is shorter than $T_{ST,WS}$ (5 to 10 hours) and longer than $T_{ST,HS}$, a cold start-up when it is longer than $T_{ST,WS}$. As in Equation (49), the inequalities in Equation (50) force the three different binary variables $\sigma_{ST,HS}$, $\sigma_{ST,WS}$ and $\sigma_{ST,CS}$ to switch their value to one when a hot, warm, or cold start-up occurs, respectively.

$$\delta_{ST}(t-1) - \delta_{ST}(t) + \sigma_{ST,HS}(t) \geq 0$$
$$\delta_{ST}(t-1) - \ldots - \delta_{ST}(t+T_{ST,HS}-1) + \delta_{ST}(t+T_{ST,HS}) + \sigma_{ST,WS}(t) \geq 0 \qquad (50)$$
$$\delta_{ST}(t-1) - \ldots - \delta_{ST}(t+T_{ST,WS}-1) + \delta_{ST}(t+T_{ST,WS}) + \sigma_{ST,CS}(t) \geq 0.$$

Note that the switch to one of the binary variables associated with a colder start-up implies that the binary variables associated with the warmer ones are also equal to one (i.e., when the warm start-up occurs both $\sigma_{ST,WS}(t)$ and $\sigma_{ST,HS}(t)$ are equal to one and when a cold start-up occurs all the three variables are equal to one).

Only hot and cold start-ups are defined for the gas turbines (GT_1 and GT_2) and the combined cycle (CC) [26]:

$$\delta_r(t-1) - \delta_r(t) + \sigma_{r,HS}(t) \geq 0$$
$$\delta_r(t-1) - \ldots - \delta_r(t+T_{r,HS}-1) + \delta_r(t+T_{r,HS}) + \sigma_{r,CS}(t) \geq 0 \qquad (51)$$
$$\text{for } r = GT_1, GT_2, CC.$$

Constraints on the maximum load ramp rates are considered for the woodchip boiler (Equation (52)), and the steam (Equation (53)) and combined cycles (Equation (54)), which are slow units because of the high thermal inertia of the combustion chamber/steam generator, whereas the remaining units are generally able to perform any load variation during $\Delta t = 1$ h (e.g., internal combustion engines and gas turbines require less than ten minutes and about half an hour to reach full load during a start-up, respectively [28,29]).

In Equations (53) and (54) different values are imposed to the maximum load ramp rates of ST and CC depending on the operating conditions, i.e., normal operation (ΔP_i^{MAX}), and hot ($\Delta P_{ST,HS}^{MAX}$), warm ($\Delta P_{WT,HS}^{MAX}$) and cold ($\Delta P_{ST,CS}^{MAX}$) start-ups. The binary variables σ are used to identify the working condition and the summations are introduced to fix the correct maximum load ramp rate during each type of start-up [26] ($\Delta t_{ST,WS}$ is the number of hours required for a warm start-up and $\Delta t_{ST,CS}$ for a cold start-up, whereas it is assumed that a hot start-up requires one hour, i.e., one time interval).

$$Q_{WB_q}(t) - Q_{WB_q}(t-1) \leq \Delta Q_{WB}^{MAX} \qquad (52)$$

$$
\begin{aligned}
P_{ST}(t) - P_{ST}(t-1) \leq\ & \Delta P_{ST}^{MAX} + \sigma_{ST,HS}(t) \cdot \left(\Delta P_{ST,HS}^{MAX} - \Delta P_{ST}^{MAX} \right) \\
& + \sigma_{ST,WS}(t) \cdot \left(\Delta P_{ST,WS}^{MAX} - \Delta P_{ST,HS}^{MAX} \right) + \sigma_{ST,CS}(t) \cdot \left(\Delta P_{ST,CS}^{MAX} - \Delta P_{ST,WS}^{MAX} \right) \\
& + \sum_{\tau=t-1}^{t-\Delta t_{ST,WS}} \sigma_{ST,WS}(\tau) \cdot \left(\Delta P_{ST,WS}^{MAX} - \Delta P_{ST}^{MAX} \right) \\
& + \sum_{\tau=t}^{t-\Delta t_{ST,WS}} \sigma_{ST,CS}(\tau) \cdot \left(\Delta P_{ST,CS}^{MAX} - \Delta P_{ST,WS}^{MAX} \right) \\
& + \sum_{\tau=t-\Delta t_{ST,WS}}^{t-\Delta t_{ST,CS}} \sigma_{ST,CS}(\tau) \cdot \left(\Delta P_{ST,CS}^{MAX} - \Delta P_{ST}^{MAX} \right),
\end{aligned}
\qquad (53)
$$

$$
\begin{aligned}
P_{CC}(t) - P_{CC}(t-1) \leq\ & \Delta P_{CC}^{MAX} + \sigma_{CC,HS}(t) \cdot \left(\Delta P_{CC,HS}^{MAX} - \Delta P_{CC}^{MAX} \right) \\
& + \sigma_{CC,CS}(t) \cdot \left(\Delta P_{CC,CS}^{MAX} - \Delta P_{CC,HS}^{MAX} \right) + \sum_{\tau=t-1}^{t-\Delta t_{CC,CS}} \sigma_{CC,CS}(\tau) \cdot \left(\Delta P_{CC,CS}^{MAX} - \Delta P_{CC}^{MAX} \right).
\end{aligned}
\qquad (54)
$$

The constraints on the minimum downtime and uptime of each energy conversion unit except PV, WT_i, and HE are expressed using the two inequalities in Equations (55)–(58) (refer to [9]), respectively. The former is introduced to limit the thermal stress in the components of the units, the latter for "good practice" (not for technological reasons).

$$\Delta t_{r,DT}(t) = \left(\Delta t_{r,DT}(t-1) + \Delta t \right) \cdot \left(1 - \delta(t) \right) \qquad (55)$$

$$\left(\delta_r(t) - \delta(t-1)\right)\cdot\left(\Delta t_{r,DT}(t) - \Delta t_{r,DT}^{MIN}\right) \geq 0 \tag{56}$$

$$\Delta t_{r,UT}(t) = \left(\Delta t_{r,UT}(t-1) + \Delta t\right)\cdot\delta(t) \tag{57}$$

$$\left(\delta_r(t) - \delta(t-1)\right)\cdot\left(\Delta t_{r,UT}(t) - \Delta t_{r,UT}^{MIN}\right) \geq 0 \tag{58}$$

for $r = WB_1, \ldots, WB_{n_{WB}}, ICE_1, \ldots, ICE_{n_{ICE}}, ST, GT_1, GT_2, CC$.

Equations (55) and (57) increase the downtime ($\Delta t_{r,DT}(t)$) and uptime ($\Delta t_{r,UT}(t)$) of the unit r by one hour (Δt) for each Δt in which the unit remains off ($\delta_r(t-1) = \delta_r(t) = 0$) or on ($\delta_r(t-1) = \delta_r(t) = 1$), respectively. The inequalities in Equations (56) and (58) assure that the unit r can turn its status (off to on, or on to off, respectively) only when the minimum downtime ($\Delta t_{r,DT}^{MIN}$) or uptime ($\Delta t_{r,UT}^{MIN}$) have elapsed. All relationships in Equations (55) to (58) include a product between unknown time-dependent variables (δ_r which is included in the decision variable set, see Section 4.5, and $\Delta t_{r,DT}$ or $\Delta t_{r,OT}$ which depend on δ_r). To keep the model linear, these equations can be linearized using a linearization method (such the Glover method [30]), which requires additional auxiliary variables, as shown in Appendix A.

The total number of start-ups of each unit is calculated, if required, as the sum of the values of the associated binary variables σ in the total period T. Again, the different binary variables υ in Equation (59) identify different types of start-ups.

$$N_{r,SU} = \sum_{t=0}^{T} \sigma_i(t) \quad N_{k,HSU} = \sum_{t=0}^{T} \sigma_{k,HSU}(t) \quad N_{ST,wsu} = \sum_{t=0}^{T} \sigma_{ST,wsu}(t) \quad N_{k,CSU} = \sum_{t=0}^{T} \sigma_{k,CSU}(t)$$
$$\text{for} \quad r = ICE_1, \ldots, ICE_{n_{ICE}}, WB_1, \ldots, WB_{n_{WB}} \tag{59}$$
$$k = GT_1, GT_2, ST, CC.$$

4.4. Objective Function

The economic profit to be maximized (Equation (20) in Section 3.2) derived from the CHP fleet of units in Figure 5 is calculated by Equation (60).

$$
\begin{aligned}
Z(\mathbf{y_D}, \mathbf{y_O}(t))_T = {} & \left[\sum_{t=1}^{8760} \left(P_{UD}(t)\cdot p_{UD,P}(t) + Q_{UD}(t)\cdot p_{UD,Q}(t) + \widetilde{F}_{GRID}(t)\cdot p_{GRID}(t) \right)\cdot\Delta t \right] \\
& + \left[\sum_{t=1}^{8760} \left(P_{PV}(t)\cdot s_{ce} + P_{PV,GRID}(t)\cdot s_{rd}(t) + \left(\sum_{i=1}^{n_{WT}} P_{WT_i,GRID}(t) + P_{HE,GRID}(t) \right)\cdot s_{oc} \right. \right. \\
& \left. \left. + \sum_{j=1}^{n_{ICE}} P_{ICE_j,GRID}(t)\cdot(s_{cv} + s_{rd}(t)) \right)\cdot\Delta t + \left[\sum_{q=1}^{n_{WB}} \beta_{WB_q}\cdot Q_{WB}^{MAX}\cdot s_{WB} \right] \right] \\
& - \left[\sum_{t=1}^{8760} \left(F_{BO}(t)\cdot c_{BO}(t) + F_{WC}(t)\cdot c_{WC}(t) + F_{CO}(t)\cdot\left(c_{CO}(t) + \alpha_{CO,CO_2}\cdot c_{CO_2}\right) \right. \right. \\
& \left. \left. + F_{NG}(t)\cdot\left(c_{NG}(t) + \alpha_{NG,CO_2}\cdot c_{CO_2}\right) + P_{GRID}(t)\cdot c_{GRID}(t) \right)\cdot\Delta t \right] \\
& - \left[P_{PV}^{MAX}\cdot c_{PV,O\&M} + \sum_{i=1}^{n_{WT}} \beta_{WT_i}\cdot P_{WT}^{MAX}\cdot c_{WT,O\&M} + P_{HE}^{MAX}\cdot c_{HE,O\&M} + \beta_{ST}\cdot P_{ST}^{MAX}\cdot c_{ST,O\&M} \right. \\
& + \sum_{j=1}^{n_{ICE}} \beta_{ICE_j}\cdot P_{ICE_j}^{MAX}\cdot\widetilde{c}_{ICE,O\&M} + \sum_{p=1}^{2} \beta_{TG_p}\cdot P_{TG_p}^{MAX}\cdot\widetilde{c}_{TG,O\&M} + \beta_{CC}\cdot P_{CC}^{MAX}\cdot\widetilde{c}_{CC,O\&M} \\
& \left. + \sum_{q=1}^{n_{WB}} \beta_{WB_q}\cdot Q_{WB}^{MAX}\cdot\widetilde{c}_{WB,O\&M} \right] - \left[E_{ES}^{MAX}\cdot\widetilde{c}_{ES,O\&M} + V_{TS}^{MAX}\cdot\widetilde{c}_{TS,O\&M} \right] \\
& - \left[\sum_{j=1}^{n_{ICE}} N_{ICE_j,SU}\cdot C_{ICE_j,SU} + \sum_{i=1}^{n_{WB}} N_{WB_q,SU}\cdot C_{WB,SU} \right. \\
& + \left(N_{ST,HSU}\cdot C_{ST,HSU} + N_{ST,wsu}\cdot(C_{ST,wsu} - C_{ST,HSU}) + N_{ST,CSU}\cdot(C_{ST,CSU} - C_{ST,wsu}) \right) \\
& + \left(N_{CC,HSU}\cdot C_{CC,HSU} + N_{CC,CSU}\cdot(C_{CC,CSU} - C_{CC,HSU}) \right) \\
& \left. + \sum_{p=1}^{2} \left(N_{GT_p,HSU}\cdot C_{GT_p,HSU} + N_{GT_p,CSU}\cdot(C_{GT_p,CSU} - C_{GT_p,HSU}) \right) \right] \\
& - \left[P_{PV}^{MAX}\cdot a_{PV} + \sum_{i=1}^{n_{WT}} \beta_{WT_i}\cdot P_{WT}^{MAX}\cdot a_{WT} + \sum_{j=1}^{n_{ICE}} \beta_{ICE_j}\cdot P_{ICE_j}^{MAX}\cdot a_{ICE} + \beta_{ST}\cdot P_{ST}^{MAX}\cdot a_{ST} \right. \\
& \left. + \sum_{p=1}^{2} \beta_{GT_p}\cdot P_{GT_p}^{MAX}\cdot a_{GT} + \beta_{CC}\cdot P_{CC}^{MAX}\cdot a_{CC} + \sum_{q=1}^{n_{WB}} \beta_{WB_q}\cdot Q_{WB}^{MAX}\cdot a_{WB} \right] \\
& - \left[E_{ES}^{MAX}\cdot a_{ES} + V_{TS}^{MAX}\cdot a_{TS} \right]
\end{aligned}
\tag{60}
$$

The optimization period T is one year, so the total number of hourly intervals is 8760. For the sake of clarity, the square brackets in Equation (60) identify the different components of economic profit presented in the Section 3.2 (Equation (20)).

The electricity is sold to both the users at the unit price $p_{UD,P}(t)$, and to the grid at the electricity market price $p_{GRID}(t)$ or at a premium feed-in tariff $p_{r,GRID}(t)$ (i.e., incentives to support generation). The latter depends on the type of energy conversion unit that generates the specific amount of electricity. The total electric power that is sent to the grid (F_{GRID}) is therefore subdivided into the different terms generated by the different units

$$
\begin{aligned}
F_{GRID}(t) \quad &= P_{PV,GRID}(t) + \sum_{i=1}^{n_{WT}} P_{WT_i,GRID}(t) + P_{HE,GRID}(t) + \sum_{j=1}^{n_{ICE}} P_{ICE_j,GRID}(t) \\
&+ P_{ST,GRID}(t) + \sum_{p=1}^{2} P_{GT_p,GRID}(t) + P_{CC,GRID}(t) + P_{ES,GRID}(t),
\end{aligned}
\tag{61}
$$

Here, premium feed-in tariffs are given to all energy conversion units using renewable sources. The electric power sold to the market ($\widetilde{F}_{GRID}(t)$ in Equations (60) and (62)) at $p_{UD,P}(t)$ is the sum of the only terms associated with fossil fuel-based energy conversion units and electric storage unit:

$$
\widetilde{F}_{GRID}(t) \quad = \quad P_{ST,GRID}(t) + \sum_{p=1}^{2} P_{GT_p,GRID}(t) + P_{CC,GRID}(t) + P_{ES,GRID}(t)
\tag{62}
$$

According to the (rather complex) Italian incentive policy, the tariff of the electricity generated by the CHP internal combustion engines and the photovoltaic unit ($P_{ICE_j,GRID}(t) \cdot \Delta t$ and $P_{PV}(t) \cdot \Delta t$) and sold to the grid has a variable term ($s_{rd}(t)$ in both cases) plus a constant term (s_{cv} and s_{ce}, respectively). Instead, the tariff for the electricity from wind turbines and the hydroelectric units ($P_{WT_i}(t) \cdot \Delta t$ and $P_{HE}(t) \cdot \Delta t$) has a constant value defined by long-term contracts (s_{oc}). Note that a constant term s_{ce} is awarded to the total amount of electricity generated by PV units ($P_{PV}(t) \cdot \Delta t$).

The installation of woodchip boilers is supported (for ten years) by an annual tax credit equal to 5% of the total purchase and installation costs of these units (i.e., s_{WB} in Equation (60) is equal to $0.05 \cdot cost_{WB}$, where $cost_{WB}$ is the total costs of WB per unit of installed capacity).

CO_2 emission allowances costs are considered only for the units that convert fossil fuels (i.e., ST, GT_1, GT_2 and CC in Figure 5). These costs are computed as additional costs associated with fuel inputs ($F_r(t)$) at the unit cost $\alpha_{r,CO_2} \cdot c_{CO_2}$, where α_{r,CO_2} is the CO_2 emission per unit of fuel and c_{CO_2} is the specific CO_2 emission allowances cost. The expenditure for the purchase of electricity from the grid ($P_{GRID}(t)$) is treated in the same way as the expenditure for consumption of primary energy at the unit price $c_{GRID}(t)$. This unit cost may differ from the selling unit price ($p_{GRID}(t)$) because of transmission costs or different types of energy contracts.

The maintenance costs of all units are calculated using constant annual costs per unit of installed capacity $\widetilde{c}_{r,O\&M}$ to avoid iterative optimization runs as discussed in Section 3.2.

Different costs are considered for the different types of start-up defined in Section 4.3 for ST, GT_1, GT_2, and CC. To evaluate correctly the total start-ups costs of these units, the costs of a warmer start-up are subtracted in Equation (60) from the costs of a colder one because, as explained in Equation (50), a colder start-up implies a warmer one.

The amortization of purchase and installation costs of the hydroelectric plant are not included because it is existing and already amortized. The interpretation of the other terms is straightforward comparing Equation (60) with Equation (20).

4.5. Decision Variables and Input Data

Table 2 lists the complete sets of the decision variables that are free to vary in the design and optimization of the fleet of energy units in Figure 1, whereas the required input data are shown in Table 3. Both tables also show the type ($\mathcal{B}$ = binary, I = integer or R = real) and units of measurement of all quantities to be used to avoid calculation errors. The index t is used to identify the quantities that vary over time, $T = 8760$ is the total length of the arrays associated with these quantities.

If the optimization concerns the operation of a fixed system configuration only, all the variables that define the design of the fleet of energy units (first two rows in Table 2) are no longer decision variables but input data (last two rows in Table 3).

Table 2. Set of the decision variables in the optimization of the design and operation of the fleet of energy units in Figure 5.

	Symbols	Type	Description	U.M.
Optimization of the design and operation	$\beta_{WT_i}, \beta_{ICE_j}, \beta_{ST}, \beta_{GT_1},$ $\beta_{GT_2}\ \beta_{CC}, \beta_{WB_q}$ for $\quad i = 1, \ldots, n_{WT}$ $j = 1, \ldots, n_{ICE}$ $q = 1, \ldots, n_{WB}$	Binary $(\mathcal{B})$	Inclusion or not of the energy conversion units (except PV and HE) in the optimum configuration	$(-)$
	P_{PV}^{MAX} E_{ES}^{MAX} V_{TS}^{MAX}	Real $(\mathcal{R})$	Maximum capacity (size) of the photovoltaic power station (PV), and electric (ES) and thermal (TS) storage units	(kWe) (kWh) (m^3)
Optimization of the operation only	$\delta_{HE}(t), \delta_{ICE_j}(t), \delta_{ST}(t),$ $\delta_{TG_1}(t), \delta_{TG_2}(t), \delta_{WB_q}(t), \delta_{CC1GT}(t),$ $\delta_{CC2GT}(t)$ for $\quad t = 1, \ldots, 8760$ $j = 1, \ldots, n_{ICE}$ $q = 1, \ldots, n_{WB}$	Binary $(\mathcal{B}^T)$	On/off status of the energy conversion units (except PV and WT_i)	$(-)$
	$P_{HE}(t), P_{ICE_j}(t), P_{ST}(t),$ $P_{TG_1}(t), P_{TG_2}(t), P_{CC}(t)$ for $\quad t = 1, \ldots, 8760$ $j = 1, \ldots, n_{ICE}$	Real $(\mathcal{R}^T)$	Electric power generated by the energy conversion units (except PV and WT_i)	(kWe)
	$P_{PV,GRID}(t), P_{WT_i,GRID}(t),$ $P_{HE,GRID}(t), P_{ICE_j,GRID}(t),$ $P_{ST,GRID}(t), P_{TG_1,GRID}(t),$ $P_{TG_2,GRID}(t), P_{CC,GRID}(t)$ for $\quad t = 1, \ldots, 8760$ $i = 1, \ldots, n_{WT}$ $j = 1, \ldots, n_{ICE}$	Real $(\mathcal{R}^T)$	Electric power generated by the energy conversion units and sent to the electric grid	(kWe)
	$Q_{ST}(t), Q_{TG_1}(t), Q_{TG_2}(t),$ $Q_{CC}(t), Q_{WB_q}(t)$ for $\quad t = 1, \ldots, 8760$ $q = 1, \ldots, n_{WB}$	Real $(\mathcal{R}^T)$	Thermal power generated of the energy conversion units (except PV, WT_i and ICE_j)	(kWe)

Table 3. Input data in the optimization of the design and operation of the fleet of energy units in Figure 5.

	Symbols	Type	Description	U.M.
Optimization of the design and operation	$P_{UD}(t)$ $Q_{UD}(t)$ for $t = 1,\ldots,8760$	Real $(\mathcal{R}^T)$	Electric and thermal power required by the users (*UD*)	(kWe) (kWt)
	$\eta_{ES,RT}, \nu_{ES,1}, \nu_{ES,2},$ $\eta_{TS,RT}, \nu_{TS,1}, \nu_{TS,2}$	Real $(\mathcal{R})$	Round-trip efficiency and oversizing/minimum capacity coefficients of the storage units	(−)
	$P_{WT}^{MAX}, P_{HE}^{MAX}, P_{ICE}^{MAX}, P_{ST}^{MAX},$ $P_{GT_1}^{MAX}, P_{GT_2}^{MAX}, P_{CC}^{MAX}$ for $j = 1,\ldots,n_{ICE}$	Real $(\mathcal{R})$	Maximum electric power output of the energy conversion units (except *PV*)	(kWe)
	$Q_{WB_q}^{MAX}$ for $q = 1,\ldots,n_{ICE}$	Real $(\mathcal{R})$	Maximum thermal power output of the woodchip boilers (*WB$_q$*)	(kWt)
Optimization of the design and operation	$\pi_{PV}(t)$ for $t = 1,\ldots,8760$	Real $(\mathcal{R}^T)$	Electric power generated by the photovoltaic power station (*PV*) per unit of nominal power (calculated from solar irradiance $G_{PV}(t)$ and ambient temperature $\theta_A(t)$, see Equation (34))	(−)
	$\check{P}_{WT_i}(t)$ for $t = 1,\ldots,8760$ $i = 1,\ldots,n_{WT}$	Real $(\mathcal{R}^T)$	Electric power that each wind turbine (*WT$_i$*) could produce (obtained from wind speed $v_{WT_i}(t)$, see Equation (35))	(kWe)
	$k_{HE,2}, k_{WB_q,2}, k_{ICE_j,2}, k_{GT_1,2}, k_{GT_2,2},$ $k_{ST,3}, k_{CC1GT,3}, k_{CC2GT,3}$ for $q = 1,\ldots,n_{WB}$ $j = 1,\ldots,n_{ICE}$	Real $(\mathcal{R})$	Ratio between minimum and maximum load of the energy conversion units	(−)
	$k_{WB_q,0}, k_{WB_q,3}, k_{ICE_j,0}, k_{ICE_j,3},$ $k_{ST,0}, k_{ST,5}, k_{ST,7}, k_{ST,9}$ for $q = 1,\ldots,n_{WB}$ $j = 1,\ldots,n_{ICE}$	Real $(\mathcal{R})$	Constant parameters which multiply the binary variables δ in the characteristic maps of energy conversion units (*ICE$_j$*, *WB$_q$* and *ST*)	(kW)
	$k_{HE,0}(t), k_{GT_p,0}(t), k_{GT_p,4}(t),$ $k_{s,0}(t), k_{s,5}(t), k_{s,7}(t), k_{s,9}(t)$ for $t = 1,\ldots,8760$ $p = 1,2$ $s = CC1GT, GG2GT$	Real $(\mathcal{R}^T)$	Variable parameters which multiply the binary variables δ in the characteristic maps of energy conversion units (*HE*, *GT$_p$* and *CC*) (calculated from available water head $h_{HE}(t)$ or ambient temperature $\theta_A(t)$)	(kW)
	$k_{HE,1}, k_{WB_q,1}, k_{WB_q,4}, k_{ICE_j,1}, k_{ICE_j,4}$ $k_{GT_p,1}, k_{GT_p,5}, k_{ST,1}, k_{ST,2}, k_{ST,6},$ $k_{ST,8}, k_{ST,10}, k_{ST,11}, k_{s,1}, k_{s,2}, k_{s,6},$ $k_{s,8}, k_{s,10}$ for $q = 1,\ldots,n_{WB}$ $j = 1,\ldots,n_{ICE}$ $p = 1,2$ $s = CC1GT, GG2GT$	Real $(\mathcal{R})$	Constant parameters which multiply an electric or thermal power output in the characteristic maps of energy conversion units (all except *PV* and *WT$_i$*)	(−)
Optimization of the design and operation	$k_{HE,3}(t), k_{GT_p,3}(t), k_{ST,4}(t), k_{s,4}(t)$ for $t = 1,\ldots,8760$ $p = 1,2$ $s = CC1GT, GG2GT$	Real $(\mathcal{R}^I)$	Variable parameters which multiply an electric or thermal power output in the characteristic maps of energy conversion units (*HE*, *CT$_p$* and *CC*) (calculated from available water head $h_{HE}(t)$ or ambient temperature $\theta_A(t)$)	(−)
	$T_{r,HS}, T_{ST,WS}$ for $r = ST, TG_1, TG_2, CC$	Real $(\mathcal{R})$	Time periods which define the type of start-ups in Equations (50) and (51)	(h)
	$\Delta t_{ST,WS}, \Delta t_{r,CS}$ for $r = ST, CC$	Real $(\mathcal{R})$	Number of hours required for start-ups	(h)
	$\Delta Q_{WB}^{MAX}, \Delta P_{r,HS}^{MAX}, \Delta P_{ST,WS}^{MAX}, \Delta P_{r,CS}^{MAX}$ for $r = ST, CC$	Real $(\mathcal{R})$	Maximum load ramp rates during normal operation and start-ups	$\left(\frac{kW}{h}\right)$
	$\Delta t_{r,UT}^{MIN}, \Delta t_{r,OP}^{MIN}$ for $r = WB_1,\ldots,WB_{n_{WB}},$ $ICE_1,\ldots,ICE_{n_{ICE}},$ ST, GT_1, GT_2, CC	Real $(\mathcal{R})$	Minimum uptime and operating time of the energy conversion units (except *PV*, *WT$_i$* and *HE*)	(h)
	$F_r^{MAX}(t)$ for $t = 1,\ldots,8760$ $r = HE, WC, BO$	Real $(\mathcal{R}^T)$	Available power associated with hydropower (*HE*), woodchips (*WC*) and bio-oil (*BO*) and natural gas (*NG*) (calculate by Equations (23) and (24) from the corresponding mass flow rates)	(kW)

Table 3. *Cont.*

Symbols	Type	Description	U.M.
F_{NG}	Real $(\mathcal{R})$	Maximum power associated with natural gas (*NG*) (calculate by Equation (24) from the corresponding maximum mass flow rates)	(kW)
$\theta_A(t), G_{PV}(t)$	Real $(\mathcal{R}^T)$	Historical or forecasted time profiles of ambient temperature and solar irradiance	(°C) $\left(\frac{kW}{m^2}\right)$
$h_{HE}(t), v_{WT_i}(t)$ for $t = 1, \ldots, 8760$ $i = 1, \ldots, n_{WT}$	Real $(\mathcal{R}^T)$	Historical or forecasted time profiles of available water head and wind speed	(m) $\left(\frac{m}{s}\right)$
$p_{UD,P}(t), p_{UD,Q}(t), p_{GRID}(t), c_{BO}(t),$ $c_{WC}(t), c_{CO}(t),$ $c_{NG}(t), c_{GRID}(t), s_{rd}(t)$ for $t = 1, \ldots, 8760$	Real $(\mathcal{R}^T)$	Variable energy prices, costs and premium feed-in tariffs	$\left(\frac{€}{kWh}\right)$
s_{ce}, s_{oc}, s_{cv}	Real $(\mathcal{R})$	Constant premium feed-in tariffs	$\left(\frac{€}{kWh}\right)$
s_{WB}	Real $(\mathcal{R})$	Annual tax credit per unit of installed capacity of the woodchip boilers (WB_q)	$\left(\frac{€}{kWt \cdot h}\right)$
$\widetilde{c}_{p,O\&M}$ for $p = $ $PV, WT, HE, ST,$ ICE, TG, CC, WB	Real $(\mathcal{R})$	Annual operation and maintenance costs per unit of installed capacity of the energy conversion units	$\left(\frac{€}{kWe \cdot h}\right)$ $\left(\frac{€}{kWt \cdot h}\right)$
$c_{ES,O\&M}, c_{TS,O\&M}$	Real $(\mathcal{R})$	Annual operation and maintenance costs per unit of installed capacity of the electric and thermal storage units	$\left(\frac{€}{kWt \cdot h}\right)$ $\left(\frac{€}{m^3 \cdot h}\right)$
$\alpha_{r,CO_2}(t)$ for $r = $ $ICE_1, \ldots, ICE_{n_{ICE}},$ ST, GT_1, GT_2, CC	Real $(\mathcal{R})$	CO_2 emission per unit of fuel energy of the units converting fossil fuels	$\left(\frac{t}{kWh}\right)$
c_{CO_2}	Real $(\mathcal{R})$	Specific CO_2 emission allowances cost	$\left(\frac{€}{t}\right)$
$C_{ICE_j,su}, C_{WB,su}, C_{CC,HSU}, C_{CC,CSU},$ $C_{ST,HSU}, C_{ST,WSU}, C_{ST,CSU}, C_{TG_1,HSU},$ $C_{TG_1,CSU}, C_{TG_2,HSU}, C_{TG_2,CSU}$ for $j = 1, \ldots, n_{ICE}$	Real $(\mathcal{R})$	Cost of the start-ups of the energy conversion units (except PV, WT_i and HE)	(€)
a_r for $r = $ all energy conversion units except HE	Real $(\mathcal{R})$	Annual amortization of purchase and installation costs of energy conversion units (except HE) per unit of installed capacity	$\left(\frac{€}{kWe \cdot h}\right)$ $\left(\frac{€}{kWt \cdot h}\right)$
a_{ES}, a_{TS}	Real $(\mathcal{R})$	Annual amortization of purchase and installation costs of the electric and thermal storage units per unit of storage capacity	$\left(\frac{€}{kWt \cdot h}\right)$ $\left(\frac{€}{m^3 \cdot h}\right)$
$\beta_{WT_i}, \beta_{ICE_j}, \beta_{ST}, \beta_{GT_1},$ $\beta_{GT_2}, \beta_{CC}, \beta_{WB_p}$ for $i = 1, \ldots, n_{WT}$ $j = 1, \ldots, n_{ICE}$ $p = 1, \ldots, n_{WB}$	Binary	Inclusion or not of the energy conversion units (except PV and HE) in the optimum configuration	(−)
p_{PV}^{MAX} E_{ES}^{MAX} V_{TS}^{MAX}	Real	Maximum capacity (size) of the photovoltaic power station (*PV*), and electric (*ES*) and thermal (*TS*) storage units	(kWp) (kWh) (m^3)

(left margin, rotated: Optimization of the operation only)

5. Examples of Numerical Applications

This Section presents two numerical applications that were solved using the general formulation and equations presented in Section 3. In particular, these applications consider two energy system configurations which include the same types of energy conversion and storage units of the general CHP fleet of energy units in Figure 5. The equations and variables used in the two optimizations are extracted from those in Section 4. The aim is to demonstrate the potential of the presented mathematical formulation.

Both numerical applications were implemented in the GAMS® environment [31] and solved using the branch-and-bound algorithm of the CPLEX® optimizer [32], which has proven to be one of the most efficient solvers of MILP problems. Other solvers recommended in the literature are LINDO© [33] (available in GAMS® environment) and Gurobi™ [34] (usually coupled with the open source programming language PYTON™ [35]).

5.1. Optimization of a Fleet of Energy Units in a District Heating Network

The fleet of energy units in Figure 9 has been designed to serve a district heating network similar to the network of the eastern part of Berlin [20,25]. The total system is subdivided into three supply areas (Figure 9). The thermal energy is required by the network (users) at different temperatures, which depend on the ambient temperature ($\theta_{UD_{in}}(t) = f(\theta_A(t)) = 80$ to $130\,°\text{C}$). No electric users are considered, and all the generated electricity is sold to the grid at the market price (no incentives/premium feed-in tariffs are considered). The optimization procedure aims at evaluating the convenience of including a thermal storage unit (thermocline tank containing pressurized water, TS_i in Figure 9) of unknown capacity in each supply area, whereas no electrical storage units are considered.

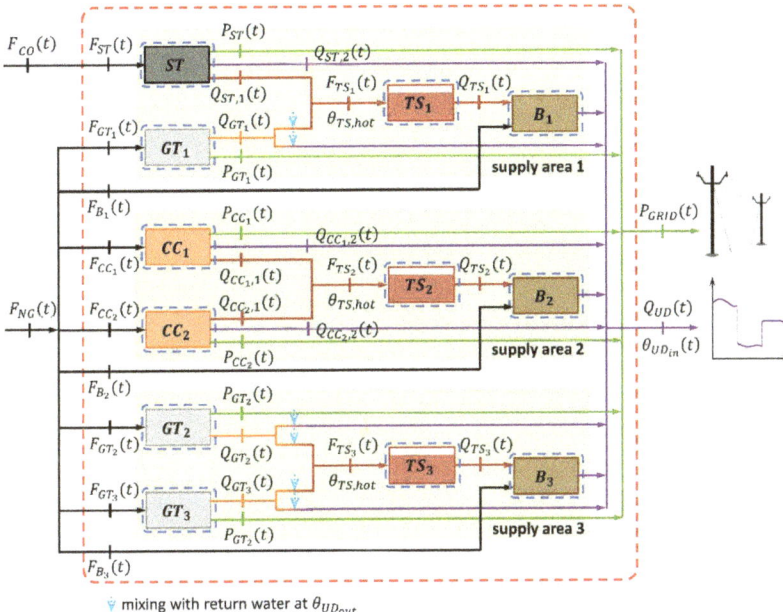

Figure 9. Fleet of large-scale energy units serving a district heating network.

The types of CHP energy conversion units are (Table 4): coal fired steam units (*ST*), natural gas fueled gas turbines (*GT*), and natural gas fueled combined cycles (*CC*). The temperatures in the hot and cold region of the storage units are considered to be constant ($\theta_{TS,hot} = 110\,°\text{C}$ and $\theta_{TS,cold} = 70\,°\text{C}$). The *ST* and *CC* units generate two different streams of pressurized hot water at different temperature: one stream ($Q_{i,1}$ at constant temperature $\theta_{i,1} = \theta_{TS,hot} = 110\,°\text{C}$) is sent to the thermal storage unit, the other stream ($Q_{i,2}$ at variable temperature $\theta_{i,2}(t) = \theta_{UD_{in}}(t) = 80$ to $130\,°\text{C}$) is directly sent to the district heating network. Instead, the gas turbines generate a single stream of pressurized water at constant temperature ($\theta_i = 130\,°\text{C}$) that is then split to be sent to the thermal storage unit or to the network. In both cases, a stream of return water at temperature ($\theta_{UD_{out}} = \theta_{TS_i,cold}(t) = 70\,°\text{C}$) is mixed with the two streams exiting the splitter to obtain the desired final temperature ($\theta_{i,1} = 110\,°\text{C}$ or $\theta_{i,2}(t) = \theta_{UD_{in}}(t) = 80$ to $130\,°\text{C}$). Each supply area includes a peak load boiler (*B*) fueled by natural gas which may generate additional thermal energy or, when needed, increase the temperature of the hot water stream discharged from the storage up to $\theta_{UD_{in}}(t)$.

Table 4. Nameplate characteristics of the energy conversion units included in the system in Figure 9 at $\theta_A = 15\,°C$. For *ST* and *CC* units, $Q_{i,tot}^{MAX}$ refers to the total thermal power output $(Q_{i,1} + Q_{i,2})$ at the maximum temperature $(\theta_{i,2} = 130\,°C)$.

Unit	Supply Area	P_i^{MAX} (MWe)	P_i^{MIN} (MWe)	$Q_{i,tot}^{MAX}$ (MWt)	Manufacturer–Model
ST	1	324.6	162.3	303.8	–
GT_1	1	188.9	94.5	258.8	Siemens–SGT5-3000E (41MAC) 50 Hz
CC_1, CC_2	2	214.1	107.1	126.5	Siemens–2 x SGT5-1000F 50 Hz
GT_2	3	46.7	23.4	96.0	Siemens–SGT-800 50 Hz
GT_3	3	50.3	25.2	53.6	General Electric–GE LM6000 PC Sprint 50 Hz

The optimization problem was formulated including Equations (31) and (32) in the model of the thermal storage units and Equations (37) to (54), (59), and (A1) to (A4) for the energy conversion units (note that Equations (A1) to (A4) are included instead of Equation (55) to (58) to obtain a linear model). The objective function is obtained by excluding from Equation (60) the terms associated with incentives (second and third row) and all revenues and costs associated with PV, wind, and ICE units. The optimization procedure was performed considering the whole year of operation divided into hourly intervals.

In the optimum configuration two thermal storage units having an optimum maximum capacity of 54, 610 and 15, 046 m^3 are included in the supply areas 1 and 2, respectively, whereas it is not convenient in terms of profit to include a thermal storage unit in the supply area 3. The benefit derived from the inclusion of the two storage units is evaluated by optimizing also the operation of the system in Figure 9 without storage units. Results showed that the storage units lead to an increase of 8.67% in the optimum profit (about 3 M€ per year), which is mainly due to the strong reduction in the use of the boilers (−43.90% of the total thermal energy generated by the boilers) in favor of a higher load factor of the CHP units (from +1.82% for the ST unit to +240.71% for the CC units).

5.2. Optimization of a Fully Renewable Fleet of Energy Units

This second numerical application considers an existing fleet of renewable energy units (Figure 10), which includes ten energy conversion units and electric and thermal users (twenty houses, two hotels, ten shops or craft workshops, and seven small industries) [10]. The type and nameplate characteristics of the energy conversion units and the peak thermal and electric power of the users are shown in Table 5. Several optimization runs were carried out considering different scenarios derived from the connection to or isolation of the system from the electric grid, and the possible bypass of the heat recovery system of the internal combustion engines. The aim is to find the optimum operation of the existing fleet of renewable energy units together with the optimum capacities of a thermal and an electric storage unit (not yet included) in the different scenarios.

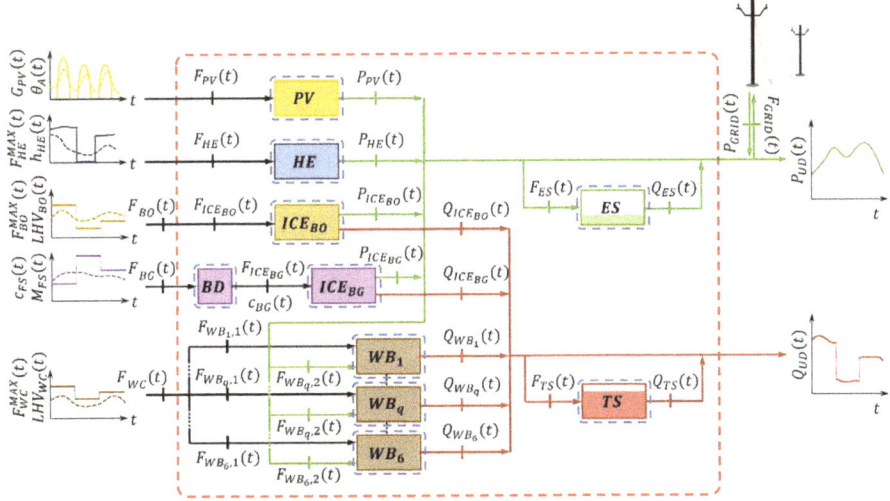

Figure 10. Fleet of medium-scale energy units in a mountain resort.

Table 5. Nameplate characteristics of the renewable energy conversion units included in the system in Figure 10 and peak thermal and electric power of the users.

Unit	Type	P_i^{MAX}(kWe)	P_i^{MIN}(kWe)	Q_i^{MAX}(kWt)	Q_i^{MIN}(kWt)	Manufacturer–Model
PV	PV power station	200.0	0	–	–	modules: BP 3230T
HE	Hydroelectric plant	17,430.0	4041.0	–	–	–
ICE_{BO}	Bio-oil CHP ICE	427.0	213.5	524.7	204.4	MAN–D 2842 LE 211
ICE_{BO}	Bio-gas CHP ICE	265.0	132.5	326.3	184.3	MAN–E 2848 LE 322
$WB_1,\dots,WB_6$	Woodchip boilers	−0.5	−0.2	195.0	56.0	ETA–Hack 200
UD	Users (total)	863.5	–	1785.0	–	–

The model of the system includes Equations (25) to (32), (36) to (40), (49), (52), and (59). As explained in Section 3.2, the generation profile of the PV power station was calculated independently from the optimization procedure. The complete model of this unit is presented in the Appendix of [10]. The objective function is obtained by excluding from Equation (60) the terms associated with wind, ST, TG, and CC units. In [10], the profit derived from the hydroelectric unit *HE* was evaluated separately from that derived from the rest of the system because of its significantly higher size (Table 5).

Results showed that the electric storage unit is required when the system is isolated from the electric grid, but the profit is negative (up to − 4189 €/day) although it includes the actual tariffs for the sale of electric and thermal energy to users. This is because of the very high installation cost of the electric storage unit (about 60 k€ for a maximum capacity of about 235 kWh). On the other hand, the inclusion of a thermal storage unit is convenient in all scenarios and its optimum capacity depends on the scenario being considered (36.9 to 110.2 m³). In particular, the inclusion of the thermal storage increases the optimum profit (+133% to +188% depending on the scenario) because the engines are free to operate at higher load factors taking advantage from the premium feed-in tariffs and the woodchip boilers are kept off for most of the time. Similarly to the *ICEs*, the hydroelectric unit *HE* works, when possible, at full load because of the premium feed-in tariffs and the low operating costs.

6. Conclusions

The paper provides general guidelines to model and optimize the design and operation of a fleet of energy conversion and storage units, which can be summarized as follows:

i) *Modelling:*

- The use of a lumped element (black boxes) schematic of the configuration of the fleet of units allows for modelling each of them using the same structure and type of equations. In particular, the behavior of each energy unit can be described with the minimum loss of information using a Mixed-Integer NonLinear programming (MINLP), where binary variables identify the on/off status of the energy conversion units fed by fossil and renewable dispatchable energy sources.

- The number of variables and equations can be kept as small as possible by considering only power streams and energy quantities, whereas mass flow rates, pressures, and temperatures are not explicit in the model. This simplification is generally acceptable except for some units, in which the compliance with the operational constraints is guaranteed by modifying the input–output relationships (characteristic maps) according to values of the only intensive or extensive variables having a strong impact on the unit behavior.

- The energy contained in the storage units can be evaluated only using their dynamic energy balance equations in which the round-trip efficiency describes the operating characteristics of the unit. For the energy conversion units, a number of steady-state characteristic maps equal to the sum of input and output energy streams minus one is sufficient to describe their behavior. The energy balance of these units is considered only when it is required to explicitly calculate their emissions/losses streams. In all other cases, the costs for emission allowances can be computed starting from the fuel consumption. Additional dynamic constraints such as maximum load ramp rate and minimum uptime and downtime are to be considered when rapid load variations could lead to malfunctions of the unit.

ii) Optimization:

- In the optimization of the system design and operation, the capacities of the storage units and the size of the energy conversion units consisting in modular components (e.g., photovoltaic power stations) are free to vary together with additional binary decision variables, which are used to include or exclude the other energy conversion units in the optimum system configuration.

- The model of energy conversion units fed by non-dispatchable renewable energy sources (sun, wind, and run-of-river hydropower) can be simulated independently from the optimization procedure and the resulting generation profile becomes an input data of this procedure.

- An objective function based on the economic profit is proposed, which includes: a) revenues from selling the generated outputs to the users or to the grid and incentives to support generation and investments; b) expenditures derived from primary energy consumption, purchase of electricity from the grid, and emission allowances, maintenance, and start-up costs; and c) amortization of the purchase and installation costs of the energy units. Different objectives, such as the maximization of the total system efficiency (on energy or exergy basis) or the minimization of the environmental impact can be considered (as alternative or additional objectives) without the need of changing the model and the choice of the decision variable (or with only minor changes) thanks to the generality of the formulation of the optimization problem.

- The MINLP optimization problem can be reduced to a MILP one with a minimum loss of accuracy by considering linear characteristic maps of the energy conversion units and applying linearization technique to the nonlinear constraints (minimum uptime and downtime). A two-step decomposition technique can be applied to further reduce the computational effort required to optimize the configuration of fleets of energy units involving a very large number of binary and real decision variables resulting from the generally long optimization period (e.g., one year). In the first step, the design of the fleet of units is optimized considering shorter periods of time, which are representative of the total period (e.g., few typical days). In the second step, the operation of the resulting optimum configuration is optimized in the total period.

All the equations required to assemble the model and formulate the optimization problem are shown in the paper in a general form.

These guidelines are finally applied to a CHP fleet of units including very different types of energy units to show how the general modelling approach allows for quite easily describing the behavior of each unit and providing the designer with a basic database of equations: energy balance of thermal and electric storage units, characteristic maps of fossil and renewable energy conversion units, and constraints on the maximum load ramp rate and minimum uptime and downtime. A complex framework of electricity market rules and incentive mechanisms are considered to include in the objective function all the main terms that should be considered to properly evaluate the economic profit of the considered type of energy system.

Funding: This research received no external funding.

Conflicts of Interest: The authors declare no conflict of interest.

Appendix A

The inequalities in Equations (55)–(58) in Section 4.3 are the only nonlinear relationships in the model of the general fleet of energy units in Figure 5 (Equations (21) to (59) in Section 4). To solve the optimization problem of such a system using a Mixed-Integer Linear Programming (MILP) method which, in general, requires a significant lower computational effort than NonLinear (MINLP) ones, these relationships are linearized by applying the Glover's method [30]:

- The inequalities in Equations (55) and (57) which calculate the time elapsed from the latest shutdown ($\Delta t_{r,DT}(t)$) and start-up ($\Delta t_{r,UT}(t)$) of the energy conversion unit r, are expressed in linear form using Equation (A1) and (A2), respectively [9]

$$
\begin{aligned}
&\Delta t_{r,DT}(t) \geq \underline{M}_{r,1} \cdot (1 - a_r(t)) \\
&\Delta t_{r,DT}(t) \leq \overline{M}_{r,1} \cdot (1 - a_r(t)) \\
&\Delta t_{r,DT}(t) \geq 1 + \Delta t_{r,DT}(t-1) - \overline{M}_{r,1} \cdot a_r(t) \\
&\Delta t_{r,DT}(t) \leq 1 + \Delta t_{r,DT}(t-1) - \underline{M}_{r,1} \cdot a_r(t) \\
&\underline{M}_{r,1} < \min(\Delta t_{r,DT}(t) + 1) \\
&\overline{M}_{r,1} > \max(\Delta t_{r,DT}(t) + 1)
\end{aligned}
\tag{A1}
$$

$$
\begin{aligned}
&\Delta t_{r,UT}(t) \geq \underline{M}_{r,2} \cdot (1 - a_r(t)) \\
&\Delta t_{r,UT}(t) \leq \overline{M}_{r,2} \cdot (1 - a_r(t)) \\
&\Delta t_{r,UT}(t) \geq 1 + \Delta t_{r,UT}(t-1) - \overline{M}_{r,2} \cdot a_r(t) \\
&\Delta t_{r,UT}(t) \leq 1 + \Delta t_{r,UT}(t-1) - \underline{M}_{r,2} \cdot a_r(t) \\
&\underline{M}_{r,2} < \min(\Delta t_{r,UT}(t) + 1) \\
&\overline{M}_{r,2} > \max(\Delta t_{r,UT}(t) + 1)
\end{aligned}
\tag{A2}
$$

for $\quad t = 1, \ldots, 8760$

$\quad r = WB_1, \ldots, WB_{n_{WB}}, ICE_1, \ldots, ICE_{n_{ICE}}, ST, GT_1, GT_2, CC.$

- The inequalities in Equations (56) and (58) which assure that the unit r can turn its status only when the minimum downtime ($\Delta t_{r,DT}^{MIN}$) or uptime ($\Delta t_{r,UT}^{MIN}$) have elapsed are expressed in linear form using Equations (A3) and (A4), respectively [9]

$$b_{r,1}(t) \geq \underline{M}_{r,3} \cdot a_r(t)$$
$$b_{r,1}(t) \leq \overline{M}_{r,3} \cdot a_r(t)$$
$$b_{r,1}(t) \geq \Delta t_{r,DT}(t-1) - \Delta t_{r,DT}^{MIN} - \overline{M}_{r,3} \cdot (1 - a_r(t))$$
$$b_{r,1}(t) \leq \Delta t_{r,DT}(t-1) - \Delta t_{r,DT}^{MIN} - \underline{M}_{r,3} \cdot (1 - a_r(t))$$
$$b_{r,2}(t) \geq \underline{M}_{r,3} \cdot a_r(t-1)$$
$$b_{r,2}(t) \leq \overline{M}_{r,3} \cdot a_r(t-1)$$
$$b_{r,2}(t) \geq \Delta t_{r,DT}(t-1) - \Delta t_{r,DT}^{MIN} - \overline{M}_{r,3} \cdot (1 - a_r(t)) \qquad \text{(A3)}$$
$$b_{r,2}(t) \leq \Delta t_{r,DT}(t-1) - \Delta t_{r,DT}^{MIN} - \underline{M}_{r,3} \cdot (1 - a_r(t))$$
$$b_{r,1}(t) - b_{r,2}(t) \geq 0$$
$$\underline{M}_{r,3} < \min(\Delta t_{r,DT}(t) + 1)$$
$$\overline{M}_{r,3} > \max(\Delta t_{r,DT}(t) + 1)$$

$$b_{r,3}(t) \geq \underline{M}_{r,4} \cdot a_r(t)$$
$$b_{r,3}(t) \leq \overline{M}_{r,4} \cdot a_r(t)$$
$$b_{r,3}(t) \geq \Delta t_{r,DT}(t-1) - \Delta t_{r,DT}^{MIN} - \overline{M}_{r,4} \cdot (1 - a_r(t))$$
$$b_{r,3}(t) \leq \Delta t_{r,DT}(t-1) - \Delta t_{r,DT}^{MIN} - \underline{M}_{r,4} \cdot (1 - a_r(t))$$
$$b_{r,4}(t) \geq \underline{M}_{r,4} \cdot a_r(t-1)$$
$$b_{r,4}(t) \leq \overline{M}_{r,4} \cdot a_r(t-1) \qquad \text{(A4)}$$
$$b_{r,4}(t) \geq \Delta t_{r,DT}(t-1) - \Delta t_{r,DT}^{MIN} - \overline{M}_{r,4} \cdot (1 - a_r(t))$$
$$b_{r,4}(t) \leq \Delta t_{r,DT}(t-1) - \Delta t_{r,DT}^{MIN} - \underline{M}_{r,4} \cdot (1 - a_r(t))$$
$$b_{r,3}(t) - b_{r,4}(t) \geq 0$$
$$\underline{M}_{r,4} < \min(\Delta t_{r,DT}(t) + 1)$$
$$\overline{M}_{r,4} > \max(\Delta t_{r,DT}(t) + 1)$$

for $\quad t = 1, \ldots, 8760$
$$r = WB_1, \ldots, WB_{n_{WB}}, ICE_1, \ldots, ICE_{n_{ICE}}, ST, GT_1, GT_2, CC.$$

In Equations (A1) to (A4) $\underline{M}_{r,1}, \overline{M}_{r,1}, \underline{M}_{r,2}, \overline{M}_{r,2}, \underline{M}_{r,3} \overline{M}_{r,3} \underline{M}_{r,4}, \overline{M}_{r,4}, a_r(t), b_{r,1}(t), b_{r,2}(t), b_{r,3}(t)$ and $b_{r,4}(t)$ are additional variable which are required to linearize Equations (55) to (58).

References

1. Ringkjøb, H.K.; Haugan, P.M.; Solbrekke, I.M. A review of modelling tools for energy and electricity systems with large shares of variable renewables. *Renew. Sustain. Energy Rev.* **2018**, *96*, 440–459. [CrossRef]
2. Frangopoulos, C.A. Recent developments and trends in optimization of energy systems. *Energy* **2018**, *164*, 1011–1020. [CrossRef]
3. Rech, S.; Lazzaretto, A. From component to macro energy systems: A common design and off-design modeling approach. In Proceedings of the ASME 2011 International Mechanical Engineering Congress and Exposition (IMECE2011), Denver, CO, USA, 11–17 November 2011.
4. Dillon, T.S.; Edwin, K.W.; Kochs, H.D.; Taud, R.J. Integer programming approach to the problem of optimal unit commitment with probabilistic reserve determination. *IEEE Trans. Power Appar. Syst.* **1978**, *6*, 2154–2166. [CrossRef]
5. Consonni, S.; Lozza, G.; Macchi, E. Optimization of Cogeneration Systems Operation—Part B: Solution Algorithm and Examples of Optimum Operating Strategies. In Proceedings of the 1989 ASME International Symposium on Turbomachinery, Combined-Cycle Technologies and Cogeneration (1989 ASME COGEN-TURBO), Nice, France, 30 August–1 September 1986.
6. Ito, K.; Yokoyama, R.; Akagi, S.; Yamaguchi, T.; Matsumoto, Y. Optimal Operational Planning of a Gas Turbine Combined Heat and Power Plant Based on the Mixed-Integer Programming. *Power Syst. Model. Control Appl.* **1988**, *21*, 371–377. [CrossRef]

7. Yokoyama, R. Optimal Sizing of a Cogeneration Plant in Consideration of its Operational Strategy. In Proceedings of the 1991 ASME International Symposium on Turbomachinery, Combined-Cycle Technologies and Cogeneration (1991 ASME COGEN-TURBO), Budapest, Hungary, 3–5 September 1991.

8. Ito, K.; Yokoyama, R.; Shiba, T. Optimal operation of a diesel engine cogeneration plant including a heat storage tank. *J. Eng. Gas Turbines Power* **1992**, *114*, 687–694. [CrossRef]

9. Christidis, A.; Koch, C.; Pottel, L.; Tsatsaronis, G. The contribution of heat storage to the profitable operation of combined heat and power plants in liberalized electricity markets. *Energy* **2012**, *41*, 75–82. [CrossRef]

10. Rech, S.; Lazzaretto, A. Smart rules and thermal, electric and hydro storages for the optimum operation of a renewable energy system. *Energy* **2018**, *147*, 742–756. [CrossRef]

11. Duran, M.A.; Grossmann, I.E. An outer-approximation algorithm for a class of mixed-integer nonlinear programs. *Math. Program.* **1986**, *36*, 307–339. [CrossRef]

12. Stojiljković, M.M.; Stojiljković, M.M.; Blagojević, B.D. Multi-objective combinatorial optimization of trigeneration plants based on metaheuristics. *Energies* **2014**, *7*, 8554–8581. [CrossRef]

13. Amusat, O.; Shearing, P.; Fraga, E.S. System design of renewable energy generation and storage alternatives for large scale continuous processes. In Proceedings of the 12th International Symposium on Process Systems Engineering (PES 2015) and 25th European Symposium on Computer Aided Process Engineering (ESCAPE 2015), Copenhagen, Denmark, 31 May–4 June 2015. [CrossRef]

14. Voll, P.; Lampe, M.; Wrobel, G.; Bardow, A. Superstructure-free synthesis and optimization of distributed industrial energy supply systems. *Energy* **2012**, *45*, 424–435. [CrossRef]

15. Yokoyama, R.; Hasegawa, Y.; Ito, K. A MILP decomposition approach to large scale optimization in structural design of energy supply systems. *Energy Convers. Manag.* **2002**, *43*, 771–790. [CrossRef]

16. Bischi, A.; Taccari, L.; Martelli, E.; Amaldi, E.; Manzolini, G.; Silva, P.; Campanari, S.; Macchi, E. A rolling-horizon optimization algorithm for the long term operational scheduling of cogeneration systems. *Energy* **2017**. [CrossRef]

17. Gao, H.C.; Choi, J.H.; Yun, S.Y.; Lee, H.J.; Ahn, S.J. Optimal Scheduling and Real-Time Control Schemes of Battery Energy Storage System for Microgrids Considering Contract Demand and Forecast Uncertainty. *Energies* **2018**, *11*, 1371. [CrossRef]

18. Rech, S.; Toffolo, A.; Lazzaretto, A. TSO-STO: A two-step approach to the optimal operation of heat storage systems with variable temperature tanks. *Energy* **2012**, *45*, 366–374. [CrossRef]

19. Bischi, A.; Taccari, L.; Martelli, E.; Amaldi, E.; Manzolini, G.; Silva, P.; Campanari, S.; Macchi, E. A detailed MILP optimization model for combined cooling, heat and power system operation planning. *Energy* **2014**, *74*, 12–26. [CrossRef]

20. Rizzetto, A. Combined Optimization of the Operation of CHP Power Plants and the Design of Thermal Storage Systems in a District Heating Network Using Dynamic Programming (MIP) Approach. Master's Thesis, University of Padova, Padova, Italy, 2011.

21. Oberndorfer, U. EU emission allowances and the stock market: Evidence from the electricity industry. *Ecol. Econ.* **2009**, *68*, 1116–1126. [CrossRef]

22. Mazzi, N.; Rech, S.; Lorenzoni, A.; Lazzaretto, A. Application of a new optimal operating strategy to a smart energy system in the de-regulated electricity market. In Proceedings of the 30th International Conference on Efficiency, Cost, Optimization, Simulation and Environmental Impact of Energy Systems (ECOS2017), San Diego, CA, USA, 2–6 July 2017.

23. Durisch, W.; Bitnar, B.; Mayor, J.C.; Kiess, H.; Lam, K.H.; Close, J. Efficiency model for photovoltaic modules and demonstration of its application to energy yield estimation. *Sol. Energy Mater. Sol. Cells* **2007**, *91*, 79–84. [CrossRef]

24. Simani, S. Overview of modelling and advanced control strategies for wind turbine systems. *Energies* **2015**, *8*, 13395–13418. [CrossRef]

25. Rech, S. Analysis and Optimization of the Configuration of a Macro Energy Conversion System. Ph.D. Thesis, University of Padova, Padova, Italy, March 2013.

26. Scarabello, G.; Rech, S.; Lazzaretto, A.; Christidis, A.; Tsatsaronis, G. Optimization of thermal power plants operation in the German de-regulated electricity market using dynamic programming. In Proceedings of the ASME 2012 International Mechanical Engineering Congress and Exposition (IMECE2012), Huston, TX, USA, 9–15 November 2012.

27. Kosman, G.; Rusin, A. The influence of the start-ups and cyclic loads of steam turbines conducted according to European standards on the component's life. *Energy* **2001**, *26*, 1083–1099. [CrossRef]
28. Schill, W.P.; Pahle, M.; Gambardella, C. Start-up costs of thermal power plants in markets with increasing shares of variable renewable generation. *Nat. Energy* **2017**, *2*, 17050. [CrossRef]
29. Wärtsilä–Enabling Sustainable Societies with Smart Technology. Available online: https://www.wartsila.com/energy (accessed on 9 January 2019).
30. Glover, F. Improved Linear Integer Programming Formulations of Nonlinear Integer Problems. *Manag. Sci.* **1975**, *22*, 455–460. [CrossRef]
31. Rosenthal, R.E. *GAMS—A User's Guide*; GAMS Development Corp.: Washington, DC, USA, 2016; Available online: https://www.gams.com/latest/docs/ (accessed on 10 March 2019).
32. IBM ILOG CPLEX A Quick Start to CPLEX Studio. Available online: https://www.ibm.com/support/knowledgecenter/SSSA5P_12.5.0/ilog.odms.ide.help/OPL_Studio/oplquickstart/topics/opl_quickstart.html (accessed on 10 March 2019).
33. LINDO Systems, Inc. User Manuals. Available online: https://www.lindo.com/index.php/help/user-manuals (accessed on 10 March 2019).
34. Gurobi Optimization, LLC. Gurobi Optimizer Reference Manual. Available online: https://www.gurobi.com/documentation/8.1/refman.pdf (accessed on 10 March 2019).
35. Python Packaging User Guide. Available online: https://www.python.org/doc/ (accessed on 10 March 2019).

MDPI

St. Alban-Anlage 66

4052 Basel

Switzerland

Tel. +41 61 683 77 34

Fax +41 61 302 89 18

www.mdpi.com

Energies Editorial Office

E-mail: energies@mdpi.com

www.mdpi.com/journal/energies